Intelligent Systems
Design and Applications

T0135174

Advances in Soft Computing

Editor-in-chief
Prof. Janusz Kacprzyk
Systems Research Institute
Polish Academy of Sciences
ul. Newelska 6
01-447 Warsaw, Poland
E-mail: kacprzyk@ibspan.waw.pl
http://www.springer.de/cgi-bin/search-bock.pl?series=4240

Robert John and Ralph Birkenhead (Eds.)
Soft Computing Techniques and Applications
2000. ISBN 3-7908-1257-9

Mieczysław Kłopotek, Maciej Michalewicz
and Sławomir T. Wierzchoń (Eds.)
Intellligent Information Systems
2000. ISBN 3-7908-1309-5

Peter Sinčák, Ján Vaščák, Vladimír Kvasnička
and Radko Mesiar (Eds.)
The State of the Art in Computational Intelligence
2000. ISBN 3-7908-1322-2

Bernd Reusch and Karl-Heinz Temme (Eds.)
Computational Intelligence in Theory and Practice
2000. ISBN 3-7908-1357-5

Rainer Hampel, Michael Wagenknecht,
Nasredin Chaker (Eds.)
Fuzzy Control
2000. ISBN 3-7908-1327-3

Henrik Larsen, Janusz Kacprzyk,
Sławomir Zadrozny, Troels Andreasen,
Henning Christiansen (Eds.)
Flexible Query Answering Systems
2000. ISBN 3-7908-1347-8

Robert John and Ralph Birkenhead (Eds.)
Developments in Soft Computing
2001. ISBN 3-7908-1361-3

Mieczysław Kłopotek, Maciej Michalewicz
and Sławomir T. Wierzchoń (Eds.)
Intelligent Information Systems 2001
2001. ISBN 3-7908-1407-5

Antonio Di Nola and Giangiacomo Gerla (Eds.)
Lectures on Soft Computing and Fuzzy Logic
2001. ISBN 3-7908-1396-6

Tadeusz Trzaskalik and Jerzy Michnik (Eds.)
Multiple Objective and Goal Programming
2002. ISBN 3-7908-1409-1

James J. Buckley and Esfandiar Eslami
An Introduction to Fuzzy Logic and Fuzzy Sets
2002. ISBN 3-7908-1447-4

Ajith Abraham and Mario Köppen (Eds.)
Hybrid Information Systems
2002. ISBN 3-7908-1480-6

Przemysław Grzegorzewski, Olgierd Hryniewicz,
Maria A. Gil (Eds.)
*Soft Methods in Probability, Statistics
and Data Analysis*
2002. ISBN 3-7908-1526-8

Lech Polkowski
Rough Sets
2002. ISBN 3-7908-1510-1

Mieczysław Kłopotek, Maciej Michalewicz
and Sławomir T. Wierzchoń (Eds.)
Intelligent Information Systems 2002
2002. ISBN 3-7908-1509-8

Andrea Bonarini, Francesco Masulli
and Gabriella Pasi (Eds.)
Soft Computing Applications
2002. ISBN 3-7908-1544-6

Leszek Rutkowski, Janusz Kacprzyk (Eds.)
Neural Networks and Soft Computing
2003. ISBN 3-7908-0005-8

Jürgen Franke, Gholamreza Nakhaeizadeh,
Ingrid Renz (Eds.)
Text Mining
2003. ISBN 3-7908-0041-4

Tetsuzo Tanino, Tamaki Tanaka,
Masahiro Inuiguchi
*Multi-Objective Programming and Goal
Programming*
2003. ISBN 3-540-00653-2

Mieczysław Kłopotek, Sławomir T. Wierzchoń,
Krzysztof Trojanowski (Eds.)
Intelligent Information Processing and Web Mining
2003. ISBN 3-540-00843-8

Ajith Abraham
Katrin Franke
Mario Köppen (Eds.)

Intelligent Systems Design and Applications

With 219 Figures and 72 Tables

 Springer

Editors:

Ajith Abraham
Oklahoma State University
Tulsa, OK, USA
Email: aa@cs.okstate.edu

Katrin Franke
Fraunhofer Institute for
Production Systems and Design Technology
Berlin, Germany
Email: katrin.franke@ipk.fhg.de

Mario Köppen
Fraunhofer Institute for
Production Systems and Design Technology
Berlin, Germany
Email: mario.koeppen@ipk.fhg.de

Associated Editors:

Damminda Alahakoon
Monash University
Victoria, Australia

Leon S. L. Wang
New York Institute of Technology
New York, USA

Prithviraj Dasgupta
University of Nebraska
Omaha, USA

Vana Kalogeraki
University of California
Riverside, USA

ISSN 16-15-3871
ISBN 3-540-40426-0 Springer-Verlag Berlin Heidelberg New York

Cataloging-in-Publication Data applied for.
Bibliographic information published by Die Deutsche Bibliothek. Die Deutsche Bibliothek lists this
publication in the Deutsche Nationalbibliografie; detailed bibliographic data is available in the Internet
at <http://dnb.ddb.de>.

Springer-Verlag Berlin Heidelberg New York
a member of BertelsmannSpringer Science+Business Media GmbH

http://www.springer.de

© Springer-Verlag Berlin Heidelberg 2003
Printed in Germany

The use of general descriptive names, registered names, trademarks, etc. in this publication does not imply,
even in the absence of a specific statement, that such names are exempt from the relevant protective laws
and regulations and therefore free for general use.

Typesetting: Digital data supplied by the authors
Cover-design: E. Kirchner, Heidelberg
Printed on acid-free paper   62 / 3020 hu – 5 4 3 2 1 0

Foreword

Soft computing (computational intelligence) is the fusion/combination of methodologies, such as fuzzy logic (FL) including rough set theory (RST), neural networks (NN), evolutionary computation (EC), chaos computing (CC), fractal theory (FT), wavelet transformation (WT), cellular automata, percolation model (PM) and artificial immune networks (AIN). It is able to construct intelligent hybrid systems. It provides novel additional computing capabilities. It can solve problems (complex system problems), which have been unable to be solved by traditional analytic methods. In addition, it yields rich knowledge representation (symbol and pattern), flexible knowledge acquisition (by machine learning from data and by interviewing experts), and flexible knowledge processing (inference by interfacing between symbolic and pattern knowledge), which enable intelligent systems to be constructed at low cost (cognitive and reactive distributed artificial intelligences).

This conference puts emphasis on the design and application of intelligent systems. The conference proceedings include many papers, which deal with the design and application of intelligent systems solving the real world problems. This proceeding will remove the gap between theory and practice. I expect all participants will learn how to apply soft computing (computational intelligence) practically to real world problems.

Yasuhiko Dote   May 2003
Muroran, Japan
http://bank.csse.muroran-it.ac.jp/

ISDA'03 General Chair's Message

On behalf of the ISDA'03 organizing committee, I wish to extend a very warm welcome to the conference and Tulsa in August 2003. The conference program committee has organized an exciting and invigorating program comprising presentations from distinguished experts in the field, and important and wide-ranging contributions on state-of-the-art research that provide new insights into 'Current Innovations in Intelligent Systems Design and Applications". ISDA'03 builds on the success of last years. ISDA'02 was held in Atlanta, USA, August 07-08, 2002 and attracted participants from over 25 countries. ISDA'03, the Third International Conference on Intelligent Systems Design and Applications, held during August 10-13, 2003, in Tulsa, USA presents a rich and exciting program. The main themes addressed by this conference are:

> Architectures of intelligent systems
> Image, speech and signal processing
> Internet modeling
> Data mining
> Business and management applications
> Control and automation
> Software agents
> Knowledge management

ISDA'03 is hosted by the College of Arts and Sciences, Oklahoma State University, USA. ISDA'03 is technically sponsored by IEEE Systems Man and Cybernetics Society, World Federation on Soft Computing, European Society for Fuzzy Logic and Technology, Springer Verlag- Germany, Center of Excellence in Information Technology and Telecommunications (COEITT) and Oklahoma State University. ISDA'03 received 117 technical paper submissions from over 28 countries, 7 tutorials, 3 special technical sessions and 1 workshop proposal. The conference program committee had a very challenging task of choosing high quality submissions. Each paper was peer reviewed by at least two independent referees of the program committee and based on the recommendation of the reviewers 60 papers were finally accepted. The papers offers stimulating insights into emerging intelligent technologies and their applications in Internet security, data mining, image processing, scheduling, optimization and so on.

I would like to express my sincere thanks to all the authors and members of the program committee that has made this conference a success. Finally, I hope that you will find these proceedings to be a valuable resource in your professional, research, and educational activities whether you are a student, academic, researcher, or a practicing professional. Enjoy!

Ajith Abraham, Oklahoma State University, USA   May 2003
ISDA'03 – General Chair
http://ajith.softcomputing.net

ISDA'03 Program Chair's Message

We would like to welcome you all to ISDA'03: the Third International Conference on Intelligent Systems Design and Applications, being held in Tulsa, Oklahoma, in August 2003.

This conference, the third in a series, once again brings together researchers and practitioners from all over the world to present their newest research on the theory and design of intelligent systems and to share their experience in the actual applications of intelligent systems in various domains. The multitude of high quality research papers in the conference is testimony of the power of the intelligent systems methodology in problem solving and the superior performance of intelligent systems based solutions in diverse real-world applications.

We are grateful to the authors, reviewers, session chairs, members of the various committees, other conference staff, and mostly the General Chair Professor Ajith Abraham, for their invaluable contributions to the program. The high technical and organizational quality of the conference you will enjoy could not possibly have been achieved without their dedication and hard work.

We wish you a most rewarding professional experience, as well as an enjoyable personal one, in attending ISDA'03 in Tulsa.

ISDA'03 Program Chairs:

Andrew Sung, New Mexico Institute of Mining and Technology, USA
Gary Yen, Oklahoma State University, USA
Lakhmi Jain, University of South Australia, Australia

ISDA'03 – Organization

Honorary Chair
Yasuhiko Dote, Muroran Institute of Technology, Japan

General Chair
Ajith Abraham, Oklahoma State University, USA

Steering Committee
Andrew Sung, New Mexico Institute of Mining and Technology, USA
Antony Satyadas, IBM Corporation, Cambridge, USA
Baikunth Nath, University of Melbourne, Australia
Etienne Kerre, Ghent University, Belgium
Janusz Kacprzyk, Polish Academy of Sciences, Poland
Lakhmi Jain, University of South Australia, Australia
P. Saratchandran, Nanyang Technological University, Singapore
Xindong Wu, University of Vermont, USA

Program Chairs
Andrew Sung, New Mexico Institute of Mining and Technology, USA
Gary Yen, Oklahoma State University, USA
Lakhmi Jain, University of South Australia, Australia

Stream Chairs
Architectures of intelligent systems
Clarence W. de Silva, University of British Columbia, Canada
Computational Web Intelligence
Yanqing Zhang, Georgia State University, Georgia
Information Security
Andrew Sung, New Mexico Institute of Mining and Technology, USA
Image, speech and signal processing
Emma Regentova, University of Nevada, Las Vegas, USA
Control and automation
P. Saratchandran, Nanyang Technological University, Singapore
Data mining
Kate Smith, Monash University, Australia
Software agents
Marcin Paprzycki, Oklahoma State University, USA
Business and management applications
Andrew Flitman, Monash University, Australia
Knowledge management
Xiao Zhi Gao, Helsinki University of Technology, Finland

Special Sessions Chair
Yanqing Zhang, Georgia State University, Georgia

Local Organizing Committee
Dursun Delen, Oklahoma State University, USA
George Hedrick, Oklahoma State University, USA
Johnson Thomas, Oklahoma State University, USA
Khanh Vu, Oklahoma State University, USA
Ramesh Sharda, Oklahoma State University, USA
Ron Cooper, COEITT, USA

Finance Coordinator
Hellen Sowell, Oklahoma State University, USA

Web Chairs
Andy AuYeung, Oklahoma State University, USA
Ninan Sajith Philip, St. Thomas College, India

Publication Chair
Katrin Franke, Fraunhofer IPK-Berlin, Germany

International Technical Committee
Andrew Flitman, Monash University, Australia
Carlos A. Coello Coello, Laboratorio Nacional de Informtica Avanzada, Mexico
Chun-Hsien Chen, Chang Gung University, Taiwan
Clarence W. de Silva, University of British Columbia, Canada
Costa Branco P J, Instituto Superior Technico, Portugal
Damminda Alahakoon, Monash University, Australia
Dharmendra Sharma, University of Canberra, Australia
Dimitris Margaritis, Iowa State University, USA
Douglas Heisterkamp, Oklahoma State University, USA
Emma Regentova, University of Nevada, Las Vegas, USA
Etienne Kerre, Ghent University, Belgium
Francisco Herrera, University of Granada, Spain
Frank Hoffmann, Royal Institute of Technology, Sweden
Frank Klawonn, University of Applied Sciences Braunschweig, Germany
Gabriella Pasi, ITIM - Consiglio Nazionale delle Ricerche, Italy
Greg Huang, MIT, USA
Irina Perfilieva, University of Ostrava, Czech Republic
Janos Abonyi, University of Veszprem, Hungary
Javier Ruiz-del-Solar, Universidad de Chile, Chile
Jihoon Yang, Sogang University, Korea
Jiming Liu, Hong Kong Baptist University, Hong Kong
John Yen, The Pennsylvania State University, USA
Jos Manuel Bentez, University of Granada, Spain
Jose Mira, UNED, Spain
Jose Ramon Alvarez Sanchez, Univ. Nac.de Educacion a Distancia, Spain
Kalyanmoy Deb, Indian Institute of Technology, India
Karthik Balakrishnan, Fireman's Fund Insurance Company, USA
Kate Smith, Monash University, Australia

ISDA'03 Technical Sponsors

 IEEE Systems, Man, and Cybernetics Society

WORLD FEDERATION ON SOFT COMPUTING

Springer

COEITT (Center of Excellence in Information Technology and Telecommunications)

Oklahoma State University

Associate Editors

Contents

Part III: Agent Architectures and Distributed Intelligence

Part IV: Intelligent Web Computing

Part V: Internet Security

Part VI: Data mining, Knowledge Management and Information Analysis

Part VII: Computational Intelligence in Management

Part VIII: Image Processing and Retrieval

Part IX: Optimization, Scheduling and Heuristics

Part X: Special Session on Peer-to-Peer Computing

Part XI: 2003 International Workshop on Intelligence, Soft computing and the Web

Part I

Connectionist Paradigms and Machine Learning

New Model for Time-Series Forecasting Using RBFs and Exogenous Data

Juan Manuel Górriz Carlos G. Puntonet[2], J.J.G. de la Rosa[1], and Moisés Salmerón[2]

[1] Departamento de Ingeniería de Sistemas y Automática, Tecnología Electrónica y Electrónica. Universidad de Cádiz (Spain) juanmanuel.gorriz@uca.es
[2] Department of Computer Architecture and Computer Technology. University of Granada (Spain) carlos@atc.ugr.es

Summary. In this paper we present a new model for time-series forecasting using Radial Basis Functions (RBFs) as a unit of ANN´s (Artificial Neural Networks), which allows the inclusion of exogenous information (EI) without additional pre-processing. We begin summarizing the most well known EI techniques used ad hoc, i.e. PCA or ICA; we analyse advantages and disadvantages of these techniques in time-series forecasting using Spanish banks and companies stocks. Then we describe a new hybrid model for time-series forecasting which combines ANN´s with GA (Genetic Algorithms); we also describe the possibilities when implementing on parallel processing systems.

Introduction

Different techniques have been developed in order to forecast time series using data from the stock. There also exist numerous forecasting applications like those ones analyzed in [16]: signal statistical preprocessing and communications, industrial control processing, Econometrics, Meteorology, Physics, Biology, Medicine, Oceanography, Seismology, Astronomy y Psychology.

A possible solution to this problem was described by Box and Jenkins [7]. They developed a time-series forecasting analysis technique based in linear systems. Basically the procedure consisted in suppressing the non-seasonality of the series, parameters analysis, which measure time-series data correlation, and model selection which best fitted the data collected (some specific order ARIMA model). But in real systems non-linear and stochastic phenomena crop up, thus the series dynamics cannot be described exactly using those classical models. ANNs have improved results in forecasting, detecting the non-linear nature of the data. ANNs based in RBFs allow a better forecasting adjustment; they implement local approximations to non-linear functions, minimizing the mean square error to achieve the adjustment of neural parameters. Platt´s algorithm [15], RAN (Resource Allocating Network), consisted in the control of the neural network's size, reducing the computational cost associated to the calculus of the optimum weights in perceptrons networks.

Matrix decomposition techniques have been used as an improvement of Platt model [22] with the aim of taking the most relevant data in the input space, for the sake of avoiding the processing of non-relevant information (NAPA-PRED "Neural model with Automatic Parameter Adjustment for PREDiction"). NAPA-PRED also includes neural pruning [23].

The next step was to include the exogenous information to these models. Principal Component Analysis (PCA) is a well-established tool in Finance. It was already proved [22] that prediction results can be improved using this technique. However, Both methods linear transform the observed signal into components; the difference is that in PCA the goal is to obtain principal components, which are uncorrelated (features), giving projections of the data in the direction of the maximum variance [13]. PCA algorithms use only second order statistical information. On the other hand,in [3] we can discover interesting structure in finance using the new signal-processing tool Independent Component Analysis (ICA). ICA finds statistically independent components using higher order statistical information for separating the signals ([4], [12]). This new technique may use Entropy (Bell and Sejnowski 1995, [6]), Contrast functions based on Information Theory (Comon 1994, [10]), Mutual Information (Amari, Cichocki y Yang 1996, [2]) or geometric considerations in data distribution spaces (Carlos G. Puntonet 1994 [17],[24], [1], [18], [19]), etc. Forecasting and analysing financial time series using ICA can contributes to a better understanding and prediction of financial markets ([21],[5]). Anyway in this paper we want to exclude preprocessing techniques which can contaminate raw data.

1 Forecasting Model(Cross Prediction Model)

The new prediction model is shown in 1. We consider a data set consisting in some correlated signals from the Stock Exchange and try to build a forecasting function $\mathbf{P}$, for one of the set of signals $\{series_1, \ldots, series_S\}$, which allows including exogenous info coming from the other series. If we consider just one series [22] the individual forecasting function can be expressed in term of RBFs as [25]:

$$\mathbf{F}(\mathbf{x}) = \sum_{i=1}^{N} f_i(\mathbf{x}) = \sum_{i=1}^{N} h_i exp\{\frac{||\mathbf{x} - \mathbf{c}_i||}{\mathbf{r}_i^2}\} \tag{1}$$

where $\mathbf{x}$ is a p-dimensional vector input at time $\mathbf{t}$, $\mathbf{N}$ is the number of neurons (RBFs) , $\mathbf{f}_i$ is the output for each neuron i-th , $\mathbf{c}_i$ is the centers of i-th neuron which controls the situation of local space of this cell and $\mathbf{r}_i$ is the radius of the i-th neuron. The global output is a linear combination of the individual output for each neuron with the weight of $\mathbf{h}_i$. Thus we are using a method for moving beyond the linearity where the core idea is to augment/replace the vector input $\mathbf{x}$ with additional variables, which are transformations of $\mathbf{x}$, and then use linear models in this new space of derived input features. RBFs are one of the most popular kernel methods for regression over the domain $\mathbb{R}^n$ and consist on fitting a different but simple model at each query point $\mathbf{c}_i$ using those observations close to this target point in order to get a smoothed function. This localization is achieved via a weighting function or kernel $\mathbf{f}_i$.

We apply/extent this regularization concept to extra series so we include one row of neurons 1 for each series and weight this values via a factor $\mathbf{b}_{ij}$. Finally the global smoothed function for the stock $\mathbf{j}$ will be defined as:

$$\mathbf{P}_j(\mathbf{x}) = \sum_{i=1}^{S} \mathbf{b}_{ij} F_i(\mathbf{x}, j) \tag{2}$$

where $\mathbf{F}_i$ is the smoothed function of each series, $\mathbf{S}$ is the number of input series and $\mathbf{b}_{ij}$ are the weights for j-stock forecasting. Obviously one of these weight factors must be quite relevant in this linear fit ($\mathbf{b}_{jj} \sim 1$, or auto weight factor).

We can use matrix notation to include the set of forecasts in an S-dimensional vector $\mathbf{P}$ ($\mathbf{B}$ in 1):

$$\mathbf{P}(\mathbf{x}) = diag(\mathbf{B} \cdot \mathbf{F}(\mathbf{x})) \tag{3}$$

where $\mathbf{F} = (\mathbf{F}_1, \ldots, \mathbf{F}_S)$ is a $S \times S$ matrix with $F_i \in \mathbb{R}^S$ and B is an $s \times s$ weight matrix. The operator $diag$ extract the main diagonal.

In order to check this model we can choose a set of values for the weight factors as functions of correlation factors between the series, thus we can express 2 as:

$$\mathbf{P}(\mathbf{x}) = (1 - \sum_{i \neq j}^{S} \rho_i)\mathbf{F}_j + \sum_{i \neq j}^{S} \rho_i \mathbf{F}_i \tag{4}$$

where $\mathbf{P}$ is the forecasting function for the desired stock j and ρ_i is the correlation factor with the exogenous series i.

We can include 4 in the Generalized Additive models for regression proposed in supervised learning [11]:

$$\mathbf{E}\{Y|\mathbf{X}_1, \ldots, \mathbf{X}_n\} = \alpha + \mathbf{f}_1(\mathbf{X}_1) + \ldots + \mathbf{f}_n(\mathbf{X}_n) \tag{5}$$

where $\mathbf{X}_i s$ usually represent predictors and $\mathbf{Y}$ represents the system output; $\mathbf{f}_j s$ are unspecific smooth ("nonparametric") functions. Thus we can fit this model minimizing the mean square error function or other methods presented in [11].

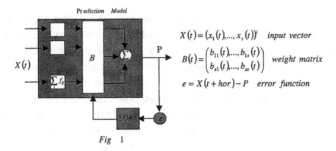

Fig. 1. Schematic representation of CPM with adaptive radius, centers and input space ANNs (RAN + NAPAPRED + LEC [21]). This improvement consists on neural parameters adaptation when input space increases, i.e. RBF centers and radius are statistically updated when dynamic series change takes place.

2 Forecasting Model and Genetic Algorithms.

CPM uses a genetic algorithm for $\mathbf{b}_i$ parameters fitting. A canonical GA is constitute by operations of parameter encoding, population initialization, crossover , mutation, mate selection, population replacement etc. Our encoding parametric system consist on the codification into genes and chromosomes or individuals as string of binary digits using one's complement representation somehow there are other encoding methods also possible i.e [14], [20],[9] or [8] where the value of each parameter is a gene and an individual is encoded by a string of real numbers instead binary ones. In the Initial Population Generation step we assume that the parameters lie in some bounded region $[0, 1]$ (in the edge on this region we can reconstruct the model without exogenous data) and N individuals are generated randomly. After the initial population $\mathbf{N}$ is generated the fitness of each chromosome $\mathbf{I}_i$ is determined via the function:

$$\aleph(I_i) = \frac{1}{\mathbf{e(I}_i)} \tag{6}$$

(To amend the convergence problem in the optimal solution we add some positive constant to the denominator) Another important question in canonical GA is defining Selection Operator. New generations for mating will be selected depending their fitness function values roulette wheel selection. Once we select the newly individuals, we apply crossover ($\mathbf{P}_c$) to generate two offspring which will be applied in the next step the mutation Operator ($\mathbf{P}_m$) to preserve from premature convergence. In order to improve the speed convergence of the algorithm we included some mechanisms like elitist strategy in which the best individual in the current generation always survived into the next.

The GA used in the forecasting function 2 has a error absolute value start criterion. Once it starts, it uses the values (or individual) it found optimal (elite) the last time and apply local search around this elite individual. Thus

Table 1. Pseudo-code of CPM+GA.

Step 1:Initialization Parameters
W = size of prediction window; M = input series maximum length; Hor = forecast horizon; N_{ind} = n° individuals of GA; Epsilon = neural increase; delta= distance between neurons; uga = activation of GA;Error = forecast error; Matrix N_{inps} = number of neural inputs of each series; Matrix n_{RBFs} = number of neurons of each series; Matrix B = Matrix Weight Vector; M_{neuron} = Neurons Parameters Matrix, radius, centres, etc.of each series $Vect_{inp}$ = Input Vector Matrix; $Vect_{Out}$ =Predicted values for each series Target = Real Data in (t+hor); P = forecast function
Step 2: Modelling Input Space.
Toeplizt A Matrix in t_o Relevant data series determination in A(Des. SVD,QR)
Step 3: Iteration.
FOR $i = 1 \rightarrow Max - Iter$ P(t) = BT*Output; Error = Target(t+hor) - P(t) *(Seek for vector B)* IF (error > uga) Execute GA (Selection, Crossover, Mutation, Elitism...) ENDIF *(Neural parametrers)* IF $(error > epsilon$ and $dist(Vect_{inp}, radius) > delta)$ Add neuron centered in $Vect_{inp}$ ELSE *(Evolution of neural networks)* Execute pruning. Update M_{neuron} (Gradient Descend Method). ENDIFELSE *(Input Space Fit)* IF $(error >> epsilon)$ Modelling Input Space. Update M_{neuron} and $Vect_{inp}$. ENDIF ENDFOR

we do an efficient search around an individual (set of $\mathbf{b}_i s$) in which there's a parameter more relevant than the others.

The computational time depends on the encoding length, number of individuals and genes. Because of the probabilistic nature of the GA- based method, the proposed method almost converges to a global optimal solution on average. In our simulation we didn't find any nonconvergent case. In table 1 we show the iterative procedure we implemented to the global prediction system including GA.

Table 2. Simulations.1 Up: Real Series ACS;Bottom: Real Series BBVA.2 Real Series and Predicted ACS Series with CPM. 3 Error Absolute Value with CPM. 4 Real Series and Predicted ACS Series with CPM+GA. 5 Real Series and Predicted ACS Series without exogenous data. 6 Error Absolute Value with CPM + GA. 7 NRMSE evolution for CPM(dot) CPM + GA(line).

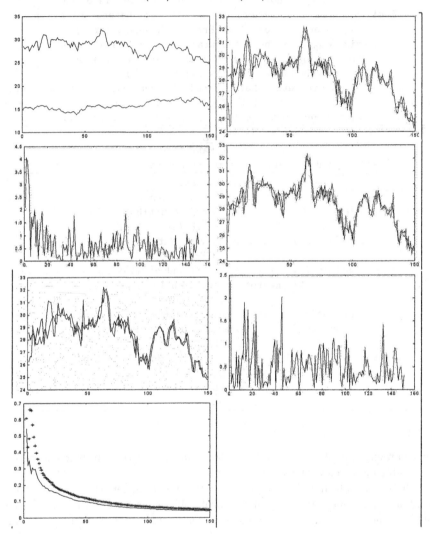

3 Simulations and Conclusions.

With the aim of assessing the performance or the CP model we have worked with indexes of different Spanish banks and companies during the same period. We have specifically focussed on the IBEX35 from Spanish stock, which we consider the most representative sample of the Spanish stock movements .

It is considered the most simple case which consists of two time series corresponding to the companies ACS ($series_1$) and BBVA ($series_2$). The first one is the target of the forecasting process; the second one is introduced as external information. The period under study covers from July to October 2000. Each time series includes 200 points corresponding to selling days (quoting days).

We highlight two parameters in the simulation process. The horizon of the forecasting process (hor) was fixed to 8; the weight function of the forecasting function was a correlation function between the two time series for the $series_2$ (in particular we chose its square) and the difference to one for the series 1. We took a 10 forecasting window ($\mathbf{W}$), and the maximum lag number was fixed to the double of W, so that to achieve a 10×20 Toeplitz matrix. In the first time point of the algorithm it will be fixed to 50. Table 1 shows the forecasting results from "lag" 50 to lag 200 corresponding to $series_1$.

Table 3. Correlation Coefficients between real signal and the predicted signal with different lags.

delay	ρ	delay	ρ
0	0.89	0	0.89
+1	0.79	−1	0.88
+2	0.73	−2	0.88
+3	0.68	−3	0.82
+4	0.63	−4	0.76
+5	0.59	−5	0.71
+6	0.55	−6	0.66
+7	0.49	−7	0.63
+8	0.45	−8	0.61
+9	0.45	−9	0.58
+10	0.44	−10	0.51

We point out the instability of the system in the very first iterations until it reaches an acceptable convergence. The most interesting feature of the result is shown in table 3; from this table it is easy to work out that if we move horizontally one of the two series the correlation between the dramatically decreases. This is the reason why we avoid the delay problem which exhibit certain networks, in sections where the information introduced to the system is non-relevant. This is due to the increase of information associated to the fact that we have enclosed only one additional time series $(series_2)$, despite the fact that neuron resources increase. At the end of the process we used 20 neurons for net 1 and 21 for net 2. Although the forecasting is acceptable we expect a better performance with more data point series.

Table 4. Dynamic and values of the weights for the GA.

b_{series}	T_1	T_2	T_3	T_4
b_1	0.8924	0.8846	0.8723	0.8760
b_2	0.2770	0.2359	0.2860	0.2634

The next step consisted in using the general algorithm including the GA. A 4 individuals (N_{ind}) population was used, with a 2×1 dimension; this is a small number because we have a bounded searching space. The genetic algorithm was run four time before reaching the convergence; the individuals were codified with 34 bits (17 bits for each parameter). In this case convergence is defined in terms of the adjustment function; other authors use another parameters of the GA, like the absence of change in the individuals after a certain number of generations, etc. We point out a sensible improvement in the forecasting results; we also note the evidence of the disappearance of the delay problem, as t is shown in table 1.

The number of neurons at the end of the process is the same as in the former case, because we have only modified the weight of each series during the forecasting process. The dynamic and values for the weights are shown in table 4.

Error behavior is shown in table 1. We note:

- We can bound the error by means of the adequate selection of the parameters bi, when the dynamics of the series is coherent (avoiding big fluctuations in the stock).
- The algorithm converges faster, as it is shown in the very beginning of the graph.
- The forecasting results are better using GA, as it is shown in table 1, where it is graphed the evolution of the normalized mean square error.

Due to the symmetric character of our forecasting model, it is possible to implement parallel programming languages (like PVM) to build a more general forecasting model for a set of series. We would launch the same number for son processes and banks; and these ones would run forecasting vectors, which would we weighed by a square matrix with dimension equal to the number of series **B**. The "father" process would have the results of the forecasting process for the calculus of the error vector, so that to update the neuron resources. Therefore we would take advantage the computational cost of a forecasting function to calculate the rest of the series.

3.1 Conclusions:

In addition to the above ideas we conclude that our new forecasting model for time-series is characterized by:

- The enclosing of external information. We avoid pre-processing and data contamination applying ICA and PCA. Series are enclosed in the net directly.
- The forecasting results are improved by means of hybrid techniques using well known techniques like GA.
- The possibility of implementing in parallel programming languages (i.e. PVM; "Parallel Virtual Machine"); and the major performance and the lower computational time achieved using a neuronal matrix architecture.

References

1. C.G. Puntonet A. Mansour, N. Ohnishi, *Blind multiuser separation of instantaneous mixture algorithm based on geometrical concepts*, Signal Processing **82** (2002), 1155–1175.
2. S. Amari, A. Cichocki, and H. Yang, *A new learning algorithm for blind source separation*, Advances in Neural Information Processing Systems. MIT Press **8** (1996), 757–763.
3. A. D. Back and A. S. Weigend, *Discovering structure in finance using independent component analysis*, Computational Finance (1997).
4. Andrew D. Back and Thomas P. Trappenberg, *Selecting inputs for modelling using normalized higher order statistics and independent component analysis*, IEEE Transactions on Neural Networks **12** (2001).
5. Andrew D. Back and A.S. Weigend, *Discovering structure in finance using independent component analysis*, 5th Computational Finance 1997 (1997).
6. A.J. Bell and T.J. Sejnowski, *An information-maximization approach to blind separation and blind deconvolution*, Neural Computation **7** (1995), 1129–1159.
7. G.E.P. Box, G.M. Jenkins, and G.C. Reinsel, *Time series analysis. forecasting and control*, Prentice Hall, 1994.
8. L. Chao and W. Sethares, *Non linear parameter estimation via the genetic algorithm*, IEEE Transactions on Signal Processing **42** (1994), 927–935.
9. S. Chen and Y. Wu, *Genetic algorithm optimization for blind channel identification with higher order cumulant fitting*, IEEE Trans. Evol. Comput. 1 (1997), 259–264.
10. P. Comon, *Independent component analysis: A new concept?*, Signal Processing **36** (1994), 287–314.
11. T. Hastie, R. Tibshirani, and J. Friedman, *The elements of statistical learning*, Springer, 2000.
12. A. Hyvarinen and E. Oja, *Independent component analysis: Algorithms and applications*, Neural Networks **13** (2000), 411–430.
13. T. Masters, *Neural, novel and hybrid algorithms for time series analysis prediction*, John Miley & Sons, 1995.
14. Z. Michalewicz., *Genetic algorithms + data structures = evolution programs*, Springer-Verlag, 1992.

15. J. Platt, *A resource-allocating network for function interpolation*, Neural Computation **3** (1991), 213–225.
16. D.S.G. Pollock, *A handbook of time series analysis, signal processing and dynamics*, Academic Press, 1999.
17. C.G. Puntonet, *Nuevos Algoritmos de Separación de Fuentes en Medios Lineales*, Ph.D. thesis, University of Granada , Departamento de Arquitectura y Tecnología de Computadores, 1994.
18. C.G. Puntonet and Ali Mansour, *Blind separation of sources using density estimation and simulated annealing*, IEICE Transactions on Fundamental of Electronics Communications and Computer Sciences **E84-A** (2001).
19. M. Rodríguez-Álvarez, C.G. Puntonet, and I. Rojas, *Separation of sources based on the partitioning of the space of observations*, Lecture Notes in Computer Science **2085** (2001), 762–769.
20. T. Szapiro S. Matwin and K. Haigh, *Genetic algorithms approach to a negotiation support system*, IEEE Trans. Syst. , Man. Cybern **21** (1991), 102–114.
21. J.M. Górriz Sáez, *Predicción y Técnicas de Separación de Senales Neuronales de Funciones Radiales y Técnicas de Descomposición Matricial*, Ph.D. thesis, University of Cádiz , Departamento de Ing. de Sistemas y Aut. Tec. Eleectrónica y Electrónica, 2003.
22. M. Salmerón-Campos, *Predicción de Series Temporales con Redes Neuronales de Funciones Radiales y Técnicas de Descomposición Matricial*, Ph.D. thesis, University of Granada , Departamento de Arquitectura y Tecnología de Computadores, 2001.
23. Moisés Salmerón, Julio Ortega, Carlos G. Puntonet, and Alberto Prieto, *Improved ran sequential prediction using orthogonal techniques*, Neurocomputing **41** (2001), 153–172.
24. F. J Theis, A. Jung, E.W. Lang, and C.G. Puntonet, *Multiple recovery subspace projection algorithm employed in geometric ica*, In press on Neural Computation (2001).
25. J. Moody y C. J. Darken, *Fast learning in networks of locally-tuned processing units*, Neural Computation **1** (1989), 284–294.

On Improving Data Fitting Procedure in Reservoir Operation using Artificial Neural Networks

S.Mohan[1], V.Ramani Bai[2]

[1]Department of Civil Engineering, Indian Institute of Technology, Madras, Chennai-600 036, Tamilnadu, India. mohan@civil.iitm.ernet.in

[2]Environmental and WaterResources Engineering Divsion, Department of Civil Engineering, Indian Institute of Technology, Madras, Chennai-600 036, Tamilnadu, India. vramanibai@hotmail.com

Abstract. It is an attempt to overcome the problem of not knowing at what least count to reduce the size of the steps taken in weight space and by how much in artificial neural network approach. The parameter estimation phase in conventional statistical models is equivalent to the process of optimizing the connection weights, which is known as 'learning'. Consequently the theory of nonlinear optimization is applicable to the training of feed forward networks. Multilayer Feed forward (BPM & BPLM) and Recurrent Neural network (RNN) models as intra and intra neuronal architectures are formed. The aim is to find a near global solution to what is typically a highly non-linear optimization problem like reservoir operation. The reservoir operation policy derivation has been developed as a case study on application of neural networks. A better management of its allocation and management of water among the users and resources of the system is very much needed. The training and testing sets in the ANN model consisted of data from water year 1969-1994. The water year 1994-1997 data were used in validation of the model performance as learning progressed. Results obtained by BPLM are more satisfactory as compared to BPM. In addition the performance by RNN models when applied to the problem of reservoir operation have proved to be the fastest method in speed and produced satisfactory results among all artificial neural network models

1. Introduction

The primary objective of the reservoir operation is to maintain operation conditions to achieve the purpose for which it has been created with least interference to the other systems. In this paper, integration of surface and groundwater available is made use of as resource available for water meeting irrigation demand of the basin. Effects of inter basin transfer of water between Periyar and Vaigai system is also carefully studied in determining release policies. The inter-basin transfer of water not having negative environmental impacts is a good concept but that should also have social and cultural consideration, price considerations and some environmental justice considerations. There are significant water-sharing conflicts within agriculture itself, with the various agricultural areas competing for scarce water supplies. Increasing basin water demands are placing additional stresses on the limited water resources and threaten its quality. Many hydrological models have been developed to problem of reservoir operation. System modeling based on conventional mathematical tools is not well suited for dealing with nonlinearity of the system. By contrast, the feed forward neural networks can be easily applied in nonlinear optimization problems. This paper gives a short review on two methods of neural network learning and demonstrates their advantages in real application to Vaigai reservoir system. The developed ANN model is compared with LP model for its performance. This is accomplished by checking for equity of water released for irrigation purpose. It is concluded that the coupling of optimization and heuristic model seems to be a logical direction in reservoir operation modeling.

2.   Earlier Works

Tang et al. (1991) conclude that artificial neural networks (ANN) can perform better than the Statistical methodology for many tasks, although the networks they tested may not have reached the optimum point. ANN gives a lower error, while the Statistical model is very sensitive to noise and does not work with small data sets. An ANN can then be trained on a particular set of the input and output data for the system and then verified in terms of its ability to reproduce another set of data for the same system. The hope is that the ANN can then be used to predict the output given the input. Such techniques have been applied, for example, to the rainfall-runoff process (Mason et al 1996; Hall and Minns 1993) and combined hydraulic/hydrologic models (Solomatine and Torres 1996). A disadvantage of these models is that a separate neural network has to be constructed and trained for each particular catchment or river basin.

3.   Study Area and Database

The model applies to Vaigai basin (N 9° 15'-- 10° 20' and E 77° 10' – 79° 15') is located in south of Tamil Nadu in India (Fig. 2.1). The catchment area is 2253 sq.km and water spread area 25.9 sq.km. The reservoir is getting water from catchments of two states namely, Tamil Nadu and Kerala. Maximum storage capacity of the dam is 193.84 Mm^3. The depth to bottom of aquifer varies from 7 to 22m. The zone of water level fluctuation varies from 2 to 15m. The river basin is underlined by a weathered aquifer, phreatic to semi confined aquifer in the alluvium and valley fills in the crystalline rock formation and semi confined to confined aquifer conditions in the sedimentary formations. The period of study is from 1969-'70 to 1997-'98 of measured historical data. Flow data from water year 1969-'70 to 1993-'94 is used for training of the neural network model for operation of the reservoir and data from water year 1994-'95 to 1997-'98 is used for validation of the model.

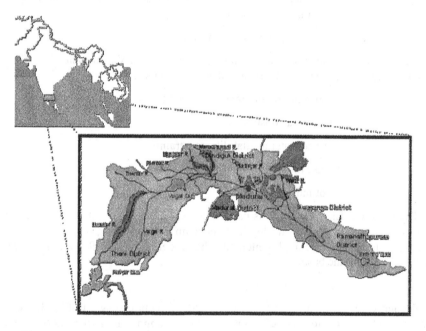

Fig. 2.1 Location of Vaigai Dam in Inter–state water transfer Periyar Dam

4. Model Development

The most challenging aspect of mathematical modeling is the ability to develop a concise and accurate model of a particular problem. Traditional optimization methods namely linear programming has considerable loss in accuracy as a linear model from a nonlinear real world problem is developed.First, a Linear programming model is developed with the objective of maximizing the net annual benefit of growing six primary crops in the whole basin deducting the annualized cost of development and supply of irrigation from surface and groundwater resources.

For maximizing net benefit for every node,

$$\text{Max} \sum_{i=1}^{n} \sum_{c=1}^{6} B_{i,c} A_{i,c} \qquad (3.1)$$

where,

c   = Crops, 1 to 6 (Single crop, First crop, Second crop Paddy, Groundnut, Sugarcane and Dry crops)

i   = Nodes, i = V, P, H, B and R

$B_{i,c}$ = Net annual benefit after production cost and water cost for crop c grown in node i in Rs./Mm^2 and

$A_{i,c}$ = Land area under crop c in node i in Mm^2

The accounting system of water, which is still in use in Vaigai system, is taken into consideration in the model with all other constraints as,

1.   Crop water requirement Constraints

2.   Land Area constraints

3.   Surface water availability constraints

4.   Groundwater availability constraints

5.   Continuity Constraints

The software, LINDO (2000) is used to solve the developed LP model. It is an optimization model with in-depth discussion on modeling in an array of application areas using Simplex method. The model is run for 29 years of inflow (1969-1997) into the basin.

Secondly, an artificial neural network (ANN) structure shown in Fig. 3.1. can be applied to develop an efficient decision support tool considering the parameters, which are non-linear in nature and to avoid addressing the problem of spatial and temporal variations of input variables.   By this work an efficient mapping of non-linear relationship between time period (T_t), Inflow (I_t) Initial Storage (S_t), Demand (D_t) and Release (R_t) pattern into an ANN model is performed for better prediction on optimal releases from Vaigai dam. Two algorithms namely, Feed Forward Error Back Propagation Network (FFBPN) with momentum correction and Back Propagation Network with Levenberg-Marguardt models are considered in this study. In an attempt to overcome the problem of not knowing at what learn count to reduce the size of steps taken in weight space and by how much, a number of algorithms have been proposed which automatically adjust the learning rate and momentum co-efficient based on information about the nature of the error surfaces.

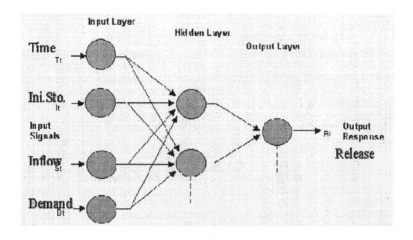

Fig. 3.1 Architecture of three layers FFNN for Reservoir Operation

The hyperbolic tangent transfer function, log sigmoid transfer function; the normalized cumulative delta learning rule and the standard (quadratic) error function were used in the frame work of the model. This calibration process is generally referred to as 'training'. The global error function most commonly used is quadratic (mean squared error) function

$$E(t) = \frac{1}{2} \Sigma \left(d_j\,(t) - y_j\,(t) \right)^2 \tag{3.2}$$

where,

$E\,(t)$ = the global error function at discrete time t.

$y_j\,(t)$ = the predicted network output at discrete time t and

$d_j\,(t)$ = the desired network output at discrete time t.

The aim of the training procedure is to adjust the connection weight until the global minimum in the error surface has been reached. The output from linear programming model; inflow, storage, release and actual demand are given as input into ANN model of the basin. Monthly values of Time period, Inflow, initial storage and demand are the input into a three layer neural network and output from this network is monthly release. The training set consisted of data from 1969-'94. The same data were used to test model performance as learning progressed. Through controlled experiments with problems of known posterior probabilities, this study examines the effect of sample size and network architecture on the accuracy of neural network estimates for these known posterior probabilities. Neural network toolbox in Matlab 6.1, Release 12, software is used to solve the developed model with 4 input variables and 1 output variable.

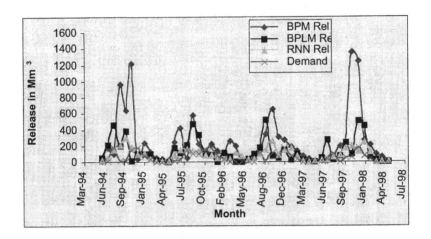

Fig. 3.2 Comparison of performance by different methods of ANN

The length of data for both input and output values is from 1 to 300. A better prediction is given by the three layers ANN model with 25 neurons in each hidden layer. This network is used to compute water release from Vaigai reservoir. The value of error is higher than the best values for the BPM neural networks shown before. It is also found that BPLM is quicker in the search for a good result. Thirdly, the special type of neural network called recurrent neural network (RNN) is used to fit the flow data for reservoir operation problem. The release produced by RNN is compared with that of other neural network models for reservoir operation. The results are compared in terms of meeting the demand of the basin. The results are shown in Fig. 3.2.

5.   Validation of the Model

Once the training (optimization of weights) phase has been completed the performance of the trained network needs to be validated on an independent data set using the criteria chosen. It is important to note that the validation data should not have been used as part of the training process in any capacity. The data of water year 1994-1997 are used for validating the developed ANN model. The validation results are shown in Fig. 3.2.

6.   Results and Discussion

The applicability of different kinds of neural networks for the probabilistic analysis of structures, when the sources of randomness can be modeled as random variables is summarized. The comparison comprehends two network algorithms (multi-layer neural networks). This is a relevant result for neural networks learning because most of the practical applications employ the neural learning heuristic back-propagation, which uses different activation functions. Further discussions on the results obtained are as below.

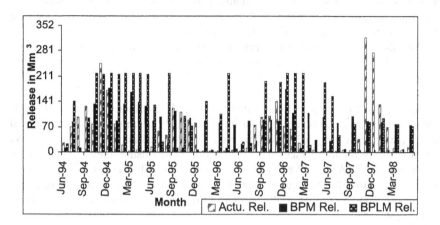

Fig. 3.3 Comparison of reservoir release by two ANN methods

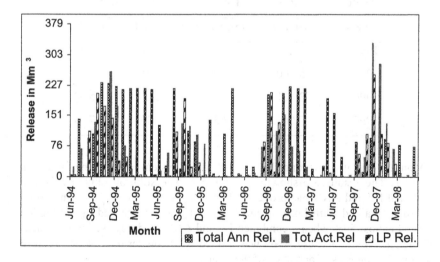

Fig. 3.4 Comparison of reservoir release by optimization and Heuristic methods

The suitability of a particular method is generally a compromise between computation cost and performance. The comparison between the results obtained from BPM and BPLM are shown in Fig 3.3. The data set resulted in a better representation, when an improved ANN model replaces back propagation with momentum. We observe that the BPLM algorithm obtains the best results more quickly than the BPM algorithm. In addition, BP using Levenberg-Marguardt algorithm needs no additional executable program modules in the source code. BPM has taken epoch number of 5000. On the contrary, the number of epochs to find the optimal solution at different tests is significantly reducing. Further, the figures showed a relatively small running time taken to find the optimal solution when an ANN replaces the linear programming models.

Further, the time series of calculated monthly flow values of water year 1994-1997 (validation periods) are plotted for ANN (BPLM) model and conventional optimization technique (LP) model. The data set resulted in a better representation when an ANN model replaces the optimization model using linear programming. The release from ANN, LP and actual values and groundwater extraction from the basin are plotted for testing performance of classical linear programming and heuristic method as applied to the problem of reservoir operation. This is exhibited in Fig. 3.4.

Table 5.1 Reservoir operation performances by LP & ANN models

Water	Result by LP		Result by ANN	
Year	% Deficit	Amt. Def.	% Deficit	Amt. Def.
1994-95	15.75	3.98	9.65	2.44
1995-96	55.00	13.90	8.70	2.20
1996-97	38.21	9.66	21.84	5.52
1997-98	72.61	18.36	31.13	7.87

Monthly deficit in meeting the demand of the system is estimated by both the methods. The results are shown in Table 5.1. The solution obtained from the combined LP and ANN has proved to better in its performance than the optimal solutions obtained from linear programming. The deficit caused by LP model is reduced almost 50% when a combined LP and ANN model is used in the place of LP model. This is because ANN was trained with courser data, which represent the complete physics of operation of the system.

7. Conclusions

In this paper, the neural network approach for deriving operating policy for Vaigai reservoir is investigated. A comparative study on the predictive performance of two different backpropagation algorithms namely backpropagation with momentum correction and backpropagation with Levenberg Marguardt algorithms are evaluated. Overall results showed that the use of neural network models in operation of reservoirs systems is appropriate. This developed model can be implemented for future operation of Vaigai system with the significance of no a priori optimization for getting release from the dam need to be formulated or tested. A time series modeling on operation using LP with ANN provides a promising alternative and leads to better predictive performance than classical optimization techniques as linear programming. Furthermore, it is demonstrated that the proposed multi-layer Feed forward network architecture with Levenberg-Marguardt learning algorithm is capable of achieving comparable or lower prediction errors, and lesser running time as compared to traditional feed-forward MLP network with momentum correction. It also proved that recurrent neural networks, owing to their feedback effect they are the fastest learning method among commonly used networks in ANN for reservoir operation.

Acknowledgements

The authors sincerely thank the timely help rendered by Dr. Ajit Abraham and essential comments made by Reviewers of isda03 conference on this paper.

References

1. Hall, M J and Minns, A. W. 1993. Rainfall-runoff modeling as a problem in artificial intelligence: Experience with a neural network. *Proceedings BHS 4th National Hydrology Symposiu*m, pp5.51-5.57.
2. Mason, J.C., Price, R.K.and Tamme, A. 1996. A neural network model of rainfall-runoff using radial basis functions. *Journal of Hydraulic Research,* 34(4), IAHR, pp. 537-548
3. Solomatine, D.P. and Torres, A.L. 1996. Neural network approximation of a hydrodynamic model in optimizing reservoir operation. Proc. 2 nd Int. Conf. on Hydroinfor-matics, Zurich, September, pp. 201-206.
4. Tang, Z., de Almeida, C., Fishwick, P.A., 1991. Time series forecasting using neural networks vs Box-Jenkins methodology. Simulation 57(5), pp. 303-310.

Automatic Vehicle License Plate Recognition using Artificial Neural Networks

Cemil Oz
UMR Computer Science Department, Rolla, MO 65401, USA or
Sakarya University Computer Engineering Department, Sakarya, TURKEY
(e-mail: ozc@umr.edu or coz@sau.edu.tr)

Fikret Ercal
UMR Computer Science Department, Rolla, MO 65401, USA
(e-mail: ercal@umr.edu)

Abstract. In this study, we present an artificial neural network based computer vision system which can analyze the image of a car taken by a camera in real-time, locates its license plate and recognizes the registration number of the car. The model has four stages. In the first stage, vehicle license plate (VLP) is located. Second stage performs the segmentation of VLP and produces a sequence of characters. An ANN runs in the third stage of the process and tries to recognize these characters which form the VLP.

Keywords: Vehicle license plate, artificial neural network (ANN), computer vision, optical character recognition (OCR).

1. Introduction

Monitoring vehicles for law enforcement and security purposes is a difficult problem because of the number of automobiles on the road today. Among the reasons for traffic accidents are high speeds, drunk driving, driving without a license, and various other traffic violations. Most important of all are the violation of traffic rules and signs. Hence, to save lives and property, it is important to enforce these rules by any means possible. Computer vision can provide significant help in this context.

Automated recognition and identification of vehicle license plates (VLP) has great importance in security and traffic systems. It can help in many ways in monitoring and regulating the road traffic. For the management of urban and rural traffic, there is a lot interest in the automation of the license plate recognition in order to regulate the traffic flow, to control access to restricted areas, and to survey traffic violations.

VLP identification is a difficult problem in computer vision and hence it is divided into several steps. The first step in the process is to locate the VLP. After that, characters in the VLP are recognized one by one using a recognition program such as a neural network. Character recognition is a popular and well-studied area in computer vision. In general, ANN and matching methods are used [15, 3, 20, 6, 7] for this purpose. In the literature, there are many studies which use different approaches for the VLP identification problem; ANN based [4, 17, 11], inductive learning method based [2], and techniques based on 2D coloration [14]. Some VLP identification studies focus only in the character recognition aspect of the problem [11, 2] while others [4, 17] also include the process of finding the location of VLP in the entire vehicle image. There is also some VLP recognition work for electronic tool collection and traffic management systems, and for the help of color blindness [5, 19, 10].

In this study, we assumed that the characteristics (e.g. a text region composed of characters) and the approximate location of the VLP are known a priori. Some other studies do not make these assumptions and use elaborate image processing techniques coupled with an ANN to locate the VLP in an image. There are also similar studies which make use of color properties of the VLP to find its location. In this study, we used properties common to Turkish vehicle license plates to find its location and identification.

2. Locating the license plate in an image

Scanned or captured VLP image has color and a complex background. We need to use the color properties and this complex background composition to locate VLP in the entire image. A VLP consists of letters and numbers. Most studies in the literature use text detection and tracking for this purpose. Most commonly, these studies are either connected component based (CC-based) [9, 21], or texture based [18, 8] or ANN based [12]. Here, we used the properties of the gray-level pixel values for finding and locating the VLP. Pixel values in gray-level change very rapidly in the VLP area of the image. Histogram analysis of these changes helps us to locate and segment the VLP.

3. Plate segmentation

3.1. Vehicle license plate properties

A standard (recreational) Turkish VLP consists of three symbols; county code, a letter code, and a numeric code. Since there are a total of 85 counties in the country, county code starts at 01 and goes up to 85 in decimal. The second part in the VLP is a three-letter alpha numeric symbol using only capital letters A-Z. The third region has a 4-digit decimal number as shown in Figure 1. License plates for military and government vehicles use a different style. This study considers only civilian VLPs.

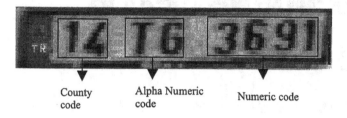

County code   Alpha Numeric code   Numeric code

Fig. 1. A sample VLP

The following steps are used in order to locate and segment the license plate image into individual characters.

First step: VLP which is located by separating it from a complex background image is converted into a binary image with background taking the pixel value of 0. Horizontal projection is used for border clearing. For each row in the image, the number of zeros (0) are counted and the resulting histogram is plotted sideways as shown in Figure 2.

Fig. 2. Horizontal projection

Second step: After the VLP is located and the background is cleaned up, the individual characters are separated from each other using vertical projection (histogram). Basically, a zero value (0) in the vertical histogram indicates a border crossing between two characters. To save processing time in the next stage, we further eliminate the blank space between characters by drawing the character boundaries as close to the characters as possible.

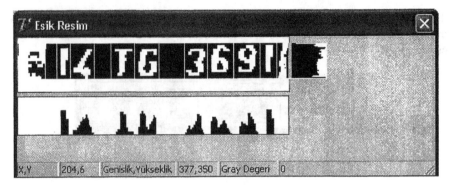

Fig. 3. VLP character segmentation

Third step: Normalization. In the original image, characters have different pixel sizes. It is important to normalize their size for further processing. In this step, every segmented character is mapped to a (19x11) pixel format. This makes it easier to design an ANN for the character recognition in the next stage.

Fourth step: In this step, each character in the license id is identified and separated from each other using a column projection technique. Then, each character is segmented into a mesh of grids containing 209 square areas (19 rows, 11 columns). The pixel data in these segments are used as input to train an artificial neural network which is designed to do the character recognition task.

4. ANN model

In this research, a backpropagation algorithm is chosen for training the ANN model. A basic structure and formulations of backpropagation is summarized as follows:

4.1 Backpropagation algorithm

Training a net involves to obtain weights so as to give output sets in responses to input sets a within an error limit. Input and output vectors make up a training pair. The backpropagation algorithm includes the following steps [19].

1-Select the first training pair and apply the input vector to the net
2- Calculate the net output
3- Compare the actual output with the desired one and find the error
4-Modify the weights so as to reduce the error.
These steps are repeated until the error within the accepted limit. In the next level output sets for test inputs are calculated, if they are the same within an

error as the expected sets then the net is called to have learnt the problem and the final weights are stored to be reused when needed.

The developed ANN has multi-layer feedforward structure and shown at Figure 4.

Variable definitions are given as follows:

L=0 input layer
L=1 hidden layer
$W_{1,ji}$ a weight between the input layer and the hidden layer.
$W_{2,tj}$ any weight between the hidden and the output layer
$i=1,2,...$ n as inputs neurons,
$j=1,2,...,m$ as the hidden neurons
$t=1,2,...,k$ as the output layer's neurons[1,13] are defined.

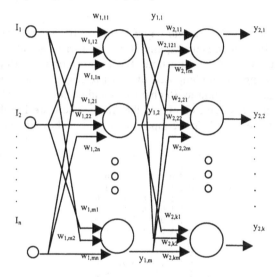

Fig. 4. A Multi-layer feedforward net structure

Thus , Output for the hidden layers is given as following.

$$y_{NET1,j} = \sum_{i=1}^{n} w_{1,ji} I_i$$

$$y_{1,j} = f_j\left[y_{NET1,j}\right] \quad j = 1,2,...,m$$

(1)

output for ouput layers are defined as following.

$$y_{NET2,t} = \sum_{i=1}^{m} w_{2,tj} y_{1,j}$$

$$y_{2,k} = f_t\left[y_{NET2,t}\right] \quad t = 1,2,...,k$$

(2)

5. ANN model design for character recognition

Characters used in the training set for the ANN are given in Table 1

Table 1. Character set.

Upper Case Letters									
A	B	C	D	E	F	G	H	I	J
K	L	M	N	O	P	R	S	T	U
V	Y	Z	Q	W	X				
Numbers									
0	1	2	3	4	5	6	7	8	9

      The proposed ANN, a multilayer feedforward network, consists of 209 input variables, 104 hidden neurons, and 36 output variables and it is designed to recognize one single character at a time. Backpropagation algorithm is used for training. The training set is composed of two files; input and output. Input file contains image data belonging to 36 characters and each character has 209 image segments which are provided as inputs to the ANN. There are 108 data sets (three data sets per character; 36*3) in the input file. Figure 5 shows one of the data sets for character "K".

```
0 0 0 0 0 0 0 0 0 0 0
0 1 1 0 0 0 0 0 1 1 0
0 1 1 0 0 0 0 0 1 1 0
0 1 1 0 0 0 0 1 1 0 0
0 1 1 0 0 0 1 1 1 0 0
0 1 1 0 0 1 1 1 0 0 0
0 1 1 0 1 1 1 0 0 0 0
0 1 1 1 1 1 0 0 0 0 0
0 1 1 1 1 0 0 0 0 0 0
0 1 1 1 1 1 0 0 0 0 0
0 1 1 0 1 1 1 0 0 0 0
0 1 1 0 0 1 1 1 0 0 0
0 1 1 0 0 0 1 1 1 0 0
0 1 1 0 0 0 0 1 1 0 0
0 1 1 0 0 0 0 0 1 1 0
0 1 1 0 0 0 0 0 1 1 0
0 0 0 0 0 0 0 0 0 0 0
0 0 0 0 0 0 0 0 0 0 0
0 0 0 0 0 0 0 0 0 0 0
```

Fig. 5. A Segmented "K" character

The output file consists of 36 outputs each representing a single letter. The output encoding is shown in Table 2.

Table 2. Output characters

			Outputs Value(36)							
			O1	O2	. . .	O26	O27	O28	. . .	O36
Upper Case Latters	1	A	1	0	. . .	0	0	0		0
	2	A	1	0	. . .	0	0	0		0
	3	A	1	0		0	0	0		0
	1	B	0	1	. . .	0	0	0		0
	2	B	0	1	. . .	0	0	0		0
	3	B	0	1	. . .	0	0	0		0
		⋮								
	1	X	0	0		1	0	0		0
	2	X	0	0	. . .	1	0	0	. . .	0
	3	X	0	0	. . .	1	0	0	. . .	0
Numbers	1	0	0	0		0	1	0		0
	2	0	0	0		0	1	0		0
	3	0	0	0		0	1	0		0
		⋮	.	.		.	.	.		.
	1	9	0	0	. . .	0	0	0	. . .	1
	2	9	0	0	. . .	0	0	0	. . .	1
	3	9	0	0	. . .	0	0	0	. . .	1

6. Character recognition using an ANN model

ANN recognizes one character at a time. It recognizes the characters stored in the test file one at a time in sequential order starting from the first character until it reaches the last character. A special control mechanism is used for generating the outputs. As explained earlier, Turkish VLP has three regions containing, from left to right, the county code, a letter code, and a numeric code as shown in Figure 1. County code starts at 01 and goes up to 85 in decimal. Therefore, first two characters must be numeric. The second region in the VLP is a three-letter alpha numeric symbol using only capital letters A-Z. The third region has a 4-digit decimal number. This knowledge of the VLP composition helps in the recognition process. When recognizing the characters in the first and third plate regions, we only need to use the outputs $O_{27},...,O_{36}$. Each time, we pick the one with the maximum value. Similarly, outputs $O_1, .., O_{26}$ are used for recognizing the characters in the second plate region. The output of ANN is streamed into a text file.

7. Test results

We tested our VLP recognition technique using the images of 40 vehicles which were chosen at random. The images had the following properties:

· Images were taken with a simple HP Digital camera and they were shot from a fixed distance.
· Images were taken with a shooting angle ranging between 0^0 -15^0.
· Vehicles were either stationary or moving at a certain speed between 0-30 km/h.

Our program recognized 38 out of 40 license plates with 100% accuracy. This result translates into a 95% recognition rate which is considered to be fairly high. The system failed to recognize only two VLPs. The reason for this failure is mainly due to the poor results obtained during the image processing phase. Basically, the quality of original images was so poor that the segmentation system failed to separate the characters correctly. And hence, ANN system could not recognize them. In general, the incorrect results were due to the recognition failure in only one character. If the image quality is poor, some characters in VLP appear to be the same. The pairs (B, E), (F, P), and (P, R) are examples of such characters. There are other characters with similar properties too. If the image quality is not good and these characters are presented to ANN, they may be confused for one another.

8. Conclusions

In this study, a VLP recognition system based on an ANN is designed and tested on real VLP images. The system has three parts: (i) localization of VLP, and (ii) segmentation of the license image into individual characters, and (iii) character recognition using a multilayer feedforward neural network. Test results show that the system can obtain almost 95% success rate on 40 test images. Generally, the failure to recognize was due to the poor image quality or the incorrect separation of the characters during the image processing phase. Recognition ability of the ANN system was very reliable.

References

1-   Abulafya, N., "Neural Networks for system Identification and Control" MSc Thesis. University of London 1995.
2-   Aksoy M. S, Cagil G, Turker A. K (2000) Number palate recognition using inductive learning. Elsevier, Robotics and Autonomous Systems, vol. 33, pp.149-153.
3-   Blumenstein M, Verna B (1998) Neural based segmentation and recognition technique for handwritten words. Gold coast, Australia.
4-   Draghici S " A neural network based artificial vision system for license plate recognition.
5-   Grattoni P, Pettiti G, Rastello M. L (1999) Experimental set-up for the characterization of automated number-plate recognizers. Elsevier, Measurement, Vol. 26. Pp.103-114.

6- Han K, Sethi I. K (1993) Handwritten signature Retrieval and Identification. Pattern Recognition Letters, 14, pp. 305-315.

7- Iee S. W (1996) Off-line Recognition of Totally unconstrained Handwritten Numerals using Multilayer Cluster Neural Network. IEEE, Transactions on pattern Analysis and Machine Intelligence 18, pp.648-652.

8- Jain A. K, Bhattacharjee S (1992) Text segmentation Using Gabor filters for Automatic document processing. Mach. Vis. Appl., vol. 5,pp.169-184.

9- Jain A. K, Yu B (1998) Automatic text location in images and video frames. Proc. ICPR, pp. 1499-1498.

10- Kou Y, Hsu J (1998) Color-blindness plate recognition using a neuro-fuzzy approach. Pergamon, Engineering Applications of Artificial Intelligence, vol.11, pp. 531-547.

11- Lee C (2003) SIMNET: A Neural Network Architecture for pattern recognition and data mining. PhD thesis, UMR.

12- Li H, Doermann D, Kia O (2000) Automatic Text Detection and Tracking in digital video, IEEE transactions on image processing, Vol. 9, No. 1, pp.147-156, January.

13- Narendra, K.S (1990) Adaptive Control Using Neural Networks" in Neural Networks for Control (W.T. Miller, R.S. Sutton, and P.J. Werbos, edds) 3 rd printing 1992 MIT.

14- Oz C, Koker R (2001) Vehicle License plate recognition using artificial neural Networks. Second International Conference on Electrical and Electronics Engineering, pp. 378-382, Bursa, Turkey.

15- Pal U, Chaudhuri B.B (2001) Machine-printed and hand-written text lines identification", pp 431-441, pattern recognition letters, Elsevier.

16- Richard P. L (1987) An Introduction to Computing with Neural Nets. IEEE ASSP, pp.2-22.

17- Wei W, Wang M, Huang Z (2001) An automatic method of location for Number-palate using color features. IEEE, pp. 782-785.

18- Wu V, Manmatha R, Riseman E. N (1997) Finding text in images. DL'97:Proc. 2nd ACM Int. Conf. Digital Libraries, Images and Multimedia, pp.3-12.

19- Zarrillo M. L, Radwan A. E, Aldeek H. M (1997) Modeling Traffic Operations at Electronic Tool Collection and traffic management Systems. Pergamon, Computers ind. Engng., vol. 33, pp. 857-860.

20- Zheng B., Qian W, Clarke L (1994) Multistage Neural network for pattern Recognition in Mammogram Screening. IEEE ICNN, Orlando, pp. 3437-3448.

21- Zhou J, Lopresti D (1997) Extracting text from www images. In Proc. ICDAR, pp. 248-252.

Weather Forecasting Models Using Ensembles of Neural Networks

Imran Maqsood[1], Muhammad Riaz Khan[2] and Ajith Abraham[3]

[1] Faculty of Engineering, University of Regina, Regina, SK S4S 0A2, Canada, E-mail: maqsoodi@uregina.ca

[2] AMEC Technologies, TTI, 400-111 Dunsmuir Street, Vancouver, BC V6B 5W3, Canada, E-mail: riaz.khan@amec.com

[3] Computer Science Department, Oklahoma State University, Tulsa, Oklahoma OK 74106, USA, E-mail: ajith.abraham@ieee.org

Abstract. This paper examines applicability of Hopfield Model (HFM) for weather forecasting in southern Saskatchewan, Canada. The model performance is contrasted with multi-layered perceptron network (MLPN), Elman recurrent neural network (ERNN) and radial basis function network (RBFN). The data of temperature, wind speed and relative humidity were used to train and test the four models. With each model, 24-hr ahead forecasts were made for winter, spring, summer and fall seasons. Moreover, ensembles of these models were generated by choosing the best values among the four predicted outputs that were closest to the actual values. Performance and reliabilities of the models were then evaluated by a number of statistical measures. The results indicate that the HFM was relatively less accurate for the weather forecasting problem. In comparison, the ensembles of neural networks and RBFN produced the most accurate forecasts.

Keywords. artificial neural networks, forecasting, model, simulation, weather.

1. INTRODUCTION

Weather is a continuous, data-intensive, multidimensional, dynamic and chaotic process, and these properties make weather forecasting a formidable challenge. Generally, two methods are used to forecast weather (a) the empirical approach and (b) the dynamical approach (Lorenz 1969). The first approach is based upon the occurrence of analogs and is often referred to by meteorologists as analog forecasting. This approach is useful for predicting local-scale weather if recorded cases are plentiful. The second approach is based upon equations and forward simulations of the atmosphere, and is often referred to as computer modeling. Because of the grid coarseness, the dynamical approach is only useful for modeling large-scale weather phenomena and may not predict short-term weather efficiently. Most weather prediction systems use a combination of empirical and dynamical techniques. However, a little attention has been paid to the use of artificial neural networks (ANNs) in weather forecasting (Moro et al. 1994; Kuligowski and Barros 1998; Maqsood et al. 2002a, 2002b).

ANNs provide a methodology for solving many types of nonlinear problems that are difficult to solve by traditional techniques. Most meteorological processes often exhibit temporal and spatial variability, and are further plagued by issues of nonlinearity of physical processes, conflicting spatial and temporal scale and uncertainty in parameter estimates. The ANNs exist capability to extract the relationship between the inputs and outputs of a process, without the physics being explicitly provided (Zurada 1992). Thus, these properties of ANNs are well suited to the problem of weather forecasting under consideration.

The objectives of this study are to: (a) examine the applicability of ANN approach by developing effective and reliable nonlinear predictive models for weather analysis in the southern Saskatchewan, Canada; and (b) compare and evaluate the performance of the developed models including MLPN, ERNN, RBFN and HFM in forecasting hourly temperature, wind speed and relative humidity for winter, spring, summer and fall seasons.

2. ARTIFICIAL NEURAL NETWORKS

2.1 Multilayer Perceptron Network (MLPN)

In this study, a fully connected feedforward type neural network (i.e., MLPN) trained with standard backpropagation algorithm is used. This network consists of one input layer, one hidden layer and one output layer. In each layer, every neuron is connected to a neuron in the adjacent layer with different weights. With the exception of the input layer, each neuron receives signals from the neurons of the previous layer linearly weighted by the interconnect values between neurons. The neuron then produces an output signal by passing the summed signal through a sigmoid function. The initial weights are usually adopted as random numbers. The training error level was set to 10^{-4}. The optimal number of hidden neurons is obtained experimentally by changing the network design and running the training process several times until a good performance is obtained (Hagan et al. 1996; Khan and Ondrusek 2000; Tan et al. 2001).

2.2 Elman Recurrent Neural Network (ERNN)

The Elman network is also known as partial recurrent network or simple recurrent network. In this network, the outputs of the hidden layer are allowed to feedback onto itself through a buffer layer. This feedback allows Elman networks to learn to recognize and generate temporal patterns, as well as spatial patterns. Every hidden neuron is connected to only one neuron of the context layer through a constant weight of value one. Hence, the context layer constitutes a kind of copy of the state of the hidden layer, one instant before. The number of context neurons is consequently the same as the number of hidden neurons. Every neuron in the hidden layer receives as input, in addition to the external inputs of the network. Inputs, output and context neurons have linear activation functions, while hidden neurons have sigmoidal activation function.

2.3 Radial Basis Function Network (RBFN)

The RBFN is embedded into a 3-layered feedforward neural network comprised of input, hidden/memory, and output neurons. RBFNs exhibit a good approximation and learning ability, are easier to train, and can converge quickly. It uses a linear transfer function for the output units and Gaussian function (radial basis function) for the hidden units.

2.4 Hopfield Model (HFM)

The Hopfield network is a single layer network with symmetric weight matrices in which the diagonal elements are all zero. Inputs are applied simultaneously to all neurons, which then output to each other and the process continues until a stable state is reached; this leads to the network output. The feedback loops involve the use of particular branches composed of unit-delay elements, which result in a nonlinear dynamical behavior by virtue of the nonlinear nature of the neurons.

3. MODELING SETUP

The weather parameters including temperature, wind speed and relative humidity recorded at the Regina Airport were collected by the Meteorological Department, Environment Canada in 2001 for developing and analyzing the intelligent based forecasting models. To examine the seasonal variations, the available weather data were split into four seasons namely winter (December – February), spring (March – May), summer (June – August) and fall (September – November). For all four models (i.e., MLPN, ERNN, RBFN and HFM), hourly seasonal data were used for training the networks. The hourly datasets of February 26, May 6, August 7 and November 10 were selected as typical days for winter, spring, summer and fall, respectively, in order to test the trained models. A Pentium-III, 1GHz processor computer with 256 MB RAM was used to simulate all the experiments using MATLAB version 5.3.

Neural networks generally provide improved performance with the normalized data. The use of original data as input to the neural network may cause a convergence problem. All the weather data sets were, therefore, transformed into values between -1 and 1 through dividing the difference of actual and minimum values by the difference of maximum and minimum values. The main goal of normalization, in combination with weight initialization, is to allow the squashed activity function to work at least at the beginning of the learning phase. Thus, the gradient, which is a function of the derivative of the nonlinearity, will always be different from zero. At the end of the each algorithm, the outputs were denormalized into the original data format for achieving the desired result.

The configuration of the neural networks depends highly on the problem. Therefore, it is left with the designer to choose an appropriate number of hidden layers and hidden-layer-nodes based on his/her experience. Thus, an appropriate architecture is required for each application using the trial and error method. In this study, training of different neural networks took about few seconds to 30 minutes with a Pentium-III, 1 GHz processor computer. The learning-rate parameter and momentum term were adjusted intermittently to speed up the convergence. In order to keep the simplicity of the modeling structure, only one hidden layer with 72 hidden-layer nodes was used for the MLPN and ERNN models. This number was arrived after analyzing 24, 48, 72, 98 and 120 neurons in the hidden layer. The architecture with 24 and 48 neurons in the hidden layer was faster in computation, but the convergence rate was very slow. The architecture with 98 and 120 neurons in the hidden layer was converging equally well as that with 72 neurons. Therefore, the architecture with 72 neurons in the hidden layer was selected. It was noted that with increased number of hidden layers, convergence rate of the MLPN and ERNN models was decreased. On the other hand, two hidden layers with 180 nodes were chosen for training the RBFN model.

MLPN, ERNN, RBFN and HFM networks were used after deciding the relevant input/output parameters, training/testing data sets and learning algorithms. To decide architectures of the MLPN, HFM and ERNN, a trail and error approach was used. Networks were trained for a fixed number of epochs, and the error gradient was observed over these epochs. Performance of the MLPN, HFM and ERNN networks were evaluated by increasing or decreasing the number of hidden nodes. Since no significant reduction in error was observed beyond forty-five hidden nodes, a single hidden layer network comprising of forty-five neurons was identified. The input and output values were scaled between -1 and +1, and the One-Step-Secant learning algorithm was used for training the MLPN and ERNN networks. The activation functions for MLPN and RNN models were chosen to be log-sigmoid and hyperbolic tangent sigmoid for hidden units, respectively, and pureline for the output units. Since, there is no exact rule for fixing the number of hidden neurons and hidden layers to avoid underfitting or overfitting in MLPN and RNN, therefore, RBFN is investigated to address this difficulty. In RBFN, the numbers of hidden layers and neurons selected by the model were 2 and 180, respectively; the Gaussian activation function was chosen for hidden units, and the pureline for the output units.

4. RESULTS ANALYSIS

The obtained results indicate that satisfactory prediction accuracy has been achieved through MLPN, ERNN, RBFN and HFM models for temperature, wind speed and relative humidity parameters during winter, spring, summer and fall seasons (Figures 1 through 4). All the obtained results were analyzed, compared and evaluated in the following subsections.

4.1 Performance Evaluation and Comparison of the Models

The MLPN can achieve useful weather forecasting results in an efficient way. Compared to the Hopfield modeling results, the MLPN exhibited lower errors. It is capable of representing nonlinear functions than the single layered perceptron. However, the learning process of the MLPN algorithm is time-consuming and its performance is heavily dependent on the network parameters like learning rate and momentum.

The ERNN, compared to the MLPN, could efficiently capture the dynamic behavior of the weather, resulting in a more compact and natural representation of the temporal information contained in the weather profile. EMS error of the ERNN was much lower than that of the Hopfield model. The recurrent network took more training time, but this depends on the data size and the number of network parameters. It can be inferred that the ERNN could yield more accurate results, if good data-selection strategies, training paradigms, and network input and output representations are determined properly.

The RBFN network can gave the overall best results in terms of accuracy and training time. The RBFN is better correlated compared to the MLPN, ERNN and HFM. The proposed RBFN network can also overcome several limitations of the MLPN and ERNN such as highly nonlinear weight update and the slow convergence rate. Since the RBFN has natural unsupervised learning characteristics and a modular network structure, these properties make it more effective for fast and robust weather forecasting.

Hopfield model shows higher values for RMSE, MAD and MAP. Figures 1 through 4 indicate that the Hopfield model overestimated most of the predicted values. Overall, the performance of the Hopfield model is reasonable. However, compared to the other models, it is less accurate for the weather-forecasting problem.

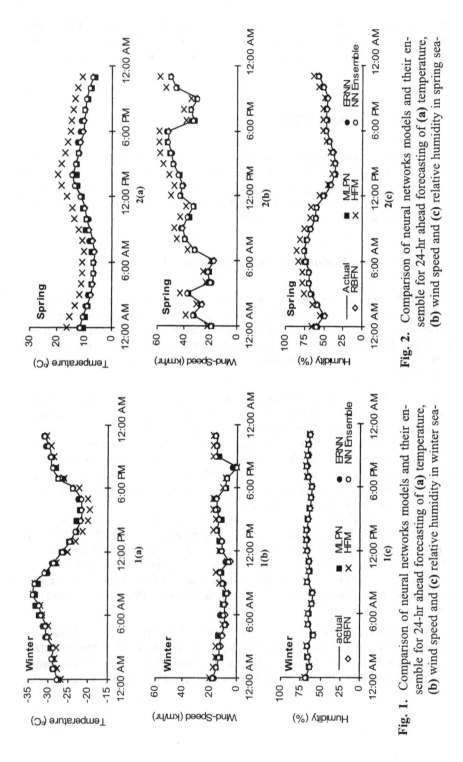

Fig. 1. Comparison of neural networks models and their ensemble for 24-hr ahead forecasting of (**a**) temperature, (**b**) wind speed and (**c**) relative humidity in winter sea-

Fig. 2. Comparison of neural networks models and their ensemble for 24-hr ahead forecasting of (**a**) temperature, (**b**) wind speed and (**c**) relative humidity in spring sea-

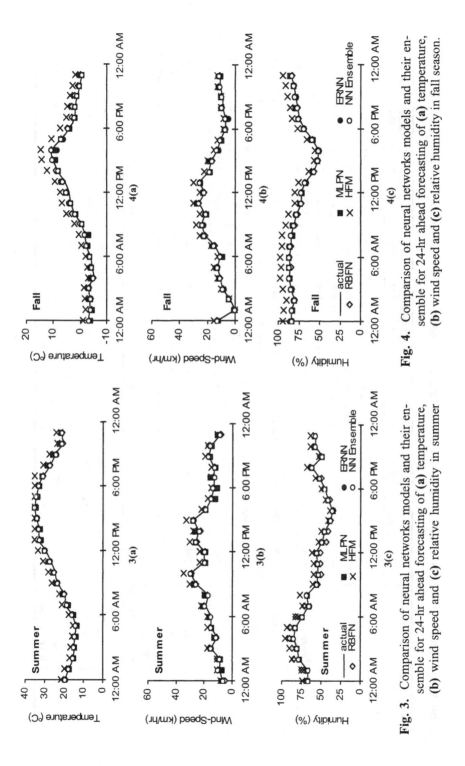

Fig. 3. Comparison of neural networks models and their ensemble for 24-hr ahead forecasting of **(a)** temperature, **(b)** wind speed and **(c)** relative humidity in summer

Fig. 4. Comparison of neural networks models and their ensemble for 24-hr ahead forecasting of **(a)** temperature, **(b)** wind speed and **(c)** relative humidity in fall season.

Table 1. Performance of MLPN, ERNN, RBFN and HFM for forecasting temperature, wind speed and humidity.

Forecasting Season	Reliability Parameter	Temperature (°C)				Wind Speed (km/h)				Humidity (%)			
		MLPN	ERNN	RBFN	HFM	MLPN	ERNN	RBFN	HFM	MLPN	ERNN	RBFN	HFM
Winter	MAD	0.5685	0.5610	0.5242	2.9894	0.8505	0.8158	0.7246	2.8120	0.6285	0.6249	0.5669	3.5575
	RMSE	0.0200	0.0199	0.0060	0.0456	0.0199	0.0199	0.0131	0.0327	0.0200	0.0199	0.0071	0.0395
	CC	0.9919	0.9883	0.9997	1.0000	0.9526	0.9032	0.9192	0.9872	0.9712	0.9738	0.9845	0.9982
	Training Time	19 min	25 min	.05 min	.03 min	10 min	8 min	.05 min	.03 min	15 min	25 min	.05 min	.03 min
	Iteration number	17678	16983	-	-	11909	4894	-	-	13742	7513	-	-
Spring	MAD	0.7958	0.7898	0.9296	3.2721	0.9012	0.8888	0.8678	3.2832	1.1975	1.1699	1.1561	7.7781
	RMSE	0.0200	0.0199	0.0032	0.1184	0.0200	0.0199	0.0044	0.1543	0.0200	0.0199	0.0054	0.1537
	CC	0.9436	0.9461	0.9992	1.0000	0.9948	0.9974	0.9975	0.9994	0.9976	0.9945	0.9965	0.9999
	Training Time	10 min	15 min	3 sec	2 sec	7 min	5 min	3 sec	2 sec	25 min	32 min	3 sec	2 sec
	Iteration number	9335	10173	-	-	5514	3271	-	-	21055	19597	-	-
Summer	MAD	0.5259	0.4699	0.4641	2.3151	0.8784	0.8794	0.8615	2.5811	1.1211	1.1349	1.2937	6.9252
	RMSE	0.0199	0.0199	0.0050	0.0650	0.0200	0.0199	0.0119	0.0365	0.0199	0.0199	0.0155	0.2029
	CC	0.9937	0.9976	0.9996	0.9897	0.9691	0.9603	0.9900	0.9892	0.9877	0.9879	0.9891	0.9852
	Training Time	17 min	21 min	.05 min	.03 min	4 min	9 min	.05 min	.03 min	19 min	30 min	.05 min	.03 min
	Iteration number	20109	12383	-	-	3904	6090	-	-	16500	18249	-	-
Fall	MAD	0.7100	0.6872	0.6739	1.0242	0.8331	0.8633	0.8258	3.0417	1.1910	1.1862	1.0765	6.8769
	RMSE	0.0199	0.0199	0.0169	0.0797	0.0199	0.0199	0.0092	0.0533	0.0200	0.0199	0.0120	0.1844
	CC	0.9902	0.9853	0.9961	1.0000	0.9909	0.9851	0.9983	1.0000	0.9962	0.9958	0.9991	1.0000
	Training Time	11 min	27 min	.05 min	.03 min	4 min	7 min	.05 min	.03 min	28 min	30 min	.05 min	.03 min
	Iteration number	10117	13523	-	-	3812	4777	-	-	20731	12614	-	-

4.2 Reliability Analysis

The optimal network is the one that should has lower error and reasonable learning time. Prediction reliability of the four models was evaluated based on a number of statistical analyses as shown in Table 1. The training time for the MLPN and ERNN models ranged from 5 to 30 minutes, while it took only a few seconds to run the RBFN and Hopfield models. The ERNN model took more training time and had relatively less number of iterations compared to the MLPN model. With the improvement of computing speed, the training time due to different algorithms may no longer be such a crucial factor if the training record is not too long and the design architecture is not too complicated. The testing accuracy could get worse if the selection of the algorithm to represent the problem is not proper.

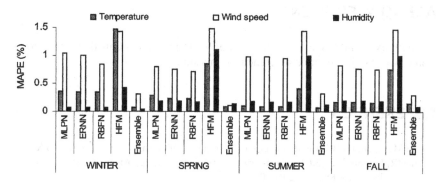

Fig. 5. Comparison of mean absolute percentage errors

Figure 5 indicates that, for winter and spring seasons, humidity was predicted with the lowest mean absolute percentage error (MAPE) values by the four soft computing models; whereas, for summer and fall seasons, temperature predictions were the most accurate with the lowest MAPE values. In comparison, humidity predictions represented the least accurate results (i.e. higher MAPE values) for all the four seasons. The different levels of MAPE associated with temperature, wind speed and humidity could be due to the variability of data patterns in the four seasons as well as the variations of the modeling structures. In overall, ensembles of neural network produced the lowest MAPE (i.e., the most accurate forecast).

5. CONCLUSIONS

Four soft computing models based on artificial neural networks were developed for hourly weather forecasting in southern Saskatchewan, Canada. Performance of the HFM was compared with those of MLPN, ERNN and RBFN. The forecasting reliabilities of these models were evaluated by computing a number of statistical measures. The modeling results indicate that reasonable prediction accuracy was achieved for most of the models. The best predictions were shown by the ensembles of the four predicted outputs. The MLPN and ERNN did equally well in forecasting temperature, humidity and wind speed. In comparison, the RBFN performed better than the MLPN and ERNN, while the HFM showed the lowest accuracy.

ACKNOWLEDGMENTS

The authors would like to thank the staff of the Environment Canada for the provision of weather information as needed for this study.

REFERENCES

Hagan MT, Demuth HB, Beale MH (1996) Neural network design. PWS, Boston, MA

Khan MR, Ondrusek C (2000) Short-term electric demand prognosis using artificial neural networks. J Elec Engg 51:296–300

Kuligowski RJ, Barros AP (1998) Localized precipitation forecasts from a numerical weather prediction model using artificial neural networks. Weather and Forecasting 13:1194–1205

Lorenz EN (1969) Three approaches to atmospheric predictability. Bulletin of the American Meteorological Society 50:345–349

Maqsood I, Khan MR, Abraham A (2002a) Intelligent weather monitoring systems using connectionist models. International Journal of Neural, Parallel and Scientific Computations 10:157–178

Maqsood I, Khan MR, Abraham A (2002b) Neuro-computing based Canadian weather analysis. In: The 2nd International Workshop on Intelligent Systems Design and Applications. Comput Intell Applic, Dynamic Publishers, Atanta, Georgia. pp 39–44

Moro QI, Alonso L, Vivaracho CE (1994) Application of neural networks to weather forecasting with local data. In: 12 IASTED Internat Conf Applied Informat 12, 68–70

Tan Y, Wang J, Zurada JM (2001) Nonlinear blind source separation using a radial basis function network. IEEE Transactions on Neural Networks 12:124–134

Zurada JM (1992) Introduction to Artificial Neural Systems. West Publishing Company, Saint Paul, Minnesota

Neural Network Predictive Control Applied to Power System Stability

Steven Ball

New Mexico Tech, Institute for Complex Additive Systems Analysis Division, Socorro NM 87801. sball@icasa.nmt.edu

Abstract

In this paper we consider the problem of power system stability with an application of predictive control systems, and in particular control systems which rely on Neural Networks to maintain system stability. The work focuses on how a hybrid control system utilizing neural networks has the capabilities to improve the stability of an electric power grid when used in place of traditional control systems, i.e. PID type controllers. Testing is done using simulation of the dynamical system with the different control schemes implemented, resulting in a measure of stability through which the controllers can be justified.
Keywords: Learning Control, Predictive Control, Electric Power Systems, Voltage Stability, Neural Network Application, Adaptive Control

1 Introduction

An electric power system can be classified as a hybrid dynamical system whose nature can be described using logic rules, discrete dynamics and continuous dynamics. Because events in an electric power system can happen without notice and time scales within the power system vary so greatly; traditional control techniques are often not adequate for controlling the stability of an electric power system during a catastrophic event, usually resulting in a failure.

Traditional power system controllers have often been unsuccessful in preventing failures, and while small power system failures typically do very little damage or even go unnoticed by consumers, large cascading failures are not uncommon [1]. In fact cascading failures happen with almost a sense of regularity, so much even that esteemed power systems authorities have alluded to power systems being of the "self-organized-critical" system type [2]. It is due to the complex nature inherent to a power system that makes it so difficult to infallibly control.

In the past power system control was often based on methods that use differential geometry to solve the non-linear characteristics associated with power systems [3]. More recently, however, the study of power system control has been based on "Direct Feedback Linearization," which allows the use of well-known linear con-

trol theory [4]. Even techniques that use super position [5] and techniques that select models that best fit the data [6] are often used in the literature.

Although these methods have been used with some degree of success we would
like to forgo some of the guess work involved in the more traditional techniques.
We can do this though the use of Neural Networks [7, 8]. Neural Networks allow
us the advantage of being able to use non-linear coupled systems without the need
to transform them into linear systems or do any other kind of hand waving to
model the system. Conversely to traditional control techniques neural networks
offer a flexible structure that can arbitrarily model a non-linear dynamical system
[9, 10]; making neural networks ideal for power system modeling and control.

This paper first discusses the power system used in this study, identifying the
plant and desired operating point of the system. Next we will discuss the artificial
neural network used in the controller including the training algorithms and testing
of the plant. Finally we will make some conclusion about the usefulness of neural
networks in power system stability.

2 Power System Representation

The system used for this study, figure 1, is a generic two bus power system.

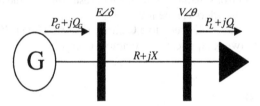

Fig. 1. Power system one-line diagram.

There are several important assumptions that go along with this model. First, it
is assumed by way of automatic voltage regulation that bus one voltage, E, is kept
constant at 1 per-unit. We also assume that the Synchronous Machine is unregu-
lated—meaning it can output as much power as demanded from it. We also let
$\delta = \delta - \theta$ to simplify calculations. Then we write the system equations using the
swing equation to model the synchronous machine, and power balance equations
to find power demand at the bus and power supplied by the generator [11].

$$\dot{\omega} = \frac{1}{M}(P_M - P_G - D_G\omega) \tag{1}$$

$$\dot{\delta} = \omega - \theta \tag{2}$$

$$P_G = E^2G - EV(G\cos(\delta) - B\sin(\delta)) \tag{3}$$

$$Q_G = E^2 B - EV(G\sin(\delta) - B\cos(\delta)) \tag{4}$$

$$0 = -BV^2 - EV(G\sin(\delta) - B\cos(\delta)) - Q_D \tag{5}$$

$$0 = -GV^2 - EV(G\cos(\delta) + B\sin(\delta)) - P_D \tag{6}$$

Noticing that (5) and (6) are algebraic equations we can use time-scale decomposition [12, 13] to model the system in the singular perturbation form of:

$$\dot{x} = f(x, u, y, \xi) \tag{7}$$

$$\xi \dot{y} = g(x, u, y, \xi) \tag{8}$$

Where g represents the algebraic equations and f represents the state equations. Furthermore, ξ is a constant that is defined as being sufficiently small such that $\xi \dot{y} \approx 0$, thus allowing the use of different time scale dynamics within a single system without the singularity-induced problems in the unreduced Jacobian [14]. So, re-writing (5) and (6) singularly perturbed and using the relationship $\delta = \delta - \theta$ we have:

$$\dot{V} = \frac{1}{\tau}\left(- BV^2 - EV(G\sin(\delta) - B\cos(\delta)) - Q_D\right) \tag{9}$$

$$\dot{\delta} = \omega - \frac{1}{D_L}\left(- GV^2 - EV(G\cos(\delta) + B\sin(\delta)) - P_D\right) \tag{10}$$

Now we have defined the system states (Eqns. 1, 9, 10) for the simple power system given in figure 1. From which we define the plant as being equations 1 and 10—the equations that come from the synchronous machine.

To further simplify the analysis of the system we assume that the voltage at a bus is set equal to 1 by way of automatic voltage regulation [15] and we assume that the transmission line is short and thereby neglect resistance in the line, or we make G=0 in the above equations. Our final system has only one input, P_m and three outputs, V, δ, and ω. Of the three outputs we would like to control V at bus 2. However, there is not a direct relationship between V and P_m therefore we would like to determine if the system is controllable.

Accessibility of a system is sufficient in determining controllability of a system. Therefore we will check accessibility of the power system to ensure controllability. This can be done by showing that the system is fully accessible using Lie Brackets as shown below, equations 11-16.

First we must write the equation in the vector form of $\dot{x} = fx + gu$, where we can say that if there is a non-zero number in g that state is controllable. So, writing the state equations in the form of:

$$\dot{x} = fx + gu \tag{11}$$

We have:

$$f = \begin{pmatrix} w - \dfrac{1}{D_L}(P_L - P_D) \\ -\dfrac{1}{M}(P_G + D_G(w)) \\ \dfrac{1}{\tau}(Q_L - Q_D) \end{pmatrix} \tag{12}$$

$$g = \begin{pmatrix} 0 \\ \dfrac{1}{M} \\ 0 \end{pmatrix} \tag{13}$$

$$u = P_m \tag{14}$$

Straight away we can see that δ is accessible. However, we must persist with Lie Bracket calculations to see if ω and V are controllable, so we write the first Lie Bracket as:

$$[g, f](x) = (f_x g - g_x f)x \equiv \begin{pmatrix} \dfrac{1}{M} \\ \dfrac{D_G}{M^2} \\ 0 \end{pmatrix} \tag{15}$$

Now we have accessibility in ω and δ, and still V is not accessible so we must write the second Lie Bracket as:

$$[g, [g, f]](x) = ([gf]_x g - g_x [gf])x \equiv \begin{pmatrix} -\dfrac{V\cos\delta}{XD_L M} - \dfrac{D_G}{M^2} \\ -\dfrac{V\cos\delta}{XM^2} + \dfrac{D_G^2}{M^3} \\ \dfrac{-V\sin\delta}{X\tau M} \end{pmatrix} \tag{16}$$

Clearly we can see that the system is fully accessible with only one control, P_m. Therefore it is valid to design a controller for this system, and we can persist with the design of the learning controller.

3 Neural Network Based Predictive Controller

Learning control architectures have properties that seem ideal for dynamical systems that combine fast and slow time scales with never-before seen events. By learning the state-space dependencies of the system the learning controller can predict the effects of novel disturbances that involve fast dynamics as well as learn how to compensate for slow-time dynamics and static non-linearities in the system [9]. Due to the behavior of this controller we will refer to it as a predictive controller in this paper.

3.1 Controller Design

The design of the predictive controller used in this paper is based on a receding horizon technique. Control predictions are optimized to determine the control signal that minimizes the performance criterion over the specified time-span [16].

A generic representation of the predictive controller is shown below in figure 2 [16].

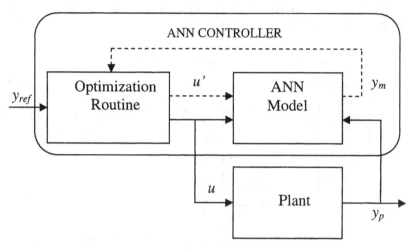

Fig. 2. Representation of a neural network controller

The optimization block outputs a control signal, u' to the neural network model. This output is used by the neural network model to determine how the plant will react to the control input. The neural network plant sends an output signal, y_m back to the optimization algorithm until the desired output, y_{ref} is achieved, at which time the final control output, u goes into the plant. As a result the plant output, y_p is achieved. The number of iterations used in this process changes the accuracy and performance of the controller substantially. Before this controller can be used a neural network must be trained and then tested to ensure that the controller is indeed capable of obtaining a valid output.

3.2 Training Process

For this system we have one output, P_m that is needed from the controller. As an input to the controller we also have one input, V. Because a neural network is adept at modeling dynamical non-linear systems and the power system is highly coupled we do not need to feed back all of the states. Further we have seen that by controlling P_m we can indeed control V (Eqns. 13, 15, 16) and therefore it is not necessary and saves computation time to only have a single input single output control system. As well as the fact that often all of the states are not available for feedback and there fore the controller would not be realistic for real life application if we had chosen to feed back all the state. We can however, expand the one state into real-time and delayed inputs which gives a better horizon for the optimization algorithm. The resulting network has four layers, one real-time input, two delayed inputs, and each hidden layer is seven neurons wide. It is a multi-layer feed-forward network that yields two outputs, P_m and P_m delayed.

The neural network model is then trained using scaled conjugate gradient back propagation as shown in figure 3, below [7].

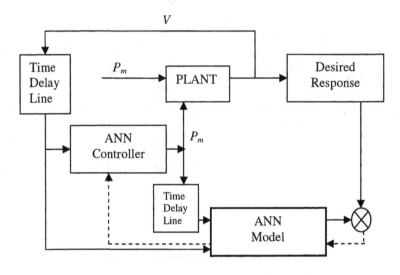

Fig. 3. Training of a neural network model.

In the literature we find that the back propagation learning algorithm is the most widely used training algorithm for feed-forward networks. Momentum is added so that the back propagation algorithm's sensitivity to small details in the error surface is decreased such that the network avoids getting stuck in local minima which would prevent the network from finding the absolute minimum. Furthermore the adaptive learning rates are used to attempt to keep the learning step size as large as possible while keeping learning stable.

Training is done for a maximum of 15000 epochs although an error of less than 10^{-6} is typically reached after only several hundred epochs. To ensure that training is adequate we train the network with 10000 samples at intervals of 0.1 seconds. The training samples are generated from several different scenarios which could happen in a power system—load changes and line outages for example. Then the network is tested to ensure that the network is working correctly.

3.3 Plant Testing

Testing is done to test the generalization capabilities of the plant. During testing the neural network is subjected to random inputs and the neural network output is compared to the actual plant output given the same input. We desire that a maximum error of 10^{-3} be achieved during testing. This will provide us with an adequate accuracy for the controller. Changing parameters of the network can have a substantial effect on the accuracy and performance of the plant. Making the hidden layer smaller makes decreases computation time while increasing the error. However making the hidden layer larger increases computation time while decreasing the error; clearly this is a tradeoff between performance and accuracy. We desire for the network to work as fast as possible, however we would also like for the accuracy to be good. Restricting the error to better than 10^{-3} we find that a hidden layer of size seven yields both relatively fast computation time and small errors. Therefore we choose this as our hidden layer size. As seen in figure 4 the plant output matches the NN output for a variety of inputs, figure 5, with a very small error, figure 6.

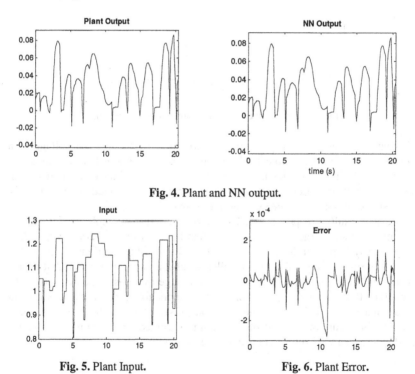

Fig. 4. Plant and NN output.

Fig. 5. Plant Input.   **Fig. 6.** Plant Error.

Clearly we can see that the results after testing indicate that the neural network representation of the plant is identical to the actual plant. Therefore we can conclude that the network was adequately trained for use in the controller.

4 Conclusion

The trained controller operates very well. In fact we can see that the neural network controller operates better than traditional PID type controllers given the same disturbance, figure 7. These results are obtained using a cost horizon of 3, a control horizon of 2, and a control weighting factor of 0.5.

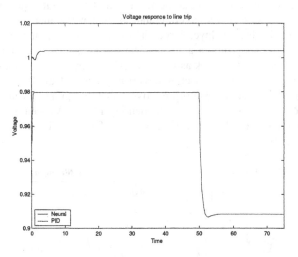

Fig. 7. Neulral network vs. PID controller comparison.

Simulation shows that the predictive controller will stabilize the voltage of a power system even when subjected to events such as random load changes and line outages. As in figure 7 it can be seen that the voltage of the controlled system is more constant and oscillates less than either an uncontrolled or traditionally controlled system—making the system more stable.

While the results are promising, they should be further explored before any judgment is made. Large power systems often have different characteristics than small power systems and therefore different results may be experienced. Also, the computation time for a predictive controller is substantially longer than that of traditional controllers and may be impractical for large systems. Therefore we cannot conclude that a neural network controller is applicable for all situations we can only say that it is applicable for the situation presented here. It should be noted that due to the nature of the controller it could be trained for any situation imaginable—leading to a very robust control for the system.

5 References

1. Thorp J, Wang H (2002), Computer Simulation of Cascading Disturbances in Electric Power Systems. Impact of Protection Systems on Transmission System Reliability, Final Report Cornell University.
2. Carreras B (2001) IEEE Evidence for Self-Organized Criticality in Electric Power System Blackouts, Hawaii International Conference on System Sciences. Maui, Hawaii.
3. Isidori, A. (1989). Nonlinear Control Systems: An Introduction. 2ed, Springer Verlag, NY, USA
4. Czajkowski S, Nachez J (2000), Linearized—Uncoupled control for a synchronous Generator. Universidad Nacional Rosario, Argentina.
5. Leach W, Superposition of Depedent Sources is Valid in Circuit Analysis. Georgia Institute of Technology. Atlanta Georgia
6. Bérubé G, Hajagos L, Testing & Modeling of Generator Controls. Kestrel Power Engineering. Toronto, Ontario, Canada.
7. G. Venaygamoorthy, R. Harley (1999), A Robust Artificial Neural Network Controller for a Turbogenerator when Line Configuration Changes. Atlanta, GA.
8. G. Venaygamoorthy, R. Harley (1998), A practical continually online trained artificial neural network controller for a Turbogenerator. Atlanta, GA.
9. White D, Sofge D (1992), Handbook of Intelligent Controls. Van Nostrand Reinhold. New Your, USA.
10. Hunt K, Sbarbaro D, Zbikowski R, Gawthrop PJ (1992), "Neural networks for control systems – a survey", Automatica, Vol 28, No. 6, pp 1083-1112.
11. Anderson P, Fouad A (1994), Power System Control and Stability. IEEE Press, NJ.
12. Kokotovic P, Khalil H, O'Reilly J (1986), Singular Perturbation Methods in Control: Analysis and Design. Academic Press.
13. Vournas C (1998), Voltage Stability of Electric Power Systems. Kluwer Academic Publishers.
14. Huang G, Zhao L, Song X (2002), A New Bifurcation Analysis for Power System Dynamic Voltage Stability Studies. IEEE.
15. IEEE/PES Power System Stability Subcommittee Special Publication (2002), VOLTAGE STABILITY ASSESSMENT: CONCEPTS, PRACTICES AND TOOLS.
16. Demuth H, Beale M (1991), Neural Network Toolbox for Use With MATLAB; User's Guide; The Mathworks Inc.

Identification of Surface Residues Involved in Protein-Protein Interaction – A Support Vector Machine Approach

Changhui Yan[1,2,5], Drena Dobbs[3,4,5], Vasant Honavar[1,2,4,5]

[1]Artificial Intelligence Research Labortory, [2]Department of Computer Science,
[3]Department of Genetics, Development and Cell Biology,
[4]Laurence H Baker Center for Bioinformatics and Biological Statistics,
[5]Bioinformatics and Computational Biology Graduate Program,
Iowa State University.
Atanasoff Hall 226, Iowa State University, Ames, IA 50011-1040, USA
chhyan@iastate.edu

Summary

We describe a machine learning approach for sequence-based prediction of protein-protein interaction sites. A support vector machine (SVM) classifier was trained to predict whether or not a surface residue is an *interface residue* (i.e., is located in the protein-protein interaction surface) based on the identity of the target residue and its 10 sequence neighbors. Separate classifiers were trained on proteins from two categories of complexes, antibody-antigen and protease-inhibitor. The effectiveness of each classifier was evaluated using leave-one-out (jack-knife) cross-validation. Interface and non-interface residues were classified with relatively high sensitivity (82.3% and 78.5%) and specificity (81.0% and 77.6%) for proteins in the antigen-antibody and protease inhibitor complexes, respectively. The correlation between predicted and actual labels was 0.430 and 0.462, indicating that the method performs substantially better than chance (zero correlation). Combined with recently developed methods for identification of surface residues from sequence information, this offers a promising approach to prediction of residues involved in protein-protein interaction from sequence information alone.

Introduction

Identification of protein-protein interaction sites and detection of specific amino acid residues that contribute to the specificity and strength of protein interactions is an important problem with applications ranging from rational drug design to analysis of metabolic and signal transduction networks. Because the number of experimentally determined structures for protein-protein complexes is small, computational methods for identifying amino acids that participate in protein-protein interactions are becoming increasingly important (reviewed in Teichmann *et al.*, 2001; Valencia and Pazos, 2002). This paper addresses the question: Given the fact that a protein interacts with another protein; can we predict which amino acids are located in the interaction site?

Based on different characteristics of known protein-protein interaction sites, several methods have been proposed for predicting protein-protein interaction sites using a combination of sequence and structural information. These include methods based on presence of "proline brackets'' (Kini and Evans, 1996), patch analysis using a 6-parameter scoring function (Jones and Thornton 1997a, 1997b), analysis of hydrophobicity distribution around a target residue (Gallet *et al.*, 2000), multiple sequence alignment (Casarai *et al.*, 1995; Lichtarge *et al.*, 1996; Pazos *et al.*, 1997), structure-based multimeric threading (Lu *et al.*, 2002), and analysis of characteristics of *spatial neighbors* of a target residue using a neural network (Zhou *and* Shan,2001; Fariselli *et al.*, 2002).

We have recently reported that a support vector machine (SVM) classifier can predict whether a surface residue is located in the interaction site using the *sequence neighbors* of the target residue. Surface residues were predicted with specificity of 71%, sensitivity of 67% and correlation coefficient of 0.29 on a set of 115 proteins belonging to six different categories of complexes: antibody-antigen; protease-inhibitor; enzyme complexes; large protease complexes; G-proteins, cell cycle, signal transduction; and miscellaneous. (Yan *et al.* 2002). The results presented in this paper show that the SVM classifiers perform even better when trained and tested on proteins belonging to each category separately, suggesting that the design of specialized classifiers for each major class of known protein-protein complexes will significantly improve sequence-based prediction of protein-protein interaction sites.

Methods

Protein complexes, proteins and amino acid residues

Proteins of protease-inhibitor complexes and antibody-antigen complexes were chosen from the 115 proteins used in our previous study (Yan *et al.* 2002). In the study described here, we focused on two set of proteins: 19 proteins from protease-inhibitor complexes and 31 proteins from antibody-antigen complexes (The proteins are available at http://www.public.iastate.edu/~chhyan/sup.htm). Solvent accessible surface area (ASA) was computed for each residue in the unbound molecule and in the complex using the DSSP program (Kabsch and Sander, 1983). The relative ASA of a residue is its ASA divided by its nominal maximum area as defined by Rost and Sander (1994). A residue is defined to be a *surface residue* if its relative ASA is at least 25% of its nominal maximum area. A surface residue is defined to be an *interface residue* if its calculated ASA in the complex is less than that in the monomer by at least 1Å^2 (Jones and Thornton, 1996). Using this method, we obtained 360 interface residues and 832 non-interface residues from the 19 proteins from the protease-inhibitor complexes and 830 interface residues and 3370 non-interface residues from the 31 proteins from the antibody-antigen complexes.

Support vector machine algorithm

Our study used the SVM in the Weka package from the University of Waikato, New Zealand (http://www.cs.waikato.ac.nz/~ml/weka/) (Witten and Frank 1999). The package implements John C. Platt's (1998) sequential minimal optimization (SMO) algorithm for training a support vector classifier using scaled polynomial kernels. The SVM is trained to predict whether or not a surface residue is in the interaction site. It is fed with a window of 11 contiguous residues, corresponding to the target residue and 5 neighboring residues on each side. Following the approach used in a previous study by Fariselli *et al.* (2002), each amino acid in the 11 residue window is represented using 20 values obtained from the HSSP profile (http://www.cmbi.kun.nl/gv/hssp) of the sequence. The HSSP profile is based on a multiple alignment of the sequence and its potential structural homologs (Dodge *et al.*, 1998). Thus in our experiments, each target residue is associated with a 220-element vector. The learning algorithm generates a classifier which takes as input a 220 element vector that encodes a target residue to be classified and outputs a class label.

Evaluation measures for assessing the performance of classifiers

Measures including *correlation coefficient, accuracy, sensitivity (recall), specificity (precision), and false alarm rate* as discussed by Baldi (2000) are investigated to evaluate the performance of the classifier. Detailed definition of these measures can be found in supplementary materials (http://www.public.iastate.edu/~chhyan/isda2003/sup.htm). The *sensitivity* for a class is the probability of correctly predicting an example of that class. The *specificity* for a class is the probability that a positive prediction for the class is correct. The false positive rate for a class is the probability that an example which does not belong to the class is classified as belonging to the class. The *accuracy* is the overall probability that prediction is correct. The *correlation coefficient* is a measure of how predictions correlate with actual data. It ranges from -1 to 1. When predictions match actual data perfectly, correlation coefficient is 1. When predictions opposite with actual data, correlation coefficient is -1. Random predictions yield a correlation coefficient of 0. We chose not to emphasize the traditional measure of prediction *accuracy* because it is not a useful measure for evaluating the effectiveness of a classifier when the distribution of samples over different classes is unbalanced (Baldi, 2000). For instance, in the antibody-antigen category there are 830 interface residues and 3370 non-interface residues in total, a predictor that always predicts a residue to be a non-interaction residue will have an accuracy of 0.80 (80%). However, such a predictor is useless for correct identification of interface residues.

Results

Classification of surface residues into interface and non-interface residues

To evaluate the effectiveness of one approach we used leave-one-out cross-validation (jack-knife) experiments on each category of complexes. For the antibody-antigen category, 31 such jack-knife experiments were performed. In each experiment, an SVM classifier was trained using a training set consisting of interface residues and non-interface residues from 30 of the 31 proteins. The resulting classifier was used to classify the surface residues from the remaining protein into *interface residues* (i.e., the amino acids located in the interaction surface) and *non-interface residues* (i.e., residues not in the interaction surface). Similarly 19 jack-knife experiments were performed for the protease-inhibitor category. The results reported in Table 1 represent averages for the antibody-antigen and protease inhibitor categories, respectively. Detailed results of the experiments are available at http://www.public.iastate.edu/~chhyan/isda2003/sup.htm .

For proteins from antibody-antigen complexes, the SVM classifies achieved relatively high sensitivity (82.3%), specificity (81.0%), with a correlation coefficient of 0.430 between predicted and actual class labels, indicating that the method performs substantially better than random guessing (which would correspond to correlation coefficient equal to zero). For proteins from protease-inhibitor complexes, the SVM classifiers performed with sensitivity of 78.5% and specificity of 77.6%, and with a correlation coefficient of 0.462. For comparison, Table 1 also summarizes results obtained in our previous study using an SVM classifier trained and tested on a combined set of 115 proteins from six categories (Yan *et al.* 2002). Note that the correlation coefficients obtained in the current study for antibody-antigen complexes (0.430) and protease inhibitor complexes (0.462), are significantly higher than those obtained for a single classifier trained using a combined dataset of all six types of protein-protein complexes (0.290).

Table 1. Performance of the SVM classifier

	Antibody-antigen complexes[a]	Protease-inhibitor complexes[a]	Six categories of complexes[b]
Correlation coefficient	0.430	0.462	0.290
Sensitivity	82.3%	78.5%	66.9%
Specificity	81.0%	77.6%	70.8%
False alarm rate	41.0%	35.7%	35.9%

[a] The SVM classifiers were trained and evaluated separately on two categories of proteins.
[b] The performance of the SVM trained and tested on a mixed set of 115 proteins from six categories (Yan *et al.* 2002)

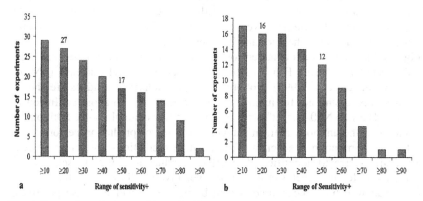

Fig. 1. Interaction site recognition: distribution of sensitivity$^+$ (sensitivity for predicting interface residues) values. The bars in graph illustrate the fraction of the experiments (vertical axis) that fall into the performance categories named below the horizontal axis. **a.** the distribution of sensitivity$^+$ values for 31 experiments in the antibody-antigen category; **b.** the distribution of sensitivity$^+$ values for 19 experiments in the protease-inhibitor category

Recognition of interaction sites

We also investigated the performance of the SVM classifier in terms of overall recognition of interaction sites. This was done by examining the distribution of *sensitivity$^+$* (the sensitivity for positive class, i.e., interface residues class). The sensitivity$^+$ value corresponds to the percentage of interface residues that are correctly identified by the classifier.

Fig. 1a shows the distribution of sensitivity$^+$ values for the 31 experiments in antibody-antigen category. In 54.8% (17 of 31) of the proteins, the classifier recognized the interaction surface by identifying at least half of the interface residues, and in 87.1% (27 of 31) of the proteins, at least 20% of the interface residues were correctly identified. **Fig. 1b** shows the distribution of sensitivity$^+$ values for the 19 experiments in protease-inhibitor category. In 63.2% (12 of 19) of the proteins, the classifier recognized the interaction surface by identifying at least half of the interface residues, and in 84.2% (16 of 19) of the proteins, at least 20% of the interface residues were correctly identified. Distributions of other performance measures for the experiments are available in supplementary materials (http://www. public.iastate.edu/~chhyan/isda2003/sup.htm).

Evaluation of the predictions in the context of three-dimensional structures

To further evaluate the performance of the SVM classifier, we examined predictions in the context of the three-dimensional structures of heterocomplexes. In the antigen-antibody category, in the "best" example (correlation coefficient 0.87, sensitivity+ 96%) 22 out of 23 interface residues were correctly identified as such (i.e., there was only 1 false negative) and 5 non-interface residue was incorrectly classified as belonging to the interface (false positive).

Fig. 2a illustrates results obtained for another example in the antigen-antibody complex category, murine Fab N10 bound to Staphylococcal nuclease (SNase) (Bossart-Whitaker *et al.*, 1995). Note that predicted interface residues are shown only for Fab N10, and not for its interaction partner (gray) to avoid confusion in the figure. The Fab N10 "target" protein shown in this example ranked 9th out of 31 proteins in the antibody-antigen category in terms of prediction performance, based on its correlation coefficient. True positive predictions are shown in red. The classifier correctly identified 20 interface residues in Fab N10 (sensitivity+ 83.3%), and failed to detect 4 of them (false negatives, yellow). Note that several residues that were incorrectly predicted to be interface residues (false positives, blue) are located in close proximity to the interaction site. In this example, the SVM classifier correctly identified interface residues from all 6 complementarity determining regions (CDRs) known to be involved in epitope recognition (Bossart-Whitaker *et al.*, 1995).

Fig. 2b and **c** illustrate results obtained for two proteins from the protease-inhibitor complex category, the "best" example (correlation coefficient 0.83) and "4th best" (correlation coefficient 0.70). In the best example (**Fig. 2b**), the target protein is a serine protease, bovine α-chymotrypsin (1acb E), in complex with the leech protease inhibitor eglin c (1acb I; Frigerio *et al.*, 1992). Only 1 interface residue in chymotrypsin was not identified as such (Gly59, yellow) and only 1 false positive residue (Leu 123 blue) is not located near the actual interface. **Fig. 2c** shows results obtained for the 4th ranked target protein in this category, porcine pancreatic elastase (1fle E) in complex with the inhibitor elafin (1fle I; Tsunemi *et al.*, 1996). In elastase, 7 interface residues were not identified (false negatives, yellow), but there were 4 false positives (blue).

Discussion

Protein-protein interactions play a central role in protein function. Hence, sequence-based computational approaches for identification of protein-protein interaction sites, identification of specific residues likely to participate in protein-protein interfaces, and more generally, discovery of sequence correlates of specificity and affinity of protein-protein interactions have major implications in a wide range of applications including drug design, and analysis and engineering of metabolic and signal transduction pathways. The results reported here demonstrate that an SVM classifier can reliably predict interface residues and recognize protein-protein interaction surfaces in proteins of antibody-antigen and protease-inhibitor complexes. In this study, interface and non-interface residues were identified with relatively high sensitivity (82.3% and 78.5%) and specificity (81.0% and 77.6%). With this level of success, predictions generated using this approach should be valuable for guiding experimental investigations into the roles of specific residues of a protein in its interaction with other proteins. Detailed examination of the predicted interface residues in the context of the known 3-dimensional structures of the complexes suggest that the degree of success in predicting interface residues achieved in this study is due to the ability of the SVM classifier to "capture" important sequence features in the vicinity of the interface.

Our previous work (Yan *et al. 2002*) used a similar approach to predict interaction site residues in 115 proteins belonging to six categories (antibody-antigen; protease-inhibitor; enzyme complexes; large protease complexes; G-proteins, cell cycle, signal transduction; and miscellaneous). In each jack-knife experiment the classifier was trained using examples from 114 proteins and tested on the remaining protein. The resulting classifier performed with specificity of 71%, sensitivity of 67%, and with a correlation coefficient of 0.29. In contrast, the results reported in this paper were obtained using separate classifiers for antibody-antigen category and protease-inhibitor category. The correlation between actual and predicted labeling of residues as interface versus non-interface residues in this case -- 0.430 and 0.462 respectively -- is substantially better than the correlation of 0.29 obtained using a single classifier trained on the combined data set from all six categories of protein-protein complexes. This indicates that there may be significant differences in sequence correlates of protein-protein interaction among proteins that participate in different broad categories of protein-protein interaction. In this context, systematic computational exploration of such sequence features, combined with directed experimentation with specific proteins would be of interest. These results also suggest that in building sequence-based classifiers for identifying residues likely to form protein-protein interaction surfaces, a 2-stage approach based on identification of the broad category of interaction the protein is likely to be involved in (say antibody-antigen versus protease-inhibitor), followed by classification of amino acid residues into interface versus non-interface classes may be worth exploring.

Because interaction sites consist of clusters of residues on the protein surface, some false positives (blue residues) in our experiments can be eliminated from consideration if the structure of target protein is known. For example, in Figure 2b, Leu 123 is predicted to be an interface residue. From the structure of the target protein, we can see that Leu 123 is isolated from the other predicted interface residues. Thus, it is highly likely that Leu 123 is not an interface residue. Thus we can remove Leu 123 from the set of predicted interface residues. Similarly two false positives in Figure 2c can be removed. Therefore the performance of the SVM classifier can be further improved if the structure of a target protein (but not the complex) is available. (If the structure of the complex is available, then there is no need to predict interface residues as they can be determined by analysis of the structure of the complex).

Recently Zhou *et al.* (2002) and Fariselli *et al.* (2002) used neural network-based approaches to predict interaction sites with accuracy of 70% and 73%. It would be particularly interesting to directly compare the results obtained in our study and theirs. Unfortunately, such a direct comparison is not possible due to differences in choice of data sets and methods for accessing performance. A notable difference between our study and the others is that the only structural information we used is knowledge of the set of surface residues of the target proteins. Knowledge of surface topology and the geometric neighbors of residues used in the other studies were not used in our study.

Several authors have reported success in predicting surface residues from amino acid sequence (Mandler 1988; Holbrook *et al.* 1990; Benner *et al.* 1994; Gallivan *et al.* 1997; Mucchielli-Giorgi *et al.* 1999; Naderi-Manesh *et al.* 2001). This raises the possibility of first predicting surface residues based on sequence information and then using the predicted surface residue information to predict the interaction sites using the SVM classifier. The classifier resulting from this combined procedure will be able to predict interaction site using amino acid

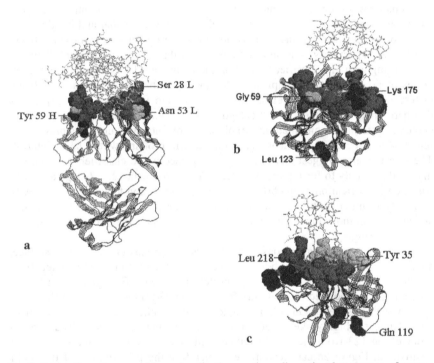

Fig. 2. Interaction site recognition: visualization on three-dimensional structures of representative heterocomplexes. The target protein in each complex is shown in green, with residues of interest shown in *space fill* and color coded as follows: red, true positives (interface residues identified as such by the classifier); yellow, false negatives (interface residues missed by the classifier); blue, false positives (residues incorrectly classified as interface). The interaction partner is shown in gray wireframe. **a.** FabN10 in the 1nsn complex; **b.** α-chymotrypsin in the 1acb complex; **c.** Elastase in the 1fle complex. Structure diagrams were generated using RasMol (http://www.openrasmol.org/)

sequence information alone. We are also exploring the use of phylogenetic information for this purpose. Other work in progress is aimed at the design and implementation of a server for identification of protein-protein interaction sites and interface residues from sequence information. The server will provide classifiers that are based on all protein-protein complexes available in the most current release PDB.

References

Baldi P, Brunak S, Chauvin Y, Andersen CAF (2000) Assessing the accuracy of prediction algorithms for classification: an overview. *Bioinformatics* **16**: 412-424

Benner SA, Badcoe I, Cohen MA, Gerloff DL (1994) Bona fide prediction of aspects of protein conformation: Assigning interior and surface residues from patterns of variation and conservation in homologous protein sequences. *J Mol Biol* **235**, 926-58

Bossart-Whitaker P, Chang CY, Novotny J, Benjamin DC, Sheriff S (1995) The crystal structure of the antibody N10-staphylococcal nuclease complex at 2.9 Å resolution. *J Mol Biol* **253**, 559-575

Braden BC, Fields BA, Ysern X, Dall'Acqua W, Goldbaum FA, Poljak RJ, Mariuzza RA (1996) Crystal structure of an Fv-Fv idiotope-anti-idiotope complex at 1.9 A resolution. *J Mol Biol***264**:137-51

Casari G, Sander C, Valencia A (1995) A method to predict functional residues in proteins. *Nat Struct Biol* **2**,171-178

Dodge C, Schneider R, Sander C (1998) The HSSP database of protein structure-sequence alignments and family profiles. *Nucleic Acids Res* **26**, 313-315

Fariselli P, Pazos F, Valencia A, Casadia R (2002) Prediction of protein-protein interaction sites in heterocomplexes with neural networks. *Eur J Biochem* **269**, 1356-1361

Frigerio F, Coda A, Pugliese L, Lionetti C, Menegatti E, Amiconi G, Schnebli HP, Ascenzi P, Bolognesi M (1992) Crystal and molecular structure of the bovine alpha-chymotrypsin-eglin c complex at 2.0 A resolution. *J Mol Biol* **225**:107-123

Gallet X, Charloteaux B, Thomas A, Brasseur R (2000) A fast method to predict protein interaction sites from sequences. *J Mol Biol* **302**, 917-926

Gallivan JP, Lester HA, Dougherty DA (1997) Site-specific incorporation of biotinylated amino acids to identify surface-exposed residues in integral membrane proteins. *Chem Biol* **4**, 739-749

Holbrook SR, Muskal SM, Kim SH (1990) Predicting surface exposure of amino acids from protein sequence. *Protein Eng* **3**, 659-665

Jones S,Thornton JM (1996) Principles of protein-protein interactions. *Proc Natl Acad Sci USA*, **93**, 13-20

Jones S, Thornton JM (1997a) Analysis of protein-protein interaction sites using surface patches. *J Mol Boil* **272,** 121-132

Jones S, Thornton JM (1997b) Prediction of protein-protein interaction sites using patch analysis. *J Mol Biol* **272**, 133-143

Kabsch W, Sander C (1983) Dictionary of protein secondary structure: pattern recognition of hydrogen-bonded and geometrical features. *Biopolymers* **22**, 2577-2637

Kini RM, Evans HJ (1996) Prediction of potential protein-protein interaction sites from amino acid sequence identification of a fibrin polymerization site. *FEBS letters* **385**, 81-86

Lichtarge O, Bourne HR, Cohen FE (1996) An evolutionary trace method defines binding surfaces common to protein families. *J Mol Biol***257**, 342-358

Lu L, Lu H, and Skolnick J (2002) MULTIPROSPECTOR: An algorithm for the prediction of protein-protein interactions by multimeric threading *Proteins* **49**, 350-364

Mandler J (1988) ANTIGEN: protein surface residue prediction. *Compute Apple Basic* **4**, 493

Mucchielli-Giorgi MH, About S, Puffery P (1999) PredAcc: prediction of solvent accessibility. *Bioinformatics* **15**, 176-177

Naderi-Manesh H, Sadeghi M, Arab S, Movahedi AAM (2001) Prediction of protein surface accessibility with information theory. *Proteins* **42**, 452-459

Pazos F, Helmer-Citterich M, Ausiello G, Valencia A (1997) Correlated mutations contain information about protein-protein interaction. *J Mol Biol* **271**, 511-523

Platt J (1998)Fast training of support vector machines using sequential minimal optimization. In B Scholkopf C J C, Burges and A J Smola editors, Advances in Kernel Methods - Support Vector Learning, p 185-208, Cambridge, MA, MIT Press

Rost B, Sander C (1994) Conservation and prediction of solvent accessibility in protein families. *Proteins* **20**, 216-226

Teichmann SA, Murzin AG, and Chothia C (2001) Determination of protein function, evolution and interactins by structural genomics. *Curr Opin Struct Biol* **11**:354-363

Tsunemi M, Matsuura Y, Sakakibara S, Katsube Y(1996) Crystal structure of an elastase-specific inhibitor elafin complexed with porcine pancreatic elastase determined at 1.9 A resolution *Biochemistry* **35**:11570-11576

Valencia A and Pazos F (2002) Computational methods for prediction of protein interactions. *Curr Opin Struct Biol* **12**:368-373

Witten I H, Frank E (1999) Data mining: Practical machine learning tools and techniques with java implementations. San Mateo, CA: Morgan Kaufmann

YanC, Dobbs D, Honavar V (2002) Predicting protein-protein interaction sites from amino acid sequence. Technical report (http://archives.cs.iastate.edu/) ISU-CS-TR 02-11. Department of computer science, Iowa State University, USA

Zhou H, Shan Y (2001) Prediction of protein interaction sites from sequence profile and residue neighbor list. *Proteins* **44**, 336--343

From Short Term Memory to Semantics-a Computational Model

Parag C. Prasad,
Doctoral Student,
School of Biosciences & Bioengi-
neering,
Indian Institute of Technology,
Bombay,
Powai, Mumbai 400 076,
India.
Phone: +91-22-25722545 extension:
4749
Email: 1) paragcp@cse.iitb.ac.in
2) paragpc@cc.iitb.ac.in
Fax: +91-22-25783480

Subramani Arunkumar,
Professor,
Department of Computer Science
and Engineering,
Indian Institute of Technology,
Bombay,
Powai, Mumbai 400 076,
India.

Author for Correspondence: Parag C. Prasad
Keywords: semantics, memory, learning

Abstract

Aphasias are disorders of language and have been shown in recent literature, to be associated with deficits of Short Term Memory (STM). Physiological STM has a semantic component that is responsible for comprehension of language. This work brings forth a new model that learns semantics of words across an STM. The approach to validating the model is through experiments that model word comprehension at the level of several sentences which when taken together convey semantics. Experimentally, the model when used to assign semantic labels to a collection of sentences, gives an accuracy that is comparable to or better than that obtained using Support Vector Machines (SVM).

1 Introduction

Wernicke's aphasia is a disorder of language in which speech is fluent and appears grammatically correct but is devoid of meaning and riddled with paraphasias in which a wrong word is substituted for the desired correct word (Dronkers et al. 2000; Mayeux and Kandel 1991). The works of Yuret (1998) and Rosenfeld (1996) inspire consideration of the semantic incorrectness of speech in Wernicke's aphasia in terms of loss of selectional information between words. Grammatical information and selectional restrictions (Yuret 1998) both are required to learn the relation between words in natural language (for example, given the verb 'eat', 'cake' is a more likely object than 'train'). The sentence "Colorless green ideas sleep furiously" which is devoid of meaning, illustrates the independence of the grammatical aspect from meaningfulness.

Studies on psychological aspects of language (Baddeley 1992; Ericsson and Kintsch 1995) also describe the existence of a Short Term Memory (STM). The variation in the normal capacity of this memory, which is 7 ± 2 items of verbal information, affects language comprehension (Just and Carpenter 1992). There is an overlap between the anatomical areas whose function is lost in Wernicke's aphasia and the areas that contribute to STM (Buchsbaum et al. 2001; Paulesu et al. 1993). This suggests that the loss of function in Wernicke's aphasia is related to a pathologically reduced STM.

Rosenfeld (1996) has shown experimentally that most of the information about a given word by a window of k words preceding it, is obtainable for k= 5, and tapers off thereafter. This value of k is within the normal physiological STM range. This suggests that a possible approach to modeling semantics is to learn a hypothesis using information over a physiological STM. Such a model, when tested with less than normal STM, would also produce semantically incorrect output.

2 Approach

Our approach is to learn representations of semantics of words using information in an STM of preceding words in input. The learnt hypotheses denote content in Long Term Memory (LTM); this follows from work (Ericsson and Kintsch 1995) that shows the identification of semantics as occurring due to interaction between content in STM with learnt information in LTM. The efficacy of learning is then tested on unseen input.

2.1 Semantics

The work by Prasad and Arunkumar (2003) proposes a linear space model of memory and suggests that the use of an appropriate learner would permit the learning of semantics using this model of memory. This work models memory as:
i. a countably infinite vocabulary V.
ii. a memory space M modeled as a c_{00} subspace of an l^2 linear space.
    In addition, it suggests the use of a learner L to learn a correct hypothesis in LTM content within M. The elements of M are countably infinite vectors with finitely many elements non-zero. The non-zero content denotes the content of memory, which could be either STM or LTM.

Sentences are comprehended one word at a time (Ericsson and Kintsch 1995; Haarmann et al. 2003). An STM vector m_{istm} of words preceding a single word W_i would code for the semantics of W_i. Every STM vector m_{istm} has associated with it a label that reflects the identity of the word whose semantics the vector encodes. Incorrect identification of a word by m_{istm} would lead to the association of an incorrect label with m_{istm}. The set of all possible labels is identical to V, and has a one to one mapping with the set of natural numbers N.

Let $S = \{s_i\}$ denote the set of all two tuples formed by taking the cartesian product of M with N. Then each s_i stands for a (memory vector, label) pair. Some s_i denote correct identification of semantics by the associated memory vector, and the others do not. Consider the identification, by STM vector m_{istm}, of word W_i with LTM encoding m_{iLTM}. Since STM and LTM act in tandem, the correct identification of W_i would use both m_{iLTM} as well as m_{istm} as input. There are two possible situations:

1. The word under consideration is correctly identified as W_i.
2. The word under consideration is incorrectly identified as W_j, j != i.

2.2 Learning

Prior to a word being correctly recognized, learning has to take place. During the process of learning, correctly labeled STM vectors are used to learn an LTM representation. This is akin to the process of learning semantics from multiple episodes of input. Learning would involve manipulation of information in memory vectors $m_i \in M$ and would have to be consistent with the properties of M.

Consider the learnt semantics of W_i and W_j in LTM, represented by $s_{iLTM} = (m_{iLTM}, W_i)$ and $s_{jLTM} = (m_{jLTM}, W_j)$ respectively, where m_{iLTM} is the LTM vector for W_i and m_{jLTM} is the LTM vector for W_j. Let $s_{istm} = (m_{istm}, W_i)$ and $s_{jstm} = (m_{jstm}, W_j)$ denote the semantics of W_i and W_j encoded by STM vectors m_{istm} and m_{jstm}, with i != j. Any given STM vector encodes for the semantics of a single word, and hence for each such vector, there is one correct tuple with a label that is consistent with correct identification. For instance, for m_{jstm}, the correct tuple would be $s_{jstm} = (m_{jstm}, W_j)$ and all tuples of the form (m_{jstm}, W_k) with j != k would indicate incorrect identification. The occurrence of m_{istm}, as an STM vector preceding W_i, means that m_{istm} should be taken as indicative of the presence of the semantics of W_i and absence of the semantics corresponding to all W_j, j != i.

Learning involves:
i. Learning an LTM representation of the association that word w_k has with all words $w_m \in V$, using observed information in STM, about w_k.
ii. Maximizing likelihood of the learnt LTM representation with respect to the observed information in STM.

The inner product $<m_i,m_j>$ can be thought of in terms of the component of m_i that lies along m_j. It is 0 when m_i and m_j are perpendicular. Maximizing $<m_i,m_j>$ would correspond to bringing m_i and m_j closer.

M is an inner product space with the inner product given by:

$$<m_i,m_j> = \sum_{<l>} m_{i(l)} * m_{j(l)} = m_i^T m_j$$

Learning the semantics of w_k encoded by $s_{kLTM} = (m_{kLTM}, w_k)$ would correspond to minimizing the distance between all $s_{kstm} = (m_{istm}, w_k)$, for all i, and maximizing the distance between all other s_{jstm} and s_{kLTM}. The learning of the semantics of w_k can be thought of as the learning of an optimal LTM vector corresponding to w_k. Maximizing $<m_{istm}, m_{kLTM}>$ minimizes the distance between s_{kstm} and s_{kLTM}; likewise, minimizing $<m_{jstm}, m_{kLTM}>$ maximizes the distance between s_{jstm} and s_{kLTM} for all j != i.

When s_{istm} is encountered in input:
1. s_{iLTM} is augmented to reflect a greater proximity to s_{istm}.
this is denoted as:

$$s_{iLTM(new)} = s_{iLTM(old)} + k1 * s_{istm} \qquad (2.2.1)$$

2. $s_{jLTM} \forall j != i$ is modified to reflect a lesser proximity to s_{istm}.
this is denoted as:

$$s_{jLTM(new)} = s_{jLTM(old)} + k2 * s_{istm} \qquad (2.2.2)$$

where k1 is a real number and k1 >= 0, k2 is a real number and k2 <= 0
+ denotes addition, and * denotes scalar multiplication.

2.3 Semantic Operations

Let F_{LABEL} be an onto function with S as its domain.
$F_{LABEL} : S->N$
$F_{LABEL}(s_i)$ corresponds to the natural number that identifies the label of s_i.
Let F_{MEM} be an onto function with S as its domain.
$F_{MEM}: S->M$
$F_{MEM}(s_i)$ is the $m_i \in M$ where m_i is the memory vector component of s_i.

Let $F_{MEMSIZE}$ be an onto function with S as its domain and the set of whole numbers W as its range;
$F_{MEMSIZE}: S->W$
$F_{MEMSIZE}(s_i)$ indicates the number of non zero entries in the memory vector component of s_i.
Let there exist a zero or null element $\Phi \in S: s_i + \Phi = \Phi \forall s_i \in S$

To permit learning, the set S supports the following operations:

2.3.1 Addition

Addition of two elements $s_i, s_j \in S$ is defined to yield a third element $s_k \in S$ where:

$s_k = s_i + s_j$, $s_i, s_j, s_k \in S$

The following cases are possible:

<u>Case 1</u>: $F_{LABEL}(s_i) = F_{LABEL}(s_j)$
Here, both elements s_i and s_j as well as their sum, code for the semantics of the same word. The new element s_k would contain information from the memory vector component of both s_i and s_j.
$s_k \in S$ is a new element such that $F_{LABEL}(s_k) = F_{LABEL}(s_i) = F_{LABEL}(s_j)$ and
$F_{MEM}(s_k) = F_{MEM}(s_i) + k1 * F_{MEM}(s_j)$
where the + and * operators are on M and
k1 is a real number greater than or equal to 0.

<u>Case 2</u>: $F_{LABEL}(s_i) != F_{LABEL}(s_j)$
The following two situations may be seen:

i) $F_{MEMSIZE} (s_i) != F_{MEMSIZE} (s_j)$
This situation occurs when one of s_i or s_j is an LTM vector and the other an STM vector, and the two code for different semantics.
$s_k \in S$ is a new element such that
$F_{LABEL}(s_k) = F_{LABEL}(\text{argmax } s_i, s_j (F_{MEMSIZE} (s_i), F_{MEMSIZE} (s_j))$ and
$F_{MEM}(s_k) = F_{MEM}(s_i) + k2 * F_{MEM}(s_j)$
where the + and * operators are on M and
k2 is a real number less than or equal to 0.

ii) $F_{MEMSIZE} (s_i) = F_{MEMSIZE} (s_j)$
Then $s_k = \Phi$
The above are commutative with respect to addition.

2.3.2 Scalar Multiplication

This is defined as follows:

The operation of scalar multiplication denoted by . is defined as:

$. : R \ X \ S \rightarrow S$

$s_j = \lambda \ . \ s_i,$ $\forall \lambda \in R$ and $\forall s_i, s_j \in S$
where $F_{LABEL}(s_j) = F_{LABEL}(s_i)$ and

$F_{MEM} (s_j) = \lambda . F_{MEM}(s_i)$

with . being the operation of scalar multiplication on M.

In S:

$$(\lambda\,\mu).\ s_i = \lambda\,(\mu.\ s_i); \ \forall\,\lambda,\mu \in R \ \forall\ s_i \in S \qquad (2.3.2.1)$$

$$\lambda\,.\ \Phi = \Phi\,; \forall\,\lambda \in R\,,\Phi \in S \qquad (2.3.2.2)$$

$$\lambda(\ s_i + s_j) = \ \lambda.\ s_i + \lambda.s_j; \forall\,\lambda \in R \ \forall\ s_i, s_j \in S \qquad (2.3.2.3)$$

$$(\lambda+\mu).\ s_i = \ \lambda.\ s_i + \mu.\ s_i; \ \forall\,\lambda,\mu \in R \ \forall\ s_i \in S \qquad (2.3.2.4)$$

$$1.\ s_i = s_i\,; \ \forall\ s_i \in S \qquad (2.3.2.5)$$

2.4 Application

The semantics that a given domain denotes, as seen from the results of language models (Rosenfeld 2000) would be different from the semantics of other domains. In data sets used for classification, each data point is a document consisting of one or more sentences, and with a label from a set $\{l_i\}$ of predefined labels. The labels denote semantically different categories. The training phase consists of learning the semantics of words that constitute each document, given its label. The testing phase consists of identifying the collective semantics of words that constitute a document whose label is not known beforehand, and mapping the identified semantics to a label in $\{l_i\}$. This approach of labelling unlabelled data is similar to that of contemporary text classification algorithms and so our results can be compared with the results of such algorithms:

Our model learns LTM representations of words. To apply it in the classification framework, the following was done:
  i. Long Term Memory representations were learnt for each class of data.
  ii. If the same word occurred in more than one class, LTM representations were learnt for it for each of the classes in which it occurred.
  iii. Text comprehension based on a semantic STM takes place one word at a time. This lets us compute a Hypothesis as follows:

In the classification framework, we aim to identify a maximum aposteriori (MAP) hypothesis H_{MAP} which is the most probable hypothesis h ε H, given the data X; H being the space of all possible hypotheses.
In the case of identifying the label of a data point based on its semantics S,
$H_{MAP} = \text{argmax}_{h \in H}\ p(h \mid S)$
$= \text{argmax}_{h \in H}\ (p\,(S|h) * p(h)\,)\,/\,p(S)$ - by Bayes Formula.
$= \text{argmax}_{h \in H}\ (p\,(S|h) * p(h)\,)$ because p(S) is a constant independent of h.
where p(h) is the prior probability of hypothesis h, which in this case is the prior probability of the corresponding class.

Let the semantics of a data point X_t be S_{Xt}. X_t consists of a sequence of words x_i, and the semantics of the x_i are identified one word at a time. Let the semantics of $x_{i.}$, be denoted as s_{STMxi}, and the learnt semantics of x_i be denoted as s_{LTMxih} where h indicates the class to which the learnt semantics belong.

$$s_{STMxi} = (m_{STMxi}, x_i) \qquad\qquad (2.4.1)$$

$$s_{LTMxih} = (m_{LTMxih}, x_i) \qquad\qquad (2.4.2)$$

The output obtained by presenting a m_{STMxi} as a test data point to each m_{LTMxih}, denotes the probability $p(s_{STMxi} \mid h)$.

The s_{STMxi} of a given x_i are taken as independent of that of other x_j j != i, that is:

$$P(S|h) = \prod p(s_{STMxi} \mid h) \qquad\qquad (2.4.3)$$

This permits the calculation of the MAP hypothesis as:

$$H_{MAP} = \text{argmax}_{h \in H} (\prod p(s_{STMxi} \mid h) * p(h)) \qquad\qquad (2.4.4)$$

3 Experiments

3.1 Data

We used used four different subsets, with two classes of data each, from the 20 Newsgroups text data set (Lang 1995) for testing purposes. This choice was motivated by the data sets used by Schohn and Cohn (2000). During processing, all words are converted to uppercase and replaced by their stems; for example computer, compute and computed all would map to the stem COMPUT. Headers that provide pointers to the true label of the document were removed during parsing.

Table 1   Data Subset Used

Data Subsets From 20Newsgroups Data Set		
Data Sub-set Label	Labels of Classes	Total Documents Per Class
DS1	1. alt.atheism 2. talk.religion.misc	1000
DS2	1. comp.graphics 2. comp.windows.x	1000
DS3	1. comp.os.ms-windows.misc 2. comp.sys.ibm.pc.hardware	1000
DS4	1. rec.sports.baseball 2. sci.crypt	1000

We carried out two kinds of experiments that we shall refer to as Experiment1 and Experiment 2.

3.2 Experiment 1

This experiment assessed the effect of different STM sizes on the accuracy, on the test set, of our model as a text classifier. We also compared our results with those published by Schohn and Cohn (2000). A training set DSi_{Train} was created for each data subset i, by randomly selecting data points. The rest of the subset was taken as a test set DSi_{Test}. We tried with different sizes of STM, and averaged results across 5 trials, with the train and test set sizes constant across trials, and a different train-test split taken for each trial.

The results of this experiment are depicted in Table 2, Table 3 and Table 4.

Table 2 shows the effect of different STM sizes on accuracy, expressed as a percentage, obtained with each of the data sets. Table 3 shows the value of STM at which the accuracy was highest for each of the data sets. Table 4 compares our results with those obtained on the same data sets using SVMs.

Table 2   Accuracy for Different STM Sizes

STM Data Set	1	2	3	4	5	6	7
DS1	79.1	82.3	85.1	84.2	84.5	83.9	84.1
DS2	89.2	89.3	90.5	92.7	92.1	91.9	91.8
DS3	88.2	89.8	92.6	91.7	92.1	91.9	91.8
DS4	98.3	98.7	99.3	99.1	98.9	98.7	98.9

Table 3   STM for Which Accuracy was Highest

Data Set	STM at which accuracy was highest on test set
DS1	3
DS2	4
DS3	3
DS4	3

These accuracy figures are compared with figures for Support Vector Machine(SVM) classifiers in the published results of Schohn and Cohn (2000) in table 4 below. All tests used a training set of 500 documents per class.

Table 4   Comparison with Accuracy of SVM Classifier

Data Set	SVM Best Accuracy	STM Model Best Accuracy
DS1	less than 80%; approximately 78% as seen on a graph	85.1 %
DS2	approximately 90% as seen on a graph	92.7 %
DS3	less than 90%; approximately 89% as seen on a graph	92.6 %
DS4	less than 100%; approximately 99% as seen on a graph	99.3 %

These results can be summarized as follows:

i. The accuracy of the STM model on the test set was highest at an STM of 3 to 4 depending on the data set used.

ii. The STM model gave an accuracy that was higher than that given by the Support Vector Machine classifier for two data sets - DS1 and DS3.

iii. The accuracy of the STM model was comparable to that of the Support Vector Machine classifier for the other two data sets.

3.3 Experiment 2

This experiment assessed the variation in the initial error of the untrained model with different STM sizes. The results are shown in figure 1 below.

The figure shows that the error of the untrained model:

i. decreases monotonically as STM increases from 1 to a value of 6 to 7, which is in the physiological range of 7 ± 2.

ii. rises marginally after this value of STM.

iii. demonstrates a steep fall initially during its monotonic decrease.

In the light of results from experiment 1, we interpret the above findings as follows:

i. The decrease in error on the training set as STM increases to the normal physiological value indicates that increase in STM to the physiological normal helps learn a progressively better hypothesis on the given data.

ii. The fact that accuracy on a test set is maximum at an STM that is lower than the physiological normal, suggests the occurrence of overfitting during training, for STM in the physiological range.

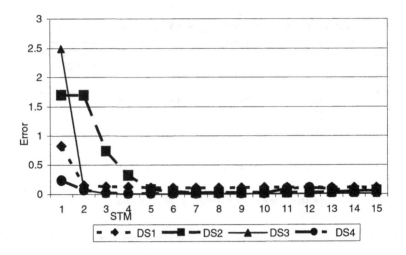

Figure 1. Initial Error Versus STM

4 Conclusion

We have presented an approach to learning semantics using information present in an STM of input. We have validated this approach using experiments that model word comprehension at the level of several sentences; these sentences when taken together, convey semantics. Our approach is consistent with physiological systems as seen in the improvement of performance with rise in STM from 1 towards the physiological normal of 5 to 9. Also, experimentally, our approach performs as well as or better than existing best techniques (SVM) on the data tested.

References

Baddeley A (1992) Working Memory. Science 255:556–559

Buchsbaum BR, Hickok G, Humphries C (2001) Role of left posterior superior temporal gyrus in phonological processing for speech perception and production. Cognitive Science 25:663–678

Dronkers NF, Redfern BB, Knight RT (2000) The Neural Architecture of Language Disorders. In: Gazzaniga MS (eds) The New Cognitive Neurosciences. The MIT Press, Cambridge, pp 949–958

Ericsson KA, Kintsch W (1995) Long Term Working Memory. Psychological Review 142:211–245

Haarmann HJ, Davelaar EJ, Usher M (2003) Individual differences in semantic short-term memory capacity and reading comprehension. Journal of Memory and Language 48:320-345

Just MA, Carpenter PA (1992) A Capacity Theory of Comprehension: Individual Differences in Working Memory. Psychological Review 99:122-149

Lang K (1995) Newsweeder: Learning to filter netnews. In: Proceedings of the 12th International Conference on Machine Learning. San Francisco, pp 331-339

Mayeux R, Kandel ER (1991) Disorders of language: The aphasias. In: Kandel ER, Schwartz JH, Jessell TM(eds) Principles of Neural Science. Elsevier Science Publishing Co. Inc, New York, pp 839-851

Paulesu E, Frith CD, Frackowiak RSJ (1993) The neural correlates of the verbal component of working memory. Nature 362:342–345

Prasad PC, Arunkumar S (2003) Learning Semantics Across a Physiological Short Term Memory. In: Proceedings of the First International Conference on Theoretical Neurobiology. National Brain Research Centre, Haryana, India

Rosenfeld R (1996) A Maximum Entropy Approach to Adaptive Statistical Language Modeling. Computer, Speech and Language 10:187-228

Rosenfeld R (2000) Two decades of Statistical Language Modeling: Where Do We Go From Here?. Proceedings of the IEEE 88:1270-1278

Schohn G, Cohn D (2000) Less is More: Active Learning with Support Vector Machines. In: Proceedings of the 17th International Conference on Machine Learning, Stanford University, pp 839-846

Yuret D (1998) Discovery of Linguistic Relations Using Lexical Attraction. Ph.D Dissertation, Massachusetts Institute of Technology

Part II

**Fuzzy Sets, Rough Sets and
Approximate Reasoning**

Axiomatization of Qualitative Multicriteria Decision Making with the Sugeno Integral

D.Iourinski and F. Modave

University of Texas at El Paso
500W. University Ave.
El Paso, Tx, 79968-0518
fmodave@cs.utep.edu , dmitrii@math.utep.edu

Summary. In multicriteria decision making (MCDM), we aim at ranking multidimensional alternatives. A traditional approach is to define an appropriate aggregation operator acting over each set of values of attributes. Non-additive measures (or fuzzy measures) have been shown to be well-suited tools for this purpose. However, this was done in an *ad hoc* way until recently. An axiomatization of multicriteria decision making was given in a quantitative setting, using the Choquet integral for aggregation operator.

The aim of this paper is to formalize the axiomatization of multicriteria decision making in the qualitative setting, using the Sugeno integral in lieu of the Choquet integral.

1.1 Introduction

Decision theory is generally divided into two main paradigms; decision under uncertain and multicriteria decision making. In decision under uncertainty, a decision maker has to choose an optimal act, that is an act leading to the preferred consequence. However, the value of the acts depend on what is the current state of the world, and this current state is priori unknown to the decision maker.

In multicriteria decision making, the state of the world is known, but the decision maker has to order multidimensional alternatives. Even though additive approaches offer a appreciated level of simplicity to such problems, they are not always appropriate to integrate complex notions such as dependency. Therefore, it seems quite natural to use non-additive techniques instead.

In particular, non-additive (or fuzzy) measures and integrals have been proposed. The axiomatization of decision under uncertainty in the non-additive setting has been initiated by Schmeidler [13], among others, then refined by [16]. In the multicriteria framework, there was a clear consensus that fuzzy integrals were useful aggregation operators [4]. However, no axiomatization was available until recently, and therefore fuzzy integrals were used in a rather *ad hoc* way. Recently, it was shown that under mild assumptions, the preferences of a decision maker in MCDM could be represented by a Choquet integral with respect to some fuzzy measure [9]. From a measurement theory (see [5] for a thorough presentation), the above approach has the drawback to require that the utility values given to the preferences are meaningful, and not only the ranking of these preferences. Therefore, it is desirable to offer a qualitative axiomatization for MCDM, based on the Sugeno integral.

The first part of this paper recalls the basics of fuzzy measures and integrals, and present more formally both paradigms of decision making. We then briefly present a result due to Dubois et al. [2] that rephrases the Savage's approach in a qualitative setting. Based on [6] we extend this result to the MCDM paradigm.

1.2 Fuzzy Measures and Integrals

In this section Ω denotes a finite set and $\mathcal{P}(\Omega)$ the set of subsets of Ω.

Definition 1. *A non-additive measure (also called fuzzy measure) on* $(\Omega, \mathcal{P}(\Omega))$ *is a set function* $\mu : \mathcal{P}(\Omega)) \to [0, +\infty]$ *such that*
$\mu(\emptyset) = 0$ *and*
if $A, B \subset \Omega$, $A \subset B$, *then* $\mu(A) \leq \mu(B)$, *that is,* μ *is a non decreasing set function.*

Note that this definition encompasses the notions of probability measures, possibility and necessity measures, the belief functions of Shafer [14] that were already known and used in the AI community. Let us now give the definitions of the main integrals of non-additive measure theory. See [1] and [15] for the original articles.

Definition 2. *Let* μ *be a non-additive measure on* $(\Omega, \mathcal{P}(\Omega))$ *and an application* $f : \Omega \to [0, +\infty]$. *The Choquet integral of* f *w.r.t* μ *is defined by:*
$(C) \int_B f d\mu = \int_0^{+\infty} \mu(\{x : f(x) > t\}) dt$
which reduces to
$(C) \int_B f d\mu = \sum_{i=1}^n (f(x_{(i)}) - f(x_{(i-1)})) \mu(A_{(i)})$ *with* $\Omega = \{x_1, \cdots, x_n\}$
The subscript $(.)$ *indicates that the indices have been permuted in order to have* $f(x_{(1)}) \leq \cdots \leq f(x_{(n)})$, $A_{(i)} = \{x_{(1)}, \cdots, x_{(n)}\}$ *and* $f(x_{(0)}) = 0$ *by convention.*

Definition 3. *Let* μ *be a non-additive measure on* $(\Omega, \mathcal{P}(\Omega))$ *and an application* $f : \Omega \to [0, +\infty]$. *The Sugeno integral of* f *w.r.t* μ *is defined by:*
$(S) \int_B f \circ \mu = \sup_{\alpha \geq 0} \{\alpha \wedge \mu(\{f \geq \alpha\})\}$
which reduces to
$(S) \int_B f \circ \mu = \bigvee_{i=1}^n (f(x_{(i)}) \wedge \mu(A_{(i)}))$
and where $\vee$ *is the supremum,* $\wedge$ *is the infimum and with the same notations and conventions as above.*

From a formal point of view, the Choquet integral and the Sugeno integral differ only with the operators used in their definition; respectively $+$, $\times$ and $\vee$, $\wedge$. Nevertheless, they are very different in essence, since the Choquet integral is more adapted to numerical problems and the Sugeno integral is better suited for qualitative problems. See for example [7] for a measurement perspective on fuzzy integrals.

1.3 Decision Making Paradigms

1.3.1 Decision under uncertainty

A decision under uncertainty problem can be written as a quadruple $(\Theta, X, A, \succeq)$, where Θ is the set of the states of the world, X is the set of consequences (we assume that consequences and results are identitcal), A is the set of acts that is the functions from Θ to X, and $\succeq$ is a preference relation on the set of acts. A possible approach to solve such problems is to assume some rationality axioms (that are understandable from a decision maker's behavior perspective) on the preference relation defined over the actions, and derive the existence of a probability measure and a utility function such that, if f and g are two actions and $f \succeq g$ means f is preferred to g,

$$f \succeq g \Leftrightarrow \int_{\Theta} u(f(\theta))dP(\theta) \int_{\Theta} u(g(\theta))dP(\theta)$$

where the integrals represent the subjective expected utility of the actions. We encounter two problems here. To have the additivity of the probability measure, we require an independence hypothesis, the sure-thing principle [12], that is unlikely to be satisfied in most practical applications, as the Ellsberg's paradox [3] shows. This result was generalized by Schmeidler [13]. It was shown that under some weakened independence hypothesis, then

$$f \succeq g \Leftrightarrow (C)\int_{\Theta} u(f(\theta))d\mu(\theta) \geq (C)\int_{\Theta} u(g(\theta))d\mu(\theta)$$

This approach despite an increase in complexity, avoids the paradoxes encountered in the additive approach.

1.3.2 Multicriteria decision making

The general multicriteria decision making problem can be represented by a tuple $(X = X_1 \times X_2 \times \cdots \times X_n, \succeq)$, where X is a multidimensional set of the consequences and $\succeq$ is a preference relation over X. We assume that each set X_i is endowed with a weak order (a preference relation) $\succeq_i$ that is given by the decision maker. This means that the decision maker is able to express her or his partial preferences. The problem consists in finding a way to aggregate these partial preferences into the global preference $\succeq$.

The approach generally used is similar in structure to what is done in uncertainty. We know that under very unrestrictive assumptions over the sets X_i, then, there exists functions $u_i : X_i \to \mathbb{R}$, such that for $x_i, y_i \in X_i$, $x_i \succeq y_i$ if and only if $u_i(x_i) \geq u_i(y_i)$. We then need to construct an operator $\mathcal{H} : \mathbb{R}^n \to \mathbb{R}$ such that for $x = (x_1, \cdots, x_n), y = (y_1, \cdots, y_n) \in X$, $x \succeq y$ if and only if $\mathcal{H}(u_1(x_1), \cdots, u_n(x_n)) \geq \mathcal{H}(u_1(y_1), \cdots, u_n(y_n))$ and the obvious choice is to set weights $\alpha_1, \cdots, \alpha_n$ in $[0, 1]$, on the criteria such that $\sum_{i=1}^{n} \alpha_i = 1$ and

$$\mathcal{H}(u_1(x_1), \cdots, u_n(x_n)) = \sum_{i=1}^{n} \alpha_i u_i(x_i)$$

Once again, this approach has a major drawback. Indeed it was shown that using an additive aggregation operator is equivalent to assuming the criteria to be independent (see [8] and [11]).

It was shown in [9] that under a set of structural conditions on $\succeq_i$ and X, the global preference $\succeq$ can be represented by a Choquet integral. This result is based on the formal equivalence between decision under uncertainty (with a finite number of states of the world) and multicriteria decision making. Indeed, these two problems are formally equivalent by identifying the set of states of the world Θ with the set of criteria $\{1, \cdots, n\}$, and the set of acts A with the set of consequences $X = X_1 \times \cdots \times X_n$ (see [9], [6] for more details).

In the remainder of this paper, we extend this result to the Sugeno case. The approach we follow is identical to the approach followed to prove the result in the case of the Choquet integral. We first identify our MCDM problem to a decision making problem under uncertainty. Then, we present the result in the uncertainty case, and the isomorphism between the two paradigms provides us with the result in a very natural way. To do so, and for the sake of completeness, we provide the reader with the key results that allow us to convert an MCDM problem into a decision under uncertainty problem.

Definition 1 *Let* $(X, \succeq)$ *be a set with weak order relation. A subset of* X, A *is said to be order-dense w.r.t.* $\succeq$ *iff for all* $x, y \in X$, $x \succ y$, *there exists* $a \in A$ *such that* $x \succeq a \succeq y$. X *is said to be order separable iff there exists an order-dense, at most countable subset of* X.

Theorem 1 *Let* $(X, \succeq)$ *be a set with a weak order. Then there exists an application* $\phi : X \to R$ *such that for all* $x, y \in X$ *the following propositions are equivalent*

1. $x \succeq y \Leftrightarrow \phi(x) \geq \phi(y)$

2. X *is order-separable.*

If ψ *is an other function verifying conditions above, then there exists a non-decreasing bijection* $f : R \to R$ *such that* $\psi = f \circ \phi$.

The proof of this result can be found in [5].

Now let us assume that there is a preference relation $\succeq_\cup$ (given by a decision maker) on the set $X_\cup = \cup_{i \in I} X_i$ such that its restriction to X_i coincides with $\succeq_i$; such a relation is called compatible with weak order $\succeq_i$ for all i. Then the commensurability hypothesis [9], [10] can be stated as follows:

(CA) Let $(X_i, \succeq_i)$ be the weak-ordered sets of a multicriteria decision making problem. Assume there exists a preference relation $\succeq_\cup$ on $X_\cup$. Then the MCDM problem satisfies (CA) iff there is a family $(\psi_i)_{i \in I}$ of representations of $(X_i, \succeq_i)_{i \in I}$ such that if $x_i \in X_i$ and $x_j \in X_j$ are such that $\psi_i(x_i) \geq \psi_j(x_j)$, is equivalent to $x_i \succeq_\cup y_i$.

This hypothesis is not very restrictive, as can be seen from the theorem below:

Theorem 2 Let $(X, \succeq)$ be an MCDM problem. Assume that for all $i \in I$, $(X_i, \succeq_i)$ is order-separable. If there is a weak-order $\succeq_\cup$ on $X_\cup$ that is compatible with $\succeq_i$ for all $i \in I$, then $(X, \succeq)$ verifies (CA).

Proof is found in [10].
Note: We confine ourselves to finite sets of consequences; thus, $(X = X_1 \times X_2 \times \ldots \times X_n, \succeq)$ satisfies CA, and there is a family of functions $\phi_i : X_i \to R$ such that $x \preceq y \Leftrightarrow \phi_i(x_i) \leq \phi_i(y_i)$

In this case the qualitative utility function can be defined as:

$$U_S(x) = \int_{Im\phi}^{Sug} x d\mu = \max_{\lambda \in Im\phi} \min(\lambda, \mu(\Phi_\lambda)) \tag{1.1}$$

where $\Phi_\lambda = \{x_i \in X, \phi_i(x_i) \geq \lambda\}$

1.4 Savage's Framework in a Qualitative Setting

We are going to use the approach first developed by Dubois, Prade and Sabaddin for a single criterion decision making [2], which, in its own turn, is based on Savage's framework [12]. In his approach, Savage uses the preference relation $\succeq$ to rank acts (decisions), where acts are functions from the state space S to the set X of consequences. In the MCDM setting, we have the sets of initial states of the world preference relation on which is given by the decision maker, and a set of multidimensional consequences $\phi(X) = (\phi_1(x_1), \phi_2(x_2), \ldots, \phi_n(x_n))$, and thus need to rank vectors from X.

Te first of Savage's axioms is very natural: it requires the relation $\succeq$ to be a complete, i.e. $\forall x, y \in X, (x \succeq y$ or $y \succeq x)$; the strict preference

$\succ$ and indifference are also defined as per the usual equivalence relation: $f \sim g \Leftrightarrow f \succeq g \wedge g \succeq f$, and $f \succ g \Leftrightarrow f \succeq g$ and not $g \succeq g$.

Sav 1[1] *Ranking:* $(X, \preceq)$ is a complete preorder.
In Savage's setting the ranking of consequences is induced by ranking the acts, where two special (somehow degenerate) types of acts play an important role: constant act and binary act. In the MCDM setting the analogue of a constant act is such a vector $x = (x_1, \dots, x_n)$ that $\phi(x_i) = \bar{x}$ for all $i \in I$. In other words, this is the state of the world in which all criteria are assessed as having the same degree of satisfaction for the decision maker. So, in the MCDM setting, $x \succeq y$, where x, y are constant vectors implies that $\phi_i(x_i) \geq \phi_i(y_i)$ and that $\phi_i(x_i) = \phi_j(x_j), \forall i, j$. To avoid degenerate situations when either only one set of parameters is always preferred or when all sets are equally desirable, Savage proposed a non-triviality axiom, which in our setting takes the following form:

Sav 5 *Non triviality:* there are $x_i, y_i \in X_i$ for some $i \in I$ such that $\phi_i(x_i) < \phi_i(y_i)$.

The ordering process starts with the ranking of constant acts. This approach suggests an inductive development of the argument, and the next natural step would be to look into the ordering of the binary acts, which in MCDM can be seen as binary tuples $(\mathbf{x}, x_j)$, where $\mathbf{x} = \times_{i \in J \subseteq I} x_i$, s.t. $\phi(x_{i \in J}) = \bar{x}, \forall i$, and $x_j \in X$. If we produce two such tuples by adjoining a criterion z to two constant vectors $\mathbf{x}, \mathbf{y}$ that are already ranked, say, $\mathbf{x} \succeq \mathbf{y}$, then the existing preference should be preserved: $(\mathbf{x}, z) \succeq (\mathbf{y}, z)$. This requirement is stated as an axiom:

WS 3 *Weak compatibility with constant vectors:* let $J \subseteq I = 1, \dots, n$ and $X_J = \times_{i \in J} X_i$, $\phi_{i \in J}(x_i) = \bar{x}$, $\phi_{i \in J}(y) = \bar{y}$ and if $\bar{x} \geq \bar{y}$ then for any $z \in X_{J^c}$ we have $(x, z) \succeq (y, z)$.

In other words, if there is such a set J that all parameters from it are mapped into the same value in R, it is preferentially independent of its complement. To see that the utility function defined in (1.1) satisfies **WS 3**, we first need a notion of pointwise preference:

Definition 2 *Pointwise preference:* $\mathbf{x} \geq_p \mathbf{y} \Leftrightarrow u_i(x_i) \geq u_i(y_i) \forall i$, where $\mathbf{x} = (x_1, \dots, x_n), \mathbf{y} = (y_1, \dots, y_n)$.

The pointwise preference is preserved by the Sugeno-based utility function:

Lemma 1. $x \geq_p y \Rightarrow \int_{Imu}^{Sug} x d\mu \geq \int_{Imu}^{Sug} y d\mu$

---

[1] To stress the parallelism between two paradigms we keep the same names and numbers of the axioms as in [2]

Proof: let $\phi(X) = \{\phi_{(0)}(x_{(0)}) < \phi_{(1)}(x_{(1)}) < \ldots, \phi_{(n)}(x_{(n)})\}$ be the reordering of valuations of parameters and $F_i = \{i \in I, \phi_i(x_i) \geq \phi_{(i)}(x_i)\}$ and $G_i = \{i \in I, \phi_i(y_i) \geq \phi_{(i)}(y_i)\}$, obviously $F_n \subseteq F_{n-1} \subseteq \ldots \subseteq F_0 \subseteq R$ and $G_i \subseteq F_i \Rightarrow \mu(F) \geq \mu(G)$ QED.

The pointwise preference is interpreted as fuzzy set inclusion in the terminology of fuzzy sets (for more details see [2]), and the lemma above just shows the monotonicity of the quantitative utility function with respect to fuzzy set inclusion.

It is now quite trivial to see that **WS 3** is satisfied by U_S: if $\mathbf{x} \geq_p \mathbf{y} \Rightarrow (\mathbf{x}, z) \geq_p (\mathbf{y}, z)$, then by Lemma 1 $U_S(\mathbf{x}, z) \geq U_S(\mathbf{y}, z)$ as desired.

The axioms introduced so far have enabled us to preserve the ranking of quite restricted subset of X: we deal only with constant vectors or compositions of a single criterion with a constant vector. We still do not have the means to give us any kind of transitivity on the preorder of X for arbitrary vectors. Such means are available when Restrictive Conjunctive/Disjunctive Dominance properties are satisfied:

RCD *Restricted Conjunctive-dominance:* $x, y \in X$ and $z = (z_1, \ldots, z_n) \in X$ such that $\phi_i(z_i) = \overline{z} \forall i; x \succ y, z \succ y \Rightarrow x \wedge z \succeq y$.

RCD states that if a constant vector z is more favorable than y and an arbitrary vector x is preferred to y, then choosing the less preferable of x and z is still better than choosing y. The qualitative utility function U_S enjoys RCD property:

Lemma 2. *Sugeno integral as an aggregation operator satisfies RCD.*

Proof: let us consider the following set $\Phi X_i = \{x_j; \phi(x_i \wedge z_i) > \phi_j(x_j)\}$. Then $\forall x_i \in X; \phi_i(x_i) \wedge \phi_i(z_i) < \overline{z}$. Hence $\forall x_i$ such that $\phi_i(x_i) > \phi_i(z_i)$ we have that $\Phi X_i = \emptyset$ and, similarly, $\forall x_i$ s.t. $\phi_i(x_i) \leq \phi_i(z_i); \Phi X_i = \Phi_{x_i} = x \in X; U_S(x) > \phi_i(x_i)$; thus $U_S(x \wedge z) = \max_{\phi_i(x_i) \leq \phi_i(z_i)} \min(\phi_i(x_i), \mu(\Phi_{x_i}))$ Now we may have 2 cases:

1. if $\phi_i(z_i) < U_S(x)$ then $\forall x_i; \phi_i(x_i) \leq \phi_i(z_i)$ and $\min(\phi_i(x_i), \mu(\Phi_{x_i})) < \phi_i(z_i)$ so $U_S(x \wedge z) = \min(\mu(\Phi_z), \phi_i(z))$ which implies that $U_S(x \wedge z) = \phi_i(z_i) = \min(U_S(x), \phi_i(z))$

2. if $\phi_i(z_i) \geq U_S(x)$ then the same reasoning gives $U_S(x \wedge z) = U_S(x) = \min(\phi_i(z_i), U_S(x))$ QED.

Restricted disjunctive dominance is a property dual to RCD, and is also satisfied by U_S:

RDD *Restricted disjunctive-dominance:* $x, y \in X$ and $z = (z_1, \ldots, z_n) \in X$ such that $\phi_i(z_i) = \bar{z} \forall i;\ x \succ y,\ x \succ z \Rightarrow x \succ z \vee y$.

RDD could be interpreted that if a vector x is preferred to both a constant vector z and a vector y, then choosing the best among y and z, still leaves x to be the preferable choice. This property is also satisfied by U_S as the lemma below shows:

Lemma 3. *Sugeno integral as an aggregation operator satisfies RDD.*

Proof: follows the same pattern as in Lemma 2.

1.5 Main Result

In this section we are going to show that a preference relation $\succeq$ on the set X that satisfies **Sav 1**, **Sav 5**, **WS 3**, **RCD**, and **RDD** can be represented by a qualitative utility function defined in 1.1.

Theorem 3 *let $(X = X_1 \times X_2 \times \ldots \times X_n, \succeq)$ be an MCDM problem, and let preference relation $\succeq$ satisfy Sav 1, Sav 5, WS 3, RCD and RDD. Then there exist a family of functions $\phi_i : X_i \to R$ and a fuzzy measure $\mu : 2^X \to R$ such that $x \succeq y \Leftrightarrow U_S(x) \geq U_S(y)$, where $U_S(x)$ is a utility function defined in 1.1.*

Proof: we proceed in the following 4 steps:

1. We can rank constant vectors in X's by looking at their $\phi_i(x_i)$ for any i.
2. Now we assume that **WS 3** and **Sav 5** are also satisfied, and then construct the fuzzy measure $\mu : 2^X \to R$, using the following strategy: if $\mathbf{x}$ is a constant vector, i.e. $\mathbf{x} = \{x_i\}_{i \in J}$, where $J \subseteq I$ and $\phi_i(x_i) = \bar{x}$, $\forall i \in J$, then $\mu(\mathbf{x}) = \phi(x_i)_{i \in J}$. Now, consider a tuple $(\mathbf{x}, y)$, where y is a single parameter, and define the measure of $\mu(\mathbf{x}, y) = \max(\mu(\mathbf{x}), \phi(y^*))$, where $\phi(y^*)$ is the highest degree of satisfaction attained by the criterion y. After that, by induction we can extend our definition to the constructions of the form $(\mathbf{y}, z)$, where $\mathbf{y}$ is some vector in X, whose measure is known, and z is a single parameter. The measure of $(\mathbf{y}, z)$ is then $\mu(\mathbf{y}, z) = \max(\mu(\mathbf{y}), \phi(z^*))$. Since the theorem 1 guarantees that the functions ϕ_i's can be chosen such that the best possible valuation of criterion x_i has $\phi_i(x_i) = 1$ and the worst has $\phi_i(x_i) = 0$, we can see that $\mu(X) = 1$ and $\mu(\emptyset) = 0$. Let $\mathbf{x} = \{x_i\}_{i \in I} \subseteq \mathbf{y} = \{y_j\}_{j \in I}$. Since the relation $\succeq$ satisfies **WS 3**, then $\mathbf{x} = \{x_i\}_{i \in I} \sim (\mathbf{x}, \mathbf{x}^*) = (\{x_i\}_{i \in I}, \{x_k^*\}_{k \in K})$, where set K is set of indices such that $\mathbf{y} - \mathbf{x} = \{x_k\}_{k \in K}$ and $\phi_k(x_k^*) = 0$, and that $\mathbf{y} \succeq (\mathbf{x}, \mathbf{x}^*)$. Thus $\mu(\mathbf{x}) \leq \mu(\mathbf{y})$, and μ is a monotonic measure.

3. Using RCD we can now rank pairs of the form (x_j, y_i^*) where $y_i^* \in X$ is such that $\phi_i(y_i^*) = \min(\phi_i(x_i), \forall i)$, i.e. the least desirable value. This will give $U_S(x_j, y_i^*) = \min(\phi_j(x_j), \mu(A))$ where $A = \{x_i; \phi_i(x_i) = \phi_i(y_i^*)\}$. Similarly, RDD gives $U_S((x_j, y_i^*) \vee (x_k, y_i^*)) = \max(U_S(x_j, y_i^*), U_S(x_k, y_i^*))$.
4. After ranking pairs of parameters we continue by induction and extend the notion from the previous step to any sets of parameters, arriving to the desired result:
$U_S(x) = \int_{Im\phi}^{Sug} x d\mu = \max_{\lambda \in Im\phi} \min(\lambda, \mu(\Phi_\lambda))$,
where $\Phi_\lambda = \{x_i \in X, \phi_i(x_i) \geq \lambda\}$ QED.

1.6 Conclusion

In this paper, we presented a qualitative, non-additive approach to multicriteria decision making, based on fuzzy measures and the Sugeno integral. The strategy to obtain the result is similar to what we proposed earlier in the case of the Choquet integral, and relies on the formal parallelism between decision under uncertainty and multicriteria decision making. In future work, it may be desirable to give a unified approach based on t-conorm integrals.

References

1. G. Choquet. *Theory of Capacities*. Ann. Inst. Fourier 5, 1954.
2. D. Dubois, H. Prade, and R. Sabbadin. Towards axiomatic foundations for decision under qualitative uncertainty. 1997.
3. D. Ellsberg. Risk, ambiguity, and the Savage axioms. *Quart. J. Econom.*, 75:643–669, 1961.
4. M. Grabisch and M. Roubens. Application of the Choquet integral in multicriteria decision making. In M. Grabisch, T. Murofushi, and M. Sugeno, editors, *Fuzzy Measures and Integrals — Theory and Applications*, pages 348–374. Physica Verlag, 2000.
5. D. Krantz, R. Luce, P. Suppes, and A. Tverski. *Foundations of Measurement*. Academic Press, 1971.
6. F. Modave. *Vers une unification des différents paradigmes de la décision: une approche basée sur les mesures non-additives et la théorie du mesurage*. PhD thesis, I.N.P. Toulouse, 1999.
7. F. Modave and P. Eklund. A measurement theory perspective for mcdm. In *Proc. of the 10th Fuzz-IEEE Conference*, Melbourne, Australia, December 2001.
8. F. Modave and M. Grabisch. Preferential independence and the Choquet integral. In *8th Int. Conf. on the Foundations and Applicatons of Decision under Risk and Uncertainty (FUR)*, Mons, Belgium, July 1997.
9. F. Modave and M. Grabisch. Preference representation by the Choquet integral: the commensurability hypothesis. In *Proc. 7th Int. Conf. on Information Processing and Management of Uncertainty in Knowledge-Based Systems (IPMU)*, Paris, France, July 1998.
10. F. Modave and V. Kreinovich. Fuzzy measures and integrals as aggregation operators: solving the commensurability problem. In *Proc. of the NAFIPS-FLINT 2002 Conference*, New-Orleans, LA, June 2002.

11. T. Murofushi and M. Sugeno. *Fuzzy measures and integrals. Theory and applications*, chapter The Choquet integral in multiattribute decision making. Physica Verlag, 1999.

12. L. J. Savage. *The Foundations of Statistics*. Dover, 2 edition, 1972.

13. D. Schmeidler. Integral representation without additivity. *Proc. of the Amer. Math. Soc.*, 97(2):255–261, 1986.

14. G. Shafer. *A Mathematical Theory of Evidence*. Princeton University Press, 1976.

15. M. Sugeno. *Theory of fuzzy integrals and its applications*. PhD thesis, Tokyo Inst. of Technology, 1974.

16. P. Wakker. A behavioral foundation for fuzzy measures. *Fuzzy Sets & Systems*, 37:327–350, 1990.

A Self-learning Fuzzy Inference for Truth Discovery Framework

Alex T H Sim [1,2], Vincent C S Lee [1], Maria Indrawan [3], and Hee Jee Mei [4]

[1] School of Business Systems, Monash University, PO Box 63B, Clayton, Wellington Road, Victoria 3800, Australia. Tel: +61 3-99052360, Email: vincent.lee@infotech.monash.edu.au

[2] Information Systems Group, Faculty of Computer Science and Information System, University Technology Malaysia, 81300 Skudai, Johore, West Malaysia.

[3] School of Computer Science and Software Engineering, Monash University, Caulfield Campus, Australia.

[4] Faculty of Education, University Technology Malaysia, 81300 Skudai, Johore, West Malaysia.

Website: http://www.bsys.monash.edu.au

Abstract. Knowledge discovery from massive business data is a nontrivial research issue. A generalized framework to guide knowledge discovery process is necessary to improve its efficiency and effectiveness. This paper proposes a framework to relate certainty factor to an absolute factor hereafter called alpha (α) factor. Alpha factor represents the magnitude of useful local knowledge that is extracted from raw data in a fuzzy inference system also its correctness in relate to a global knowledge. The concept of alpha, α, is explained with mathematical illustration and a research design. Three specific case studies are included to illustrate the use of the proposed self-learning framework for true discovery.

1 Introduction

Previous research [1] has shown that experts expressed the strength of their belief in the terms that were neither logical nor mathematically consistent. It is due to this deficiency, that MYCIN project team was unable to use classical statistical approach to solve machine reasoning; instead they set up a certainty factor, *cf*, a metric to measure expert's belief.

Ever since then, expert system developer can either choose to use Bayesian approach or certainty factor to solve problem. If certainty factor approach is chosen to measure of belief or disbelief with the help of logic operators of 'OR' and 'AND', the measurements obtained are often non-robust.

Solutions provided by machine reasoning such as fuzzy logic or rough set techniques are generally for solving specific problems within narrow boundaries. From knowledge inference perspective, these techniques solve a narrow domain problem, give ideas on its possible solutions, predict outcomes but do not infer further useful knowledge that would be useful knowledge for another problem domain. In other words, problems are solved without knowledge revelation.

The aim of this paper is to propose a practical framework that can be used integrate knowledge learned from solving one problem to facilitate the process of solving similar problems. This paper is organized in four sections. Section 2 discusses and states the methodology and most importantly the S-FIT framework of this research. Section 3 states a research design base on its methodology and some specific case studies to illustrate the use of S-FIT framework. Section 4 provides conclusions drawn and further works.

2   The Self-learning Fuzzy Inference for Truth Discovery Framework

Most experts state their feeling of confident based on their own understanding or insight on a specific problem domain. Hypothetically, all experts state their level of confident, cf, in relative to each other or in measure to the substances of data, information and knowledge in their possession. An Expert's cf value is subjective and it seldom holds the value of +1 (absolute true) or -1 (remain unknown). Therefore, each value of cf_i is really a subset to a hypothetical truth.

We believe that knowledge induced while solving problem is related to a temporary unknown global true. We propose to use alpha, α to represent this global true. The alpha factor denotes the degree of truth being reached by current knowledge induction. In practice, α-factor is a measure against valid statistical truth. At its peak, alpha factor takes the value of +1, which means that the inferred (knowledge) is of absolute true.

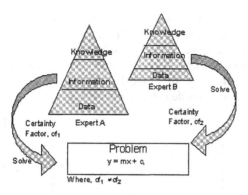

Figure 1: Relative Nature of Certainty Factor

In the experiment, to be followed in section 3, the task is to solve a series of numbers that meet a mathematical equation. With limited data and information, expert A might conclude that there is an algorithm that can represent a series of numbers exactly, with a certainty factor of cf. However, even with a high cf value, the expert opinion is no guaranteed to be correct. It is relatively true to his or others understanding. To cater for self-learning capability, an Alpha factor is introduced so that we do not stop from finding a global true or induced fact in a relative realm.

$$cf \notin \alpha \dots\dots\dots\dots\dots\dots\dots (1)$$

In section 3, three examples on how cf is different from α are illustrated. Here are some interesting observations:

i)   Expert provides cf value based on his initial belief, which is in relative to an earlier belief. For example, if an expert makes an initial conclusion with 99% certainty, then most likely, any conclusion that follows which is close to the initial certainty becomes true.

ii) Expert seeks an algorithm or higher representation and logic that can explicitly summarize and explain phenomena. From observations, if expert knows how disturbance factor, d is applied and its effect on the equation, then expert's cf can vary greatly. In other words, truth effects cf. For example, if experts know the relationship between disturbance factor, d and the equation, y = mx+c then expert can provides confident answer.

iii) Expert did not judge well in scenario with multiple relationships or criteria, which is in consistent with findings by Shortliffe and Buchanan, the developers of MYCIN [2].

In this paper, we demonstrate that truth by its nature is absolute, it is either true or false but with an alpha factor, which holds a value in range of [0,1]. Certain percentage of truth can be revealed through a framework of data collection, categorizing for information, meaning or feeling discovery. This kind of framework is able to reveal certain percentage of a global truth with a degree of discovery (alpha factor), the flow of which is depicted in Figure 2.

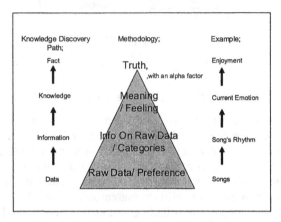

Figure 2: Truth Discovery Framework

'Raw Data' is the available data on a problem domain. 'Information on Raw Data' is the processed, categorized information on 'Raw Data'. 'Meaning' explains the 'Information on Raw Data'. Whenever, meaning on specific subject is hard to related or confirmed, 'feeling' is used with a degree of discovery. 'Truth' is a generalized, new knowledge explaining 'meaning' or 'feeling'. In IF-THEN rules,'Truth' can also be viewed as the consequence and 'meaning' as the antecedent in a fuzzy system.

Classical design on fuzzy rule based expert system is to elicit the "Knowledge" to match "Data", which causes rules to fire, solving problem without inducing further re-usable or meaningful knowledge from the experience of solving a problem. The S-FIT framework, enable us to induce "fact" with a degree of discovery, alpha factor, in relative to discrete global true. Secondly, S-FIT provides a paradigm shift from traditionally 'data' to 'knowledge' as its fuzzy antecedent. Also, a shift from 'knowledge' to inferred new knowledge – 'truth' as its fuzzy consequence.

3   Experiment and Discussion

Base on the proposed S-FIT framework, we consider the following research design to induce a fact of 'user enjoyment', which is rather common in type-1 fuzzy logic and knowledge of 'user emotions', which is an example on the new knowledge that can be inferred from S-FIT framework from user song selection. User is to select from 8 titles of song. Song selection is the raw data. In general songs can be categorized by its rhythms into the category of 'Romance', 'Blue', and 'Contemporary' and 'Pop'. These are in fuzzy terms. Each song can have its degree of membership in more than 1 category.

According to the S-FIT framework in Figure 2, we are to obtain knowledge, which means 'meaning' from our former categorized information on songs. With a fuzzy system, we use fuzzy linguistic term and take in imprecise terms to perform categorization. Whenever meaning is hard to reveal from songs category (information on raw data), we hypothetically assume a relationship between this 'information' and the 'knowledge' based on feeling.

In this prototype research, we predefine a two contradicting feeling which are 'Feel love' and 'Feel bore'. When user chooses a song, a degree of membership is assigned to its correspondence linguistic variable of feeling based on categorized song information.

Till then, we have hypothetically assumed there is true relationship between songs categories and feeling. We do not have to prove it now, but assuming a hypothetical true on it with a degree of discovery in related to true, alpha factor. This alpha value carries the meaning, a degree of correctness in relate to statistical true in a real world problem.

$$\mu_\alpha(x): X \to [0,1] \quad \ldots\ldots\ldots\ldots\ldots (2)$$

A research survey can later be carried out to survey the listening pattern of each song categories to its feeling to verify the correctness of alpha value. Truth or new knowledge from the problem domain is then revealed by using a summarized linguistic term, which is capable of explaining a range of feeling used in this experiment (i.e. Feel love, fell bore). In this case, summarised linguistic term can be "enjoyment".

Till here, we have completed in designing a prototype experiment. In comparing with conventional fuzzy system to S-FIT fuzzy system, we find that conventional fuzzy system provides a degree of truth on 'feeling' but normally not to induce new knowledge such as 'enjoyment' from a problem domain as described. 'Enjoyment' is inferred from 'meaning' or 'assumed meaning –feeling' which is normally the consequence of conventional fuzzy logic rules. Adding to it, conventional system (i.e. FLS) does not indicate on a result's degree of correctness. This has caused fine tuning works on conventional fuzzy system both to be slow and difficult. S-FIT system infers new knowledge and gives its result with an alpha factor. This factor is an estimate or an index, which is related to statistical correctness by a further research carried out on a later time.

In this experiment, enjoyment comes with an alpha factor, which is a relative measurement to our ability to seek for a whole truth, alpha. It is always with a good data collection and effective information gathering that meaningful and closer globally knowledge can be inferred with higher degree of alpha. Three case studies are given as follows to illustrate the use of S-FIT for truth discovery.

Case study 1:

A series of random selected and paired data x and y is extracted from a large data population of W. Expert is to confirm whether these paired data sample can be represented exactly by an algorithm. Hence, we can use this single algorithm to generate new instances of W. W globally holds numerical values for variable y, m, x, and c. This knowledge of W is of course hidden from expert in this case study.

Table 1. Sample data for case study 1

y	x
(55.0599)	(5.01)
(44.0699)	(4.01)
(33.0799)	(3.01)
(22.0899)	(2.01)
(11.0999)	(1.01)
-	-
11.0999	1.01
22.0899	2.01
33.0799	3.01
44.0699	4.01
55.0599	5.01

From observation, expert after sometimes, might suspect for a linear relationship between the given value of x and y through equation of $y = mx + c$. But, without aided tools, it is hard to give a definite answer since expert is not sure of the consistency value of the slope, m or the constant value, c. There might have two linear equations hidden within. However, since, the value x and y repeat itself in symmetrical manner, the probability that the data can be represented by a single algorithm is high. Also, an assumed m with value around 11, seems to relate x and y. So, expert endorsed the fact that series of x and y, most probably is presented by a single equation and population of W will hold the value m around 11 and c all around 0.

By using spreadsheet and chart, we know that a single algorithm represents the above study wholly. The variable, m holds a value 10.99 and variable, c holds a value 0.

Table 2. Sample data plot for case study 1

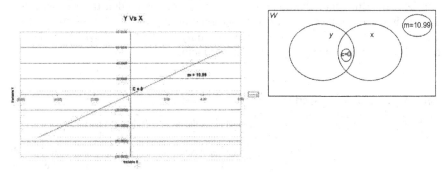

Therefore, we are able to denote the concept of certainty factor and alpha factor as shown in table 3.

Table 3. Illustration of the concept of certainty factor and alpha factor

Solving y = mx + c	Expert Judgment without electronic means;	Judgment using FIT frame-work
	Result: x and y are related. cf = 0.99	Meaning: x and y are related. Fact: W can be represented by $y = (10.99)x + 0$, with $\alpha = 1.0$

Note: $\alpha = 1.0$. It is because we have mathematical proof that information on x and y is only and fully related through an algorithm.

Case study 2:

Base on case study (1), if we are now to modify the sample data to be $y = mx_d + c$ where x_d is a manipulation of $(x + d)$. 'd' is a small disturbance value. And, expert are asked whether x and y still matches the former straight line equation. In another words, in case study (1), we have paired data of x and y, now we have paired data x_d and y. Can x_d and y be represented by a same, former algorithm of $y = mx + c$?

Table 4. Sample data for case study 2

y	x (not shown)	x_d
(88.0299)	(8.01)	(8.02)
(77.0399)	(7.01)	(7.02)
(66.0499)	(6.01)	(6.02)
-	-	-
66.0499	6.01	6.02
77.0399	7.01	7.02
88.0299	8.01	8.02

$d = -0.01$, where x<0
$d = +0.01$, where x>0
$d = 0$, where x=0

If expert knows the nature on how disturbance factor is applied, the answer will be 'no'. But, since this manipulation knowledge is hidden from expert, and expert has been trained to believe a true relation between x and y with $cf=0.99$ therefore any relation close to a former certainty factor of 99% will be assumed true, of course this time with lower certainty factor.

Table 5. Comparison of Expert System and S-FIT Model judgment

Solving $y = mx_d + c$	Purely Expert Judgment without electronic means;	Judgment using FIT framework
	Result: x_d and y is having a same relation as its former x and y, therefore x_d and y are related. $cf = 0.98$	Meaning: x and y are related. Fact: W cannot be represented by the former equation $y = (10.99)x + 0$, with $\alpha = 0.0$.

Note: $\alpha = 0.0$. It is because we have no mathematical proof that information on x and y is only and fully related through an algorithm. x and y are related but through a deferent equation or logic. Note that $\alpha \neq -1.0$ because the relationship of x and y is known to be uncorrelated but not unknown.

Case study 3:

In this case study, we repeat our case study (1) using following data

Table 6. Sample data used in case study 3

y	m	x
(55.0599)	10.99	(5.01)
(44.0699)	10.99	(4.01)
(33.0799)	10.99	(3.01)
(22.0899)	10.99	(2.01)
(11.0999)	10.99	(1.01)
-	10.99	-
11.0999	10.99	1.01
22.0899	10.99	2.01
33.0799	10.99	3.01
44.0699	10.99	4.01

that really consists of 2 sets of linear equations shown as below,

Table 7. Plot of X, Y variables' relationship using two data sets

y	x	y_2	x_2	These relations can be plotted as
(55.1698)	(5.02)	(48.2400)	(4.02)	
(33.1898)	(3.02)	(24.2400)	(2.02)	
(11.2098)	(1.02)	-	-	
-	-	24.2400	2.02	
11.2098	1.02	48.2400	4.02	
33.1898	3.02			
55.1698	5.02			

Normally, real world problems are more complex and have not categorised into meaningful groups. This is represented by the sample data, which are not categorised into meaningful groups as in Table 7 but simply sorted by values of x. Based on the same reasoning as case study (1), experts are asked for conclusion. A comparison of the results between expert system and S-FIT model judgement are given in Table 8.

Table 8. Comparison of Expert System and S-FIT Model judgment

Solving y = mx + c	Expert Judgment without electronic means;	Judgment using FIT frame work
	Result: x and y is related. cf = 0.99	Meaning: x and y is related but by more than an algorithm. Fact: W can be represented by $y = (10.99)x + 0$, with $\alpha = 0.5$

Note: $\alpha = 0.5$. It is because we later prove by mathematic that information on x and y can be related through 2 algorithms. The meaningful relation between x and y is considered true but not wholly in related to global true, alpha.

4   Conclusion and Future Research

New knowledge can be inferred from raw data using S-FIT framework. In our example, the fact of "enjoyment" is revealed after using S-FIT on user song selections.

With enough data collection and a good information categorization, we will be able to seek for a new revealed knowledge that will have its potential in marketing. Product build on this concept not only be able to derive at how much customer feel for each songs but also to show a generalized important truth on what is a customer current state of emotion, and with the current state of emotion, how much this customer is entertained with a product. Referring back to figure 2, this answer provides a higher understanding on the problem domain. This information is definitely precious to marketing. It can be strategically used for critical success especially on internet selling even web based training. A strategic business marketer might consider to e-mail in needed product brochure to potential customer at their right moment using knowledge revealed (current state of feeling) and its global understanding on customer (degree of enjoyment).

This experiment believes that higher knowledge can be induced by constantly relating new inferred knowledge to hypothetical global true, alpha. The inferred new knowledge is normally generalized enough to prove its usefulness across different problem domain. S-FIT can also be used to improve self-learning in fuzzy logic system. Further works on improving self-learning base on this proposed framework are being carried out.

Future works are directed on developing efficient implementation algorithm, search progression and logic representation based on this proposed S-FIT framework.

Acknowledgement

Alex Sim acknowledges the funding support from the university of Universiti Teknologi Malaysia (UTM), Malaysia and Monash, Australia for this research.

References

[1]   Negnevitsky, M. (2002). *Artificial Intelligence: A Guide to Intelligent Systems*, Pearson Education Limited, England.

[2]   Shortliffe, E . H. and Buchanan, B. G. (1975). A model of inexact reasoning in medicine, *Mathematical Biosciences*, vol. 23, no. ¾, pp. 351-379.

[3]   Au, W. H. and Chan, Keith C.C. (2001). Classification with Degree of Membership: A Fuzzy Approach, *IEEE Transaction on Fuzzy Systems*.

[4]   Reynolds, R. G. and Zhu, S. (2001). Knowledge-Based Function Optimization Using Fuzzy Cultural Algorithms with Evolutionary Programming, *IEEE Transaction on Systems, Man, and Cybernetics*, Vol. 31, No.1, pp 1-18.

Exact Approximations for Rough Sets

Dmitry Sitnikov [1], Oleg Ryabov [2], Nataly Kravets [1], Olga Vilchinska [3]

[1] Kharkov State Academy of Culture, Ukraine
[2] National Institute of Advanced Industrial Science and Technology, Japan
[3] Kharkov National University of Radio Electronics, Ukraine

Abstract

Classical topological definitions of rough approximations are based on the indiscernibility relation. Unlike classical approaches in this paper we define rough approximations in an algebraic way. We do not use any binary indiscernibility relation but only unary predicates in terms of which an arbitrary predicate should be described. The terms "exact upper approximation" and "exact lower approximation" have been introduced to stress the fact that there can exist a variety of approximations but it is always possible to select the approximations that cannot be improved in the terms of the approximation language. These new definitions are compared to the classical ones (which use an equivalence relation) and are shown to be more general in the sense that the classical definitions can be deduced from them if we put some restrictions on our model. The process of generating logic rules based on the exact approximations is considered. We also introduce an algebraic definition of a local reduct (a minimal set of predicates describing a rough set) for any subset of the universe.

Introduction

The rough set concept is a relatively new mathematical approach to vagueness and uncertainty in data [1, 2]. The indiscernibility relation and approximations based on this relation form the mathematical basis of the rough set theory. Classical definitions of lower and upper approximations were originally introduced to describe some topological properties of rough sets [3]. These definitions were proposed with reference to an indiscernibility relation, which was assumed to be an equivalence relation. Various generalized definitions of rough approximations have been developed, the majority of them dealing with more general types of relations. In particular, some authors considered a tolerance relation (reflexive and symmetric) as a basis for approximations, which allowed them to express weaker forms of indiscernibility [4]. In some papers researchers proposed interesting definitions of rough approximations based on a similarity relation [5].

The authors of the above papers follow the classical topological way of defining ambiguity concepts. They start from introducing an indiscernibility relation, define some special properties for it and then lower and upper approximations appear in a natural way.

Usually rough set approximations are defined as follows. Consider a finite set of objects U called the universe and a binary relation I over U called the indiscernibility relation (it is normally an equivalence or tolerance relation). The relation I is introduced to demonstrate the fact that our knowledge about elements of the universe is limited and we sometimes cannot discern them. The indiscernibility relation allows us to consider clusters of similar elements (so called atoms) instead of single elements, which we are unable to identify. Let $X \subseteq U$.

The I-lower and the I-upper approximation of X are defined as follows: $I_*(X) = \{x \in U: I(x) \subseteq X\}$, $I^*(X) = \{x \in U: I(x) \cap X \neq \varnothing\}$, where $I(x)$ is the set of all the objects indiscernible with x. One can say that the lower approximation consists of the elements that definitely belong to X, the upper approximation represents the elements that may belong to X.

The concept of a boundary region is introduced to identify the elements about which one cannot say whether or not they belong to X. The boundary region is defined as the difference between the upper and lower approximations. This region consists of the elements that belong to $I^*(X)$ but do not belong to $I_*(X)$.

In order to demonstrate the possibility of reducing the number of features that are used to describe objects, the concept of reducts has been introduced in the rough theory [6]. If instead of one indiscernibility relation there is a set of such relations $I = \{I_1, I_2, ..., I_n\}$, the intersection of these relations is also an equivalence relation. A minimal subset I^r of I such that $\cap I = \cap I^r$ is called a reduct of I (in general there can be several reducts for one indiscernibility relation). We would like to stress that according to the classical definition a reduct depends only on the relations $I_1, I_2, ..., I_n$ and does not depend on the set X that should be described with the help of the indiscernibility relations.

In this paper we do not use any binary indiscernibility relation but only unary predicates in terms of which an arbitrary set of objects can be described. Our approach is rather algebraic than topological. We define lower and upper approximations for sets of objects in an algebraic way using Boolean operations. These new definitions are compared to the classical ones (which use an equivalence relation) and are shown to be more general in the sense that the classical definitions can be deduced from them if we put some restrictions on our model. Using binary codes allows us to quickly calculate the approximations of a set in accordance with the new definitions. We also suggest an algebraic definition of a reduct for any subset of the universe.

An algebraic definition of rough approximations

Suppose we are given a finite nonempty universe $U=\{a_1, a_2, ..., a_n\}$. Consider also a set of unary predicates (functions that take on their values from the set $\{0,1\}$) defined on U: $P_1(t), P_2(t), ..., P_k(t)$, which we will call coordinates.

The predicates P_1, P_2, ..., P_k can be interpreted as characteristic functions for some properties of objects of the universe. In this case an object a_i has the property P_j if and only if $P_j(a_i) = 1$. Following the basic concepts of the rough set theory we should describe an arbitrary set $X \subseteq U$ in terms of the coordinates. Since there exists a one-to-one correspondence between all the predicates defined on U and all the subsets of U, instead of a set $X \subseteq U$ we can consider a predicate $X(t)$ that equals 1 if and only if $t \in X$. Thus we should give a description of an arbitrary predicate $X(t)$ in terms of the predicates P_1, P_2, ..., P_k. In this connection it seems natural to consider some logic or algebraic language that would allow us to discover links between the predicates $X(t)$ and P_1, P_2, ..., P_k. In this paper we suppose that only Boolean operations can be applied to the predicates. We will say that the approximation language consists of the unary predicates P_1, P_2, ..., P_k and the Boolean operations. It is necessary to stress that in the general case the approximation language can include other types of predicates and operations.

Consider the set Φ of all possible formulae constructed with the help of the Boolean operations conjunction (&), disjunction (V) and negation ($\neg$) applied to the predicates P_1, P_2, ..., P_k. For example: $(P_1 \& P_2 \& \neg P_3) V P_4$, $(P_1 V (P_2 \& \neg P_3)) \& P_4$, $\neg (P_1 \& P_2 \& ... \& P_4)$ etc. On calculating all the formulae belonging to Φ, we will obviously obtain a set of predicates, which we will denote Λ. We must note here that different formulae (not necessarily equivalent ones) can correspond to the same predicate. Let us consider a simple example. Suppose $U = \{a_1, a_2\}$; $P_1(a_1)=1$, $P_1(a_2)=1$; $P_2(a_1)=1$, $P_2(a_2)=0$. On calculating the formulae $\varphi = P_1 \& P_2$ and $\pi = P_1$ we get the same predicate P_1 although these formulae are not logically equivalent (neither of them can be obtained from the other by equivalent transformations), nevertheless in this particular case their values are equal.

If the predicate $X(t)$ belongs to Λ, it means that $X(t)$ can be expressed in terms of the coordinates and the set X corresponding to the predicate $X(t)$ can be called crisp with respect to the coordinates. If the predicate $X(t)$ does not belong to Λ, this predicate can not be expressed in terms of the coordinates and we should describe it approximately. The following definitions allow us to do it.

Definition 1. If $A(t) \in \Lambda$ and $\forall t \in U$ $A(t) \to X(t)$ then we say that $A(t)$ is a lower approximation for $X(t)$.

Definition 2. If $B(t) \in \Lambda$ and $\forall t \in U$ $X(t) \to B(t)$ then we say that $B(t)$ is an upper approximation for $X(t)$.

Definition 3. If a predicate $I_*(t)$ is a lower approximation for the predicate $X(t)$ and for any lower approximation $A(t)$ of this predicate $\forall t \in U$ $A(t) \to I_*(t)$ then we say that $I_*(t)$ is an exact lower approximation for $X(t)$.

Definition 4. If a predicate $I^*(t)$ is an upper approximation for the predicate $X(t)$ and for any upper approximation $B(t)$ of this predicate $\forall t \in U$ $I^*(t) \to B(t)$ then we say that $I^*(t)$ is an exact upper approximation for $X(t)$.

Comparison with the classical definitions

Let us compare the classical definitions of lower and upper approximations to the definitions of exact approximations introduced in this paper. In the classical rough set theory an indiscernibility equivalence relation is the mathematical basis for rough approximations. Suppose we are given an equivalence relation I, which is used for constructing approximations of a set X [2]. The relation I defines a set of equivalence classes, where each class corresponds to a predicate taking on a value of 1 for the elements belonging to the class and a value of 0 for the other elements of the universe. Thus we have a set of predicates $P_1(t)$, $P_2(t)$, ..., $P_k(t)$ that satisfy the following conditions corresponding to the well-known properties of equivalence classes:

$$\exists t\ P_i(t),\ i = 1, 2, ..., k, \tag{1}$$

$$\forall t \in U\ \ P_1(t) \vee P_2(t) \vee ... \vee P_k(t), \tag{2}$$

$$\forall t \in U\ \ \neg(P_i(t)\ \&\ P_j(t));\ i, j = 1, 2, ..., k;\ i \neq j. \tag{3}$$

It can be deduced from the classical definitions of rough approximations that

1. the lower approximation for a set X is the union of all the equivalence classes that are subsets of X,
2. the upper approximation for X is the union of all the equivalence classes that have a non empty intersection with X.

Using logic terms we can re-formulate these definitions as follows:

Definition 5. The lower approximation for a predicate X(t) is the disjunction of all the predicates $P_i(t)$ for which $\forall t \in U\ P_i(t) \rightarrow X(t)$.

Definition 6. The upper approximation for a predicate X(t) is the disjunction of all the predicates $P_i(t)$ for which $\exists t \in U\ P_i(t)\ \&\ X(t)$.

Let us show that definitions 5 and 6 are equivalent to definitions 3 and 4 accordingly (exact upper and lower approximations in definitions 3 and 4 correspond to the upper and lower approximations in definitions 5 and 6) given conditions Eq.(2) and Eq.(3) are satisfied. Obviously it is sufficient to prove that for an arbitrary set X the approximations obtained according to these definitions are the same predicates.

Consider an arbitrary predicate X(t), $t \in U$. On calculating any Boolean formula with the predicates $P_1(t)$, $P_2(t)$, ..., $P_k(t)$ we obtain the disjunction of some of these predicates, since the conjunction of two different predicates $P_i(t)$ and $P_j(t)$ equals 0 for any $t \in U$ and the negation operation applied to the disjunction of some predicates $P_i(t)$ gives the disjunction of the rest of the predicates (see conditions Eq.(2) and Eq.(3)). Since

1. any lower approximation is the disjunction of some of the predicates $P_i(t)$ for which $\forall t \in U\ P_i(t) \rightarrow X(t)$,
2. the exact lower approximation for X(t) is the disjunction of all the lower approximations obtained according to definition 1 (Consider the predicate $I_i(t)$

that is the disjunction of all the lower approximations for X(t). It is obvious that for any lower approximation A(t) the following property is true: $\forall t \in U \; A(t) \rightarrow I_*(t)$. Thus $I_*(t)$ is an exact lower approximation),

we can state that the exact lower approximation obtained according to definition 3 is equal to the lower approximation obtained according to definition 5. We have shown that definitions 3 and 5 are equivalent under the assumptions made. Since

1. the disjunction D of all the predicates $P_i(t)$ for which $\exists t \in U \; P_i(t) \; \& \; X(t)$ is an upper approximation in terms of definition 2,
2. any other upper approximation F is the disjunction of some predicates $P_i(t)$, which necessarily includes all the predicates from D

we can state that the exact lower approximation obtained according to definition 4 is equal to the lower approximation obtained according to definition 6. We have demonstrated the fact that definitions 4 and 6 are equivalent under the assumptions made.

Note that the new definitions are equivalent to the classical ones only if the predicates $P_i(t)$ satisfy conditions Eq.(2) and Eq.(3) (i.e. in the case where the predicates $P_i(t)$ describe equivalence classes). If it is not so, nevertheless, definitions 3 and 4 still can be used to approximately describe rough sets in terms of the coordinates.

Calculating the exact approximations for a predicate

Verifying all possible Boolean formulae to calculate exact approximations is a time consuming procedure. Nevertheless, there exists a way of obtaining the approximations for a predicate that allows us to quickly write down necessary formulae. Consider table1:

Table 1. Information representation in the form of binary codes.

	a_1	a_2	...	a_n
P_1	δ_{11}	δ_{12}	...	δ_{1n}
P_2	δ_{21}	δ_{22}	...	δ_{2n}
...	...	...	...	...
P_k	δ_{k1}	δ_{k2}	...	δ_{kn}
X	λ_1	λ_2	...	λ_n

where $\delta_{ij}, \lambda_j \in \{0,1\}$, if $\delta_{ij} = 1$ then $P_i(a_j) = 1$, if $\delta_{ij} = 0$ then $P_i(a_j) = 0$, if $\lambda_j = 1$ then $X(a_j)=1$, if $\lambda_j = 0$ then $X(a_j) = 0$.

Suppose that the predicate X should be described in terms of the coordinates P_1, P_2, ..., P_k. Let us find the exact upper approximation for X. For this purpose consider the columns of the table that contain 1 for the predicate X and write down the corresponding disjunctive normal form. A simple example is given in table 2:

Table 2. An example.

	a_1	a_2	a_3	a_4	a_5
P_1	1	0	0	1	0
P_2	0	0	1	1	1
P_3	0	1	0	0	0
X	0	1	0	1	1

For this example we will get the following formula:

$$I^* = (\neg P_1 \& \neg P_2 \& P_3) \lor (P_1 \& P_2 \& \neg P_3) \lor (\neg P_1 \& P_2 \& \neg P_3) \tag{4}$$

Consider now the columns that contain 0 for the predicate X and write down the corresponding conjunctive normal form. For this example:

$$I_* = (\neg P_1 \lor P_2 \lor P_3) \& (P_1 \lor \neg P_2 \lor P_3) \tag{5}$$

Eqs. (4) and (5) produce the following results:

Table 3. Exact approximations.

	a_1	a_2	a_3	a_4	a_5
P_1	1	0	0	1	0
P_2	0	0	1	1	1
P_3	0	1	0	0	0
X	0	1	0	1	1
I^*	0	1	1	1	1
I_*	0	1	0	1	0

In the general case the predicates I^* and I_* can be represented as follows:

$$I^* = (\lambda_1 \& P_1{}^*\delta_{11} \& P_2{}^*\delta_{21} \& ... \& P_k{}^*\delta_{k1}) \lor (\lambda_2 \& P_1{}^*\delta_{12} \& P_2{}^*\delta_{22} \& ... \tag{6}$$
$$\& P_k{}^*\delta_{k2}) \lor ... \lor (\lambda_n \& P_1{}^*\delta_{1n} \& P_2{}^*\delta_{2n} \& ... \& P_k{}^*\delta_{kn}),$$

$$I_* = (\lambda_1 \lor P_1{}^*(1-\delta_{11}) \lor P_2{}^*(1-\delta_{21}) \lor ... \lor P_k{}^*(1-\delta_{k1}))\&(\lambda_2 \lor P_1{}^*(1- \tag{7}$$
$$\delta_{12}) \lor P_2{}^*(1-\delta_{22}) \lor ... \lor P_k{}^*(1-\delta_{k2})) \& ... \& (\lambda_n \lor P_1{}^*(1-\delta_{1n}) \lor$$
$$P_2{}^*(1-\delta_{2n}) \lor ... \lor P_k{}^*(1-\delta_{kn})),$$

where $P^*\delta = P$ if $\delta = 1$ and $P^*\delta = \neg P$ if $\delta = 0$ for any predicate P.

Let us show that the predicates I^* and I_* are the exact upper and lower approximations. It is obvious that the predicate I^* is an upper approximation for the predicate X in terms of definition 2. If one removes from Eq. (6) any conjunction where $\lambda_i = 1$, the resulting formula will not be an upper approximation, as the predicate I^* will have 0 in a column where X has 1. (We suppose here that in case there are several conjunctions identical to the one removed all of them should be removed). For example if one removes the conjunction $(\neg P_1 \& P_2 \& \neg P_3)$ from Eq. (4), then $I^*(a_5)$ becomes 0 whereas $X(a_5) = 1$. It means that the approximation I^* cannot be improved and, therefore, I^* is the exact upper approximation. The predicate I_* is obviously a lower approximation for X in terms of definition 1. If one removes from Eq. (7) any disjunction where $\lambda_i = 0$, the resulting formula will not be a lower approximation as the predicate I_* will have 1 in a column where X has 0. (We suppose that if there are several disjunctions identical to the one removed all of them should be removed). For example if one removes the disjunction $(P_1 \lor \neg P_2 \lor P_3)$ from Eq. (5), then $I'(a_3)=1$ whereas $X(a_3)=0$. It means that the approximation I_* cannot be improved and, therefore, I_* is the exact lower approximation.

Approximation-based logic rules

Consider an example of generating logic rules with the help of Eqs. (4) and (5). Following traditional rough set concepts we can say that rules based on the exact upper approximation may exist in the data set, and rules based on the exact lower approximation must exist in the data. Transform the expression on the right side of Eq.(5) to get: $I_*=P_3 \lor (\neg P_1 \lor P_2)$ & $(P_1 \lor \neg P_2)=P_3 \lor (P_1 \& P_2) \lor (\neg P_1 \& \neg P_2)$. We can now formulate the following exact rules:

1. An element *belongs* to the set X *if*

a) property P_3 is true for this element

OR

b) properties P_1 and P_2 are true for this element

OR

c) neither property P_1 nor P_2 are true for this element.

This rule says that if one of the conditions a), b) or c) holds, an element belongs to the set X.

Let us simplify the expression on the right side of Eq. (4) to get: $I^* = (\neg P_1 \& \& \neg P_2 \& P_3) \lor (P_1 \lor \neg P_1) \& (P_2 \& \neg P_3) = (\neg P_1 \& \neg P_2 \& P_3) \lor (P_2 \& \neg P_3)$. This formula allows us to formulate the following approximate rules:

2. An element *may belong* to the set X *if* neither property P_1 nor P_2 are true *AND* property P_3 is true for this element.
3. An element *may belong* to the set X *if* property P_2 is true *AND* property P_3 is not true for this element.

Note that the rule 2 can be removed from the set of the rules since, according to the rule 1, if property P_3 is true then an element belongs to X.

Defining reducts

Following the algebraic approach to describing rough sets, we suggest a new definition of a reduct that differs from the classical definition. It seems natural to consider minimal subsets of predicates that describe rough sets with the same precision as the whole set of predicates does. We say that a set Γ of predicates is a local reduct for a set X if:

1. Predicates from Γ produce the same exact upper and lower approximations of X as the exact approximations obtained with the help of the whole set of predicates.
2. Γ is a minimal set of predicates satisfying the above condition.

Thus, a local reduct is a minimal set of coordinates that allows obtaining the same exact approximations of a set X as the whole set of coordinates does. Note that local reducts depend on the sets for which they are constructed.

Conclusion

In this paper a new approach to defining rough set approximations has been considered. The main idea of this approach is to describe a predicate in terms of other predicates in an algebraic way, i.e. to find formulae that represent the most exact approximations of the predicate. Although in this paper only Boolean functions have been investigated, this approach can be applied to other logic and algebraic structures that may be used to describe rough concepts. The terms "exact upper approximation" and "exact lower approximation" have been introduced to stress the fact that there can exist a variety of upper and lower approximations, but it is possible to select approximations that cannot be improved in the terms of the approximation language. Such approximations describe a rough set in the most precise way. Logic rules generated with the help of the exact approximations can be used to classify new elements, i.e. to define whether or not an element belongs to a set. We have also introduced an algebraic definition of a local reduct, which is a minimal set of predicates with the help of which one can obtain the same exact approximations as by using the whole set of predicates.

References

[1] Pawlak Z (1982) Rough sets. International Journal of Computer and Information Sciences, vol 11, pp 341-356.
[2] Pawlak Z (1995) Vagueness and uncertainty: a rough set perspective. Computational Intelligence, vol 11 (issue 2), pp 227-232.
[3] Pawlak Z (1996) Rough sets, rough relations and rough functions. Fundamental Informaticae, vol 27, no 2/3, pp 103-108.
[4] Polkowski L, Skowron A, Zytkow J (1995) Tolerance based rough sets. In: Lin T, Wildberger A (eds) Soft computing: Rough Sets, Fuzzy Logic, Neural Networks, Uncertainty Management, Simulation councils, Inc., San Diego, pp 55-58.
[5] Slowinski R, Vanderpooten D, (1997) Similarity relation as a bases for rough approximations. In: Wong PP (ed) Advances in Machine Intelligence and Soft-Computing, vol. IV, Bookwrights, Raleigh, NC, pp 17-33.
[6] Skowron A, Rauszer C (1992) The Discernibility Matrices and Functions in Information Systems. In: Slowinski R (ed) Intelligent Decision Support – Handbook of Advances and Applications of the Rough Set Theory. Kluwer Academic Publishers, Dordrecht, Boston, London, pp 311-362.

Correlation Coefficient Estimate for Fuzzy Data

Yongshen Ni[1]   John Y. Cheung[2]

1,2 School of Electrical and Computer Engineering, University of Oklahoma
202 W. Boyd, CEC 219 , Norman, OK 73019 jcheung@ou.edu 405-325-4324

Abstract . Correlation coefficient reflects the co-varying relationship between random variables and has become an important measurement in many applications. However conventional methods assume the data set as crisp value which isn't proper under some situations. In this paper, fuzzy set theory is introduced to deal with observations in fuzzy environment; also we proposed a method to calculate correlation coefficient of fuzzy data. Then some data sets are tested to check the validity of the proposed algorithm. Realization of this method can help decision-maker have a better knowledge of the data characteristics and therefore very useful in many data mining applications.

Keywords . Fuzzy sets, correlation coefficient, fuzzy data, stochastic simulation

1. Introduction

In real world statistical analysis application, it's very useful to know whether random variables co varies or not, so that people may make a decision more wisely based on the knowledge of the inherent association between different data. Correlation coefficient in all its forms has been put forward to provide a measure of linear predictability between random variables.

Pearson's correlation coefficient definition plays a critical role in crisp estimation area. His formula is defined as follows:

Definition1:

$$R_{xy} = \frac{1}{N-1} \Sigma \left(\frac{x - \bar{x}}{S_x} \right) \left(\frac{y - \bar{y}}{S_y} \right)$$

N: number of observations

$\bar{x}$: Mean of x coordinates of the data

$\bar{y}$: mean of y coordinates of the data

S_x: standard deviation of $x - \overline{x}$

S_y : standard deviation of $y - \overline{y}$

Pearson's correlation coefficient ranges from -1 to 1. The closer of the absolute value is to 1; the higher is the linearity between two random variables, 0 means no correlation at all. It not only tells us the magnitude, but tells us the co-varying direction; minus sign means random variables change in opposite directions.

The above definition assumes all data points are clear-cut measurement; it is a useful tool to analyze conventional crisp data. However, classic statistical theory can't handle subjective data, such as observations described by linguistic terms etc, or some situations under which precise estimation is almost impossible; those data have the characteristic of vagueness. We call them random fuzzy data.

Fuzzy set theory [1] can be introduced to solve the problem; we describe the basic idea as follows:

Definition 2

Given a universal set X, fuzzy set A is defined by a function

A: F(X) -> [0, 1]

This function F is called membership function and A is characterized by this function. In the discussion of this paper, we assume membership function is a fuzzy number.

According to first decomposition theorem, we have $A = \bigcup_{\alpha \in [0,1]} \alpha \cdot^{\alpha} A(x)$, so

we can first apply algorithm to each a-cut level, after exploring data characteristic at each a-level, the results could be summarized to go back to the original fuzzy data set.

2. Related work

Much research work has been done in this field. Pedrycz put his focus on exploring relationship on a local level of the data instead of the global level, his method works well in analyzing information granule in which case local information is more important, in his method data points has to be objective crisp value, although the meaning of the observations is subjective, that is the reason why fuzzy correlation coefficient is needed . Watanabe uses the concept " fuzzy random variable (FRV)" and computed the correlation coefficient with pre-designed membership function at each a level. Both methods worked with crisp data and derived a membership function of correlation coefficient which is a fuzzy number with the domain from -1 to 1 [2, 3].

Ding-An Chiang and Lin picked up random samples from a data set (x, y), use their fuzzy set membership function value (μ_x, μ_y) to compute correlation coefficient of fuzzy set (X,Y) [4, 5]. Chaudhuri and Bhattacharya [6] also proposed a formula to measure fuzzy set association, all those methods only can tell us correlation relationship between two intuitional fuzzy sets instead of original data set.

Liu extended application to fuzzy data analysis; his paper gives us a fuzzy number correlation coefficient with domain from -1 to 1. The model is based on a pair of nonlinear programs with bounded constraints; commercial nonlinear programming solver is turned to when multiple fuzzy observations are involved [7]. Nonlinear optimization is an untraceable problem mathematically; the algorithm usually becomes inefficient when the size of dataset is large. Moreover, although a fuzzy number correlation coefficient reflected the uncertainty of the data, decision-maker may still need a crisp value to have a summarized knowledge of the data association. He or she will take different action depending on whether this value exceeds his or her threshold or not.

In next section, we will propose a stochastic sampling based method to estimate correlation coefficient for fuzzy dataset. Then some datasets are tested by using this algorithm by comparing our simulation result with Pearson's results it will be very convenient for us to check the validity of the proposed method.

3. Fuzzy correlation coefficient algorithm description

In this section we will build a practical model to estimate correlation coefficient between two fuzzy variables.

Since input data is fuzzy, according to Extension Principle, the correlation coefficient which is a function of input data is supposed to be fuzzy too.

According to the reason we have discussed in section 1, fuzzy data can be characterized by its membership function. We have the following definition.

Definition 3:

$r_{xy}(a)$ is defined as correlation coefficient at each a level .

Because every fuzzy data has its own membership function (in both x and y direction), two intervals are generated in both directions for each observation if we take a cut to the membership functions.

$([X_a^L, X_a^U]_j [Y_a^L, Y_a^U]_j)$ j=1,2,…,m

m: the cardinality of data set Λ

X_a^L : lower bound of a-level X coordinate of jth fuzzy data

X_a^U : upper bound of a-level X coordinate of jth fuzzy data

Y_a^L : lower bound of a-level Y coordinate of jth fuzzy data

Y_a^U : upper bound of a-level Y coordinate of jth fuzzy data

Now each of our data has been converted to a rectangle area, we need to use those data to compute correlation coefficient $r_{xy}(a)$. Obviously our $r_{xy}(a)$ is no longer a single value, but an interval

$r_{xy}(a) = [\min r_{xy}(a), \max r_{xy}(a)]$

Stochastic simulation algorithms such as likelihood weighting often give fast and effective approximations to complicated large size problem. So in this interval computing case, stochastic simulation is a good choice.

We assume actual value of the data follows uniform distribution in the above defined rectangle area.

Hence, at each a level, for every fuzzy observation data points (X_i, Y_i) are generated randomly ($i=1,2\ldots n$) , we will get one sample space of r_{xy} (a). If the number of samples is big enough, we will obtain an interval to represent the correlation coefficient at a level.

$$r_{xy} (a) =[\min r_{xy} (a), \max r_{xy} (a)]$$

After the membership function of correlation coefficient is induced, it's not difficult to derive a crisp value which reflects the association between the fuzzy data (this step is usually referred as defuzzification).

$$d_{,p}(r) \equiv \frac{\sum_{k=1}^{n} r(\alpha)\Delta\alpha}{\sum_{k=1}^{n}\Delta\alpha}$$

In general, the data interval included in the computing with bigger a is a subset of data interval included in the calculation with smaller a, so the above defined correlation coefficient provides us a complete view of the data characteristics with different uncertainty. By defuzzification, we may get a summarized conception about the association between fuzzy data set.

Fuzziness of the data is also an index people may feel interested in for data mining, here we define a fuzziness index for fuzzy data [1].

$$f : A(x) \rightarrow R^+$$

A(x) is membership function of fuzzy set, f function gives a value which reflects the degree to which the bound of A is not sharp. We use the following formula

$$f(A) = b - a - \int_a^b |2A(x)-1| \, dx = b - a - \sum_{x=a}^{b} |2A(x)-1| * \Delta x$$

a : Lower bound of support of membership function

b : Upper bound of support of membership function

When the interval Δx is taken small enough, the above equation holds so that we can use summation to take place of integration. f(A) ranges from 0 to b-a. If dataset is crisp, the f(A) equals to zero. If A(x) equals to 0.5, f(A) arrives the peak value b-a which means dataset has biggest uncertainty. This definition is intuitively reasonable. Bigger value indicates more imprecision of data measurement.

Now we summarize this algorithm.

Initialize $\alpha=0.0$, carrying out the following steps:

1. Fuzzifying each data with proper membership function.

2. Take a cut on membership functions (in both directions) and get a rectangle area for each fuzzy data.

3. Do stochastic simulation on the data to obtain r_{xy} (a) , here we use Pearsons' formula .

4. $\alpha = \alpha + \Delta\alpha$, continue this process till $\alpha=1.0$, go to (2)

5. Defuzzifying the correlation coefficient membership function. Obtain a summarized crisp estimation value.

6. Compute fuzziness of the correlation coefficient

4. Mathematical properties of the proposed correlation coefficient

The proposed measure satisfies the following four properties.

Property 1. Correlation coefficient is a function of a-cut.

We compute correlation coefficient of fuzzy data at each a level , and only use those sampled data points which fall into the corresponding a -cut interval of data membership function. Obviously the result is a function of a. When a equals to 1, we get back to our Pearson's estimation, with a decreasing, dataset has more uncertainty. If a decreases to 0, we will get widest interval for $r_{xy}(a)$.

Property2. Correlation coefficient $r_{xy}(a)$ ranges from -1 to 1

At each a level, we use Pearson's formula to compute the corresponding measure, so just like the property of Pearson's formula, $-1 \leq r_{xy}(a) \leq 1$.

Property 3. Correlation coefficient is a fuzzy number.

To be a fuzzy number, it has to meet [8]:
   1. Normal fuzzy set.
   2. Its a -cut is a closed interval for every $\alpha \varepsilon [0, 1]$.
   3. The support must be bounded
   From property 2, it's trivial that any a-cut of correlation coefficient is a closed interval $\in$ [-1, +1], as for the first condition, since we have restricted in this paper to discuss fuzzy number dataset, it's obvious the correlation coefficient which is computed with those fuzzy data is also a normal set. Since all three conditions are satisfied, our proposed measure is a fuzzy number.

Property 4. Correlation coefficient can get back to a crisp value by defuzzification.

Decision-maker usually will take action by comparing objective data with his or her own threshold. So it's important to get our fuzzy measure back to a crisp value so that it may help better. In many applications, this value should be close to its crisp dataset counterpart.

5. Simulation studies

We will do some tests in this section to verify the above properties. Suppose we want to know the correlation coefficient of a set of N fuzzy observations, each observation has triangle membership function for convenience

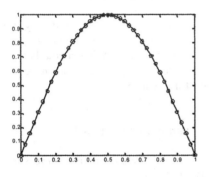

Fig. 1 Data set 1

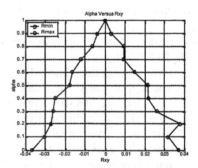

Fig. 2 Membership function of Rxy

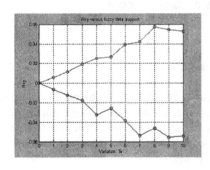

Fig. 3 a=0.5 r_{xy} versus variation.

Fig. 2-1 Membership function of r_{xy}. (X is crisp)

Figure 1 is the first dataset we will test with the proposed algorithm, If all data are crisp, the correlation coefficient with Pearson's formula is 0. Now we fuzzify these data, assume membership function of these data is a symmetric triangle for both x and y coordinates (support of membership function is 10% of the maximum measurement of dataset), the points appearing on the curve are symmetric centers of each membership function, finally we will get a sequence of fuzzy data. Figure 2 shows the simulation result, it's a membership function of correlation coefficient for fuzzy dataset 1. From this figure we can find correlation coefficient is equal to 0 when a=1 which means there is no uncertainty at all in data set. In that case fuzzy data is boiled down to crisp data, therefore fuzzy estimation is boiled down to classic Pearson's correlation coefficient. With the decrease of a, uncertainty of data increases so that correlation coefficient becomes more spread distributed.

Now we take a=0.5, then change the support of membership function from 0 to 20% of maximum value of the data. Figure 3 shows when the support is zero, which means the data is crisp data, we will get one value for correlation coefficient i.e. Pearson's value. Then when support increases, the lower bound of correlation coefficient becomes smaller, upper bound becomes larger. This result isn't surprising since with membership function support wider, the interval into which data points are included computing also wider, so we have more combinations of x and y which results in computed value is expanded.

If for two random variables x and y, only y is fuzzy, while x is still crisp, we will obtain correlation coefficient as Figure 2-1 , fuzziness of this measure is 0.0372, which is smaller than the case both x and y are fuzzy (see table 1). The result is easy to understand since in this case data are less fuzzy intuitively.

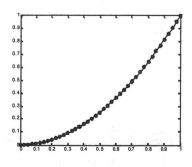

Fig.4 Dataset 2

Fig. 5 Membership function of r_{xy}

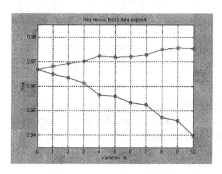

Fig. 6 a=0.5 r_{xy} versus variation

Figure 4 is the dataset we will test next. Just follow the same step we used in the first simulation. Figure 5 is the membership function of correlation coefficient for this data set. The plot shows correlation coefficient value is 0.967 when a=1 which is the same value as Pearson's result. Correlation coefficient ranges from 0.93 to 0.98 when a=0 since we have fuzzified those data in figure 4.

Figure 6 is r_{xy} versus variation curve, the interpretation of this plot is similar to the first dataset. The tails of this plot is expanded with the increasing support; the point on which support is equal to 0 indicates no uncertainty, the data set is once again boiled down to crisp.

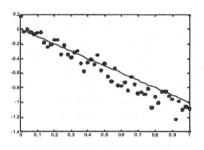

Fig. 7. Dataset 3.

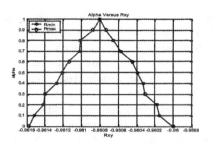

Fig. 8 Membership function of rxy.

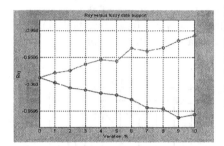

Fig. 9 a=0.5 r_{xy} versus variation

Figure 7 is a dataset which is a straight line with noise (operating on data with same steps). Figure 8 is the corresponding membership function. Figure 9 is r_{xy} versus variation curve. We find the resulting value is close to -1 which is within our expectation.

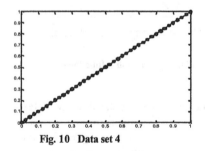

Fig. 10   Data set 4

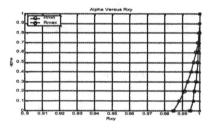

Fig. 11 Membership function of r_{xy}

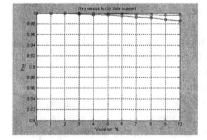

Fig. 12 a=0.5 r_{xy} versus variation

Figure 10 gives us a perfect straight line data set. Figure 11 and figure 12 displayed very narrow band for the resulting value. Straight line looks not sensitive to the measurement fuzziness.

We summarize the simulation result as the following table.

Table 1. Simulation results of four data sets

Data set No.	Pearson's value	Defuzzified value	Fuzziness of measurement (5%)	Fuzziness of measurement (10%)
1	0	0.0120	0.0582	0.1184
2	0.9668	0.9580	0.0155	0.0354
3	-0.961	-0.9523	0.0017	0.0021
4	1	0.9891	0.0024	0.008

The first column of the above table lists the four dataset we have just tested; second column is Pearson's correlation coefficient for each dataset if they are crisp. The third column provides their defuzzified value. The last two columns compare fuzziness of correlation coefficient when the support of membership function for fuzzy data is $\pm 5\%$, $\pm 10\%$ respectively.

From table1 defuzzified correlation coefficient value of fuzzy data set is close to their crisp observations counterparts. Since it reflects the inherent data characteristic within fuzzy context, it can provide us better insight of the data relationship. Fuzziness increases when we fuzzify data with a wider support membership function. We also find the shape of fuzzy dataset relationship impacts the fuzziness of correlation coefficient.

Exploring the simulation result, we find correlation coefficient membership function is in essence a nonlinear mapping from membership function of fuzzy data according to extension principle. Since our membership function of dataset is a fuzzy number, the correlation coefficient is also a fuzzy number with domain in the interval [-1, +1]. It's trivial to find +1 means two random variables co-vary perfect positively, while -1 means they will change in perfect opposite direction. The membership curves reflect the fuzzy characteristic of correlation coefficient and Also from R_{xy} versus Variation curves; we can observe the span of estimated value is wider with increasing data membership function support which is intuitionally reasonable.

6. Conclusions

In this paper, we introduced fuzzy set theory to handle some vague data in real world application. The conception of fuzzy correlation coefficient is proposed, also with stochastic sampling method which can be used to compute the correlation coefficient. This method can help us analyze data in fuzzy environment; the simulation results have supported the idea of this paper.

Stochastic simulation method permits great flexibility in developing improved sampling technique. When the size of data increases, the computation of this method is still acceptable. The result of the simulation shows fuzzy correlation estimation is in fact generalized crisp correlation estimation.

7. Reference

1. George J. Klir/Bo Yuan, *Fuzzy Sets and Fuzzy Logic.* Englewood Cliffs, NJ:Prentice-Hall, 1995
2. Witold Pedrycz, Michael H. Smith, *Granular Correlation Analysis in Data Mining*, 1999 IEEE
3. Norio Watanabe and Tadashi Imaizumi, *A Fuzzy Correlation Coefficient for Fuzzy Random Variables,* IEEE International Fuzzy Systems Conference Proceedings, pp.Π1035-Π1038(1999.8)
4. Ding-An Chiang, Nancy P. Lin, *Correlation of fuzzy sets,* Fuzzy Sets and Systems, 102(1999) 221-226
5. Ding-An Chiang, Nancy P. Lin, *Partial correlation of fuzzy sets*, Fuzzy Sets and Systems 110(2000) 209-215
6. B.B. Chaudhuri, A. Bhattacharya, *On correlation between two fuzzy sets*, Fuzzy Sets and Systems 118(2001) 447-456
7. Shiang-Tai Liu, Chiang Kao, *Fuzzy measures for correlation coefficient of fuzzy numbers*, Fuzzy Sets and Systems 128(2002) 267-275
8. Pedrycz, Witold.; Gomide, Fernando, *An Introduction to Fuzzy Sets* : Analysis and Design. Cambridge, Mass. MIT Press, 1998

Part III

**Agent Architectures and
Distributed Intelligence**

A Framework for Multiagent-Based System for Intrusion Detection

Islam M. Hegazy, Taha Al-Arif, Zaki. T. Fayed, and Hossam M. Faheem
Computer Science Department,
Faculty of Computer and Information Sciences,
Ain Shams University

Abstract. Networking security demands have been considerably increased during the last few years. One of the critical networking security applications is the intrusion detection system. Intrusion detection systems should be faster enough to catch different types of intruders. This paper describes a framework for multiagent-based system for intrusion detection using the agent-based technology. Agents are ideally qualified to play an important role in intrusion detection systems due to their reactivity, interactivity, autonomy, and intelligence. The system is implemented in a real TCP/IP LAN environment. It is considered a step towards a complete multiagent based system for networking security.

Keywords: Intrusion detection, Multiagent-based systems, Security

1 Introduction

The majority of threats on TCP/IP networks come from a small number of security flaws. System administrators report that they have not corrected these flaws because they do not know which of them are the most dangerous [7]. One of the recent security applications is the intrusion detection system (IDS). Intrusion detection can be defined as the problem of identifying individuals who are using a computer system without authorization and those who have legitimate access to the system but are abusing their privileges. The IDS is a computer program that attempts to perform intrusion detection with two approaches, rule-based approach and/or anomaly-based approach [3]. The rule-based IDS searches for attacks' signatures in the network traffic. The anomaly-based IDS searches for abnormal behavior in the network traffic [6, 7]. IDSs can be classified as host-based or network-based. Host-based IDSs base their decisions on information obtained from a single host, while network-based IDSs obtain data by monitoring the traffic of information in the network to which the hosts are connected [3]. So IDS has became a corner stone in security mechanisms because system administrators do not have to decide which security flaws are the most dangerous since the IDS will monitor the network traffic and decide which are the most dangerous flaws.

The agent is a program module that functions continuously in a particular environment, able to carry out activities in a flexible and intelligent manner that is responsive to changes in the environment, and able to learn from its experience [3, 8]. The agent should be autonomous. It should take actions based on its built-in knowledge and its experience [8]. The multiagent system is a system that consists of multiple agents that can interact together to learn or to exchange experience. In order to be flexible, the agents should achieve reactivity, pro-activeness, and social ability. Reactivity means that the agent is able to perceive their environment, and respond in a timely fashion to changes that occur in it in order to satisfy its design objectives. Pro-activeness means that the agent is able to exhibit goal-directed behavior by taking the initiative in order to satisfy their design objectiveness. Social ability means that the agent is capable of interacting with other agents in order to satisfy its design objectiveness [1].

Agents can be classified into four categories: simple reflex agents, agents that keep track of the world, goal-based agents, and utility-based agents. The simple reflex agent perceives the input from its environment and interprets it to a state that matches its rules. The agent that keeps track of the world keeps an internal state of the past input because its actions need to be taken in correlation between the past states and the new state. The goal-based agent needs to know some information about its goal because the percepts are not enough to determine the action to be taken. Sometimes knowing the goal to the agent is not sufficient to take the right action, especially when there are conflicting goals, so the utility-based agent maps the percept state into a number that determines the degree of achieving the goal [8].

The IDS consists of several agents working together. Since the attacks grow everyday, the signatures also grow so the agents can learn the new signatures or detect the abnormal traffic resulted from the new attacks. During this paper, modeling of three kinds of attacks, the Denial of Service (DoS) attack, the ping sweep attack, and the secure coded document theft is presented. The DoS is to attack a certain service so that it cannot be available to legitimate users to use. Ping sweep is a pre-attack methodology to know the online hosts to attack later [6, 7]. The secure coded document theft is to read a document that you are not authorized to read.

Building the IDS using the agent technology has several advantages. They can be added or removed from the system without altering other system components. This is because they are running independently. The agents can be reconfigured or upgraded to newer versions without disturbing the rest of the system as long as their external interface remains the same. An agent may be a member of a group of agents that perform different simple functions but can exchange information and derive more complex results than any one of them may be able to obtain on their own [3].

The paper is organized as follows: section 2 describes the suggested multiagent framework, section 3 explains software methodology and agent structure, section 4 explains the system operation, section 5 shows how to simulate the IDS operation, and section 6 augments some concluding remarks.

2 Multiagent System Framework

The proposed intrusion detection system consists of four main modules, as shown in Fig. 1, the sniffing module, the analysis module, the decision module and the reporting module. The sniffing module is responsible for gathering packets from the network. The analysis module is responsible for analyzing the packets. The decision module is responsible for taking actions relevant to the severity of the detected attack. The reporting module generates reports and logs.

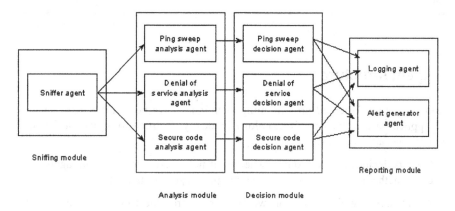

Fig. 1. System modules

The sniffing agent can be implemented as a simple reflex agent; that is, it listens to the network line and gives out the TCP/IP packets only. The analysis agents can be implemented as agents that keep track of the environment because some of the analysis types require looking back at the past packets. The decision agents can be implemented as goal based agents or utility-based agents because they have to know the goals of the system to take their appropriate decisions. The alert agents can be implemented as simple reflex agents because they need not to know anything about the past or the goals; they just give out reports or logs. Table 1 describes the type of each agent, its input and its output.

Table 1. Agents description

Agent name	Type	Percepts	Output
Sniffing agent	Simple reflex agent	Network traffic	The TCP packets
Secure code analysis agent	Keeping track of the world agent	The TCP packets	The secure code analysis
DoS analysis agent	Keeping track of the world agent	The TCP packets	The DoS analysis
Ping sweep analysis agent	Keeping track of the world agent	The TCP packets	The Ping sweep analysis

Secure code decision agent	Goal based agent	The secure code analysis	The secure code decision
DoS decision agent	Goal based agent	The DoS analysis	The DoS decision
Ping sweep decision agent	Goal based agent	The Ping sweep analysis	The ping sweep decision
Logging agent	Simple reflex agent	The TCP packets	Log file
Alert generator agent	Simple reflex agent	Decisions	Alert file

3 Agent structure

Although object-oriented programming techniques have been utilized for several AI applications, agent-based technology is ideally qualified to replace it for several reasons among them:

Objects decides for themselves whether or not to execute an action, agents decide whether or not to execute an action upon request from another agent.

Agents have flexible behavior, while the standard object model has nothing to say about such types of behavior.

A multiagent system is multithreaded, where each agent has at least one thread of control while the standard object model has a single thread of control in the system [1].

Hereinafter, agents' structures are explained in details.

The sniffing agent is the first agent to work in the system. As shown in Fig. 2, it connects to the network and begins to read the packets moving around. It has a buffer, to hold necessary packet information required by other agents. The sniffing agent reads a packet, if there is a free space in the buffer, then adds it to the buffer. If the analysis agents request new data, the sniffing agent will send the data in the buffer. The sniffing agent will work infinitely until the user stops the system.

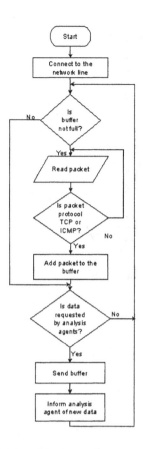

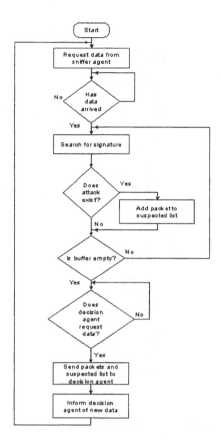

Fig. 2. Sniffing agent flowchart   **Fig. 3.** Analysis agent flowchart

The analysis agent, as shown in Fig.3, requests the buffer from the sniffing agent. The analysis agent builds a list of suspected packets. It searches through the buffer for signatures of attacks; if there is a known signature, it will put the packet into the list. When it finishes searching the buffer, it will wait until the complementary decision agent requests the suspected list of packets.

The decision agent starts working by requesting the data and the suspected list of packets from its complementary analysis agent, as shown in Fig. 4. The decision agent then will calculate the severity of the attack and take the necessary action according to the level of severity. It will also build a list of the decisions taken. Severe attacks cause the agent to disconnect the attacker and send an alert to the system administrator; otherwise it may only send an alert to the system administrator. Decisions and alerts are then forwarded to the alert generator agent. The decision agent forwards the suspected list of packets as well as the data when requested by the logging agent.

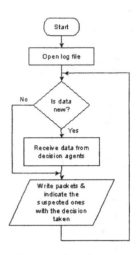

Fig. 5. Logging agent flowchart

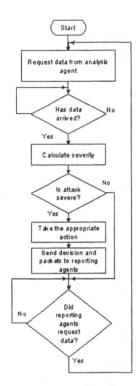

Fig. 4. Decision agent flowchart

The logging agent as shown in Fig. 5, opens the logging file and requests the buffer and suspected list of packets from the decision agents. The logging agent then writes the buffer into the log file, indicating the suspected ones. Else if the decision agent has no suspected list, the logging agent will only write the data of packets information into the log file. The logging agent will terminate when the user stops the system and consequently, it will close the log file.

When the alert generator receives the decisions list, it will send an alert to the system administrator. This alert can be a message on the screen or a message to a centralized machine or an alert file. The alert generator will terminate when the user stops the system. This is explained in Fig. 6.

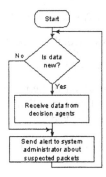

Fig. 6. Alert generator flowchart

4 System operation

The sniffing module, with its agent, percepts the network traffic passing through the host and gives out the TCP/IP and ICMP packets. The free packet capture library, winpcap, has been utilized.

The DoS analysis agent searches for DoS signatures in the packets such as the land attack where the source address and the destination address are the same, and the Xmas tree attack where the control bits are all set. The ping sweep analysis agent searches for the ping sweep attack signature where one source address floods many destination addresses with certain packets to discover the online hosts to attack. The secure code analysis agent searches for a secret code in the data portion of the packets where the source address is not authorized to read this document. The analysis agents perceive the TCP/IP packets fields, analyze them, and give out their analysis.

Corresponding decision agents perceive the results of the analysis agents and then start their operations. The decision agents issue the necessary decisions, take the appropriate actions, and yield the type of attack.

The logging agent percepts the TCP/IP fields, extracted from the packets, and then writes them into a log file. The alert generator agent percepts the type of attack and the action taken from the decision agents and writes them in an alert file.

5 Simulation

In this section a simulation of an intrusion detection system that can detect DoS, ping sweep, and secure document theft is presented. Suppose that computer A attacks computer C with the Xmas tree attack, the multiagent-based IDS is installed on computer B and monitoring the network.

The following scenario (Xmas Tree attack) will take place:

Computer *A* attacks computer *C* with the Xmas tree attack.

Computer *C* receives the defected packets.

The IDS running on computer *B* sniffs the packets on the network and analyzes it. The multiagent system will analyze all the incoming packets and will discover that computer *C* is being attacked with the Xmas tree attack.

The IDS initiates an alert and sends a reset packet to the attacker to disconnect it from the network.

Computer *A* will receive the reset packet and will be disconnected and cannot continue the attack.

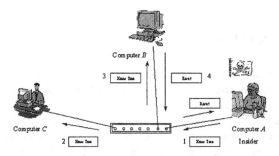

Fig. 7. Simulation of an attack

Of course the detection of the attack is a real-time process such that the attacker is instantly terminated and an alert is generated onto the console machine B. It is highly recommended that the IDS machine is fully dedicated to the intrusion detection mission. High processing speed (P-IV with 2GHz or higher), large memory (1GB), and sufficient amount of available disk storage are highly recommended.

6 Concluding remarks

Agent-oriented techniques are being increasingly used in a range of networking security applications. However, if they are to enter the intrusion detection system design it is vital to clarify why they are suitable to implement such systems as compared to other technologies such as object-oriented. This paper has sought to justify why agent-oriented approaches are appropriate for developing complex software systems such as IDS. An internal structure of each agent in the system has been addressed. A simplified simulation process has been introduced. This framework is considered a step towards a complete multiagent-based system for networking security.

References

1. Gerhard Weiss (2000) Multiagent Systems: A modern approach to distributed artificial intelligence. The MIT press, Cambridge.
2. Guy G Helmer, Johnny S K Wong, Vasant Honavar, Les Miller (1998) Intelligent agents for intrusion detection. In: The IEEE Information technology conference. Syracuse, NY, pp 121-124.
3. Jai Sunder Balasubramaniyan, Jose Omar Garcia-Fernandez, David Isacoff, Eugene Spafford, Diego Zamboni (1998) An architecture for intrusion detection using autonomous agents. In: 14th IEEE Computer security applications conference. Scottsdale, Arizona.
4. Marc Dacier (1999) Intrusion Detection vs. Detection of Errors Caused by Intentionally Malicious Faults. In: 29th Annual international symposium on fault-tolerant computing, Madison, Wisconsin.
5. Sami Saydjari (1999) The detection of novel unseen attacks. In: 29th annual international symposium on fault-tolerant computing, Madison, Wisconsin.
6. Stephen Northcutt, Judy Novak, Donald McLachlan (2001) Network Intrusion Detection. New Riders, USA.
7. Stephen Northcutt, Mark Cooper, Matt Fearnow, Karen Frederic (2001) Intrusion Signatures and Analysis. New Riders, USA.
8. Stuart Russell, Peter Norvig (1995) Artificial Intelligence a modern approach. Prentice Hall, New Jersey.
9. Teresa F Lunt (2000) Foundations of intrusion detection. In: 13th IEEE Computer security foundations workshop, Cambridge.

An Adaptive Platform Based Multi-Agents for Architecting Dependability

Amar Ramdane-Cherif, Samir Benarif and Nicole Levy

PRISM, Université de Versailles St.-Quentin, 45, Avenue des Etats-Unis, 78035 Versailles Cedex, France

Abstract. Research into describing software architecture with respect to their dependability proprieties has gained attention recently. Fault-tolerance is one of the approaches used to maintain dependability; it is associated with the ability of a system to deliver services according to its specifications in spite of the presence of faults. In this paper, we use a platform based on multi-agents system in order to test, evaluate component, detect fault and error recovery by dynamical reconfigurations of the architecture. An implementation of this platform on Client/Server architecture is presented and some scenarios addressing dependability at architectural level are outlined. In this paper, we discuss the importance of our approach and its benefits for architecting dependable systems and how it supports the improvement of dependability and performance of complex systems.

Keywords. dependability and performance, software architecture, software reliability, multi-agents system, architecting dependable systems, software agent.

1 Introduction

The dependability of systems is defined as the reliance that can justifiably be placed on the service the system delivers. The dependability has become an important aspect of computer system since everyday life increasingly depends on software [1] [2] [3]. Although there is large body of research in dependability, architectural level reasoning about dependability is only just emerging as an important theme in software engineering. This is due to the fact that dependability concerns are usually left until too late in the process of development. In addition, the complexity of emerging applications and trend of building trustworthy systems from existing, untrustworthy components are urging dependability concerns be considered at the architectural level.

In [4] the researches focus on the realisation of an idealized fault-tolerance architecture component. In this approach the internal structure of an idealized component has two distinct parts: one that implements it's normal behaviour, when no exceptions occur, and another that implements it's abnormal behaviour, which deals with the exceptional conditions. Software architectural choices have profound influence on the quality attributes supported by system. Therefore, architecture analysis can be used to evaluate the influence of the design decisions on important quality attributes such as maintainability, performance and dependability [5]. Another axe of research is the study of fault descriptions [6] and the role of event description in architecting dependable system [7]. Software monitoring is a well-know technique for observing and understanding the dynamic behaviour of programs when executed, and can provide for many different purposes [8] [9]. Besides dependability, other purposes for applying monitoring are: testing, debugging, correctness checking, performance evaluation and enhancement, security, control, program understanding and visualization, ubiquitous user interaction and dynamic documentation. Another strategy is used, like a redundant array of independent component (RAIC) which is a technology that uses groups of similar or identical distributed components to provide dependable services [10]. The RAIC allows components in redundant array to be added or removed dynamically during run-time, effectively making software components "hot-swappable" and thus achieves greater overall dependability. The RAIC controllers use the just-in-time component testing technique to detect component failures and the component state recovery technique to bring replacement components up-to-date. The approach in [11] advocates the enforcement of dependability requirements at the architectural design level of a software system. It provides a guideline of how to design an architectural prescription from a goal oriented requirements specification of a system.

To achieve high dependability of software, the architectures must have the capacity to react to the events (fault) and to carry out architectural changes in an autonomous way. That makes it possible to improve the properties of quality of the software application [12]. The idea is to use the architectural concept of agent to carry out the functionality of reconfiguration, to evaluate and to maintain the quality attributes like dependability of the architecture [13]. Intelligent agents are new paradigm for developing software applications. More than this, agent-based computing has been hailed as "the next significant break-through in software development, and "the new revolution software". Currently, agents are the focus intense interest on the part of many sub-fields of computers science and artificial intelligence. An agent is a computer system situated in some environment, and that is capable of autonomous action in this environment in order to meet its design objectives. Autonomy is a difficult concept to pin down precisely, but we mean it simply in the sense that the system should be able to act without the direct intervention of humans (or other agents), and should have control over its own actions and internal state. It may be helpful to draw an analogy between the notion of autonomy with respect to agents and encapsulation with respect to object-oriented systems. In this paper, we propose a new approach which provide a platform based agents. This platform will monitor the global architecture of a system and maintain or improve its performance and dependability. It will achieve its functional and non functional requirements and evaluate and manage changes in such architecture dynamically at the execution time.

This paper is organized as follows. In the next section, we will introduce the platform based multi-agents. Then a strategy to achieve fault tolerance by our platform will be presented. In section four, we describe an example showing the application of our platform on Client-Server architecture and its benefits are outlined through some scenarios about the dependability. Finally, the paper concludes with a discussion of future directions for this work.

2 The platform multi-agents

Autonomous agent and multi-agents systems represent a new way of analysing, designing, and implementation complex software systems. The agent-based system offers a powerful repertoire of tools, techniques, and metaphors that have potential to considerably improve the way which people conceptualize and implement many type of software systems. Agents are being used in increasingly wide variety of applications.

The platform based multi-agents is specifically developed to evaluate and reconfigure software architecture, it is constituted by three layers: superior, intermediate and reactive layers. To control the architecture by the platform, this architecture is decomposed into localities which depend on the zones. The first layer of the platform is the superior layer which is the decisional part of the platform. The superior layer includes evolved agents, which integrate decisional functionality. Each superior agent manages a lot of intermediate agents, it controls a zone of a software architecture. The second layer is the intermediate layer, it links superior layer to reactive layer. The intermediate agent in this layer controls a number of reactive agents; it supervises a locality of a software architecture. The finale layer is the reactive layer composed of reactive agents. These agents realise a basic action, and response at simple stimulus perception/action (Fig.1).

For this platform based multi-agents system we have developed an organization as well as specific planning. This platform includes different characteristics to draw from multi-agents model some systems for example: i) the multi-agents actor system (capacity of the agent to create another agents), ii) the blackboard system (inputs information are centralized towards a knowledge base) and iii) the physically distributed or centralized system (planning distributed or planning centralized).

3 The platform and fault tolerance

The basic strategy to achieve fault tolerance in a system can be divided into two steps. The first step called error processing, is concerned with the system internal state, aiming to detect errors that are caused by activation of faults, the diagnostic of the erroneous states, and recovery to error free states. The second step, called fault treatment, is concerned with the sources of faults that may affect the system and include: fault localization and fault removal. The communication between components is only through request/response messages. Upon receiving a request for a service, the components will react with a normal response if request is successfully processed or an external exception, otherwise. This external exception may be due to the invalid service request, in which case it is called an interface exception, or due to a failure in processing a valid request, in which it is called a failure exception. The error can propagate through connector of software architecture by using the different interactions between the components. Internal exceptions are associated with errors detected within a component that may be corrected, allowing the operation to be completed successfully; otherwise, they are propagated as external exceptions.

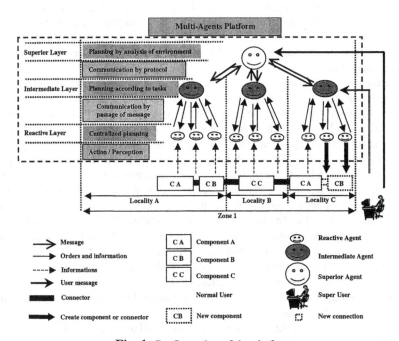

Fig. 1. Configuration of the platform

3.1 Monitoring system

System monitoring consists in collecting information from the system execution, detecting particular events or states using the collected data, analysing and presenting relevant information to the user, and possibly taking some (preventive or corrective) actions. As the information is collected from the execution of the program implementation, there is inherent gap between the levels of abstraction of the collected events, states of the software architecture. For event monitoring, there are basically two types of monitoring systems based on the information collection: sampling (time-driven) and tracing (event-driven). By sampling, information about the execution state is synchronously (in a specific time rate), or asynchronously (through direct request of the monitoring system). By tracing, on the other hand, information is collected when an event of interest occurs in the system. Tracing allows a better understanding and reasoning of the system behaviour than sampling. However, tracing monitoring generates a much larger volume of data than sampling. In order to reduce this data volume problem, some researchers have been working on encoding techniques. A more, common and straightforward way to reduce data volume is to collect interesting events only, and not all events that happen during a program execution. The second approach may limit the analysis of events and conditions unforeseen previously to the program execution. Both state and event information are important to understand and reason about the program execution. Since tracing monitoring collects information when events occur, state information can be maintained by collecting events associated to state change. With a hybrid approach, the sampling monitoring can represent the action of collecting state information into an event for the tracing monitoring. Not all events with state information should be collected, but only the events of interest. Integrating sampling and tracing monitoring and collecting the state information through events reduce the complexity of the monitoring task. The monitoring system needs to know what are the events of interest, what events should be collected.

3.2 Detection of faults with the platform based agents

We will use a monitoring system based on the agents, by implementing our platform, described above, on the top of the architecture. Each component will be supervised by a reactive agent, by sampling or tracing. The reactive agents will use sampling on architecture and collect information on the state of the components within each interval of time predefined or limited by the user. Another type of detection in reactive agent is the tracing, in this case, the component generates an external exception in the form of an event, this event will be collected and will be transmitted towards the intermediate agent, this event will be thereafter analyzed, identified and then sent by this agent towards the agent of the superior layer in order to establish plans to correct the errors. In other words, the signals are collected by the agents of the reactive layer, which transmit them immediately to the intermediate agent of their locality. This agent analyzes this information using its knowledge base containing the description of the errors. Thus, it will sort information coming from the reactive agents and send only the error messages towards the agent of the superior layer of its zone. According to the detected errors the superior agent establishes the plans in order to solve the errors coming from architectural level.

3.3 The treatment process

After the phase of detection, the platform identifies the type of error and establishes the plans in order to achieve at architectural level the necessary reconfigurations to correct the faults. This treatment process uses tow types of plans:

1. the first plans consist to reconfigure architecture connections for finding temporary solution of fault (disabled component or connector). In the detection phase, the information travel up through the layers of the platform in order to arrive to the superior agent, in this decisional layer the treatment process begins by establishing plans. The superior agent chooses the best solution to maintain dependability of the architecture. The platform can reconfigure connections of the architecture to isolate the disabled components (if the platform can't create new components). The superior agent distributes the plans to the intermediate agent on the locality of fault. When the intermediate agent receives the plans, it distributes directives to the reactive agents. The reactive agents delete the connection of disabled component and create new connection to isolate it.

2. the second plans recover errors by addition or changing disabled component or connector. If the platform has the possibility to create new component in order to recover errors at architectural level, the superior agent distributes plans to the intermediate agent. This agent distributes directives to reactive agents, and the reactive agents work together in order to delete the disabled component and it's connection and create new component and it's new connection.

4 Implementation of the platform on Client/Server architecture

We realise Client/Server architecture based on sockets, the client can connect to server by entering it's user name and host name or IP address. The principal server (server A) and the second server (server B) deliver the same services. We used the second server and the dispatcher in order to reduce charge on principal server. The dispatcher orients the connections of the clients on server that have most lowly charge, the architecture is implemented on Windows system (Fig. 2). A lot of kinds of faults can occur on this Client/Server architecture: software fault like error of program, or hardware like electrical cut or breakdown of server machine. We realise this architecture and the platform based multi-agents and create different scenarios of fault. We can see the intervention of our platform to recover faults, these faults will be corrected dynamically by our platform without disturbing the clients. With this approach, the platform evaluates and maintains automatically a high level of dependability.

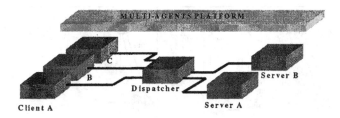

Fig. 2. Client/Server architecture

4.1 Scenario 1: Breakdown of the second server (Recover errors by reconfiguration of connections)

In this scenario the second server breakdown, the role of the platform is to intervene automatically and recover the fault at architectural level by the dynamic reconfigurations:

Fig. 3.a:
~ The user parameters the frequency of monitoring on the components.
~ The reactive agents do sampling on architecture and collect the state of each component, then this information is transferred to the intermediate layer. The intermediate agents analyze this information and identify if the faults occur on the architecture.

Fig. 3.b:
~ The second server is disabled, then, the reactive agent of this component collects information about it's state, and sends it to it's intermediate agent, which identifies the fault, and informs it's superior agent.
~ The superior agent identifies the kind of fault and establishes the plans and distributes them to the intermediate agent in locality of error. The intermediate agent disposes of plans, and controls its reactive agents in order to recover error.

Fig. 3.c:
~ The reactive agent specialized on the connection, deletes the disabled connections of the second server. The reactive agent that controls the disabled server component, deletes it.
~ The reactive agent reconfigures the dispatcher to orient all connections to the principal server.

Fig. 3.d:
~ All new client connection requests to the second server will be oriented to the principal server. The intermediate agent and the superior agent will be respectively informed about the new changes with the aim of storing the current state of the architecture.

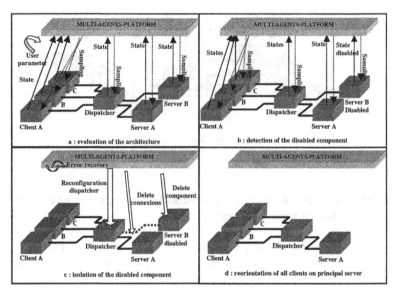

Fig. 3. Breakdown of the second server

4.2 Scenario 2: Breakdown of the dispatcher (recover errors by changing disabled component)

In this scenario, the platform changes dynamically the disabled component by new component:

Fig. 4.a:
˜ The platform monitors the architecture by sampling and collecting the states of all components.

Fig. 4.b:
˜ The state information of disabled component detected by reactive agent, is transmitted to the intermediate agent.
˜ The intermediate agent detects the error and transmits it to the superior agent, which identifies the kind of fault.

Fig. 4.c:
˜ The superior agent develops the plans in order to recover the fault, and sends them to the intermediate agent.
˜ The reactive agents delete all the client connections and the two servers connections to the dispatcher.

Fig. 4.d:
˜ The reactive agents create a new dispatcher and its connection.
˜ Now, the dispatcher can orient all client connections to both principal server and second server. The intermediate agent and the superior agent will be respectively informed about the new changes with the aim of storing the current state of the architecture.

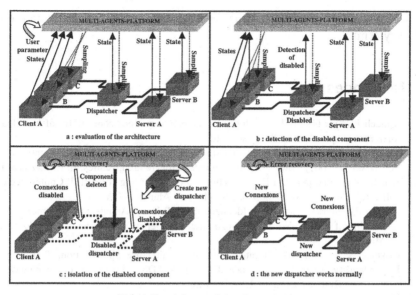

Fig. 4. Breakdown of the dispatcher

5 Conclusion

Dependability has become an important aspect of computer systems since everyday life increasingly depends on software. Although there is a large body of research in dependability, architectural level reasoning about dependability is only just emerging as an important theme in software engineering. The aim of the architecting dependable systems is to provide architectural principles involved first in building dependable systems and second in evaluating the architectures of these systems. This paper presents a new platform based multi-agents, which advocates automatically the achievement and the enforcement of dependability requirements of a software system at the architectural level. The conception of the specific mutli-agents system produces the dynamic and evolutionary platform so that this platform can constantly reach and follow the evolution of the software architecture system. In this paper we have developed our generic platform and we have applied and implemented it on the Client-Server architecture. We show by some scenarios the dynamic reconfigurations related to the improvement of performance and dependability through the structuring investigation of fault-tolerant component-based systems at architectural level. Our approach can be extended to deal with other architectural non-functional quality attributes in the context of developing complex and reliable systems.

6 References

1. Randell B, Xu J (1995) The evolution of the recovery block concept, In software fault tolerance, chapter 1. John Wiley sons ltd
2. Sloman M, Kramer J (1987) Distributed systems and computer networks. Prentice hall
3. Sotirovski D (2001) Towards fault tolerance software architectures. In R. Kazman, P. Kruchten, C. Verhoef, and H. Van Vliet, editors. Working IEEE/IFIP Conference on software architecture workshop, pages 7-13, Los Alamitos, CA
4. Asterio P, Guerra C (2002). An Idealized Fault-Tolerant Architectural Component, In proceeding of WADS: Workshop on Architecting Dependable Systems. Orlando, USA 25 May
5. Gokhale S S, et al. (2002) Integration of Architecture Specification, Testing and Dependability Analysis, In proceeding of WADS: Workshop on Architecting Dependable Systems. Orlando, USA 25 May
6. De-Lemos R, et al. (2002) Tolerating Architecture Mismatches, In proceeding of WADS: Workshop on Architecting Dependable Systems. Orlando, USA 25 May
7. Dias M S, Richardson D J (2002) The role of Event Description in Description in Architecting Dependable Systems. In proceeding of WADS: Workshop on Architecting Dependable Systems. Orlando, USA 25 May
8. Shroeder B (1995) On-line monitoring, IEEE Computer, vol. 28, n. 6, June, pp. 72-77.
9. Snodgrass R (1988) A Relation approach to monitoring complex systems, ACM Trans. Computer Systems, vol. 6, n. 2, May, pp. 156-196.
10. Liu C, Richardson D J (2002) Architecting dependable systems through redundancy and just-in-time testing. In proceeding of WADS: Workshop on Architecting Dependable Systems. Orlando, USA 25
11. Brandozzi M, Perry D E (2002) Architecture prescription for dependable systems, In proceeding of WADS: Workshop on Architecting Dependable Systems. Orlando, USA 25 May
12. Bass L, Clements P, Kazman R (1998) Software architecture in practice, SEI Series, Addison-Wesley. January
13. Ramdane-Cherif A, Levy N, Losavio F (2002) Dynamic Reconfigurable Software Architecture: Analysis and Evaluation. In WICSA'02: The Third Working IEEE/IFIP Conference on Software Architecture. Montreal, Canada, August 25-31

Stochastic Distributed Algorithms for Target Surveillance

Luis Caffarelli[1], Valentino Crespi[2], George Cybenko[2], Irene Gamba[1], and Daniela Rus[3]

[1] Department of Mathematics, University of Texas at Austin
   {caffarel,gamba}@mail.ma.utexas.edu
[2] Thayer School of Engineering, Dartmouth College
   {vcrespi,gvc}@dartmouth.edu
[3] Computer Science Department, Dartmouth College gvc@dartmouth.edu

Summary. In this paper we investigate problems of target surveillance with the aim of building a general framework for the evaluation of the performance of a system of autonomous agents. To this purpose we propose a class of semi-distributed stochastic navigation algorithms, that drive swarms of autonomous scouts to the surveillance of grounded targets, and we provide a novel approach to performance estimation based on analysing sequential observations of the system's state with information theoretical techniques. Our goal is to achieve a deeper understanding of the interrelations between randomness, resource consumption and ergodicity of a decentralized control system in which the decision–making process is stochastic.

Key words: UAV Systems, decentralized control and multi-agent systems.

1 Introduction

Employment of Agent-based technology in the design of distributed control systems is notoriously limited by the lack of a general unified model for performance estimation. The difficulty in this matter concerns both the qualification of candidate solutions and the quantification of their performance.

In this paper we deal with problems related to target surveillance[4]. Informally, we want to deploy a number of flying UAV's (Unmanned Autonomous Vehicles) with the main objective of monitoring a set of targets located on ground at fixed positions. We design a distributed agent-based navigation control system that allows the scouts to provide full coverage of the sites and, at the same time, confound the enemy to prevent its counter measures, with the

---

[4] Research partially supported by the DARPA TASK program under grant number F30602-00-2-0585.

least possible consumption of fuel per target visit. The importance of those problems relies upon the recent increasing demand for large fault-tolerant autonomous systems composed of fully automated, unmanned, easy-to-deploy and cheap vehicles.

The main purpose of this investigation is to understand nontrivial interrelations between various elements that characterize a distributed multiagent system. In particular we study how randomization affects metrics like energy consumption and the degree of unpredictability of vehicle trajectories. We provide an original application of techniques from Information Theory to the analysis of sequential observations of the system's state in order to quantify the quality of the proposed solutions.

Unlike in classical methods, based on solving a global optimization problem to provide an optimal solution we look for distributed solutions under conditions in which traditional techniques, like dynamic programming, do not apply. Our idea of approaching these problems using agent-based technologies is motivated by

- the need for decentralization in control systems characterized by hundreds of autonomous agents;
- requirements of robustness and fault tolerance with respect to abrupt increase or decrease of the number of UAV's in play;
- the need for a chaotically nondeterministic evolution of the system's state, dictated by the necessity for protection against confidentiality attacks (trajectories or throughput rates prediction).

Methodologically, we follow a three-stage process:

1. Under the assumption that the resources available to the UAV's are unbounded (e.g., they have full knowledge of the global state), we define a family of initial solutions, typically by applying techniques from classical control theory and dynamic systems theory. In this case we define a family of randomized potential functions for the navigation of the UAV's that combines, in an original fashion, a stochastic decision–making procedure with an artificial potential technique (examples of applications of artificial potentials and artificial fields in UAV's navigation and robot path planning are contained in [1, 2] and in [3, 4, 5, 6], respectively).
2. Global resources are replaced with local resources compatibly with the constraints of the given decentralized environment. During this stage we look for phase transitions in order to identify portions of the parameter space in which the Objective is fulfilled.
3. Finally, we tune the parameters of the artificial potential functions for efficient control of the scouts. This allows the system to evolve with enough nondeterminism at a reasonable cost. The resulting algorithms (implemented using a Montecarlo D/E numerical simulation) drive each scout (UAV) along the steepest descending direction of a stochastically determined potential. Instances of our methodology can be found in [7, 2].

Summarizing, we formalize our Surveillance Problem (Section 2) along with the metrics we measure (Section 3). Then we introduce a new class of navigation algorithms[5] (Section 4) and conduct an experimental analysis of their performance (Section 5). We execute a phase-transition analysis in order to identify the conditions under which a fair trade of resources can take place and finally we discuss fuel vs "space aperiodicity" trade-offs. Considerations on future extensions of this work conclude the paper (Section 6).

2 Problem Definition

We want to design a distributed algorithm that drives a swarm of UAV's to surveil targets disseminated on a given battlefield under the following assumptions consistent with the available technology:

- The vehicles can access global information by communicating with a satellite, in general, under bandwidth limitations. The sort of information that each vehicle can collect, despite such limitations, includes the coordinates of all the targets on the ground and the statistics of the monitoring level of any target at a specific time.
- The vehicles are able to fly at different altitudes so they never collide.
- The vehicles have unlimited maneuvering capabilities in the ray of curvature and in the acceleration, i.e., they can perform sharp turns.

We formulate our Surveillance problem in 2D as follows:

Instance: A battlefield consisting of a terrain and a set of m targets located at fixed positions.

Problem: Deploy n scouts (UAV's) $\{s_1, s_2, \ldots, s_n\}$ with the following Objective and Optimization requirements:

1. **Objective:** guarantee that each target receives a proper share of visits over time (e.g., equal coverage).
2. **Optimization/Trade-off:** drive scouts along unpredictable trajectories minimizing the consumption of fuel per target visit (see Section 3).

In this paper we will adopt the following terminology. We denote with $\mathbf{x}_i(t) \in \mathbf{R}^2$ the position of UAV s_i at time t and with $\mathbf{y}_j \in \mathbf{R}^2$ the position of target $j \in [m]$.

3 Objectives, Metrics and Measurements

We are going to provide a quantitative analysis of our algorithms, based on the following metrics:

---

[5] See http://actcomm.thayer.dartmouth.edu/task/demos.html

1. *Energy consumption per target visit of UAV's.*
   In our model we assume that energy and fuel consumption of a single UAV is proportional to the arc length of its trajectory. Thereby, in our analysis we estimate the lengths of those trajectories by adding lengths of segments of polygonal curves that approximate the actual curves.

2. *Predictability and stability of the trajectories.*
   This metric is important as if the potential enemy knows in advance which target is going to be visited in a certain time frame, it will be able to adopt undesirable counter measures. Let us consider the following event generation process: each time a target is visited by any scout we record the corresponding target id. Then as the system evolves we will be able to collect a (potentially infinite) sequence of target id's. Our idea is to study the compressibility of such sequences in order to estimate the entropy of the stochastic source that has generated it. Maximum entropy would then correspond to total unpredictability of the next target to be visited or, equivalently, to space aperiodicity of the trajectories[6]. We measure space aperiodicity by applying information theoretical tools. In particular we run our simulations recording the sequence of target id's as they are visited. Then we use Lempel-Ziv [8]-based Universal coding schemes to compress the string of such recordings and finally we compute the compression ratio. These values provide us with a measure of the entropy of the source.

3. *Application performance with respect to the Objective.*
   We measure of how well the algorithms ensure that targets receive a proper share of attention (in this case uniform) from the scouts. Formally, let $\theta_u = \frac{1}{m}\mathbf{1}_m$ be the uniform distribution over the set of targets, representing equal number of visits (on average) per target in a time unit. Then we compute stable (stationary) statistics of the portions of surveillance per target, say θ, and evaluate its distance from θ_u according to some suitable distance metric. In this paper we consider what in the theory of mixing Markov Chains is called *the variation distance* $d(\theta_u, \theta) = \frac{1}{2}\sum_j |\frac{1}{m} - \theta_j|$. We also compute the information entropy[7] of θ: $H(\theta) = \sum_j \theta_j \log_2 \frac{1}{\theta_j}$ that should provide us with a measure of the imbalance of θ with respect to uniformity.

As we will observe experimentally, energy consumption and predictability of the trajectories are metrics that cannot be optimized at the same time. Intuitively, optimization of fuel results into the scouts moving deterministically and this in turn makes their trajectories very predictable. So, in general we look at methods to obtain trade-offs between energy and space aperiodicity, under the condition that targets are visited on equitable basis.

---

[6] We focus on space aperiodicity since time periodicity can be easily broken by randomizing the speed of the scouts (within their capabilities).

[7] This should not be confused with the entropy of the source of target id's that we mentioned before.

4 Our Algorithms

4.1 Randomized Potentials

As anticipated our algorithms are based on artificial potentials to be minimized through a gradient descent method combined with a stochastic decision–making rule.

Each target j has a dynamic mass $m_j = M(p_j)$ that generates a gravitational potential V_j. This mass depends in general upon a dynamic priority $p_j(t)$. At each time t each drone is subjected to the linear superposition of a stochastically selected subset of all the potentials. Formally, scout i in position $\mathbf{x}$ at time t resents of the influence of potential

$$V^{(i)}(\mathbf{x},t) = \sum_{j=1}^{m} \sigma_{i,j} \cdot V_j(\|\mathbf{x} - \mathbf{y}_j\|, t) = \sum_{j=1}^{m} \sigma_{i,j} \cdot (-1)\frac{M(p_j(t))}{\|\mathbf{x} - \mathbf{y}_j\|^\alpha}, \qquad (1)$$

where $V_j(r,t) = -\frac{m_j(t)}{r^\alpha}$ and $\sigma_{i,j} \in \{0,1\}$ are Bernoulli random variables.

Targets are given dynamic priorities that increase with time and are reset when they are visited by any scout. Each scout s_i needs to make a selection of one or more targets through a randomized assignment of K ones to $\sigma_{i,j}$, for $j \in [m] = \{1, 2, \ldots, m\}$. The value $K = \sum_j \sigma_{i,j} = \#\{\sigma_{i,j} = 1 \mid j \in [m]\}$ can be fixed a priori or free to range from 1 to m and, as we will see in a moment, it is very important for the control of the level of randomness incorporated in the algorithm.

So, at time t scout i either has been assigned a target (binary) vector $\sigma_i = (\sigma_{i,j})$ defining a subset of targets that exert attraction to it, or has to select a new target vector by sampling from a probability distribution that is defined through the dynamic priorities and the Euclidean distances $\{d(s_i, l_j) = \|\mathbf{x}_i - \mathbf{y}_j\| \mid j \in [m]\}$. The selection of a target vector is irrevocable and the selecting scout has to reach one of the constituent targets before making a new selection. The quantity $p_j(t)$, the dynamic priority of target j at time t can be defined as:

$$\begin{cases} p_j(t+1) := p_j(t) + \delta \cdot w_j & \text{target } j \text{ not visited at time } t \\ \\ p_j(t+1) := 0 & \text{target } j \text{ visited at time } t \end{cases}$$

The vector $\mathbf{w} = \{w_1, w_2, \ldots, w_m\}$ is intended to be a vector of static priorities.

With these ingredients we can define a set of n basic probability distributions (one per scout) over the set of targets:

$$q_{i,j}(t) = \frac{p_j(t) \cdot d(s_i, t_j)^{-\eta}}{Q_i}, \quad Q_i = \sum_j p_j(t) \cdot d(s_i, t_j)^{-\eta}, \qquad (2)$$

where $\eta = \alpha + 1$ and α is the parameter that occurs in the formula of the gravitational potentials generated by the targets. And from those a set of n probability distributions:

$$P_{i,j}(\beta) = \frac{q_{i,j}^{\beta}}{N_i(\beta)}, \quad N_i(\beta) = \sum_j q_{i,j}^{\beta} . \tag{3}$$

Observe that, as $\beta \to 0^+$, $P_{i,j}(\beta) \to \frac{1}{m-1}$, whereas as $\beta \to \infty$ we can prove the following lemma.

Lemma 1. *Let $\alpha = 1$ and $u_{i,j} = p_j/d_{i,j}^2$. Define $I_i = \arg\max\{u_{i,k} \mid 1 \le k \le m\}$ as the set of indices of the maximal elements in the sequence $(u_{i,k})_k$. Then as $\beta \to \infty$:*

$$P_{i,j}(\beta) \to P_{i,j} = \begin{cases} \frac{1}{|I_i|} & \text{if } j \in I_i, \\ \\ 0 & \text{otherwise.} \end{cases}$$

Proof. Let $j_0 \in I_i$. Then substituting 2 in 3 and rearranging terms we obtain:

$$P_{i,j}(\beta) = \frac{u_{i,j}^{\beta}}{\sum_k u_{i,k}^{\beta}} = \frac{u_{i,j}^{\beta}}{u_{i,j_0}^{\beta}(|I_i| + \sum_{k \notin I_i}(\frac{u_{i,k}}{u_{i,j_0}})^{\beta})}$$

which clearly tends to $1/|I_i|$ for $j \in I_i$ and to 0 otherwise. We observe in fact that terms $\frac{u_{i,k}}{u_{i,j_0}}$ in the summation are less than 1.

Finally the K ones are sampled from the distribution $P_{i,j}(\beta)$ as follows. Suppose we have an urn containing m numbers $1, 2, \ldots, m$, then we extract the K values $j_1, j_2, \ldots, j_K$ one by one without re-insertion. At stage u the probability of extracting j_u is defined as $P\{j_u\} = P_{i,j_u}(\beta)/\{1 - \sum_{k=1}^{u-1} P_{i,j_k}(\beta)\}$.

4.2 The Simulation

Our algorithms are based on a gradient descent method simulated with a Discrete/Event mechanism[8]. Each agent updates its position following the steepest decreasing direction of its potential [9]. This is precisely the opposite of the gradient vector of the potential function. So, according to this method, each scout i computes the vector $-\nabla V^{(i)} = -\sum_j \sigma_{i,j} \nabla V_j$ and moves along its direction. One way to implement this is through the following rule:

$$\mathbf{x}_i(t+1) = \mathbf{x}_i(t) - \alpha^{\frac{1}{\alpha+1}} \cdot \frac{\nabla V^{(i)}(\mathbf{x}_i(t))}{\|\nabla V^{(i)}(\mathbf{x}_i(t))\|^{\frac{\alpha+2}{\alpha+1}}} . \tag{4}$$

Since

---

[8] We have produced Matlab, C and Java implementations.

$$\nabla V_j(\|\mathbf{x} - \mathbf{y}\|) = V_j'(\|\mathbf{x} - \mathbf{y}\|)\frac{\mathbf{x} - \mathbf{y}}{\|\mathbf{x} - \mathbf{y}\|}\ ,$$

V_j' has constant sign and $V^{(i)}(\mathbf{x}_i) \approx V_j(\|\mathbf{x} - \mathbf{y}_j\|)$, when scout i is sufficiently close to target j (and $\sigma_{i,j} = 1$), it is not hard to see that

$$\alpha^{\frac{1}{\alpha+1}} \cdot \frac{\nabla V^{(i)}(\mathbf{x}_i(t))}{\|\nabla V^{(i)}(\mathbf{x}_i(t))\|^{\frac{\alpha+2}{\alpha+1}}} \approx \mathbf{x}_i - \mathbf{y}_j$$

so when a scout reaches the proximity of one of its targets its new gradient adjusts gracefully avoiding overshooting.

The algorithm may be described as follows. At the beginning, say at time $t = 0$, all the agents, deployed uniformly at random over the region of interest, compute their current destinations, together with the needed travel times using random dynamic priorities and so purely random target vectors. This is accomplished by applying the position update rule (4) and assuming that the velocities are constant in norm (and here for simplicity equal to v for all). This first step produces:

- a description of the system state that consists of the current agent positions and velocities $X = \{\mathbf{x}_i\}$, $V = \{\mathbf{v}_i\}$ and
- a discrete set E of events (t, i) signifying that agent i will reach its current designated destination at time t.

Then the simulation proceeds endlessly as follows:

1. Extraction of the event (t, j) that is meant to occur first, i.e., the one that minimizes the element t. Let t_0 be such time;
2. The positions $X(t)$ of all the drones are let evolve by computing iteratively $X(t + \Delta t); t \leftarrow t + \Delta t$; in accordance to the assumed kinematics (rectilinear uniform motion along the directions $\mathbf{v}_i$), until either $t + h \cdot \Delta t > t_0$ or one or more targets are visited by any scouts;
3. In the first case, the event fires and the position update rule is applied to the agent j associated with the extracted event (t, j) to compute its new designated destination and, as a consequence, its new velocity $\mathbf{v}_j'$ and travel time t_j':

$$\mathbf{x}_j' = \mathbf{x}_j - \alpha^{\frac{1}{\alpha+1}} \cdot \frac{\nabla V^{(j)}(\mathbf{x}_j)}{\|\nabla V^{(j)}(\mathbf{x}_j)\|^{\frac{\alpha+2}{\alpha+1}}}, \mathbf{v}_j' = v\frac{\mathbf{x}_j' - \mathbf{x}_j}{\|\mathbf{x}_j' - \mathbf{x}_j\|}, t_j' = \frac{\|\mathbf{x}_j' - \mathbf{x}_j\|}{v} \quad (5)$$

This in turn causes the generation of a new event $(t_0 + t_j', j)$ to be included in the set E.
4. In the second case, all the scouts that are crossing over targets will have to sample a new target vector, apply the position update rule, compute speeds and travel times and cast the appropriate events in E.

A pseudo-code description of our Discrete Event Simulation algorithm would be the following[9]:

```
MainProgram
Input: α, β, δ, w
begin
  E ← ∅;  t ← 0
  for j ∈ Targets AND i ∈ Scouts do
    (xᵢ, yⱼ, pⱼ, σᵢ,ⱼ) ← RandomValues
    mⱼ ← 1
    UpdateRule(i, t)
  endo
  while E ≠ ∅ do
    (t₀, i₀) ← min(E)
    E ← E \ {(t₀, i₀)}
    while t + Δt < t₀ do
      X(t) ← X(t + Δt);  t ← t + Δt
      [VisitedTrgts, VisitingScts] ← CheckVisits(X(t), Y)
      GenerateEvent(VisitingScts, VisitedTrgts )
      p ← p + δ · w
    endo
    X(t) ← X(t₀);  t ← t₀
    [VisitedTrgts, VisitingScts] ← CheckVisits(X(t), Y)
    GenerateEvent(VisitingScts, VisitedTrgts)
    UpdateRule( i₀, t )
    p ← p + δ · w
  endo
end
```

$$\text{GenerateEvent}(A, B)$$
$$\textbf{begin}$$

$\quad$ **for** $i \in A$ AND $j \in B$ **do**

$\qquad p_j \leftarrow 0$

$\qquad \sigma_i \leftarrow \text{ExtractTgtVector}(K, \beta, \mathbf{p})$

$\qquad \text{UpdateRule}(i, t)$

$\quad$ **endo**

end

$$\text{UpdateRule}(i, t)$$
$$\textbf{begin}$$

$\quad \mathbf{x}_i' = \mathbf{x}_i - \alpha^{\frac{1}{\alpha+1}} \cdot \dfrac{\nabla V^{(i)}(\mathbf{x}_i)}{\|\nabla V^{(i)}(\mathbf{x}_i)\|^{\frac{\alpha+2}{\alpha+1}}};$

$\quad \mathbf{v}_i' = v \dfrac{\mathbf{x}_i' - \mathbf{x}_i}{\|\mathbf{x}_i' - \mathbf{x}_i\|}; \quad t_i' = \dfrac{\|\mathbf{x}_i' - \mathbf{x}_i\|}{v}$

$\quad \mathbf{x}_i \leftarrow \mathbf{x}_i'; \ \mathbf{v}_i \leftarrow \mathbf{v}_i'$

$\quad E \leftarrow E \cup \{(t + t_i', i)\}$

end

---

[9] Here we have adopted the following terminology: $K = \sum_j \sigma_{i,j}$, $\mathbf{m} = (m_i)$, $M(\mathbf{m}) = (M(m_i))$, $\mathbf{p} = (p_i)$, $\mathbf{w} = (w_i)$, $\sigma_i = (\sigma_{i,j})$, $X(t) = (\mathbf{x}_i(t))$, $Y = (\mathbf{y}_i)$, $V(t) = (\mathbf{v}_i(t))$ and $X(t + \Delta t) = X(t) + V(t) \cdot \Delta t$.

5 Experiments

Let us first discuss the role and importance of parameters K, the number of ones that occur in the target vectors σ_i and β, the exponent that regulates the relation between the sample probabilities $P_{i,j}$ and the dynamic priorities p_j (the feedback control). We observe the following facts.

- The values of both K and β control in some sense the amount of randomness that is incorporated in the target vector selection.
- For $K = m - 1$ (all ones except for the target last visited that is zero), we obtain a totally deterministic algorithm where no random choice is made at all. Moreover, for α large enough, the gradient direction will tend to align along $\nabla \sup_j V_j$. This implies that scouts would always be headed towards their closest targets.
- At the other extreme is the case $K = 1$, where scout i selects only one target j with probability $P_{i,j}(\beta)$. This means that any target is selectable and reachable in principle as long as the corresponding probability is nonzero. So, arbitrary target vectors imply the capability of scouts to penetrate through any "cloud" of targets and reach the farthest ones.

We have verified experimentally that the distribution of the proportions of surveillance realized by the algorithm stabilizes relatively quickly (after 6000 steps). Besides all our experiments have been conducted in stationary regime.

5.1 Space aperiodicity.

To measure this metric we exploited the incompressibility of random sequences of numbers. We have considered many choices of n scouts and m targets and run hundreds of simulations for thousands of steps. For each run we have recorded the sequence of targets as they were visited and measured the compressibility of those sequences using the Lempel-Ziv coding scheme. Given a sequence of target recordings S and its compressed representation $c(S)$, we define $r = 100 \cdot (1 - l(c(S))/l(S))$ ($l(x)$ is the length of string x). So, $r = 0$ means no compression at all (no regularities) whereas $r \approx 100$ means high compression (high regularity).

5.2 K varying: phase-transition analysis of the reachability phenomenon.

In this experiment we set $\beta = 1$ and let K vary from 1 to $m - 1$. We observed a breakdown of $H(\theta)$ as K grows above 2 (see Fig. 2). This confirms our theoretical intuition that "fat" target vectors (numerous multiple destinations) drastically limit the reachability of the peripheral targets as scouts restrict their patrols to targets located nearby the center of mass of the system.

5.3 $K = 1$: space aperiodicity vs fuel consumption trade-off.

In order to maintain a "fair" distribution of surveillance: $\theta \approx \frac{1}{m}1$ (fulfillment of the Objective), we set $K = 1$ and study the energy consumption and space aperiodicity as β ranges from 0 to ∞. Our idea here is to trade energy with space aperiodicity under the invariant condition that all the targets are to be visited on equitable basis. We observed that as β tends to infinity the probability distribution $P_{i,j}(\beta)$ tends very often to a $(0,1)$-vector $P_{i,j}$ (see Lemma 1, in case $|I_i| = 1$) with only one component set to 1. This means that as β grows the choice of the next target to visit becomes more and more of a deterministic function of the inter-target Euclidean distances and their associated dynamic priorities. And this results into a lower fuel consumption per target visit (see Fig 1). The intuition behind this is that, in presence of total determinism, scouts would tend to travel to the closest targets with consequent saving of fuel. This saving would be attained at the expenses of a more chaotic behavior desired to confound the enemy. Figure 1 displays values of fuel consumption and space aperiodicity, $n = 5$ scouts, $m = 8$ targets arranged on a fixed deployment, $\alpha = 1$ and constant masses. The compression rates have been computed on 10000 target recordings and the results averaged over 50 trials.

The resulting data reveal that we can reasonably trade energy for space aperiodicity by tuning parameter β. The two curves $E(\beta)$ and $r(\beta)$ are indeed monotonic (decreasing the first and increasing the second) as expected. In particular we can see that $\beta = 2$ allows the lowest consumption of energy while keeping the compression ratio optimal ($r = 0$).

We also would like to observe that our trade is most meaningful for $\beta < 6$ as for higher values no appreciable saving of energy can be attained. This was expected, as our algorithms were not designed to compete with the best global deterministic solutions. In fact, the special case $n = 1$ (one scout) would be roughly equivalent to the Euclidean TSP (Traveling Salesman Problem) that is notoriously NP-hard [10]. Moreover the validity of the triangular inequality guarantees approximability with performance (error) ratio ≤ 2, whereas our algorithms in pure deterministic regime would not be guaranteed to attain a bounded performance (error) ratio.

6 Conclusions and Future Work

We have introduced a new class of semi-distributed algorithms for a multi-agent surveillance system and conducted an experimental analysis to synthesize an optimized solution. Now, we would like to investigate the mathematical relation between the quantities E, r and H. Besides, our algorithms still require knowledge of the dynamic priorities of all the targets at any time and so the next step will be to investigate systems in which scouts have only limited access to that information and need to exchange their "experiences" when

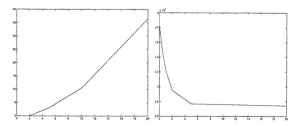

Fig. 1. Trade-off Analysis: $D = 5$: $r(\beta)$ (left) and $E(\beta)$ (right).

they come at close range. Finally, our phase-transition analysis leads to possible formalizations of the system as a percolation process since we may see agents as particles that float around the targets. Then one problem is to study the circulation of those particles (e.g., whether or not they remain imprisoned inside potential wells). This direction is interesting but must be pursued with care as it may lead to intractable problems. Statistical mechanics provides examples of multi-agent systems whose performance analysis of their emergent behaviors requires the computation of NP-hard quantities (e.g. partition functions in Ising models for ferromagnets) [11].

References

1. Valentino Crespi and George Cybenko. Agent-based Systems Engineering and Intelligent Vehicles and Road Systems. *Darpa Task Program white paper.* http://actcomm.thayer.dartmouth.edu/task/, April 2001.
2. Valentino Crespi, George Cybenko, Daniela Rus, and Massimo Santini. Decentralized Control for Coordinated flow of Multiagent Systems. In *Proceedings of the 2002 World Congress on Computational Intelligence. Honolulu, Hawaii,* May 2002.
3. J.C. Latombe. *Robot Motion Planning.* Kluver Academic Publishers, 1991.
4. Elon Rimon and Daniel E. Koditschek. Exact Robot Navigation using Artificial Potential Functions. *IEEE Transactions on Robotics and Automation, October 1992,* 8(5):501–518, 1992.
5. Leena Singh, Herry Stephanou, and John Wen. Real-time Robot Motion Control with Circulatory Fields. In *Proceedings of the International Conference on Robotics and Automation,* pages 2737–2742, 1996.
6. D. Keymeulen and J. Decuyper. The Fluid Dynamics Applied to Mobile Robot Motion: The Stream Field Method. In *Proceedings of the International Conference on Robotics and Automation,* pages 378–386, 1994.
7. Valentino Crespi, George Cybenko, and Daniela Rus. Decentralized Control and Agent–Based Systems in the Framework of the IRVS. *Darpa Task Program paper.* http://actcomm.thayer.dartmouth.edu/task/, Apr 2001.
8. Thomas M. Cover and Joy A. Thomas. *Elements of Information Theory.* Wiley Interscience, 1991.
9. D.P. Bertsekas and J.N. Tsitsiklis. Gradient Convergence in Gradient Methods. *SIAM J. on Optimization,* 10:627–642, 2000.

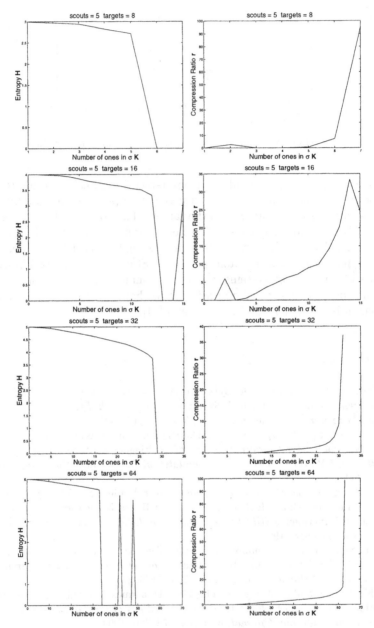

Fig. 2. Phase transition analysis: $D = 5$, $T \in \{8, 16, 32, 64\}$.

10. Micheal R. Garey and David S. Johnson. *Computers and Intractability*. W.H. Freeman and Company, 1979.
11. Mark Jerrum and Alistair Sinclair. *Polynomial Time Approximation Algorithms for the Ising Model*, volume 443. Springer Verlag, 1990.

What-if Planning for Military Logistics

M. Afzal Upal[1]

OpalRock Technologies
42741 Center St. Chantilly, VA 20152
afzal@opalrock.com

Abstract. The decision makers involved in military logistics planning need tools that can answer hypothetical ("what-if") questions such as how will a detailed logistics plan change if the high level operational plan is changed. Such tools must be able to generate alternative plans in response to such questions, while maintaining the original plan(s) for comparison. This paper reports on the work performed to add this capability to the multiagent planning and execution system called Cougaar. We state the what-if planning problem, describe the challenges that have to be addressed to solve it, discuss a solution that we designed, and describe the limitations of our approach.

1 Introduction & Background

Conventional wisdom in the AI has been that the classical AI planning problem is hard enough without the additional complexity of generating and managing multiple plans. Military logistics planning in particular, is known to be a notoriously difficult problem because of (a) the millions of different object types and thousands of heterogeneous interacting organizational units involved, (b) the complex continual interplay between planning, execution, and replanning, and (c) stringent performance requirements [1, 2, 3]. Work on DARPA's Advanced Logistics Project (ALP) and its successor Ultra*Log has produced a multi-agent, hierarchical planning, execution monitoring, and replanning system called Cougaar (Cognitive Agent Architecture) that aims to address this problem [3, 4]. Cougaar was designed to take advantage of the efficiencies available in the military logistics domain. The result is a robust system that can produce a level-5 (to the 'eaches' and 'bumper numbers') detail logistics plan for a given high level Operations Plan (Oplan) within minutes [3]. However, Cougaar does not allow the decision makers to specify hypothetical changes to the Oplan (such as "what if another tank unit needs to be moved from Fort Anderson to Fort Davis?") and see the alternative plan(s) produced to satisfy the new Oplan. This paper reports on the work that IET's Ultra*Log team performed to add this capability to Cougaar.

---

[1] This work was performed during author's tenure at Information Extraction & Transport Inc. 1901 North Fort Myer Drive, Arlington, VA 22209, under a DARPA contract.

1.1 Cougaar

A Cougaar society closely mirrors the organizational structure of military units engaged in real-world logistics planning. Each Cougaar agent represents an organizational unit that can interact with other organizational units. Figure 1 shows the 4-agent minitestconfig society [3].

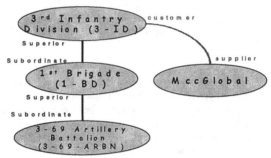

Figure 1: Cougaar minitestconfig society.

The agent interactions are governed by the roles that the interacting agents are playing at the time. For instance, a superior agent may pass requirements (such as "report for service" and "generate a logistics plan for this mission") down to its subordinates and the subordinate agents may pass the status back up to the superior. A customer agent, on the other hand, may requests a service from its supplier (such as "transport this for me" and "supply me with this"). Agent roles in such interactions are dynamic and time-phased (i.e., an agent A may be a supplier for another agent B only during the month of January). Internally each agent has a blackboard (on which various Cougaar objects, such as all the *assets* the agent owns, are published) and a number of *plugIns* that define that organization's behavior. Each agent has a limited knowledge of the society. It only knows about the agents it is related to. An agent subordinates and suppliers are represented as assets on its blackboard as are any physical assets that it owns such as trucks, tanks, and guns.

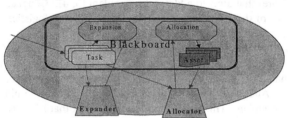

Figure 2: Internal Architecture of a Cougaar Agent.

Cougaar users and agents specify their requirements in terms of *tasks*. Similar to classical AI planning tasks, a Cougaar task consists of a verb such as *move* and an object on which the verb operates such as *Truck5931*. However, a Cougaar task may also specify the preferences of the specifying agent about resolution of the task such as "I prefer a resolution of this task that minimizes execution time". The preferences may be specified as a linear value function over the Cougaar defined world *aspects* (i.e., relevant features of the world such as cost, duration, start time, and end time). Planning at the agent level means finding suitable assets from among the assets published on the agent's blackboard that can be *allocated* to the tasks on the agent's blackboard in accordance with each task's preferences. Two types of plugIns play a crucial role in planning at the agent level; *expander/aggregator* plugIns, and *allocator* plugIns. The basic task of an expander is to expand the given tasks along with its preferences into subtasks while an aggregator merges subtasks into an aggregated task whose preferences are an aggregation of the preferences of the merged subtasks. An allocator's role is similar to that of a traditional AI scheduler. Given a list of tasks and assets, an allocator finds the best schedule such that an objective function (sum of all task preferences in case of Cougaar allocators) is optimized. An agent's plugIns are activated when a new object they subscribe to is published to that agent's blackboard. Typically, when a new task is published, either an expander/aggregator, or an allocator plugIn subscribes to it. If an expander subscribes to it, it expands the task into subtasks which are then published to the blackboard. An allocator can allocates a physical or an organizational asset to the task for some time. If an organizational asset A is allocated to a task T, the Cougaar infrastructure routines (known as *logicProviders*) copy the assigned task T to the allocated agent A's blackboard causing A's subscribing plugIns to become active and the planning process continues. Planning for a task stops when a physical asset is allocate to that task.

Example1: Figure 3 shows the plan created when the task transport(3-69ARBN, Ft. Stewart, Ft. Irwin) (denoting, "transport 3-69 ARBN from Ft. Stewart to Ft. Irwin") is published to the agent 3ID's blackboard. 3ID's expander expands this task into two subtasks T1 and T2 both of which are allocated to the MccGlobal agent. MccGlobal's allocator then allocates these tasks to the physical asset Truck1.

Cougaar is able to efficiently compute a plan given a high level task. However, it did not allow its users to specify hypothetical changes to the Oplan (such as "what if another tank unit needs to be moved from Fort Anderson to Fort Davis?") and be able to see the alternative plan produced to satisfy the new O-plan, and then be able to compare the alternative plans. IET was contracted by DARPA to add this capability to Cougaar. Our solution had to conform to the following guidelines:

1. Minimize changes to the Cougaar plugIns written by military logistics domain modeling experts.
2. Minimize the amount of memory needed to perform maintain multiple plans i.e., create a minimal number of new Cougaar objects needed to support what-if planning, and
3. Minimize the time needed to perform what-if planning.

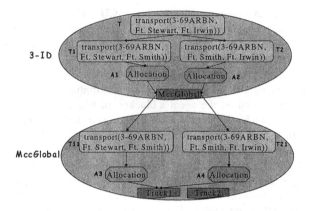

Figure 3: Plan generated by Cougaar's minitestconfig society when the task of transporting 3-69-ARBN from Ft Stwart to Ft. Irwin is posted to the agent 3ID.

2   What-if Planning in Cougaar

We assumed that a user specifies a what-if by performing the following two steps.
1. Creating a new *what-if world*.   The new world is a copy of the old world i.e., it has exactly the same state as the world from which it was created.
2. Adding/deleting/modifying Cougaar objects to/from an agent's blackboard in the what-if world created in Step 1.

The what-if planning problem then is to allow the Cougaar society to compute a plan in the what-if world while maintaining the original world state and the original plan to be compared with the what-if plan [9]. A naïve solution to this problem is to clone the Cougaar society i.e., make two copies of the entire Cougaar society, leave one copy containing the original plan alone, propagate user's modification in the other copy and let it replan. The two plans can thus be maintained separately and presented to the user. The problem with this approach is that it is space inefficient. It copies everything; the agent relationships, and complete contents of each agent's blackboard (including those objects that may not even change in the branch world). We adopted a more space efficient approach that we describe next with the help of the following example.

Example 2: Suppose that the user wishes to specify a what-if in a very simple society that contains only one object (namely the object p). We will refer to this state of the society as the Parent World.   The user creates a Child world by branching from the Parent world and adding an object c to the Child world.   Further suppose that the worlds Grand Child 1, and Grand Child 2 are created by branching from Child and by adding objects g1 and g2 respectively.

Implementing what-if using the society-cloning approach would mean cloning the entire society configuration three times.   This would involve making three extra copies of p and two extra copies of c leading to worlds with the following visible objects.

Parent World = {p}

Child = {p, c}

Grand Child 1 = {p, c, g1}

Grand Child 2 = {p, c, g2}

We adopted the *delta* approach[2] to the creation and storage of new Cougaar objects in the new what-if world. The intuition behind the delta approach is similar to that of many software version control systems that a departure from a given world should be efficiently expressible—and its effects efficiently computable. The adoption of the delta approach reflected our hope that small changes in requirements would have small effects on the resulting logistics plan—that much of the plan should be reusable. The delta approach takes a lazy approach by copying Cougaar objects only when it needs to modify them. For instance, in the above example, the delta approach requires that we do not make three copies of the object p (one for Parent's each descendent world) unless p changes in one of the descendent worlds. If that happens then, and only then, do we make a copy of p for that world. The disadvantage of the delta approach is the overhead involved in managing the objects. This involves object storage (additions, deletions, modification) and retrieval.

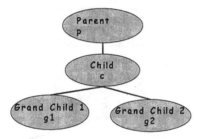

Figure 4: An example what-if world tree.

2.1 Multiple World maintenance

There are two general ways of implementing the delta approach in Cougaar.

1. **Flat object management scheme.** In this storage scheme, records are indexed by objects. Each record consists of an object and the list of the worlds the object is visible in. For instance, in Example 2, flat-tagging scheme would store the objects in a single container shown in Figure 5. Since the object p is visible in all the worlds, its visible-worlds-list contains the names of all the worlds. The object g2, however, is only visible in one world and hence only contains the name of one world.

---

[2] This approach (originally called "internal delta") was advocated by BBN in an earlier design of a proposed what-if facility in Cougaar [2].

2. **Hierarchical object management scheme**. In this scheme records are stored by the worlds in a hierarchical fashion. Each world in the hierarchy only contains those objects that are present in that world and not in its parent world. A child world is said to *inherit* objects from its parent world. For instance the context of Example2, this would mean storing only the object c in the world Child because the other object visible (namely p) in that world is present in its parent and can be inherited from it.

p	Parent, Child, Grand Child 1, Grand Child 2
c	Child, Grand Child 1, Grand Child 2
g1	Grand Child 1
g2	Grand Child 2

Figure 5: In the flat object management scheme each object has an associated visible-world-list.

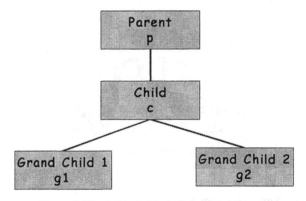

Figure 6: Hierarchical object storing scheme.

We anticipated that object retrieval by worlds will be the most common operation performed in Cougaar with what-if facility (*what-if Cougaar* henceforth). Retrieving an object *O* from the flat scheme in a given world *W* involves searching to see if the W is present in *O*'s visible-worlds-list. Retrieving an object *O* from the hierarchical scheme in the world *W* is just as efficient if the object is stored in *W*. If *O* is inherited, then the world tree has to be traversed possibly up to the root of the tree to find the object. This is less efficient than the flat object storage scheme. Adding an object has the same cost in both worlds. However, adding a new branch/world is more costly in a flat scheme because we each object's visible-worlds-list may have to be updated. The hierarchical scheme does not suffer from this problem because whenever a new branch is created, only the new branch-only objects are stored in the new branch and all the other objects can be left untouched. Thus hierarchical object storage scheme can be expected to be more efficient in those situations in which users frequently pose nested what-if questions and few inherited object retrievals are required. The major problem with the hierarchical scheme as stated so far is that it does not allow an inherited object to be deleted or modified. For instance, there is no way of deleting p from Grand Child 1 because it has no representation there. We addressed this problem by designing a modified form of hierarchical object storage scheme. In the modified scheme, each object has the list of descendent worlds it is invisible in. When an inherited object O is deleted from a world W, the world W is added to O's invisible-worlds-list. For instance, deleting p from Grand Child 1 would result in the new tree shown in Figure 7.

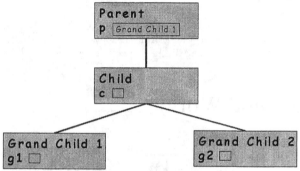

Figure 7: Hierarchical object storage scheme with invisible-worlds-list. The invisible-worlds list for each object is shown as a square box on the right of the object.

We adopted the hierarchical scheme assuming frequent nested what-if queries by users. We had no way of assessing the number of inherited object retrievals and we left it as an empirical question to be investigated during the evaluation phase.

2.3 Dependency Management

The highly interrelated nature of Cougaar objects, such as tasks, expansions, allocations, and assets makes dependency management an important feature of Cougaar processing. Standard Cougaar (i.e., Cougaar without what-if facility) uses object-links to handle these dependencies such as those between a task and an allocation. Dependency management is even more challenging in Cougaar with what-if facility because as new worlds are created and objects are added, removed, and changed, the relationships between objects must be preserved across what-if worlds to maintain proper plan structure. For instance, suppose the user wishes to specify a what-if when the Cougaar minitestconfig society is in the state shown in Figure 3 and specified in Example 1 (referred as the Parent world hereafter). The user creates a child world by branching from the Parent world and adding an asset (Truck3) to MccGlobal agent. Further suppose that the new asset is a better able to satisfy the preferences specified on the tasks T11 and T21. In this case, MccGlobal's allocator will delete the allocations A3 and A4 and reallocate asset Truck3 to T11 and T22 creating the allocations A31 and A41 as shown in Figure 8. Deletion of A3 and A4 from child world will cause them to become invisible in the child world. However, we must do more than this because if the state of the deleted objects A3 and A4 changes in the parent world then the new objects created because of this change must also be prevented from being inherited by the child world. For instance, if the user removes the asset Truck1 in the parent world (shown in Figure 3) then new allocations A32 and A42 of tasks T11 and T21 to Truck2 will be created as shown in Figure 9. However, these new objects must not be inherited by the child world because it already has allocations for those two tasks and having two allocations for the same task is an inconsistent Cougaar state.

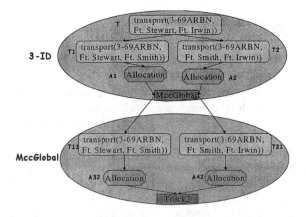

Figure 8: Child world of the world shown in Figure 3 created by addition of a new asset (Truck3).

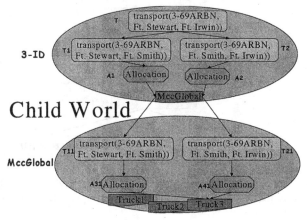

Figure 9: Parent world (Figure 3) after deletion of an asset Truck2.

Our solution involved keeping track of the object whose addition/deletion/modification caused Cougaar to engage in planning (for instance, in the above example, creation of the new objects A32 and A42 is caused by the deletion of objects A1 and A2) and initialize the worlds-invisible-in list of the new objects created during that planning to the worlds-invisible-in list of the object that caused planning to occur. This means that since, in this example the causing objects A1 and A2 are not visible in the child world, the new objects created because of a change in the state of A1 and A2 must also not be visible in the child world.

3 Implementation

IET has completed the implementation and preliminary testing of the what-if facility in Cougaar in Cougaar's minitestconfig society. The Java source code along with the user's guide is available under the open source license from IET's website[3]. It describes the user interface that IET has designed to allow the users to create new branches from the any given world (starting from the default "reality" world), and add/modify/delete selected types of objects (such as tasks and assets) to/from the blackboard of any agent. We have also modified the standard Cougaar interface to allow users to view the plans in each branching world. We are planning on further testing this facility in larger Cougaar societies and integrating this facility with our plan evaluation and probabilistic reasoning facility in Cougaar [3] to create a powerful what-if plan generation, multiple plan comparison, and optimal plan selection and execution facility in Cougaar.

4 Experiments & Results

We performed empirical experiments to study the performance of What-if Cougaar. The experiments were designed to answer the following two questions.

1. How much extra time and memory (compared to Standard Cougaar) is needed by What-if Cougaar in non-what-if planning situations because of the extra book-keeping involved in maintaining the object inheritance mechanism (even though object inheritance are not actually needed in these situations)?
2. How much savings in time and memory did we achieve by implementing the delta approach for what-if planning rather than the clone approach (described earlier)?

In order to answer the first question, we generated a number of planning scenarios in Cougaar's minitestconfig society [3], ran what-if Cougaar and standard Cougaar on them and measured the lapsed Java System Clock time and memory used by the Java Virtual Machine. The results [5] show that on average What-if Cougaar runs 18% slower than Standard Cougaar, and uses 50% more memory, on average, than that used by the Standard Cougaar. The results with respect to time confirmed our

---

[3] http://www.iet.com/Projects/UltraLog/

expectations, but we did not anticipate the large amount of memory needed by What-if Cougaar. We believe that the overhead of what-if processing appears to be larger because of the small size of the minitestconfig society problems we used in the preliminary experiments. We anticipate that the overhead will not increase proportionately to the size of the problem and that on larger problems, it will form a smaller proportion of the overall memory needs. We are planning experiments to confirm this hypothesis.

To answer the second question, we generated what-if scenarios by posing what-if questions in the test situations generated to answer the first question. We computed the amount of time and memory used by our implemented delta-based what-if Cougaar and compared it with the amount of time and memory that would be needed by a clone approach. These numbers were computed by running Standard Cougaar N times on a problem with N number of what-ifs. The total amount of time and memory used by all the runs of Standard Cougaar was then compared to the amount of time and memory used by a single run of the delta-based What if Cougaar in which N what-ifs were posed to compute the savings in time and memory obtained by implementing the delta-based approach rather than the clone-based approach. The results detailed in [5] show that on average savings of 12% in time and 43% in memory usage are obtained by using the delta approach rather than the clone-based approach to what-if.

5 Conclusion

The decision makers involved in military logistics planning need tools that can answer their hypothetical ("what-if") questions such as how will a plan change if the operational plan is changed in this way. Such tools must be able to generate alternative plans in response to such questions, while maintaining the original plan(s) for comparison. This paper reports on the work that IET performed to add this capability to the multiagent planning, execution monitoring and replanning architecture called Cougaar.

References

1. BBN 2001a. Introduction to the Advanced Logistics Project Model of Military Logistics, Technical Report, Bolt Beranek & Neumann Inc.
2. BBN 2001b. What-if Planning in Cougaar, presented at the 1ˢᵗ Cougaar Developers Workshop held on February 5, 2001.
3. BBN 2002. The Cougaar Architecture Document. Bolt Beranek & Neumann Inc. (http://www.cougaar.org).
4. Fowler, N., Cross, S.E. and Owens, C. 1995. The ARPA-Rome Knowledge-Based Planning and Scheduling Initiative, *IEEE Expert*, 10 (1), 4-9.
5. IET 2002b. The Ultra*Log Experiment Report, Technical Report, Information Extraction & Transport Inc. (http://www.iet.com/Projects/UltraLog).

Effects of Reciprocal Social Exchanges on Trust and Autonomy

Henry Hexmoor and Prapulla Poli

Computer Science & Computer Engineering Department
Engineering Hall, Room 313, Fayetteville, AR 72701
{hexmoor, ppoli}@uark.edu

Abstract

In this paper we present effects of reciprocal exchanges on trust and autonomy of peer level agents. We present models of trust and autonomy and show changes to autonomy and trust in an abstract problem domain. As social exchanges increase, average agent autonomy is increased. Autonomy and trust are more susceptible to certain number of social ties among agents mirroring the principle of peak performance.

1. Introduction

A proper balance between trust and autonomy is necessary in modeling an agent society. The two notions are inter-related. This paper explores the variations of trust and autonomy when agents are engaged in reciprocal exchanges [6]. The objects of exchange in this paper are requests for action and free agreements to perform actions, i.e., favors. In reciprocal exchanges, agents initiate an exchange without knowing whether, when, or to what extent the other will reciprocate in the future. Thus, agents offer benefits unilaterally and the exchange does not depend on negotiation. Any expectation of return is implicit and not reasoned about. Exchange networks are states of interaction among a group of agents engaged in reciprocal exchange. These states change over time with each interaction. Exchange networks and our proposed trust networks based on reciprocal exchanges are dynamic. Exchanges among agents may lead to familiarity and a social relation. However, general social ties captures a broader sense of closeness between partners and exchanges may or may not contribute to social distance between individuals. For example, giving to a call for a charitable cause does not necessarily bring the giver and receiver any closer. Later in this paper we will warrant a social network to be derived from an exchange network under further assumption about social conditions.

A main position of this paper is that different levels of reciprocal exchange contribute to different trust and autonomy values, which are used as part of the decision to perform a task or to delegate it. Another point in this paper that average autonomy and trust among individuals in a group is highest when there is a certain number of social connections in the groups. Just as there is a curvilinear relationship between the number of ties in the organization's structure and their performance, too many or too few ties degrade trust and autonomy.

The social connections among the agents in a social network influence the willingness of an agent to delegate a task. Jean Ensminger describe cattle herding behavior of East African Orma where cattle owner make sure a close relative be on the cattle camp [3]. Such trust of the kin has motivated owners to adopt young men. This is a form of generalized reciprocity. With increased relationship, the owners keep loose account of services and payments. They may reward their loyal hired help with paying them bridewealth after many years of loyal service or to marry the hired herder to one of their own daughters. We present a model of trust that combines tit-for-tat reciprocity (process-based trust) with this generalized reciprocity. Although we agree on competence being important in forming trust, we feel that Castelfranchi's definition [2] could apply to complete strangers and does not take the interpersonal sense of trust we are interested in. The "willingness" component does not capture trustee's attitude towards trusted. Interpersonal trust is also a function of familiarity and benevolence. Interpersonal trust, which is the focus of this paper, differs from system and institutional trust. System trust is an individual's trust in the reliability of social structures in its environment. Institutional trust is the trust that exists among individuals due to their participation in social norms and values of various institutions they are members of.

Social exchanges affect interpersonal trust based on the types of connections an agent has on another and the variation in the connections. Usually, trust levels accumulate and diminish gradually unless there are radical changes in agent attitude toward one another, such as major changes in benevolence [1]. Another conceptualization is that, trust is not a precursor to delegation but one between collaborating individuals who communicate. Trust is in the degree of belief in validity of messages. In this notion of trust, capability and benevolence of trustee is not in question but the agents' social interaction is taken into consideration. We have modeled benevolence among agents when agents successfully accept and complete a delegated task. The change in benevolence is our model of alliance among agents. This is combined with balance of reciprocity. We conceptualize that agents strengthen their ties as they interact about their assigned tasks and delegate tasks to others who are benevolent. Strengthening ties between agents in social ties increases their interpersonal trust. In summary, we suggest that agent X's trust in agent Y about task T (we will denote that by Trust (X, Y, T)), is partly a function of agent's X's perception of agent Y's benevolence towards it, partly a function of agent X's perception of agent Y's capability toward task T, and partly due to balance of tit-for-tat. This approach to conceptualizing trust lends itself to formulating delegation between two individuals, which requires trust between delegator and delegee [7,2].

An agent's autonomy toward a task is affected by its capability and the sense of freedom it receives from other agents [5, 4]. This sense of freedom can be approximated by a combination of factors such as social ties (both the number of ties and strength of ties) and trust it receives from others. An agent's trust in others generates benevolence and return of trust. Therefore, an agent who trusts is likely to experience autonomy. Network of social exchange might have asymmetries. In addition to direct exchange, agents might experience indirect exchange via other agents.

In the remainder of this paper we will begin by elaborating our model of trust, autonomy, and delegation. In section three, we present implemented simulation we have used for our experimental results. In section four, we will describe a series of experiments that show effectiveness of approach. In section five, we draw conclusions about relationships between trust and autonomy.

2. A Model of Trust, Autonomy, and Delegation

Our model of trust is aimed at capturing a precondition to the formation of intentions to delegate a task, i.e., asking for a task to be done by another agent. An agent's assessment prior to delegation may include an analysis of risk and utilities, creating an intermediate notion of trusting value, prior to adoption of an intention. In most applications, trust has the consequence of reducing the need for the trusting agent to supervise or monitor the trusted agent.

The variety of definitions has added to the confusion about, and misconceptions of trust. In multi-agent systems, trust has been related to models of other social notions such as autonomy, delegation, dependence, control, and power, which influence interactions between agents. In this paper, we treat trust as a dyadic relation, i.e., the amount of trust each agent has on other agents. We define Trusting value to be the amount of trust an agent has on other agents with respect to a particular task. The Trusting value can be computed for multiple agents or for individual agents. This value among the agents is subject to many factors such as benevolence of resources, general reputations, competencies, reciprocity among agents, histories or prior experiences, environmental factors: cultural factors and organizational factors.

Although all these factors are relevant in conception of trust among humans, we assume an agent environment where only competency, benevolence, and reciprocity are more central. The following Eq. (1) summarizes Trusting value:[1]

Trusting value(A, B, t) = (1/3)* [capability(B, t)+benevolence(B, A, t)+10* DH(A,B)]   (1)

Here A, B are agents and t is the task to be performed by agent B. Capability(B, t) is the agent B's ability to perform a task t and we assume both A and B perceive the same value. benevolence(B, A, t) is agent B's (i.e. trustee's) level of well wishing towards agent A (i.e. trusted) in performing a task t. This value may be a result of many factors such as kinship (i.e., thick relationship), social ties (i.e., a variety of familiarity, relationships), positive regard, alliance, coalition, team allegiance, predictability, shared values and norms, dependence, even commitment. Argua-

---

[1] Coefficients 10 and 1/3 are used to equalize the three effects of factors and to normalize Trusting value to be in the range 0-10.

bly, this value is difficult to perceive. However, since we are building agents we can artificially have control for this component and approximate this quality in biological organisms.

Assuming agents relate at an interpersonal level, using values on exchange networks, we define a value that reflects harmony in delegation. DH(A, B) is the number of times agents A and B have agreed to the delegation request from one another after internally weighing all other considerations divided by the sum of number of times agents A and B have made a delegation request. DH is considered typically to increase with time. Delegation of a task is an opportunity to increase delegation harmony. The DH value range from 0.0 to 1.0. When this value is 1.0 these agents have been in perfect harmony honoring one another's delegation request. Value of 0.0 is when they have never agreed on delegation or have never interacted. DH is only computed when agents interact at the interpersonal level. This condition is established when agents are considered to be in a social network explicitly. In the beginning of this paper we stated that many relationship types do not fit this condition such as when someone is making an anonymous purchase from a generic store.

Autonomy value for an agent is the amount of trust the agent has for itself to perform a task and is computed by the following Eq. (2.2).[2]

Autonomy value (A, t) = (1/3)*[capability(A, t) +Average(T)+ 1/(n-1)*(Balance of recip-   (2)
rocity)]

If we do assume agents relate at an interpersonal level, using values on exchange networks, we define a value that reflects a balance of reciprocity they have with other agents. Balance of reciprocity for an agent A is counting two values and subtracting two values:

- Add the number of times delegated tasks by agent A has been agreed upon divided by the number of such agents.
- Add the number of times agent A has made a delegation request regardless of accepting that request divided by the number of such agents.
- Subtract the number of times agent A has agreed to a delegation request by another agent divided by the number of such agents making the request.
- Subtract the number of times agent A has been asked for delegation regardless of whether A has agreed to work on the task divided by the number of such agents.

Considering solely an agent's self-ability in performing the tasks might not evaluate an agent's autonomy reasonably in an exchange network. A good rate of exchange among the agents will increase an agent's autonomy significantly. Hence, the two values in the balance of reciprocity that relate to an agent's self-ability were subtracted.

---

[2] Coefficients 1/(n-1) and 1/3 are used to equalize the three effects of factors and to normalize Autonomy value to be in the range 0-10.

Capability(A, t) is the agent A's ability to perform a task t. Average(T) is the average trust of all the agents on agent A and is measured by

$$1/(n-1) \sum_{i=1}^{n} T_i$$

where $T_1, T_2 \ldots T_n$, are the trusting values of the agents on agent A on a particular task t. The amount of trust an agent has on itself determines its competence for performing a task. We call this *autonomy* value of the agent as trusting value of an agent on itself. Said differently, the autonomy Eq. (2) of an agent is same as the trusting value of the self-agent. Obviously, Eq. (1) affects Eq. (2). When there is good harmony among the agents the exchange rate among them would increase. This affects the autonomy, as trust is one of the factors. To have an equal range for all the factors in the equation the factors were multiplied by the required number. The range of all the factors is set from 0 to 10.

Autonomy is compared with the trusting values of all the agents to determine which agent should perform a task. Every agent has an individual task assigned to perform. This autonomy of an agent to perform the pre-defined task is compared with the autonomy of the overall tasks determined. The agent performs a task for which the autonomy is highest. When multiple agents determine to perform a unique task, an agent whose autonomy is higher with respect to the task performs the task. For agents with equal autonomy their capabilities with respect to the task are compared and the agent with the higher capability performs the task. If the agent's capabilities are equal, the task is performed by one of the agents selected randomly.

The following two cases illustrate how the Autonomy value is affected by the variations in capability, benevolence and social exchanges among the agents.

Case 1: When there exists no social exchange among the agents, they tend to increase their capability, as there is no exchange even though there exists trust among the agents. Their autonomy value varies depending on the change in the capability of the agent.

Case 2: When social exchanges exist among the agents, both social exchanges and capability affects the autonomy value, as they are used in deriving it. i.e. the agent's capability increases or decreases depending on the success or failure of the tasks. Even the trusting values among the agents vary.

Trust and autonomy of the agents tend to increase much in an exchange network where there exists good harmony among the agents. Experimental results in section 4 supports this.

1. Initialize the values of capability matrix (C[][]) to random values between 0 to 10.
2. Initialize the values of Benevolence (B[][][]) to 0.0 initially.
3. While (tasks remain) {   /* main body of the algorithm*/
4.   for all agents and tasks {   /* trusting values */
5.   if (a = b)   /* a, b – variables stand for agents*/
        TV[t][a][b] = (1/3)*[C[a][t] + average(T) + (Balance of reciprocity)/(n-1)]
6.   else TV [t][a][b] = (1/3)*[C[a][t] + B[t][a][b]+10*DH(a,b)]
7.   A[a][t] = (1/3)*[C[a][t] + average(T) + (Balance of reciprocity)/(n-1)] /*autonomy */
8.   compare A[][] with TV[][][] to find the suitable agents performing task t

9. compute the number of tasks being executed per iteration and unsuccessful attempts
10. C⟦⟧ = C⟦⟧ + i   /*Update C⟦⟧ with success */
11. C⟦⟧ = C⟦⟧ – i   /*Update C⟦⟧ with failure */
12. B⟦⟧⟦⟧ = B⟦⟧⟦⟧ + i   /*Update B⟦⟧⟦⟧ with success */
13. B⟦⟧⟦⟧ = B⟦⟧⟦⟧ – i   /*Update B⟦⟧⟦⟧ with failure */
14. AA[a][t] = A[a][t]/(n*n)   /*average autonomy*/
15. ATV[t][a][b] = TV[t][a][b]/((n*n)*(n-1)) /*where a!=b ; average Trusting values */
        } /* for loop*/
     } /*while loop */

Fig. 1. Algorithm to calculate the average trust and average autonomy

- AA⟦⟧ is an average autonomy of all the agents with respect to tasks and is used in plotting the graph.
- Average(T) is the average trusting values of all other agents with respect to self agent.
- TV⟦⟧⟦⟧ is the matrix that holds the trusting values of agents with respect to tasks.
- B⟦⟧⟦⟧ is the benevolence matrix of the agents.
- B⟦⟧⟦⟧ is zero initially but varies with time.
- ATV⟦⟧⟦⟧ is the average trusting values of all agents except the self-trusting values with respect to the tasks and is used in plotting the graph.
- DH is the delegation of harmony among the agents.
- Balance of reciprocity is defined in the equation 2 above.
- n is the number of agents.
- i is the increment.

3. A Simulated Testbed

In our implementation simulation, N agents considered N tasks repeatedly, i.e. each agent has its own task, which is same in each time period. This does not mean that each agent has to perform the assigned task. Agents may perform tasks assigned to other agents. The tasks are performed basing on the amount of knowledge an agent have on other agents in the environment and harmony that each agent have with the other agents. The ranges of capability and trust are between 0.0 and 10.0. In our simulation we assume in general, agents perform certain tasks and develop trust, capability, benevolence and contacts among them in exchanging the tasks. In the algorithm, the aim is to focus on the performance of agents in a social exchange network. The pseudo code for our simulation is shown in Fig. 1.

The success or failure of an agent can be determined by comparing the capability values of an agent with a randomly generated number ranging 0 to 10. If the random number has a value greater than the capability value of an agent, it is considered as a failure and if the number is lesser, then it is considered as a success. The above range is set because, when an agent has a capability say x units in performing a task t. The agent can perform a task only when the task requires units less than or equal to its capability (i.e., <= x). If the required unit for a task is more than the agent's capability then the agent might not be able to perform the task which leads to failure. An agent may perform one task each time and no two agents can do the same task in same time unit. The success of a task is dependent only on the capability of the agent (as compared to the random number). The capability, trust and relations among the agents are updated with the success or failure in performing the tasks. Before the values are updated the average autonomy and trusting values of the agents are calculated to observe a relation between the two with respect to the social exchanges among the agents. The rate of successful tasks and unsuccessful attempts is measured as a factor of time with respect to social exchanges.

4. Experiments and Discussions

This section presents results of using our abstract simulation of agents and tasks. Two different sets of experiments were performed. In the first set of experiments the results were observed for 25 units of time. In each time unit the average autonomy, average-trusting values, number of successful tasks and the number of unsuccessful attempts were noted. In the first set of experiments, autonomy and trust of agents in two different social networks were considered and compared.

The average trust and average autonomy of the agents were computed with and without social exchanges among the agents in a network. In the case where we consider social exchanges among the agents, there exists certain delegation of harmony among the agents. Figs. 2 and 3 show the results of agents with and without social exchanges. From Fig. 2 we observe that the average trust and average autonomy values without any social exchanges (marked Tb and Ab in Fig. 2) were very low. These values change slightly over time. There was a gradual increase in both curves that reached a constant level after a certain time. The average trust and average autonomy values where social exchanges take place were significantly greater (marked Tn and An in Fig. 2). There was a gradual increase in both curves that might reach a constant level or vary over time. This is because of the exchanges and delegation of harmony that agents have among them. The average autonomy climbed with time and reached almost a constant high level. The fluctuations in the curve may also be due to variations in capability of the agents. The magnitudes of the average trust and autonomy with social exchanges were consistently greater than without any exchange among the agents. Benevolence value in both cases started at zero but changed as agents successfully delegated tasks to one another.

Figure. 3 show the results of cumulative task executions over time. It is observed that the number of successful tasks (Sn in Fig. 3) where we considered social exchanges was higher than when agents have no exchange among them.

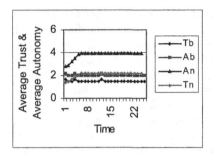   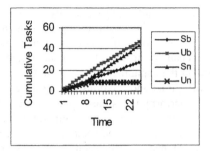

Fig. 2. Average trust and average autonomy   **Fig. 3.** Cumulative tasks with respect to time
with respect to time

In addition, the number of unsuccessful attempts (Un in Fig. 3) is very low with social exchanges. Sb and Ub in Fig.3 show that when there were no social exchanges it results in the lowest number of successful tasks being performed (Sb in Fig. 3). At time t = 25 the number of tasks were 27 successful and 48 failed respectively where there were no exchanges among the agents. With an exchange among the agents, the number of tasks was 43 successful and 8 failed respectively.

The second sets of experiments were performed to observe the effects of social exchanges in varying size of agent community. In one experiment, we considered that there were no social exchanges among the agents i.e., the agents have no delegation of harmony. While in the other we considered there exists high delegation of harmony among the agents, a maximum level. In both cases, a given population size was run for 25 time units. Average autonomy and trust values at the 25th unit of time were plotted for different population sizes.

Figures. 4 and 5 show autonomy and trust in various population sizes. Fig. 4 shows the average trust and average autonomy of the agents where there is no social exchange among the agents. It is observed that the average autonomy (shown as A in Fig. 4) and average trust (shown as T in Fig. 4) of the agents decreases with larger populations. In a network where the number of agents is three, the average trust and average autonomy were 5.61 and 5.81 respectively after 25 time units. With eight agents, the values were 2.83 and 3.5 respectively. These values clearly show that the average trust and average autonomy of the agents were lower in larger communities where there exist no exchanges among the agents in the network; there is no delegation of harmony. Autonomy and trust is low and almost stable in populations of 23 or more.

Figure. 5 show the results of the average trust and autonomy of the agents when there is high delegation of harmony among the agents (i.e., DH = 10). It is observed from Fig.5 that the average autonomy (shown as A in Fig. 5) of the agents increased with larger number of agents and the average trust (shown as T in Fig. 5) among the agents increased with larger number of agents. Average autonomy and average trust of the agents reached almost a constant level with populations greater than 23. The average trust and average autonomy values were 4.25 and 5.25 respectively when the network had three agents. When the network consisted of eight agents, the values were 4.8 and 6.22 respectively.

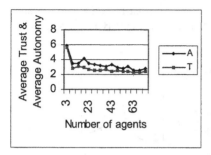

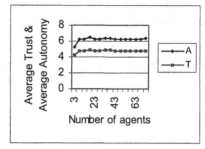

Fig. 4. Average trust & average autonomy for each population size, were the values at 25th time unit. No delegation of harmony i.e., DH = 0

Fig. 5. Average trust and average autonomy for each population size, the values at 25th time unit. Delegation of harmony is high i.e. DH = 10.

We have been talking about exchange network. The term "exchange network" defines a network representation of agents where directed arcs denote the number of times an agent assigns a task to a particular agent and the number of times the assigned agent agree to the delegated task. We observe that trust network affects exchange network. The exchange rate in the network is high among those agents that have high and equal trust for each other and fewer among those agents that have low and equal trust for each other. Also, the chances of successful delegation will be high when an agent assigns a particular task to an agent that has more trust on it i.e., if agent A has more trust on agent B than B on A then agent A delegating the task for agent B will be more likely to occur when agent B assigns a task to A. As we have discussed in section 2, a social network is derived from an exchange network. Fig. 7 shows an exchange network of three agents A, B and C. From Fig. 7 we see that the exchange rate between agent A and agent C is high. Fig. 8 shows a social network where we have allowed all agents to be in personal relationship, i.e., the number of ties = 3.

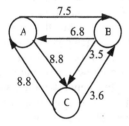

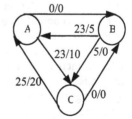

Fig. 6. A trust network corresponding to Fig. 2 at run = 25

Fig. 7. An exchange Network

Figure. 6 show a graph that represents trust among agents. In Fig. 7 just as the exchange rate between agent A and agent C is high, the trust between them is high and equal. Also, in the trust network, agent A has more trust on agent B and hence agent A delegates the tasks assigned by agent B. As trust between agent B and agent C is low and almost equal, exchange rate between agents B and C is zero.

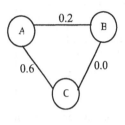

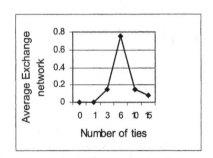

Fig. 8. A social network corresponding to the exchange network in Fig. 7

Fig. 9. Average Exchange networks for 6 agents after 25 units of interaction at 0, 1, 3, 6, 10,15 ties

5. Trust, Autonomy and Exchanges Mirror Performance in a Social Network

Experiments were performed to observe how varying ties in the social network affects average trust, average autonomy, average exchanges and performance of agents. Here, ties are considered to be established among agents who are assumed to relate at an interpersonal level. In this section we use DH(A, B) values we described in section 2 to approximate strength of ties between agents A and B. As we described in section 2, we used values on exchange networks, to update harmony in delegation, i.e., DH(A, B) values.

Figure. 9 show an average exchange network for a network of 6 agents. The maximum number of bi-directional ties in a network of n agents is (n*(n-1))/2. For n = 6 the maximum number of ties is 15. The average exchanges among the agents in Fig. 9 are average of all the DH values between the agents in the network. The values are taken after the agents have had 25 time units of interaction.

The average trust and average autonomy of the agents is low when there exists few ties among the agents. The trust and autonomy among agents is high when there exists moderate ties among the agents within the network. The trust and autonomy levels increase gradually with increase in ties from low to moderate ties among the agents. I.e., it reaches a peak and then it drops. In a network with many ties among the agents, trust and autonomy of the agents is once again low. This pattern is mirrored in performance. The number of successful tasks performed by the agents will vary same as trust and autonomy of the agents with varying ties in the network. I.e. the exchange is more common in a moderate network with which the trust and autonomy of the agents increases.

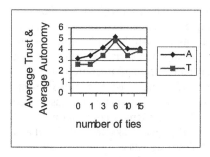

Fig. 10. Average trust and Average
Autonomy

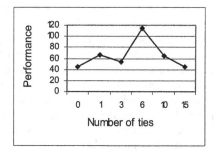

Fig. 11. Number of tasks completed

Figures. 10 and 11 support our discussion. Fig. 10 shows the average trust and average autonomy of the same network of 6 agents as in Fig. 9. We show the variations of autonomy and trust against the number of ties we permit among the agents ranging from 0 to (n*(n-1))/2. The average trust and average autonomy of the agents at the 25th time unit of each run is plotted for different number of ties in the network. The number of ties is the number of interpersonal relationships allowed to flourish.

We observe from the graph that the average trust and average autonomy curves gradually increase with increase in the ties from low range to medium range and then decrease as the ties increase further. From this we infer that the average trust and average autonomy of the agents has a peak in the middle range of ties.

The performance curve of the agents with increasing number of ties among the agents in a network is shown in Fig. 11. It is observed that the number of successful tasks performed by the agents is high when the agents are moderately connected when compared to the situation where the agents are either not connected or highly connected. This is a standard proposition well explored in sociology. We have shown that autonomy and trust values mirror that pattern. This is significant in that such peaks are not only good for performance, but afford the highest trust and autonomy levels.

6. Conclusion and Future work

A simple model of autonomy and trust that relied on social exchange network where there exists delegation of harmony among the agents was presented. This model is deliberately kept simple to illustrate the role of social exchanges that plays in the relationship between autonomy and trust. An agent trusts a second agent in consideration of delegating a task to it if the second agent is capable of performing the task and the amount of relation it has towards the first agent. An agent experiences autonomy with respect to a task if it is capable of performing and it is trusted by other agents with regard to the task along with the balance of exchange it has with other agents. There are many other parameters that affect trust and autonomy. These parameters in general have to do with the relationship among agents and their interactions. Our simple model can be easily extended to include other parameters. Presenting many parameters will have obscured our observations. We have seen from our experiments that when social exchanges among the agents in a social network are weak, their autonomy and trust are lower than that of the agents with high exchanges among the agents. Also, when there exists harmony in a social network, the trust and autonomy of the agents increase when compared with that of the trust and autonomy where there is no social network. When there exists no harmony in a social network, fewer tasks are completed and vise versa. The latter makes sense in terms of fewer agents considering task delegation. The human interaction with the agents may be considered in developing a better relation among the two social notions trust and autonomy along with the factors that exists in our model.

Acknowledgements

This work is supported by AFOSR grant F49620-00-1-0302.

References

1. A Abdul-Rahman, S Hailes (2000) Supporting Trust in Virtual Communities. In: Proceedings *Hawaii International Conference on System Sciences 33, Maui, Hawaii, 4-7 January-2000.*
2. C Castelfranchi & R Falcone (1998) Principles of trust for MAS: Cognitive anatomy, social importance, and quantification. In: Proceedings of *the Third International Conference on Multi-Agent Systems.* Paris, France, pp 72-79
3. J Engminger (2002) Reputation, Trust, & the Principal Agent Problem. In: *Trust in society,* K S Cook (ed) Vol II in the Russell sage foundation series on trust.
4. H Hexmoor and J Vaughn (2002) Computational Adjustable Autonomy for NASA Personal Satellite Assistants. In: *ACM SAC-02,* Madrid.
5. H Hexmoor (2001) Stages of Autonomy Determination. In: *IEEE Transactions on Man, Machine & cybernetics-Part C* (SMC-C),Vol. 31, No. 4, pp 509-517, November-2001.
6. L D Molm (1997) *Coercive Power in Social Exchange.* Cambridge, Cambridge University Press.
7. J S Sichman, Y Demazeau, and R Conte, C Castelfranchi (1993) A Social Reasoning Mechanism On Dependence Networks. In: Proceedings of *the Eleventh European Conference on Artificial Intelligence.*

Part IV

Intelligent Web Computing

Real Time Graphical Chinese Chess Game Agents Based on the Client and Server Architecture

Peter Vo, Yan-Qing Zhang, G.S. Owen and R. Sunderraman

Department of Computer Science
Georgia State University
P.O. Box 4110
Atlanta, GA 30302-4110 USA

Abstract. The client and server architecture is currently widely used in industry; therefore, it is worthwhile to perform further investigations into its usefulness in different applications. To accomplish this, this paper will demonstrate its appropriateness by implementing an Internet based Chinese chess game using client and server architecture. This paper also has value for developers who would like to develop similar applications, such as Western chess, Internet Relay Chat, etc. In implementing the game application, the server program is developed with Java technology while the client program is implemented in C++ with the help of MFC to facilitate the development of a 2D graphical user interface. In the future, the client program can be modified to be a web application. Importantly, the Internet-based graphical Chinese chess agent system can be used to teach students to understand intelligent agents and game playing in an artificial intelligence class, networks, graphics and other relevant techniques in other computer science classes.

1 Introduction

Computer technology has been extensively developed and a large quantity of software has been written to solve complex business and entertainment issues. More and more features, such as increased business functionality and graphical user interfaces (GUIs), have been added to these software systems. To accommodate user demands and maintain system efficiencies in a timely design and implementation enhancement cycle, software developers frequently have questions related to maintainability, usability, flexibility, interoperability, and scalability.

In a monolithic application model, with business logic, database access, and GUI all combined in one module none of the 5 essential considerations is satisfied. While client server architecture has been developed relatively recently in the history of computer technology it has been proven that client server architecture is an essential part of any enterprise level application, and it satisfies the 5 considerations.

A GUI is a basic component for any visual application ranging from pre-school educational software to sophisticated professional enterprise applications. For pre-school students, it can be used to display simple lessons and educational and entertaining games. Currently GUIs are 2D with some exploratory research into 3D GUIs. For business applications 2D graphics is a basic component for the complex GUI and is used, rather than 3D graphics, for both computational and design simplicity and for better system performance.

Since "client/server" is a well-known successful architecture, and 2D graphics is a basic and popular method for building a GUI application, these techniques are employed in this work on a Chinese chess application. This agent system consists of 3 components. The first is the server program, where game logics are validated and players' information is manipulated. The second is the client, which is the GUI for interacting with the server. The third is a Chinese chess auto-player program, which acts as an intelligent agent.

The rest of the paper is organized as follows. Section 2 discusses the Chinese chess software system in terms of its functional design specification as well as the detailed design. Section 3 discusses the Chinese chess auto-player agent with its functional design specification and the detailed design. Section 4 shows major technical merits and applications in education, artificial intelligence, networks, graphics, etc. Finally, Section 5 gives conclusions and future work.

2. The 2D Client-Server-based Chinese Chess Software

The client/server architecture shown in Fig. 1 is a logical extension of modular programming technology in which a monolithic piece of software is separated into 2 constituent components. One of the two is called the "Server" and the other is called the "Client". The server in this architecture is defined as a service provider, and the client is defined as a service requester. In essence, the client/server architecture does not require that the client and the server have to be on the same memory address space, or on the same machine. This feature is a solution to many enterprise applications. On the other hand, the separation greatly benefits both development and maintenance of the software.

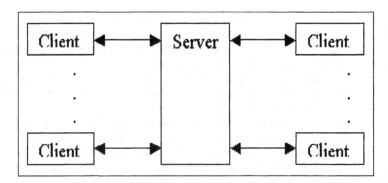

Figure 1 The Chinese chess game architecture

A client program is a service requester that sends each request to a server program and, if necessary, waits for the data response from the server. The most common role of a client is to manage the user interface components of the application. Because of this popular role of the client program, it is also referred to as the presentation layer of the client/server architecture. Before sending a request to a server, a client usually validates data entered by its users and performs some proper business logic.

A server is a program playing a role as a service provider. It is always running and waiting for a client request. Whenever it receives a request from a client, it performs the specified task and returns the result, if any, to the client. The most common role of a server is to manipulate the database for the application. It is also very common that a server program provides logic and computations for the application.

Every Internet application can be developed using client/server technology. The Internet Chinese Chess system is an example of this and the client/server architecture is a perfect model for this application. The integral part of the Internet Chinese chess application is the server, which resides on a fast computer capable of handling many simultaneous connections. Before any client connects to it, the server is running and waiting for client connections to be established. When there is at least one client connecting to it, the server also listens for any request from other potential clients. For any specific request, the server performs the task and returns the result to the corresponding client. For this application, the client is responsible for displaying the visual representation of the chess board to the players and providing a GUI to facilitate its user's interaction with the server. Figure 1 demonstrates the relationship between the server and clients.

The client program is developed using Visual C++. The server program is developed using Java [4, 5, 8] and SQL [13], therefore it can be run on any operating system with JVM installed.

Each instance of the client program runs on a player's computer. The client program's goal is to facilitate and fascinate players in communicating and playing games with others online. The client GUI comprises 1 menu bar, 1 toolbar, and 1 splitter window. Figure 2 shows the client program GUI. The menu bar consists of File, Edit, View, and Help submenus. The File submenu has 3 menu items: Print, Print Setup, and Exit. The Edit submenu has 6 menu items: Undo, Redo, Copy, Cut, Paste, and Delete. The View submenu has 2 menu items: Toolbar and Status bar. Finally, the Help menu contains only 1 menu item: About. The toolbar consists of 16 buttons. Each button corresponds to a command option supported by either the server or a client. Finally, the splitter window consists of 2 panes - an upper pane and a lower pane. The upper pane is for displaying the output response from the server while the lower pane is for the user to issue their text commands to interact with the server.

2.1 Server Detailed Design

The first time it runs, the server needs to create 2 tables: players-info and games-info. The players-info columns are: name, password, score, balance, email. The games-info columns are: game-name, game-num,, e.g., game0, game1, ... , and game9.

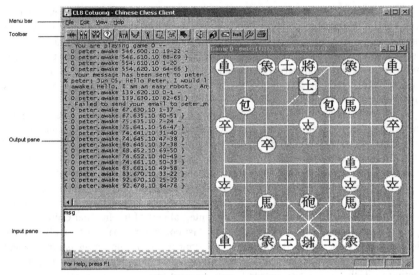

Figure 2 The client GUI

Anytime the server starts, it creates one and only one database object and connect to the database for synchronization. It also creates new types necessary for the server to operate: "Player-Record", "Game-Negotiation", and "Game-Record".

To speed up performance, lists are implemented using a Hash Table [1].

The server waits for a connection by infinitely looping and checking for new connections. When there is a connection request from a client, the server accepts the connection and spawns a new thread to communicate with that client. After the new thread has started, the server goes back to its waiting state and is ready to accept a new connection [7]. The following state diagram in Fig. 3 sketches the server's connection handling.

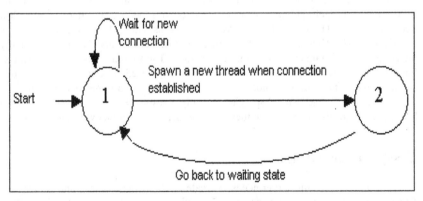

Figure 3 A sketch of server's connection handling

Algorithm:
1. Create a new socket and wait for a new connection.
2. Establish the connection when there is a connection request.
3. Create a new thread (player-thread) to communicate with the client. The main thread then goes back to 1.
4. The player-thread will then check for the player's existence in the database. If the player is already in the database, it will validate the player's user id and password. Otherwise, it will guide the new player to register as a new member and collect all necessary information required to fill out the Player-Record. Then it saves all information in the Player-Record except for thread-id into the players-info table. The Player-Record will then be added to g-players-records in memory to be ready for fast collection of player information. Some fields in Player-Record do not have any value for a first time player or a new login player. The server uses some default values as described below:

> Score: 1500,
> obs-number: -1,
> game-number: -1,
> idle-seconds: 0,
> balance: 0.

2.2 Client Detail Design

The client is the GUI of the application. From the players' point of view, the client facilitates their interactions with the server by providing an output window, an input window, and a chessboard representation. The input/output windows and the chessboard are the elements we are primarily concerned with and we used the Visual C++ [2, 12] wizard to generate the main window (CMainFrame), the menu bar (CMenu), and the toolbar (CToolBar) [10, 15].

2.2.1 CMainFrame

CMainFrame encapsulates the main window of the client. It contains the menu toolbar, and the splitter window, which has two child views – the input and output view. Any event generated when user clicks on a menu item or a toolbar button is handled inside this class.

2.2.2 Create the Splitter Window

We declare member variable m_splitter, m_input, m_output in CMainFrame. Then override CMainFrame::OnCreateClient() and create a splitter window and attach it to m_splitter. After the creation of m_splitter, invoke its CreateView() and GetPane() to assign the m_input and m_output to the result view. From now on, the m_input and m_output are the handles to the input window and output window. Whenever the client receives input from the server, it updates the text in the m_output window. The client interprets the text in the current line of m_input window as a command to interact with the server.

2.2.3 Connect to the Server

Without a connection, the user cannot interact with the server at all. The connection to the server can be done with the window socket protocol. With the help of MFC, a socket connection is no longer a difficult problem. The steps necessary to establish a connection are described below:

1. Create a member variable in CMainFrame name m_socket.
2. Create the socket whenever user requests to connect to server. Then invoke AsyncSelect(FD_READ | FD_CLOSE) to register for input and close events
3. Override the socket's OnReceive() to be informed when there is some input to be read.
4. Override the socket's OnClosed() to be informed when the connection is closed.

3. The 2D Internet-based Chinese Chess Auto-Player Agent

There are times no one is connected to the chess server, for example, when the server is first started. The first player who connects to the server will be there alone. In this case, the player cannot play any game or chat with other online players except for querying players' information. That would not be a lot of fun at all. Therefore, the server should have an intelligent agent that can imitate a player to play chess with others.

3.1 Intelligent Software Agent

An intelligent software agent, also known as a software agent, is a piece of software that acts as robot. An intelligent agent is an agent that can make some autonomous decisions. It can handle some tasks involving in making decision and act as smart as human in some situations [6, 14]. For example, an agent can be designed to watch for a specific stock symbol's price. I can make some decision that if the stock price goes up to some value, it automatically sells the stocks. On the other hand, it may notify the owner when the stock price goes down to a very low value. In the game industry an intelligent agent in a game is usually called a Non-Player Character or NPC.

3.2 Auto-player's Functional Design Specification

The auto-player is a separate process that runs on the server machine. It is equivalent to a client and a player on that client. Therefore, after the auto-player was logged onto the server, from another player's point of view it is a player. The auto-player also greets each player anytime he/she logs onto the server. Then, it automatically offers the player a game. If the player accepts the game, it will play with him/her as a real player. For this simple intelligent agent, it thinks only 3 levels in depth while playing a game.

The auto-player GUI is comprised of a menu bar, a toolbar, and a splitter window. The menu bar comprises of 2 submenus: "File" and "Help". The File menu has 2 menu items - connect, exit - while the Help menu has only menu item - about. The toolbar has 4 buttons in order: Connect, Who, Refresh, and Activate. Finally, the splitter window has 2 panes an upper and a lower pane. The upper pane is for output text sent from the server. The lower pane is for the user (usually the server's administrator) to interact with the server by issuing command lines.

3.3 Auto-player's Design

The GUI detail design of our Chinese chess auto-player is very simple and similar to that of the client program. However, the auto-player has less command buttons since the robot will not need those buttons. The detailed design we need to focus on for the robot is the board model used to perform the best move calculation.

3.4 Min-Max Technique

Suppose that our robot thinks 3 levels in depth. This means that the robot will pick the best move sequence in all possible move sequence. Let's suppose that at level 1, the robot has N1 possible moves. For each move in N1 of the robot, its opponent may have N2 possible moves. Finally, each move in N2 of the opponents, the robot in turn may have N3 possible moves (thus there are roughly N1*N2*N3 possibilities). At level 1 the robot wants to select the best move for itself. However, at level 2, when its opponent moves, it wants the opponent to have the worst move. For its turn again, it wants to select the best move. This process of computation is called Min-Max technique. Beside the Min-Max technique, we can use Alpha-Beta prune with $O(n\log n)$ sorting algorithm [3] to cut off the calculation. Moreover, we can utilize artificial intelligent to help the robot make decisions closer to a human's decisions [9].

3.5 Robot's Board Model

Although there are many tricks to make a robot compute a best move faster, all techniques are built up from the Min-Max search in a game tree. For this reason, our simple robot's best-move calculation is designed and implemented based on the Min-Max search.

The best move calculation is a fairly complex and lengthy computation. Consequently, it needs the following data structures and variables.
   a.   Data type "Move" consists of the following members:
          From(integer): move's "from position",
          To(integer): move's "to position",
          Score(integer): move's score.
   b.   Global variables:
          g-Best-Moves[]: a list of "Move" objects

g-Robot-Score: robot's score at any time,

g-Regions[]: an array of 90 squares in the Chinese chess board. The values of the elements identify the pieces in the board. For example, if g-Regions[1] = 'red rook', then a red rook is occupying square 1 in the chess board.

g-robot1[]: robot's score at level 1,

g-robot2[]: robot's score at level 2,

g-robot3[]: robot's score at level 3.

c.   Best move calculation algorithm

    a.   Get all robot's possible moves

    b.   For each move 'i' in 1., make-a-move

    c.   Get all opponent's possible moves

    d.   For each move 'j' in 3., make-a-move

    e.   Get all robot's possible moves

    f.   For each move 'k' in 5., make-a-move

    g.   Get robot's score and store in g-robot3[k]

    h.   Unmake-a-move (this will put back the move in 6)

    i.   End "for loop with variable k"

    j.   Get max of g-robot3[k] and assign the max value to g-robot2[j]

    k.   Unmake-a-move (this will put back the move in 4)

    l.   End "for loop with variable j"

    m.   Get min of g-robot2[j] and assign the max value to g-robot1[i]

    n.   Create an instance of "Move" with information consists of g-robot1[i] and the corresponding move information. Then add this instance to g-Best-Moves[].

    o.   Unmake-a-move (this will put back the move in 2)

    p.   End "for loop with variable i"

    q.   Navigate through g-Best-Moves[] to search for the "Move" object that have the best score. The move associated to that score is the best move that the robot should move.

d.   Make-a-move: is a function responsible for updating g-Regions[] and g-Robot-Score. It calculates g-Robot-Score by "adding to/subtracting from" g-Robot-Score the piece's value at the "moved-to" position if it is a "robot's move/opponent's move" respectively.

e.   Unmake-a-move: this function can be called only if make-a-move function has been called. The purpose of this function is to put back the changes made by a call to a previous make-a-move. Thus, "make-a-move(), make-a-move(), unmake-a-move(), un make-a-move()" will not alter the board model at all.

4. Technical Merits and Applications

Currently, there are many websites supporting games including Chinese chess. This paper discusses a MS Windows based implementation of the game with an auto-player. Therefore, it has some advantages and some disadvantages compared to the existing ones.

Some advantages of this work compared to the previous ones are:

    a.   The client is a MS Windows based program. Therefore, its response time is very much faster (possibly up to 3 times) than the web clients.

    b.   The auto-player amuses even the first login player.

    c.   Player can move the piece by clicking and dragging the piece naturally.

Main applications include (1) education in artificial intelligence, for example: intelligent agents, game playing, etc., these topics are not easy for students to understand, and (2) other educational applications in computer science such as networks (client-server architecture), and (3) other classes.

In general, distant students can use the online agents similar to the game agent to play with them, learn new knowledge, and create novel techniques conveniently and efficiently. Instructors can also use the smart Web agents to provide students with vivid demonstrations and online experiments like inline games. In the future, real-time online 3D graphics-based teaching and learning systems will play an important role in terms of quality, efficiency, visualization and intelligence.

5. Conclusions

Client/server is a perfect model for internet/networking enterprise applications. It is especially good for applications that support many simultaneous users. In this model, the server program is installed on a fast separate machine with an operating system that can support a large number of processes and threads. The client program has many instances running on different machines and asking for services from the server. This model has also been proven as a cost-effective solution to enterprise level applications. Although client/server is just one of many approaches to distributed computing, it has been frequently used and has dominated many other enterprise solutions.

A 2D GUI plays an important role in any application. For pre-school students, it can be used to display simple lessons and entertaining and educational 2D games. For other uses, especially in business, 2D graphics forms the basis of a GUI for all applications to increase performance and usability.

Currently, the application connects to the database (MSDE) using ODBC-JDBC driver, which is fairly slow. If we can find a JDBC driver for it, we can re-implement the database object for better performance. The application handles connections from clients using the multi-threading feature of JDK 1.3.1. Although threading is cheap, it is still more expensive compared to non-threading. JDK 1.4 supports Socket Channel that we can utilize for our multi-connection application without creating many threads. This will greatly improve our system scalability. Importantly, the Internet-based graphical Chinese Chess agent system can be used to teach students to understand intelligent agents and game playing in an artificial intelligence class, computational intelligence [9] class and networks and graphics in other computer science classes conveniently and efficiently.

Acknowledgements

The authors would like to appreciate the support by NSF under Grant IIS-9980130 and the ACM SIGGRAPH Education Committee.

References

1. Alfred V. Aho, John E. Hopcroft, and Jeffrey D. Ullman.: Data Structure and Algorithms, Addison-Wesley, Reading, MA (1987).
2. Marshall P. Cline, Greg A. Lomow.: C++ FAQs, Addison Wesley, Reading, MA (1995).
3. Thomas H. Cormen, Charles E. Leiserson, Ronald L. Rivest.: Introduction to Algorithms, MIT Press, Cambridge, MA (1990).
4. David Flanagan.: Java in a Nutshell, O'Reilly, Sebastopol, CA (1997).
5. David Flanagan.: Java Examples in a Nutshell, O'Reilly, Sebastopol, CA (1997).
6. Tim Finin and Yannis Labrou.: UMBC Agent Web.< http://agents.umbc.edu/> (2002).
7. Elliotte Rusty Harold.: Java Network Programming, Oreilly, Sebastopol, CA (1997).
8. Philip Heller and Simon Roberts.: Java 2 Developer's Handbook, SYBEX, Alameda, CA (1999).
9. J-S. R. Jang and C.-T. Sun.: Neuro-Fuzzy and Soft Computing, Prentice-Hall, Inc., Upper Saddle River, NJ (1997).
10. David J. Kruglinski.: Inside Visual C++ 6, Microsoft Press Redmond, WA (1997).
11. Joseph O'Neil.: JavaBeans Programming from the Ground Up, Osborne/McGraw-Hill, Berkeley, CA (1997).
12. Herbert Schildt.: C++ from the Ground Up, Osborne McGraw-Hill, Berkeley, CA (1994).
13. Ryan K. Stephens, Ronald R. Plew, Bryan Morgan, and Jeff Perkins.: Teach Yourself SQL in 21 Days, SAMS, Indianapolis, IN (1997).
14. Katia Sycara.: Carnegie Mellon University's Robotics Institute. <http://www-2.cs.cmu.edu/~softagents/intro.htm> (2002).
15. Viktor Toth.: Visual C++ 4 Unleashed, Sams Publishing, Indianapolis, IN (1996).

DIMS: an XML-based information integration prototype accessing web heterogeneous sources

Linghua FAN [1,2]
[1] LAMIH UMR-CNRS n° 8530, Université de Valenciennes et du Hainaut Cambresis(UVHC), Le Mont Houy , F59304 VALENCIENNES CEDEX 9, France
Jialin CAO
[2] School of Mechanical & Electronic Engineering and Automation, Shanghai University149,Yanchang Road, Shanghai 20072, China
Rene SOENEN [1,3]
[3] Centre Universitaire de Nîmes, Site des Carmes, Pl. G. Peri, 30021 NIMES Cedex 1, France

Abstract: The goal of information integration is to provide a uniform interface to a multitude of distributed, autonomics, heterogeneous information sources available online (e.g., databases, XML or HTML files from the WWW). Distributed Information Management System (DIMS) is an XML-based information integration system for accessing these web sources. It utilizes some efficiencies tools, such as Fatdog's XQEngine, Jarkarta Lucene, Jtidy, JDBC or JDBC-ODBC drivers, to wrapper heterogeneous web sources into standard XML data and uses mediator to translate the user's query to relational wrapper and integrate these results. Using materialized view to speed query response, using metadata to easy add and delete sources, etc. DIMS provide a new innovation and flexible way to design and implement. Our system uses UML as a method to design our software, and provide a prototype can be using independence platform implementing on Java language.

Keywords: information integration, mediator, wrapper, XML, GAV, LAV, GLAV

1. Introduction

With the rapid growth of the available data sources on the Internet, significant attention has been received on integrating huge distributed, autonomous and heterogeneous information to build a new unified web application. The key technology of these applications is how to homogeneous the difference structures and semantics of the information provided by the enterprise or the Internet. The need to access, retrieve, and manage information from a variety of sources and applications using different data model, representation and interfaces has created a great demand for tools supporting data and systems integration[1]. Integration of complementary information produced from the popular Web and the rapid growth Internet, providing useful service for the large amount of the Web users, become a very important and more recent research subject!

Traditionally, there are some projects based on centralized, client-server or distributed systems, which with multiple independent sources producing and managing their own data. Recently, integration technologies are needed for accessing repositories, applications and legacy systems located across the corporate intranet or at partner companies on the Internet, because of the Web rapid growth the more and more users on the Internet, and the emergence of E-Commerce. Modern applications require integrated access to various information sources (from traditional DBMS to semi-structured Web "repositories"), fast deployment and low maintenance cost in a rapidly evolving environment. The main contribution of information integration system is that users can focus on specifying what data they want rather on describing how to obtain it. An information integration system relieves the user from the burden of finding the relevant data sources, interacting with each of them separately, then combining the data they return.

Because of its flexibility, there is increase interest in using XML as a middleware model. DIMS accepts this XML model to be an XML-based integration system. In many applications, it's never the case that all sources are available, they may be offline, or network connectivity may not be available. The query response time for information mediators, particularly web-based mediators is often very high, mainly because to answer most queries a large number of Web data sources must be fetched over the Web. To improve our system's performance and quality, we accept a mix approach. Based on mediator-wrapper architecture, we will allow some useful, important or frequently used information materialization [2].

The remainder of this paper is organized as follows. Sector 2 presents basic background of information integration and XML. Sector 3 shows DIMS architecture. Discussion of DIMS mediator and wrappers' detailed design. The implementation of DIMS and some case study will be presented in sector 4. In sector 5, we will discuss and compare some current information integration systems. At last, in sector 6, we conclude the paper with future research opportunities.

2. Background of information integration and XML

Information integration (a.k.a data integration/information mediation) systems harmonize data from multiple sources into a single coherent representation. Frequently, there exist a variety of databases or other information sources that contain related information. We have the opportunity to combine these sources into one. Information integration is the process of taking multiple query results and merging them into a single response to the user. The goal is to provide an integrated view over all the data sources of interest and to provide a uniform interface to access all of these data. The access to the integrated data is usually in the form of querying rather than updating the data. However, heterogeneities in the schemas often exist; these incompatibilities include differing types, codes or conventions for values, interpretations of concepts, and different sets of concepts represented in different schemas.

The data sources to be integrated may belong to the same enterprise or may be arbitrary sources on the web. Most of the time, each of the sources is independently designed for autonomous operation. Also, the sources are not necessarily databases; they may be legacy systems (old and obsolescent systems that are difficult to migrate to a modern technology) or structured/unstructured files with different interfaces. Information integration requires that the differences in modeling, semantics and capabilities of the sources together with the possible inconsistencies be resolved. The users can access these data sources in the information integration system as if they were accessing one large database. We can see detail in **Fig. 1.**

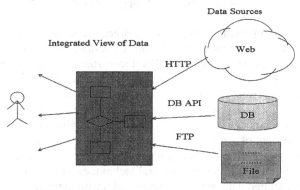

Fig. 1. Information integration architecture

Early approaches involved "federation", where each database would query the others in the terms understood by the second. More recently, information integration can be either *virtual* or *materialized*. In the first case, the integration system acts as an interface between the user and the sources [3], and is typical of multi-databases, distributed databases, and more generally open systems. In virtual integration query answering is generally costly, because it requires accessing the sources. It is referred to as *virtual warehouse* or *mediator*, also known as *"lazy"* or *"query-driven"* approach, gathers, cleanses, and integrates data only after a query has been issued. In the second case, the system maintains a replicated view of the data at the sources, and is typical, for example, both in information system reengineering and *data warehousing*. In materialized data integration, query answering is generally more efficient, because it does not require accessing the sources, whereas maintaining the materialized views is costly, especially when the views must be up-to-date with respect to the updates at the sources (view refreshment). This approach we also call the *eager* approach, or the *materialized integration*. It is an efficient mechanism to support frequently asked queries as long as the data is available in the warehouse.

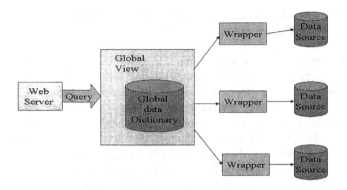

Fig. 2. Mediator-Wrapper architecture

Mediator-based information integration architecture has evolved from an initial proposal by Gio Wiederhold in [5] and elaborated through the intelligent information integration(I3) program [6] For a mediated system, two most common types of modules are called wrappers and mediators **(Fig. 2.).** An issue with such mediators, particularly Web based mediators is that the speed of any mediator application is heavily dependent on the remote sources being integrated, with often a very large amount of time being spent in retrieving data from the remote sources. The concept of a mediator has been proposed as a good basis for giving integrated views of multiple heterogeneous data sources. Currently research in mediation technology has proposed several techniques for describing, integrating or accessing structured data, semi-structured and unstructured date on the Internet. These systems differ widely in the capabilities of mediators and in the capabilities of wrappers. Wrapper deals with data model and technical heterogeneity, while mediators bridge structural and semantic heterogeneity.

XML[7] is becoming widely used for the development of Web applications that require data integration (Web portals, e-commerce, etc). Although fashion surely accounts for some of XML's popularity, it is also justified on technical grounds. XML enables easy wrapping of external sources and declarative integration, thus allowing fast development and cheap maintenance of application. XML is a versatile markup language, capable of labeling the information content of diverse data sources including structured and semi-structured documents, relational databases, and object repositories.

There are several advantages in building this application with XML. First, due to its flexible data model, XML can represent both structure and semistructured information. Second, it is easy to convert any data into XML, and to do so in a generic fashion (i.e., independently of the source schema). Third, several languages support declarative integration of XML data (E.g., MSL[12], XML-QL, XQuery[8] or YATL[4]). Finally, being a standard, XML facilitates interoperability.

Comparing the non-XML mediator system hard-coded wrappers and integration specification complex and hard-coded, XML integration system has some fast wrapping tools and a fast integration based on declaration XML languages. However, it is not like non-XML mediator system has many robust system (DISCO, Garlic, HERMES etc), and efficient distributed query process techniques, XML integration system has few academic prototype (TSIMMIS[12], YAT[4], MIS), and only recent work on query processing techniques.

3. Architecture of DIMS

For standard mediator-wrapper architecture, differences in the sources' data models are resolved by wrappers that translate the raw data into a common generic data format XML on which mediator would directly define the integrated views using a users' query and decompose the users' query into relational sub-query language of wrappers. We extend the architecture by adding metadata management and materializing some useful and often retrieved information to make our system more flexibility and extensibility (see **Fig. 3.**).

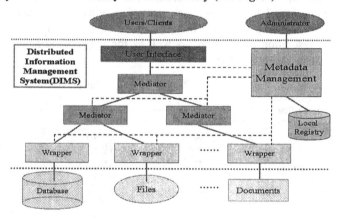

Fig. 3. DIMS architecture

3.1. DIMS mediator architecture

The mediator architecture deals with the problem of integration of heterogeneous information. The mediator, which is the query processing core of DIMS, has to decompose application requests into an efficient series of requests targeted to the sources. These requests have to be compatible with the query capabilities of the underlying sources. The main task of a mediator inside DIMS is to answer queries against its schema by using only queries executable by wrappers. Finding sequences of such wrapper queries is called query planning. Query planning relies on knowledge about the relationships between elements of the mediator schema and elements of the wrapper schemas. Such relationships are called schema correspondences and are expressed using a correspondence specification language. Previously developed correspondence specification language either use Local as View (LAV) or Global as View (GAV) approach [9]. We have studied these two approaches and we will accept a new hybrid approach GLAV to create our prototype. In GLAV, flexible definition is allowed for mediator schema. This can be achieved by making unified schema independent of particular details of sources. Between mediator schema and local data sources schema, will be defined a semantic mapping, efficiency is achieved when long paths are necessary to retrieve information.

Our mediation architecture is depicted in **Fig. 4.**. The mediator consists of a Query Processor and a Result Integrator. The Query Processor responsible for decomposing a require for data into the sub queries which are submitted to the relevant data sources. Within the Query Processor, the Query Translation parses and translates the incoming queries (which are formulated in a high-level language) into the internal format based on XML. The Query Decomposition performs the query rewriting of the client query into one or more and specific distributed sub queries. The Result Integrator responsible for integrating, cleaning, and reconciling the result data that is returned from the wrappers. Specifically, the Result Fusion joins related data based on a set of merging rules. Data restructuring is done by the Cleaning & Reconciliation and includes the removal of duplicate information, resolution of conflicts, as well as the aggregation and grouping of information into high-level units.

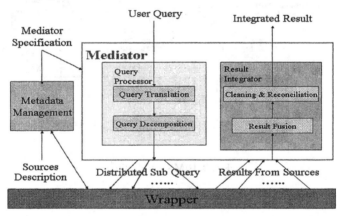

Fig. 4. The DIMS mediator architecture

3.2. DIMS wrappers architecture

The wrappers have generic interface, common to all the wrappers, and a specific part that performs the translation between the data sources models and the mediator one. A wrapper is able to execute a query sent by the mediator and give back the results. Specific wrappers can be added, they have to complain to a generic interface. For each type of source, there is a specific wrapper. Wrappers encode the knowledge to access different sources and export some information about the sources like their schema data and query processing capabilities.

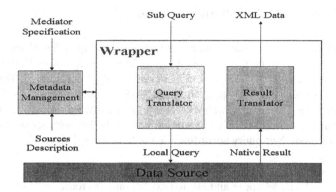

Fig. 5. The DIMS wrapper architecture

Fig. 5 shows our wrapper general architecture. The wrapper is composed of Query Translator and Result Translator. The Query Translator receives the query from the mediator, translators the query to the local query for data source, and transmits the query to data source. Data source returns the query result with its proprietary format. The Result Translator harmonizes the result as XML format, and checks results against the XML syntax.

3.3. DIMS metadata management

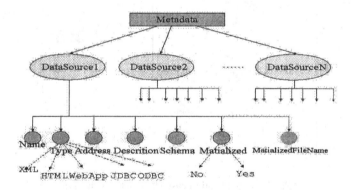

Fig. 6. The metadata structure

In DIMS system, we design a XML file named Metadata.xml to store our metadata. Use this file, we can easy add, delete or update user's data source to DIMS system. For materialized data sources, using this metadata file, we can easy and quick find, access and retrieve these type data sources. In the following, we can see his structure. All the data sources, we give the same metadata structure, see **Fig. 6.**, use these information, we can easy maintain the system data sources' and improve system's flexible and extensible.

4. DIMS implementation and case study

A DIMS prototype implementation has been developed based on Java and XML language and some useful tools, such as Model design tools Rational Rose, Java IDE JBuilder, XML query implementation Fatdog's XQEngine, Text search tool Jarkarta Lucene etc. The application of DIMS we simple example as an online books search and a weather search. For online books search, we define a mediator schema: Name, Author, Publisher, Year and keywords. For weather search, we try only let user input their query according the rules we have defined.

We find some XML files, HTML files and TEXT files. According our research, we can create XML Wrapper and Html/Text Wrapper to retrieve XML, HTML or TEXT documents and get the result represent in XML Data model for our system. Otherwise we can tidy HTML files to be a well-formed HTML through Tidy tool. The convert its well-formed HTML to XML file. For all DBMS data sources, we can utility JDBC/ODBC wrapper based on different JDBC or JDBC-ODBC driver to access and retrieve information. For better understanding DIMS system's function, we design a typical application case UML activity diagram using Rose Rational **Fig. 7.**.

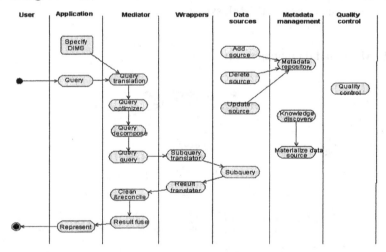

Fig. 7. An application case UML activity diagram

5. Related work

At least two decades of prior research exists in the general area of information integration. Sheth, in a recent overview [11], classified information integration into three generation approaches. In our architecture, similar to second generation approaches like TSIMMIS [12], using mediators and wrappers, but integrated views are defined using combination GLV and LAV. And with the web data, we are faced with more distributed, more autonomy, and more heterogeneity among the accessible information, information sources, and users. Sheth considers that the new generation of information systems should be able to solve semantic heterogeneity to make use of the amount of information available with the arrival of Internet and distributed computing. Our system tends to solve this type problem using XML as standard.

Two basic approaches are known for the mediator architecture [9].   According to the first one, called GAV, the global schema is constructed by several layers of view above the schemas exported by pre-selected sources. Queries are expressed in terms of the global schemas and are evaluated similarly to the conventional federated database approach. GAV is a direct extension of a central database engine to the distributed case. It is used in many projects, such as TSIMMIS, GARLIC, DISCO, IRO-DB and HERMES. Another approach, known as LAV, The LAV approach (or source-centric) allows correspondences that relate a single relation of a wrapper schema to a view on the mediator schema. This is exactly contrary to the GAV approach. In LAV, every mediator relation may appear in many views corresponding to different wrapper relations. Every such correspondence contributes to the total extension of the mediator relation. In general, after a user poses a query, the mediator decides how to use the source views to answer the query. This process is also known as answering queries using views. LAV is a relatively new approach. It was first described in 1994, and is, for instance, used in the Information Manifold and in Infomaster. It has its strength in environments with frequent evolution of sources, such as the web. However, query planning with LAV rules is considerable more complex than with GAV rules.

6. Conclusion and future research

We have introduced a solution to information integration, which allows end-users to access and retrieve information from multiple distributed, heterogeneous data sources through a consistent, integrated view. DIMS uses a combined approach, based on mediator-wrapper architecture, selecting useful and frequently access information materialized, for enhanced query performance and increased reliability. Specifically, we add metadata management to add or delete or update application data sources, use novel wrapper and mediator technologies to reduce human involvement as much as possible. Given the popularity of the Web and multitude of legacy system, DIMS is designed to integrated distributed, autonomous, and heterogeneous data sources that provide XML-based, semi-structure data, different database structure data, and HTML/TEXT document unstructured data. Within DIMS, data is united represented as XML standard. Given the rapid evolution of XML and its technologies, our next version of the prototype will use XML schema as their validate standard and data model. We will optimize the mediator process and improve the qualities of integrated results, using information quality to resolve inconsistencies among heterogeneous data sources. Other plan, include the support more complex query in the mediator and more materialized maintenance procedures, which rely on source monitors for determining when the materialized data sources needs updating rather than on user demands. We will continue report on our progress in future conferences and workshops.

ACKNOWLEDGMENTS

The authors would like to thank all the teachers in LAMIH/SP of UVHC for supporting their works and the colleagues in Shanghai University for their cooperation. We also would like to thanks our anonymous reviewers for their helpful comments and suggestions on the paper.

REFERENCES

[1]: AIT – IMPLANT (Implanting CSCW and CAD Innovations into the User Enterprise) ESPRIT PROJECT N° 24771 documents

[2]: Linghua Fan, Jialin Cao, René Soenen, Model and techniques to specify, develop a system: a mediated approach, SCI 2002, 6th World Multiconference on Systemics, Cybernetics and Informatics. Orlando, Florida, USA; ISBN 980-07-8146-3

[3]: Amit Sheth and James A. Larson. Federated database systems for managing distributed, heterogeneous and autonomous databases. ACM Computing Survey, 22(3):183--236, 1990.

[4]: Vassilis Christophides, Sophie Cluet, Jerome Simeon, On Wrapping Query Languages and Efficient XML Integration, to appear in SIGMOD'2000, Dallas, Texas, may 2000.

[5]: G. Wiederhold. Mediators in the architecture of future information systems. IEEE Computer, 25(3):38--49, 1992

[6]: Wiederhold, Gio: "Interoperation, Mediation, and Ontologies"; Proceedings International Symposium on Fifth Generation Computer Systems(FGCS94), Workshop on Heterogeneous Cooperative Knowledge-Bases, Vol.W3, pages 33-48, ICOT, Tokyo, Japan, Dec. 1994

[7]: http://www.w3.org/XML/

[8]: http://www.w3.org/XML/Query

[9]: Marc Friedman, Alon Levy, and Todd Millstein, Navigational Plans for Data Integration, The Sixteenth National Conference on Artificial Intelligence (AAAI-99), Orlando, Florida, July 18-22, 1999.

[10]: Byron Choi, Mary Fernandez, Jérôme Siméon, "The XQuery Formal Semantics: A Foundation for Implementation and Optimization", May 31, 2002

[11]: Sheth, A. Changing Focus on Interoperability in Information Systems: from System, Syntax, Structure to Semantics, in Interoperating Geographic Information Systems, M. F. Goodchild, M. J. Egenhofer, R. Fegeas, and C. A. Kottman (eds.), Kluwer, (1998)

[12]: S. Chawathe, H. Garcia-Molina, J. Hammer, K. Ireland, Y. Papakonstantinou, J. Ullman, and J. Widom. "The TSIMMIS Project: Integration of Heterogeneous Information Sources". (ps) In Proceedings of the 100th IPSJ Anniversary Meeting, Tokyo, Japan, October 1994.

A Framework for High-Performance Web Mining in Dynamic Environments using Honeybee Search Strategies

Reginald L. Walker

Tapicu, Inc., P.O. Box 88492
Los Angeles, California 90009
rwalker@tapicu.com

Summary. The methodology for the knowledge discovery in databases architecture outlines possible approaches taken by search engines to improve their IR systems. The conventional approach provided the requester with query results based on the user's knowledge of respective IR systems. This paper proposes the use of an information sharing model based on the information processing methodology of honeybees and knowledge discovery in databases as opposed to the traditional IR models used by current search engines. The major limitation of IR-based systems is their dependency on human editors which is reflected in static sets of query terms and the use of stemming. Experimental results are presented for data clustering component (Web page indexer) of the Tocorime Apicu search engine which is based on the information sharing model.

1 Introduction

The purpose of the Tocorime Apicu [1] information sharing indexing (ISI) approach to indexing/ranking and clustering Web pages is to show the feasibility of implementing unique aspects of the search strategies of honeybees coupled with knowledge discovery in databases (KDD) using stochastic optimization techniques [17] of evolutionary computation (EC) for this component of the information sharing (IS) system. An extensive discussion of the information sharing model was presented in the author's Ph.D. dissertation [21].

The use of unsupervised clustering of Web documents provides an alternative to the supervised clustering used by some search engines which requires domain expert to generate a taxonomy of the document set [15, 20]. Supervised clustering is mainly used by engines that employ human editors, such

---

[1] The word *Tocorime*, meaning "spirit" [6], comes from an ancient Amazon Indian language. *Apicu* comes from the Latin *apis cultura*, meaning "honeybee culture" or "the study of honeybees." The phrase *Tocorime Apicu* is used as "in the spirit of bee culture."

as Yahoo and Open Directory Project, to partition a dataset into disjoint categories. The generation of a partial taxonomy of the World Wide Web is becoming extremely difficult for any human or group of humans due to the growth rate of Web documents. Other shortcomings associated with the use of human categorization systems are taxonomies differing across search engines, and needing to be continuously updated as new categories emerge and others diminish in importance.

This paper focuses on coupling the honeybee information sharing model with KDD to develop a framework which could search for hidden knowledge within a collection of Web documents. The collection of Web documents were randomly retrieved by the Tocorime Apicu HTML resource discovery (HRD) system [18]. The Tocorime Apicu browser reporting interface (BRI) will be responsible for presenting the clustered results. The structure of the paper is as follows. Section 2 presents related work. Section 3 presents the Web page indexing system. Section 4 presents the information processing methodologies of honeybees. Section 5 presents the high-performance Web mining model. Section 6 presents the computational results, and a conclusion in Section 7.

2 Related Work

Search engine responses to queries [9] were analyzed to assess the size of the WWW and its coverage by selected engines. The engines in this study were limited to engines that provided full text searches (AltaVista, Excite, HotBot, InfoSeek, Lycos, and Northern Light). Two important questions from this study includes the following: whether the centralized architecture of current search engines can keep up with the expanding number of documents, and if whether the search engines can regularly update their databases to detect modified, deleted, and relocated information.

An ANTS algorithm for performing parallel searches of data warehouses [10] focused on minimizing the query response time by reducing the number of disk pages to be accessed. The data warehouse is partitioned into subsets that represent data clusters (or views) that satisfy at least query from a predefined workload, or query set. This approach falls within the Ant Colony Optimization (ACO) class, which, in this case, uses the collective behavior of simulated ants to solve the vertical fragmentation problem by (VFP) determining the optimal set of fragments (data clusters) to develop distinct views of the data warehouse.

Data clustering using artificial agents [12] was based on *Macrotermes* termites that were used as models for ant-like heuristics, this enabling the emulation of such processes as ants forming piles of items such as dead bodies (corpses), larvae, or grains of sand. The artificial agents were designed to solve problems for unsupervised clustering and data exploratory analysis for image retrieval. The goal of this approach was to reduce communication among agents by replacing coordination and agent hierarchy through direct communication by indirect interactions. The ant-like heuristics were based on the scale of dissimilarity d between objects in the space of object attributes.

An unsupervised categorizing strategy [15] was developed based on k-means clustering that tests the effects of stemming. The focus of this effort was to develop a framework for developing a standardized dataset that can be used to assess the clustering results across search engines.

An evolutionary learning approach used a modified similarity measure based on creating HTML tag weight for a retrieval engine [8] known as SCAIR (SCA Information Retrieval). The engine represented documents and queries as keyword vectors. Documents were ranked according to the similarities between the documents and query vectors. Genetic algorithms (GA) were used to find the optimal feature tags, which, in turn, used to improve the similarity results.

3 The Web Page Indexing System

The purpose of the Tocorime Apicu information sharing indexing (ISI) approach to indexing/ranking and clustering Web pages is to show the feasibility of implementing unique aspects of the search strategies of honeybees using stochastic optimization techniques[17] of evolutionary computation (EC) for unsupervised clustering. These optimization techniques form the basis of a regulatory mechanism for sharing information—migration of Web pages—in the Tocorime Apicu ISI system.

The goal of continuously applying the regulatory mechanism is to improve the subclustering of Web pages in this distributive application this leading to disjoint nodes for chosen sets of search queries known as probe sets. Given a node's workload distribution of n documents, the regulatory system attempts to minimize the function [10, 14]

$$F = \sum_{i=1}^{n} \sum_{\substack{j=1 \\ i \neq j}}^{n} s(D_i, D_j) \tag{1}$$

where $s(D_i, D_j)$ is the similarity between documents i and j. The value of F is computed iteratively when performing searches by applying static or dynamically chosen probe sets to the subclusters of pages residing on each indexer node.

4 The Information Processing Methodologies of Honeybees

The social hierarchy of honeybees [4, 5] is a result of environmental influences [1]—genetic potentiality, social conditions (regulations) of the colony, and those which are mostly ecological and physiological. Social insects [3, 12] are characterized by: 1) cooperation among adults in brood care and nest construction, 2) overlapping of at least two generations, and 3) reproductive division of labor—the queen specializes in egg laying and the workers in brood care. Hive-to-hive communication by way of drone congestion areas provides the bee population with a gene flow, thus maintaining the genetic homogeneity of the population. Therefore, populations can be differentiated, leading to the isolation of subpopulations which, in turn, results in the process of speciation.

The sharing of information plays a vital role in the survival of a bee colony. Information sharing between bees is facilitated via auditory, tactile, and chemical means. In honeybee colonies, individuals migrate from one subpopulation to another for several diverse reasons including crowding, changes in environmental conditions, limitations on colony activities, or members becoming disoriented.

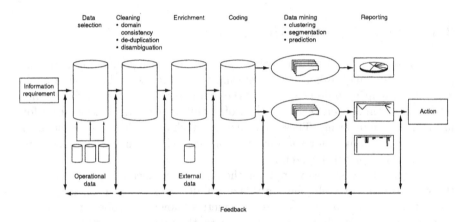

Fig. 1. The KDD architecture (courtesy of Adriaans and Zantinge 1996)

5 Supplementing the KDD Model with an Adaptive Honeybee Search Strategies

5.1 The KDD Model

The methodology for the KDD architecture [2, 11] outlines possible approaches taken by search engines to improve their IR systems (as shown in Figure 1). The conventional approach provided the requester with query results based on the user's knowledge of respective IR systems. Since a typical user often has a limited knowledge of the structural and search methodologies that pertain to individual search engines, the user represents a significant limitation to the current search engines. The components comprising the KDD model include:

1. data selection
2. cleaning—reducing the file complexity
3. enrichment
4. coding
5. data mining—the discovery phase
6. reporting

The benefits of incorporating the KDD model's methodologies include provisions for: 1) long-term vision, 2) mechanisms for updating documents on distributive computers, 3) mechanisms for connecting Web pages, and 4) mechanisms for interpreting the relations between a diverse collection of files, also known as the formulation of hypertext [8, 13]. The KDD approach, however, derives queries from the resulting databases built by the search engines; the IR system, in turn, organizes the database and presents the user with useful information. Thus, the incorporated structure and search methodologies of KDD systems do not require in-depth knowledge by end-users. The KDD IR system does, however, require an intelligent tool [2] coupled with a methodology that eliminates repeated queries and provides useful data back to the end-user for an automated IR system.

5.2 Extending the KDD Model

Table 1 presents a comparison of the components associated with the honeybee IS model, KDD model, and the Tocorime Apicu information sharing (IS) model. Each component generates feedback for its predecessor by using IS protocols. Each of these components was developed with the capability of acting as an independent mechanism, since the only form of communication is via shared IS protocols.

Phase 1 uses selected Web documents retrieved by the HTML resource discovery (HRD) system. Phase 2 results are created via the data parsing (cleaning) process which is a component of the indexing mechanisms. Phase 3 incorporates periodic updates by all components of this search engine. Corrupted and/or badly formed Web pages are then removed by either foragers or indexers during Phase 4. Phase 5 encompasses the ISI system. Phase 6 results in the browser reporting interface (BRI) system interacting with users as well as the indexing mechanisms.

The HRD Web probes, scouts, and foragers [18] are responsible for retrieving external data. In the first pass, the raw (external) data files are reduced to canonical HTML files that contain no padding or HTML header information [8], such as the date of last modification or date of expiration. The canonical HTML file is equivalent to the information presented to the user by her/his chosen Web browser. The raw external file contains information that

Table 1. Comparison of the honeybee IS model, the KDD model, and the Tocorime Apicu IS model.

Phases of KDD model	Honeybee IS model	KDD model	Tocorime Apicu IS model
1	Sharing externally located information within the bee colony	Data selection	HRD system 1) Collecting Web foraging results 2) Sharing Web results within the search engine structure
2	1) Cell cleaning 2) Garbage collection 3) Information sharing resulting from cleaning other bees	Cleaning 1) Domain consistency (removal of pollutants) 2) De-duplication of records 3) Disambiguation (Application of pattern recognition algorithms)	Removal of: 1) Duplicate Web pages 2) Stop words 3) Duplicate non-stop words 4) Non-contributory tags and their attributes 5) Non-contributory attributes 6) Header information 7) Comments Possible correction of common typographical errors
3	Information sharing within colony	Enrichment	Periodic buildup of initial data set with Web pages provided by information sharing between Web probes, scouts, foragers, and indexers
4	Removal of debris	Coding	Removal of corrupted/badly formed Web pages by the Web foragers
5	Information sharing via 1) Pattern recognition techniques 2) Clustering	Data mining techniques 1) Clustering 2) Segmentation 3) Prediction	ISI system 1) Nearest-neighbor techniques 2) Adaptive load-balancing 3) Stochastic optimization techniques of evolutionary computation
6	Protecting contents	Reporting	BRI system

can be used by the advanced mechanisms, which in themselves, are components of some popular browsers.

The second pass on the canonical HTML file occurs during the tag parsing process which utilizes a probe set. This is a component of the ISI Web page parser. The data to be used in these mechanisms is filtered during the first pass in order to assure that the appropriate wrappers exist. Wrappers used by the HRD Web prototypes filter out documents that may not be in English or follow incorrect HTML format. Currently, the search engine is unable to perform any language translations.

6 Computational Results

6.1 Execution Environments

Earlier studies of the ISI system [16, 19] were limited to 1024 pages and presented the feasibility of incorporating honeybee search strategies. These results focus on the second pass of the two-pass ISI parser [18] which has been expanded and tested using HP Pavilions with three 866 MHz (30 Gigabytes hard drive) and one 800 MHz (30 Gigabytes of memory) Pentium III processors, 128 MB SDRAM, and Intel Pro/100+ Server Adapter Ethernet cards, connected via two D-Link DSH-16 10/100 dual speed hubs with switches through a 144 Kbps router. The indexer tests are using Red Hat Linux release 7.0 (Guiness). The dataset tested consists of 1771 Yahoo Business Headline documents [22] supplemented with 22415 HRD raw (parsed for the correct format) data files supplied by the HRD system [18] during the data enrichment and cleaning phases (see Table 2).

The methodology used in clustering (retrieval) calculations [19]—computing the stochastic measurements—was based on: 1) generating the canonical format of the raw Web page—an application-specific document of structural information, and 2) applying the retrieval algorithms—computing the *raw fitness, standardized fitness*, and *adjusted fitness* (the *normalized fitness*).

This study relied on a set of 47 strings stored in a static probe set—the approach used by operational IR systems—in order to compute the associated stochastic measures for each Web page. A static probe set, which does not have adequate *a priori* knowledge about randomly chosen Web documents

Table 2. Selected Web documents retrieved by the HTML resource discovery (HRD) system.

Retrieval Period	ISP response results	Web forager dispatcher				Totals
		Node 0	Node 1	Node 2	Node 3	
15 Oct 2001	Raw HTML pages	1422	1127	1003	1112	4664
— 28 Jan 2002	Access forbidden pages					
	—Firewall pages	96	37	72	48	253
	—Web mail pages	324	288	345	328	1285
	—403 forbidden pages	7	4	5	5	21
	—404 not found	1	1	2	3	7
	Useful raw HTML pages	994	797	579	728	3098
28 Jan 2002	Raw HTML pages	3526	2302	2819	2448	11095
— 13 May 2002	Access forbidden pages					
	—Firewall pages	220	61	55	43	379
	—Web mail pages	190	573	104	226	1093
	—403 forbidden pages	23	19	22	11	75
	—404 not found	8	6	4	4	22
	Useful raw HTML pages	3085	1643	2634	2164	9526
16 Sep 2002	Raw HTML pages	1923	1558	1716	1459	6656
— 04 Mar 2003	Access forbidden pages					
	—Firewall pages	110	75	71	69	325
	—Web mail pages	486	514	534	540	2074
	—403 forbidden pages	37	13	53	17	120
	—404 not found	12	6	3	7	28
	Useful raw HTML pages	1278	950	1055	826	4109
Totals	Raw HTML pages	6871	4987	5538	5019	22415
	Access forbidden pages					
	—Firewall pages	426	173	198	160	957
	—Web mail pages	1000	1375	983	1094	4452
	—403 forbidden pages	67	36	80	33	216
	—404 not found	21	13	9	14	57
	Useful raw HTML pages	5357	3390	4268	3718	16733

that supplement the dataset, must be developed by a human editor. The 47 search strings used in this study were derived from randomly chosen phrases from a disjoint collection of Wall Street Journal (WSJ) articles [7] which have no interdependencies with the randomly chosen dataset of Web pages.

6.2 Unsupervised Dataset of Web Pages

The initial dataset used in a search engine case study was 512 pages [16] followed by Versions A, B, C, and D of the ISI system [19] which were limited to 1024. Both of these studies used subsets of the 1771 Yahoo pages. The hash table used to store the dataset for indexing purposes was limited to 1024 elements. These five studies presented feasibility results reflective of the utilization of honeybee search strategies as a clustering and indexing mechanism.

The experimental results in this paper were derived from a study aims at assessing the impact of Web page scalability on the ISI system. The HRD system returned 22415 Web pages from randomly selected location throughout the Internet. These 22415 pages resulted from the first 4 phrases of the KDD model. The HRD Web scouts located a total of 28727 raw data files that contained various forms of HTML documents written in a host of languages.

The HRD Web foragers were responsible for coding (phase 4) the raw data files for the indexers by executing the first pass of the two pass HTML parser. The foragers eliminated 6312 raw data files due to incorrect formats, corrupted files, and/or non-English files. The resulting file types are presented in Table 2. The resulting 22415 files were further partitioned into firewall pages, Web mail access pages, 403 (access) forbidden pages, 404 (file) not found pages, and useful raw HTML pages. The percentages of theses files that comprised the dataset are 4.2%, 19.9%, 1.0%, 0.2%, and 74.7%, respectively.

The current versions of the ISI system to be tested is Versions E and F which differ by the structure of the probe set. These versions differ from their predecessors in two distinct manners. The dataset size will be 22415 (+ 1771 Yahoo pages) which is a 22-fold increase in the dataset size and a hash table that has at most 32768 elements.

Before the parallel version of Versions E and F are tested, the sequential version is being used with a pseudo-node that initially contains the 1771 Yahoo pages at the beginning of the simulation. The pseudo-node is used so to emulate a two-node indexer cluster coupled with the stochastic regulatory mechanism for information sharing between the two nodes. The UNIX function /usr/bin/time was used to capture the runtime resource usage for four distinct nodes executing the sequential version of Version E.

The hash table size was increased by powers of 2 starting from a hash structure size of 1024. The study was terminated with a hash size of 32768. The output generated by the UNIX command provides run-time information related to how the application software utilizes the major system resources—CPU and memory. Each of the nodes utilizes the system resources differently

Table 3. Execution results for a pseudo-node with 1771 pages and a node with 4664 pages.

Hash table size	OS Usage	Node 0	Node 1	Node 2	Node 2
1024	CPU timing (elapsed)	86:56:06	76:42:33	89:34:06	91:44:48
	CPU timing (user)	95494.90	102740.60	95582.33	108248.18
	CPU timing (system)	3151.72	3419.16	8358.86	4611.25
	CPU utilization	4%	7%	5%	8%
	page faults (major)	20217567	21219010	23348805	22271294
	page faults (minor)	18063326	18227509	18422797	18473458
	page swaps	891266	992168	1106580	1135227
2048	CPU timing (elapsed)	88:16:57	77:17:10	86:38:36	81:42:51
	CPU timing (user)	94494.37	101677.49	93302.16	107019.31
	CPU timing (system)	2989.58	3396.43	7722.58	4240.17
	CPU utilization	3%	6%	4%	8%
	page faults (major)	19699071	22141370	19809162	21803346
	page faults (minor)	17783977	18532508	17798479	18690136
	page swaps	714667	1138898	720854	1240631
4096	CPU timing (elapsed)	84:27:22	74:35:15	83:34:03	79:23:29
	CPU timing (user)	92226.53	99259.45	91692.19	105041.97
	CPU timing (system)	2855.57	3249.32	7870.62	4296.15
	CPU utilization	3%	6%	4%	8%
	page faults (major)	22210976	21335491	22199305	22540909
	page faults (minor)	18308356	18065741	18352411	18211708
	page swaps	980466	851953	1006141	893464
8192	CPU timing (elapsed)	88:36:29	76:50:48	86:09:27	81:53:56
	CPU timing (user)	90262.82	97639.74	89977.98	103455.31
	CPU timing (system)	2827.84	3391.01	7874.10	4639.60
	CPU utilization	2%	5%	3%	7%
	page faults (major)	24346468	24240287	23128138	25611445
	page faults (minor)	18814675	18473646	18757880	18469424
	page swaps	1408180	1189773	1366295	1181784
16384	CPU timing (elapsed)	81:23:12	71:49:00	80:15:21	75:14:24
	CPU timing (user)	88549.63	95136.15	87770.73	100379.20
	CPU timing (system)	2622.25	2935.13	7293.53	3833.38
	CPU utilization	1%	4%	3%	6%
	page faults (major)	20429131	20144997	20584363	19423002
	page faults (minor)	17940436	17724845	17957506	17687554
	page swaps	859842	728641	872209	682569
32768	CPU timing (elapsed)	84:23:08	74:58:50	83:52:38	76:51:55
	CPU timing (user)	91939.85	98812.46	91174.06	103942.82
	CPU timing (system)	2644.12	3037.01	7486.09	3875.40
	CPU utilization	2%	5%	4%	7%
	page faults (major)	21064694	22040748	22070392	19517012
	page faults (minor)	18638159	18401523	18588344	18302823
	page swaps	1189082	1019620	1149957	940127

based on internal fluctuations in each nodes hardware. The usage of the UNIX *time* function is not accessible when using MPI for the parallel versions.

6.3 Experimental Results

As the table size was increased, the performance of Version E with various hash table sizes showed improvement in all areas of resource usage (see Table 3). These improvements occur on a computer-by-computer basis in which each machine responded uniquely to the table size variations. The elapsed CPU times for all the nodes showed decreases that ranged from approximately 5 hours for node 0 to approximately 16 hours for node 3. Node 2 had a decrease of approximately 9 hours. Each node experienced a resource usage increase for the hash table sizes of 8192 and 32768. These increases reflect the distribution of the pages within the hash table which reflects collisions within each element. Additionally, the elapsed time for nodes 0 and 1 showed an approximately one hour increase for a table size of 2048 where node 2 experienced a 3 hour decreased. However, node 3 experienced a 10 hour decrease for the same table size.

The user CPU times did not reflect the timing trends shown in the elapsed times. The CPU time measurements for elapsed time was hours:minutes:seconds, and seconds for the user and system times. The user CPU timings for each node decreased by approximately 2000.0 seconds (5 hours) as the hash table size was increased. The approximate difference for each node was approximately 8000.0 seconds as the table size varied from 1024 to 32768. The system CPU times showed timing fluctuations consistent with the elapsed CPU times. Nodes 0 and 3 showed decreases in system times consistent with elapsed times. As the elapsed time decreased, decreased was experienced in the system time. Nodes 1 and 2 did not follow any noticeable trend and varied in what appears to be an inconsistent manner.

The most interesting component of this study was the variations in the percentage of CPU utilization. The percentages of utilization ranged from 1% to 8%. Node 0 started with 4% and reached a low of 1%. Node 1 had a similar percentage range of 7% down to 4%. Both of these nodes decreased by approximately 3%. Nodes 2 and 3 decreased by approximately 2%. All of the nodes showed decreases until the tale size was increased to 32768. The percentages of CPU utilization do not appear to reflect any single resource measurement but appear to result from a combination of measurements.

The other factor that is evident from the CPU utilization percentages reflects the efficient usage of memory which encompasses major and minor page faults and page swaps. The major and minor page faults as well as the page swaps fluctuate as the hash table sizes increased. The ratio of the hash table size to the dataset size (4664 pages) was 0.22, 0.44, 0.88, 1.76, 3.51, and 7.03, respectively. The factor that appears to affect the memory utilization as well as the CPU utilization is the distribution of pages throughout the hash table. A reduction in the number of collisions within each table element results in a decrease in all memory and CPU measurement. As the dataset grows, there are limitations on the size of the hash table. The underlying goal of a good hash function is to distribute the elements of the dataset as equally as possible.

7 Conclusion

The Tocorime Apicu IS system has incorporated suggested approaches for improving the IR systems of current search engines with the search strategies of honeybees. Earlier studies of this model showed the benefits of unsupervised clustering of Web pages using adaptive probe sets. The experimental results showed that the model will scale as additional Web pages are added to the dataset. Also, the data structure used to facilitate the unsupervised clustering of Web pages has a large impact on the efficiency of this adaptive implementation.

The CPU utilization percentages showed steady decreases which on first notice appear to indicate an application program that is inefficient. The measurement that is most noticeable in parallel applications is the reduction in elapsed CPU time, as opposed to the percentage of CPU utilization across all nodes in a parallel application. The speedups reflected in the sequential versions should transfer to the parallel version of this application.

Acknowledgments

The author wishes to express his gratitude to the reviewers whose detailed and useful comments helped tremendously to improve the quality of this paper. This work was supported by Honeybee Technologies and Tapicu, Inc.

References

1. H.A. Abbass. MBO: Marriage in Honey Bees Optimization A Haplometrosis Polygynous Swarming Approach. In *Proceedings of CEC 2001*, pages 207–214. IEEE, Piscataway, NJ, 2001.
2. P. Adriaans and D. Zantinge. *Data Mining*. Addison-Wesley, Harlow, England, 1996.
3. E. Bonabeau, A. Sobkowski, G. Theraulaz, and J. Deneubourg. Adaptive Task Inspired by a Model of Divsion of Labor in Social Insets. In D. Lundh, B. Olsson, and A. Narayanan, editors, *Biocomputing and Emergent Computation*, pages 36–45. World Scientific, 1997.
4. J.B. Free. *The Social Organization of Honeybees*. (Studies in Biology no. 81) The Camelot Press Ltd, Southampton, 1970.
5. J.B. Free. *Pheromones of Social Bees*. Comstock Publishing Associates, Ithaca, New York, 1987.
6. P. Fritsch. Five Mellow Guys Follow Their Dream: A 'Tall Ship' in Brazil. *The Wall Street Journal*, CXLII(35):1, Friday, February 18, 2000.
7. Wall Street Journal. *(Western Edition)*. Dow Jones and Company, 200 Liberty St., New York, February 2003.
8. S. Kim and B. Zhang. Evolutionary Learning of Web-Document Structure for Information Retrieval. In *Proceedings of CEC 2000*, pages 1253–1260. IEEE, Piscataway, NJ, 2000.
9. S. Lawrence and C.L. Giles. Searching the World Wide Web. *Science*, 280:98–100, 1998.

10. V. Maniezzo, A. Carbonaro, M. Golfarelli, and S. Rizzi. An ANTS Algorithm for Optimizing the Materialization of Fragmented Views in Data Warehouses: Preliminary Results. In E.J.W. Boers et al., editor, *EvoWorkshop 2001, LNCS 2037*, pages 80–89. Springer-Verlag, Berlin, 2001.

11. S. Parthasarathy and M. Ogihara. Exploiting Dataset Similarity for Distributed Mining. In *IPDPS 2000 Workshops, LNCS 1800*, pages 399–406. Springer-Verlag, Berlin, 2000.

12. V. Ramos, F. Mugo, and P. Pina. Self-Organized Data and Image Retrieval as a Consequence of Inter-Dynamic Synergistic Relationships in Artificial Ant Colonies. In A. Abraham, J. Ruiz del Solar, and M. Koppen, editors, *Soft Computing Systems: Design, Management and Applications*, pages 500–509, IOS Press, Amsterdam, December 2002.

13. G. Salton, J. Allen, C. Buckley, and A. Singhal. Automaic Analysis, Theme Generation, and Summarization of Machine-Readable Texts. *Science*, 264:1421–1426, 1994.

14. G. Salton, A. Wong, and C.S. Yang. A Vector Space Model for Automatic Indexing. *Communications of the ACM*, 18:613–620, 1975.

15. M.P. Sinka and D.W. Corne. A Large Benchmark Dataset for Web Document Clustering. In A. Abraham, J. Ruiz del Solar, and M. Koppen, editors, *Soft Computing Systems: Design, Management and Applications*, pages 881–890, IOS Press, Amsterdam, December 2002.

16. R.L. Walker. Search Engine Case Study: Searching the Web Using Genetic Programming and MPI. *Parallel Computing*, 27(1/2):71–89, March 2001.

17. R.L. Walker. Applying Evolutionary Computation Methodologies for Search Engine Development. In L. Wang, K.C. Tan, T. Furhashi, J. Kim, and X. Yao, editors, *SEAL'02: Proceedings of the 2002 Asia-Pacific Conference on Simulated Evolution and Learning*, pages 208–213, Singapore, November 2002. Nanyang Technological University Press.

18. R.L. Walker. Simulating an Information Ecosystem within the WWW. In A. Abraham, J. Ruiz del Solar, and M. Koppen, editors, *Soft Computing Systems: Design, Management and Applications*, pages 891–900, IOS Press, Amsterdam, December 2002.

19. R.L. Walker. Using Nearest Neighbors to Discover Web Page Similarities. In H.R. Arabnia, editor, *PDPTA'02: Proceedings of the 2002 International Conference on Parallel and Distributed Processing Techniques and Applications*, pages 157–163. CSREA Press, June 2002.

20. R.L. Walker. Comparative Study of the Information Retrieval Systems of Current Search Engines. In W. Abramowicz and G. Klein, editors, *BIS 2003: Proceedings of the 2003 Business Information System Conference*, June 2003. To appear.

21. R.L. Walker. *Tocorime Apicu: Design of an Experimental Search Engine using an Information Sharing Model*. Ph.D. Dissertation, University of California, Los Angeles, 2003.

22. Yahoo. Yahoo Web Page. Yahoo Inc. Santa Clara, CA, November 1998.

Part V

Internet Security

Real-time Certificate Validation Service by Client's Selective Request

Jin Kwak, Seungwoo Lee, Soohyun Oh, and Dongho Won

School of Information and Communications Engineering,
Sungkyunkwan University, Korea
{jkwak, swlee, shoh, dhwon}@dosan.skku.ac.kr

Abstract. The application of PKI(Public Key Infrastructure) was enlarged with the development of Internet. The certificate using services to validate the public key are increasing according to this. To verify the certificate, the client must confirm the availability of certificate's current status first. Various methods to validate the certificate have been proposed so far and the most of them are CRL-based. But those CRL-based methods have many problems because of the CRL's periodicity. Therefore, the CA in the field that requires frequent modification needs to provide the latest certificate status information to the client in a real-time. In this paper, we propose a new model which can offer the timely certificate status information to the client in a real-time, we called RCVM. Also, we define the MITP(Modified Information Transmission Protocol), and the selective request and response message. The RCVM is that client does not incur the overhead of certificate status validation and the certificate status validation service provides information about a selective request of the client.

Key words: PKI, certificate validation, real-time, timeliness, selective request/response

1 Introduction

With the rapid progress of research to construct a secure network, a lot of new services have been proposed. Those services are expanded with the adaptation of PKI that provides entity authentication, data integrity and non-repudiation about the transaction. These certificate using services with PKI are mostly used for an entity authentication. This is achieved provide secure network by means of using a certificate. A certificate is digital document that associate an individual or entity with its specific public key. A certificate is a data structure containing a public key, pertinent details about the key owner, and optionally, some other information, all digitally signed by a trusted third party, usually called a CA(Certification Authority).

When the client A wants to send a message to the client B, s/he does not attach the public key to the message, but the certificate instead. The client B receives the message with the certificate and then checks the signature of the third party on the certificate. If the signature was signed by a certifier that s/he trusts, the client B can safely accept that the certificate is trusted[5]. Therefore, a certificate must be validated before verifying it and the certificate status information is important in the fields that use lots of certificates. Also, as especially the network structure came to be complex, certificate validation service needed information provide such as path validation, trusted time-stamp, and additional information[6].

Existing CRL(Certificate Revocation List)-based methods are proposed to verify the availability of certificate. But the CRL-based methods cannot provide the current certificate status information[10]. Because OCSP(Online Certificate Status Protocol) and SCVP(Simple Certificate Validation Protocol) has been proposed to overcome this problem. But OCSP and SCVP do not contain any concrete explanation about the operation and mechanisms about checking certificate status validation, only define the format and structure of a message between the server and the client.

In this paper, we describe existing certificate validation methods and then propose a new model which can provides certificate status validation by client's selective request, RCVM(Real-time Certificate Validation Model). Also, we define the MITP(Modified Information Transmission Protocol), and the selective request/response message. This paper is organized as follows. In section 2, we review the existing certificate validation methods. In section 3, we present modules and protocols for providing certificate validation service in the RCVM. In section 4, we explain the characteristics of proposed in this paper. Finally, in section 5, we bring to a conclusion.

2 Certificate Validation Methods

Certificate validation is the process of verifying that a certificate is valid or not. This process checks that the certificate is within its validity period, and performs an integrity check based on the digital signature of the CA. Certificates are valid until they expire, if certificates are not revoked. There are many things can happen that require the revocation of a certificate before the expiry date, such as the secret key lost or destroy, the key compromised, change of certificate holder's information, and so on[4][5].

The traditional certificate validation methods is to periodically publish a CRL, CRL is widely used method to validate the certificate status. The CA signs to the certificate including the serial number, revoked reason and make it public. Then the client downloads the CRL and search for obtaining the certificate status information. CRL contains a serial number and reason of all revoked certificates. It is the general method to show the revoked reason. But, the main problem with this method is that the list can be large for whole CRL, and the communication overload because of all clients downloading the whole CRL. It can't be providing the current certificate status because it is issued periodically[2][3].

Methods relevant to certificate validation are OCSP and SCVP. The OCSP is proposed to provide the current certificate status information. It is composed of client and server, and proposed by IETF RFC2560[9]. It is a protocol that provides certificate status information to a client without using CRL and it is used online between the server and the client. If a client connects to a server and requests a certificate status that s/he need, the server searches that information and digitally signs it[10]. Then the server sends it to the client. The client can obtain the certificate status information using the OCSP. SCVP is to reduce the overload of certificate path validation from the client. SCVP uses a simple request and response protocol, designed run over HTTP. The SCVP client sends a specific certificate in the inquiry to be validated. The server performs the validation processing and digitally signs it. Then the server sends it to the client[8]. OCSP and SCVP are only proposed protocols. They don't contain any concrete explanation about the operation and only structure of a message between the server and the client.

3 Real-time Certificate Validation Model

3.1 RCVM

The propose RCVM is a system that provides certificate validation process to client without additional overhead. RCVM consist of CA, repository(CA repository), CVS(Certificate Validation Server), MIP(Modified Information Provider) that sends information to Cdb(CVS database), several modules such as CPaM(Certification Path Manager), CPoM(Certification Policy Manager), MITP(Modified Information Transmission Protocol), and Client. Fig. 1 shows the proposed RCVM architecture.

- CA(Certification Authority)
  - Issuing and managing the certificates of the CVS and Client.
  - Issuing and managing the whole CRLs and modified status information of certificate.
  - When certificate is modified, CA informs it to the Cdb using MIP.
  - Modified status information may include the certificate's serial number, distinguished name(DN), revoked reason and etc.
- Repository (CA repository)
  - Storing whole CRLs created and issued by CA
  - Storing and updating the modified information of certificate.
- MIP(Modified Information Provider)

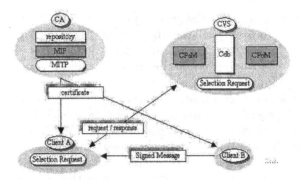

Fig. 1. Architecture of RVCM

- MIP is a transmission module that provides new modified certificate information from CA to Cdb.
- It cannot delete or insert any information issued by the CA and only can transmits the information to the CVS.
- MIP sends it to the CVS instantly. MIP uses the MITP negotiated between the MIP and the CVS.
- MITP(Modified Information Transmission Protocol)
  - MITP is a protocol to provide new information about certificate status to CVS from CA.
  - It is a mutual agreement protocol between the CA and the CVS.
  - It must guarantee real-time property as definite validity time.
- CVS(Certificate Validation Server)
  - CVS stores the latest certificate status information and path/policy information to own Cdb.
  - CVS checks in Cdb and then provides it to the client.
  - If the service request is excessively concentrated on the CVS, it possible to run into state of denial of service. Then the CVS can dispersion of request to other CVS.
- Cdb(CVS database)
  - Cdb only includes modified certificate information, not the whole CRL.
  - It includes certificate path validation and policy information.
  - It has the same certificate status information issued by the CA and stored in CA repository.
- CPoM(Certification Policy Manager)
  - CPoM manages certification policies and policy mapping information.
  - It stores certification policy related information to Cdb.
- CPaM(Certification Path Manager)
  - CPaM builds all certification paths that consist of CA certificates.
  - It verifies certification path and managed all related path information.
  - It stores certification path related information to Cdb.

In the RCVM, by using MIP, the CA provides modified certificate information to CVS. CPaM and CPoM provide validation information to CVS. By using MITP, CVS may obtain timely latest modified certificate information in a real-time and it provides the timeliness of the selective request of the client without any overload. The client sends the selective request to the CVS, and then the client may obtain validation information from CVS.

Real-time: It means that the definite transmission and response time are guaranteed. When CA produces new modified certificate information and provides it to the CVS, CA can guarantee a definite transmission time and the CVS can also guarantee a definite response time for the client's request.
Timeliness : The response to the client's request must be timely. If the modified certificate information is issued periodically, it cannot provide the latest information of certificate to the client.

3.1.1 Modified Information Transmission Protocol

The MITP is a new transmission protocol that provides the latest information from the MIP to the CVS. MIP receives the newest information created by CA, and then transmits it to the CVS at once. It can't delete or insert any information. It is the module that can transmit the modified certificate information to the CVS. The MIP provides the latest certificate information to the client in a real-time using the MITP. The ASN.1 definition of the MITP is given in Appendix A, and the meaning of each field is described as follows[1][7].

CertModifiedMessage field is composed of *header, body* and *signatureValue*. *CertModifiedheader* field is explanations of the *sender* and *recipient* about modified information generated by the CA. The sender is a CA and the recipient is a CVS.
CertModifiedBody field include substantial modified certificate information. This is composed of the *certStatus* that presents the status of each certificate and the *certStatusInfo* that presents the additional information of each certificate. *producedAt* is the time at which the CA signed this message.
CertStatusInformation field is composed of the *certSerial* that indicates each certificate's serial number, the *reason* that indicates the revoked reason, the *revocationTime* that indicates the revoked time, and the *crllocation* that supports the searching for the CRL.

3.1.2 Selective request

Selective request message is the message that client, who using the RCVM, selective requests the information to the CVS. This message sends the certificate information to the server including signature of requestor, so availability of the certificate is held back before receiving the Selective response of the

CVS. The ASN.1[7] definition of the Selective request message is given in Appendix B, and the meaning of each field is described as follows.

version field specifies the version of the message and it's initial value is DEFAULT 0. *requestorName* field indicates that the general name of requestor. *reqCertList* field is the information that specifies the certificates in the request. This is composed of the *CertID, IssuerSerial, PkCert* and *CertHash.* The *CertID* composed of hash algorithm OID(Object Identifier), hash value, and a serial number of the certificates. *hashAgorithm* is used to hash the issuer's DN(Distinguished Name) and public key. *certSerialNumber* contains the serial number of certificate. *issuerSerial* indicates the proper number of issuers. *PkCert* indicates the public key certificate.

selectionRequests field is the request of selective request by the client. The selective request is the request of the information by the client's needs. If s/he needs validation of certificate status, s/he selects *certStatus*. Also, s/he needs validation of certification path or valid time, s/he selects *certPath* or *validTime.*

requestTime field is the time at which the client transmits the request message.

requestExtensions field indicates beforehand mutual agreement between CVS and client and this field is used to request an additional information.

signature field is composed of *sigAlgorithm* used to make a signature and *signatureValue* that represents the BIT STRING.

Signed request message may include in the *certs* field of signature element certificate that assist the CVS to verify the requestor's signature.

3.1.3 Selective response

Selective response is the response message that includes a result of the selective request. It is done by the CVS it receives the request message from the client. Response message sent to the client also contains the responder's signature. CVS checks the information of certificate and provides it to the client. The ASN.1[7] definition of the selective response message is given in Appendix C, and the meaning of fields are described as follows.

version field specifies the version of the message, initial value is DEFAULT 0. *responderName* field indicates that the general name of responder.

validationResult field is the validation result that specifies the certificates for the client's selective request. This is the construction based on *certStatus* in MITP. This is composed of the *certStatus, certPathStatus,* and *validTime.*

certStatus field indicates that the result of the validation about each certificate status, Table 1 shows the result of the certificate status validation.

certPathStatus field indicates that the result of the path validation of each certificate, Table 2 shows the result of the certificate path validation *response* field includes additional information of response message. This is composed of

Table 1. Result of the certificate status validation

Value	Explanation
valid	validity ofcertificate
internalError	CVS reached an inconsistent internal state
finished	certificate is not valid any more
revoked	certificate is revoked before the term of validity
tryLater	CVS is operational but unable to return a status for the requested certificate, it can be used to indicate that the service exists, but it is temporarily unable to respond
unauthorized	client who is not authorized to the CVS

the *responderID* indicates the identifier of the responder, *producedAt* indicates the time at which the responder signed the response, reason indicates the reason of revoked certificate, *revocationTime* is the time of certificate revocation, *crllocation* gives support to refer to CRL, and *responsedataExtensions. responseTime* field is the time at which the CVS transmits the response message. *responseExtensions* field indicates that beforehand mutual agreement between CVS and client, this field used response about additional information. *signature* field is signature of CVS.

Table 2. Result of the path status validation

Value	Explanation
valid	certificates at a minimum satisfies the path validation rules
invalid	certificate does not satisfy one or more necessary to the production of a "*valid*"
unknown	CVS Server has no knowledge of the subject certificate

All the above information is digitally signed with the private key of the CVS, and the signature can be verified the public key contained in the public key certificate issued by the CA

3.2 Operation of the service

The operation of the RCVM that proposed in this paper is showed fig. 2. Service is provided by this sequence as follows.

Generation of new modified certificate information(2): First, the user request issuing certificate to CA, CA checks user's identity and other information. If there are no problem, CA issues certificate to clients*(1)*. At the same time, CA generates the information of the user, certificates, certificate status and CRL and stores them in CA's own repository. The new modified certificate information contains the latest information of certificate such as

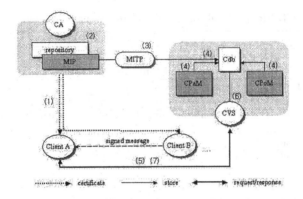

Fig. 2. Operation of service

the certificate's serial number, DN, current status and the reason of modification/revocation etc. Then the CA stores it to repository and sends it to MIP at the same time after the generation of new information.

Transmission of the modified certificate information(3): After receiving the new information from the CA, MIP sends it to the Cdb instantly. MIP uses the MITP negotiated between the MIP and the CVS, the MITP provides the reliability of the information. The MIP only can transmit the new information to the Cdb and cannot delete or insert any information to the new information issued by CA.

Modified certificate information storing to the Cdb(4): CVS stores the new information to it's own Cdb. The new modified certificate information transmitted from the MIP only contains the new information and it does not contain the whole CRL.

Certification path information storing to the Cdb(4): CPaM builds all certification paths that consist of CA certificates. It managed all related path information. It stores certification path related information to Cdb.

Certification policy information storing to the Cdb(4): CPoM manages certification policies and policy mapping information. It stores certification policy related information to Cdb.

Selective request of the certificate validation information(5): The user can validate the certificate by requesting the certificate information that s/he wants to validate the CVS and validate the replying message from the CVS. The selective request message only contains the requested information, there is less communication overload. Selective requests are certificate status, certification path and time-stamp information. Client can select not only one but also all request type.

Search/extraction of the information(6): If the CVS receives the selective request message from the user, it searches the information stored in Cdb. Because the new information in Cdb is equivalent to the latest information issued by CA, it can provide the timely modified certificates information to the client.

Response of the validation result(7): CVS provides the result of reference to the client with the format of Selective response. Response message contains the information that can present the current certificate status and selective requested. Response message can provide not only the certificate status but also the revocation reason, selected information, revoked time and the location of CRL in the case of client's request. Selective response provides the real-time property. Because the proposed service obtains the latest information using the MITP, it can provide the timely information to the client.

4 Characteristics

The proposed service provides new information to the client in real-time by using the MIP. On defining the new protocol, the service can transmit the new information in real-time and provide the timeliness of the selective request of the client without any overload.

Timeliness : Unlike the CRL based services, the proposed service can send the modified information to the CVS instantly. So it can provide the timely response and latest information to the client.

Real-time: The new information issued by the CA is transmitted to the CVS through the MIP instantly. The response time to the client's request can be guaranteed in the proposed service by searching the status information in the CVS. So the proposed service can respond to the client in a real-time.

Independence: If the clients only connect to the CVS to validate the certificate status, s/he can get the CA's modified information. The CA also needs only additional module to provide the new information to the CVS. Therefore the independent operation of the CA and the CVS is possible.

Decentralization : If there are several requests at the same time, it is possible that the response server cannot provide the service. But the proposed service can provide the service by dispersing the selective request messages of the clients with several CVSs.

Selectivity : The client selective requests the specific certificate's status information or other information. Also CVS provides the specific certificate's status information and additional information. Therefore, the client can acquire the necessary information selectively about the request.

Lightness : The proposed service minimizes the length of the modified information provided to the client by the CA and the CVS. So it can provide modified information and selective request of client without overloading the client and the CA.

As stated above, the proposed RCVM can provide those characteristics to the CA, CVS and the client. And it can be applied not only to certificate validation but to the additional service using extension fields. Especially, it is apt to a hierarchy structure of PKIs.

5 Conclusions

PKI needs a real-time certificate information demanded present point about status of certificate and real-time offer about information of certificate. So, this paper has presented a model of real-time certificate validation service in PKI environment. By proposed new protocol, it provides additional information such as path validation or time-stamp. For this, we analyze the problem of the existing methods and proposed a new model that the CA generates new modified information of certificate and the information provides to the CVS. CVS provides the information of the request to the client based on this modified information.

The proposed Real-time Certificate Validation Service by Client's Selective Request, reduces communication overload of both the CA and the client. Also this service can provide the information timely for the user wants. And it is suitable for the user who needs information about certificate validation as well as path validation and trusted time-stamp at the point of time. RCVM can give an additional support to the client. Also, it is to make useful applications to a hierarchy structure of PKIs.

References

1. C.Adams and S.Farrell, Internet X.509 Public Key Infrastructure Certificate Management Protocols, RFC2510, 1999
2. J.Author, Certificate Revocation Paradigms, Technical Report, Cybernetica Estonia, 1999
3. D.A.Cooper, A Model of Certificate Revocation, Proceeding of the 15th Annual Computer Security Applications Conference, 1999
4. P.Gutmann, PKI:It's Not Dead, Just Resting, IEEE Computer, vol.35, no.8, pp.41-49, Aug. 2002.
5. R.Housley, W.Ford, W.Polk, and D.Solo, Internet X.509 Public key Infrastructure Certificate and CRL Profile, RFC 2459, 1999
6. ISO/IEC 9549-8, Information technology Open System Interconnection The Directory : Authentication Frame Work, X.509, 1997
7. ISO/IEC 88240-1, Information technology-Abstract Syntax Notation One (ASN.1) : Specification of Basic Notation, 1997
8. A.Malpani, P.Hoffman and R.Housley, Simple Certificate Validation Protocol, draft-ietf-pkix-scvp-09.txt, Jun. 2000.
9. M.Myers, R.Ankney, A.Malpni, S.Galperin, and C.Adams, Internet X.509 Public Key Infrastructure Online Certificate Status Protocol -OCSP, RFC 2560, 1999
10. M.Naor and K,Nissim, Certificate Revocation and Certificate Update, In Proceeding of the 7th USENIX Security Symposium, 1998, pp. 217-228.

Appendix A : Definition of MITP

```
CertModifiedMessage :: = SEQUENCE {
    Header          CertModifiedHeader
    body            CertModifiedBody
    signatureValue  BIT STRING OPTIONAL }
CertStatusHeader :: = SEQUENCE {
    Sender          GeneralName
    Recipient       GeneralName }
CertModifiedBody ::= SEQUENCE {
    certStatus      CertStatus
    certStatusInfo  CertStatusInformaton
    producedAt      GeneralizedTime}
CertStatusInformation ::= SEQUENCE {
    certSerial      INTEGER
    reason          RevokedReason
    revocationTime  RevokedTime
    crllocation     CRL location OPTIONAL }
```

Appendix B : Selection request

```
SELRequest ::= SEQUENCE {
    version             INTEGER DEFAULT 0
    requestorName       GeneralName
    reqCertList         SEQUENCE OF RequestCertList
    selectionRequests   SelectionRequests
    requestTime         GenelizedTime OPTIONAL
    requestExtensions   EXPLICIT Extensions OPTIONAL
    signature           Signature OPTIONAL }
RequestCertList ::= CHOICE {
    CertID          CertID
    IssuerSerial    [0] IssuerSerialNumber
    PkCert          [1] Certificate
    CertHash        [2] OCTET STRING }
SelectionRequest ::= CHOICE {
    certStatus      [0] CertStatus
    certPath        [1] CertPathStatus
    validTime       [2] ValidationTime }
CertID ::= SEQUENCE {
    hashAlgorithm    AlgorithmIdentifier
    issuerNameHash   OCTET STRING -- Hash of issuer's DN
    issuerKeyHash    Certificate -- Hash of issuer's Public key
    certSerialNumber CertificateSerialNumber }
Signature ::= SEQUENCE {
```

```
       sigAlgorithm          AlgorithmIdentifier
       signatureValue        BIT STRING
       certs                 EXPLICIT SEQUENCE OF Certificate OPTIONAL }
```

Appendix C : Selection response

```
SELResponse ::= SEQUENCE {
     Version                INTEGER DEFAULT 0
     responderName          GeneralName
     validationResult       ValidationResult
     response               SEQUENCE OF ResponseData
     responseTime           GeneralizedTime OPTIONAL
     responseExtensions     EXPLICIT Extensions OPTIONAL
     signature              Signature OPTIONAL }
ValidationResult ::= CHOICE {
     certStatus         [0] CertStatusResult
     certPathStatus     [1] CertPathStatusResult
     validtime          [2] ValidationTime }
CertStatusResult ::= ENUMERATED {
     valid              (0)
     internalError      (1)
     finished           (2)
     revoked            (3)
     trylater           (4)
     unauthorized       (5) }
CertPathStatusResult ::= ENUMERATED {
     valid              (0)
     invalid            (1)
     unknown            (2) }
ResponseData ::= SEQUENCE {
     responderID               ResponderID
     producedAt                GeneralizedTime
     reason                    RevokedReason
     revocationTime            Time
     crllocation CRL           CRL location OPTIONAL
     responseDataExtensions    EXPLICIT Extensions OPTIONAL }
Signature ::= SEQUENCE {
     sigAlgorithm       AlgorithmIdentifier
     signatureValue     BIT STRING
     certs              EXPLICIT SEQUENCE OF Certificate OPTIONAL }
```

Internet Attack Representation using a Hierarchical State Transition Graph

Cheol-Won Lee, Eul Gyu Im[1], and Dong-Kyu Kim[2]

[1] National Security Research Institute
62-1 Hwa-am-dong, Yu-seong-gu
Daejeon, 305-718, Republic of Korea
{cheolee,imeg}@etri.re.kr
[2] Department of Computer Engineering
Ajou University, Suwon 442-749
dkkim@madang.ajou.ac.kr

Summary. Internet attacking tools become automated and advance quickly, so an attacker can easily deploy attacks from distributed hosts to acquire resources as well as to disrupt services of a target host. One of the best feasible ways to study internet attacks and their consequences is to simulate attacks. In this paper, we introduced a new approach to express attack scenarios for simulation. Our approach allows relations between states to be expressed in graphs, so that users can identify relations between states and find new scenarios.

Key words: Internet Attack Modeling, Attack Scenario, Attack Simulation, State Transition Graph

1 Introduction

Cyber attacks are increasingly taking on a grander scale with the advance of attack techniques and attack tools, such as Nessus Scanner [1], Internet Security Systems(ISS) Internet Scanner [2], COPS security checker [3], N-Stealth [4], and so on. Attacks may be launched from a wide base against a large number of targets with the intention of acquiring resources as well as disrupting services [5] of a target host. Research on cyber attacks and their effects is fundamental to defend computer systems against cyber attacks in real world [6, 7, 8]

Experiments of attacks in real environments have some limitations because 1) experiments may disrupt services of systems, and 2) scales of target systems are getting bigger and bigger. Therefore, simulations become one of the best feasible ways to study cyber attacks and their consequences. There are several ways to represent internet attacks in simulations: a tree-structure, Petri-Net, and a state transition diagram. Each approach has its own advantages and disadvantages.

Among them, state transition diagrams are widely used to express attack scenarios where a transition represents a single action or step taken by attackers a state represents an intermediate or final situation of attackers. A scenario is defined by a set of transitions and states. If new scenarios are added, the number of transitions and that of states continuously grow in a diagram. If a newly added state is different from other states even though it is quite similar to several states, it is added to the diagram and many similar states exist separately.

Previous work focused on expressing a single scenario, and a new approach needs to be developed to express many scenarios together. In this paper, we propose a hierarchical state transition graph to represent Internet attacks. With the hierarchical structure, relations between states, such as similarity and inclusion can be expressed, and a hierarchical state transition graph reduces the number of states and transitions by merging related states. In addition, users can find new paths in the graph because relations between states are expressed in the graph. These new paths generate new attack scenarios for users.

The rest of this paper are organized as follows: Section 2 addresses previous work done by others. We will explain mechanisms to express attack scenarios and their pros and cons. Our approach and a design of a simulator are explain in Section 3 and Section 4, followed by summary in Section 5.

2 Related Work

There are several ways to model cyber attacks. One of them is to use tree structures [7, 9]: each node as a state, each branch as an action and embedded actions as children of a node. In the Attack Tree model [9], the final goal is expressed in the root node and sub nodes shows attacks that must be succeeded to achieve the goal. Each node has a logical operation which is an AND or an OR operation. If a node has an AND operation, the attack of the node is successful when all the attacks in sub nodes are succeeded. If a node has an OR operation, the attack of the node is successful when any attack in sub nodes are succeeded.

Figure 1 shows an example of a scenario represented in the Attack Tree model. Advantages of this approach are that 1) it is action-oriented and 2) it modularizes actions very well. But with this approach, dynamic selections of actions and loops of actions cannot be expressed.

Petri-Net [10] is used to represent internet attacks. In Petri-Net approaches, a token moves around nodes to indicate executions of attack scenarios. If a node has a token, that node represents a current state of a scenario. One of advantages of this mechanism is that it can express attacks that hap-

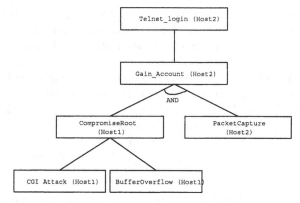

Fig. 1. An attack scenario in the Attack Tree model

pen simultaneously. One of problems with Petri-Net is that some graphs are hard to be implemented as the size of a graph increases.

Vigna, et al. used state transition diagrams to express internet attacks [11]. They proposed the State Transition Analysis Tool (STAT) tool suite. In STAT, scenarios are expressed as state transition graphs and they are processed in STAT core-based architecture. STAT has states and transitions but no tokens. Each state or transition has assertions, and for any incoming event, assertions of transitions are examined. If assertions are matched with the event, the corresponding transition occurs. If new attack scenarios are found, then a new diagram must be drawn since only one scenario is expressed in each diagram. USTAT and NetSTAT are variants of STAT [12, 13].

The above mechanisms lack to express relations between states. So, if a state is slightly different from another state, they are handled separately and it is not easy to figure out their similarities. But, to find similar attacks and cluster them together, there should be mechanisms to express relations between states.

Bordeleau, et al. also proposed a hierarchical state graph [14]. In this approach, the *scenario Interaction* relationship, the *scenario dependency* relationship, and the *scenario clustering* relationship are identified, and the hierarchy of scenarios is built based on these relationships. Their approach is to group commands of attach scenarios, whereas our approach is to group states into a bigger or higher state.

3 Hierarchical State Transition Graph (HSTG)

The integration of a new scenario $S1$ into an existing graph G requires pairwise comparisons between states in $S1$ and those in G. We identify three important types of inter-state relations:

Similarity: There can be similarity relation among states because there are many variants for an Internet attack. Therefore, there should be a mechanism that can extract common factors of states and express them.

Inclusion: If a state A is a specific case of a more general state B, A is a subset of B, and B *includes* A. If two states, C and D are similar, we can create a new state, E that contains common properties of two states, C and D. In this case, E includes both C and D, and this relation spawns a hierarchy of states. This will be addressed later.

Transition: Attack scenarios are expressed as a sequence of transitions between states in a state transition graph. Therefore, there are transition relations.

These relations can be expressed in a hierarchical state transition graph in HSTG. In HSTG, common aspects of states are extracted, and a hierarchical structure is built using these common characteristics. A hierarchical state transition graph that has multiple levels of states can reduce the number of states as well as the number of transitions by merging similar states, and as a result the overall computation overheads will be reduced.

In HSTG, a state can have one or more sub states, and this state can be a sub state of another state. Any state can transit to any other states regardless of its level. Internet attack scenarios can be expressed using HSTG, and if a new vulnerability was found and a new transition is possible because of this new vulnerability, the newly created states and transitions can be easily merged into the existing HSTG.

One of advantages of HSTG is that overall computation overheads are reduced because of the reduced number of states. In addition, when a new state is added to the graph, the number of comparison between states is $O(lgN)$ whereas the number of comparison in non-hierarchical graphs is $O(N)$. In HSTG, a parent state is created based on relations of states, so that relations among states can be recognized in the graph.

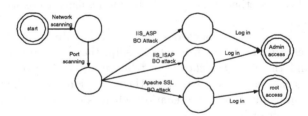

Fig. 2. An Example of a non-Hierarchical State Transition graph

Both Figure 2 and Figure 3 show a buffer overflow attack scenario. The sequence of this example scenario is as follows:

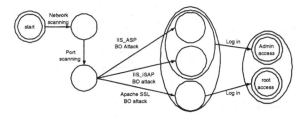

Fig. 3. An Example of a Hierarchical State
Transition graph

Network scan: To determine a target host, some information must be gathered
first. In this step, an attacker scans ip addresses to find out live hosts.
How to implement this function is beyond the scope of this paper, and
application programmers develop actors for this step.

Port scan: An attacker scans ports of some live hosts to find out open ports.
Based on information about open ports and alive hosts, a target host or
target hosts are selected.

Buffer overflow attack: Using vulnerabilities of web servers, an attacker tries
to access accounts in a target host. We used the following vulnerabilities available from CERT Coordination Center: IIS ASP Buffer Overflow
identified as CAN-2002-0079 which is a CVE name, where CVE stands for
Common Vulnerabilities and Exposures, IIS ISAPI Buffer Overflow identified as CAN-2001-0241, and Apache mod_SSL Buffer Overflow identified
as CAN-2002-0082.

Gain an admin or root privilege: Since buffer overflows allow attackers to execute arbitrary codes, attackers can gain a user privilege.

Figure 2 shows a non-hierarchical state transition graph. After a target
host is selected, one of buffer overflow attacks is executed depending on envi
ronments of the target host. For example, if the target host runs an IIS web
server, IIS_ASP or IIS_ISAP is used to attack the host. But, in this graph, relations between states, more specifically states about buffer overflow attacks,
cannot be expressed.

In HSTG shown in Figure 3, similar states are grouped into a bigger state.
For example, a buffer overflow attack using Apache SSL and a buffer overflow
attack using Apache mod_cookies grouped into a state called Apache buffer
overflow attack, and the Apache buffer overflow attack state and IIS web
server buffer overflow attack are grouped into a state called buffer overflow
attack.

Table 1 shows the overall comaprison of HSTG with other approaches.
The main advantages of HSTG are that it can express relationships between
unit attacks and it can group unite attacks into a higher state.

	tree-based	PetriNet	STAT	HierState	HSTG
express a scenarios?	O	O	O	O	O
express multiple scenarios?	X	O	O	O	O
express relationships between unit attacks?	X	X	X	X	O
grouping of unit attacks	O	X	X	X	O

Table 1. Comparison of HSTG with other approaches

3.1 Non-hierarchical representation of HSTG

HSTG can be expressed in a non-hierarchical form with additional relations.

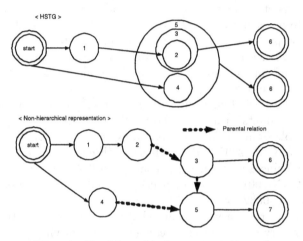

Fig. 4. Non-hierarchical representation of HSTG

Figure 4 shows a HSTG and its equivalent non-hierarchical graph. To convert an HSTG to non-hierarchical form, we need to add a special relation, called *parental relation*. *Parental relation* allows an unconditional transition from a child state to a parent state since the parent state is a super set of the child state.

3.2 Formal Definition of HSTG

Based on the non-hierarchical representation of HSTG, the following is a formal definition of HSTG.

Definition 1 An hierarchical state transition graph or HSTG is a tuple $G = (S, \tau, S_0, S_s)$, where S is a set of states, $\tau \subseteq S \times S$ is a set of transitions, $S_0 \subseteq S$ is s set of initial states, and $S_s \subseteq S$ is a set of states.

The size of S_0 must be greater than one. In other words, there should be at least one initial states. S_s denotes a set of states where the intruder has achieved his goals, and it cannot be empty. An attack scenario is defined as a sequence of states and its final state is in S_s.

3.3 Definition of a State

For HSTG, a state is defined as follows:

```
State ::= [initialization]
    state StateId {Annotation} [':' pStateId] '{'
        [Children cStateIds*]
        [Assertion]
        [CodeBlock] '}'
```

The above definition is based on state definition in STATL [15] and new fields are added to support hierarchies of states. To express hierarchical relations, parents and children can be able to be specified. Parent relation is expressed using a similar method as inheritance in the object-oriented design. In the above definition, 'pStateId' followed by a colon represents an optional parent id of a state. Children states can be specified in the 'Children' field, and each state can have zero or more children states and must have one parent state.

Relation	Merging Method
Case 1 : $A \cap B = \phi$	Two states have no common factors. Since a newly added state has no relation with existing states, the new state is added independently.
Case 2 : $A \supset B, A \neq B$	The state B is a subset of the state A, i.e. the state B has all the characteristics of the state A. In this case, the state A will be a parent of the state B.
Case 3 : $A \cap B \neq \phi$	Two states have some common characteristics. In this case, a new state C which is $A \cup B$ is created, and A' and B' are added as children of the state C, where A' is $A - (A \cap B)$ and B' is $B - (A \cap B)$

Table 2. Adding and Merging States

3.4 Adding a New State

When a new state is created, it is compared with states in the existing graph and added to the graph. The comparison process is done in a top-down manner, so a new state is first compared with top-level states as explained in Table 2. When a state is added to another state as a child state, the newly added state is compared with other children. This process is done until no comparison is required. Every state has a parent state, and top-level states have a virtual parent.

3.5 Adding a new scenario

When a new scenario is added to the graph, each state in the new scenario compared with existing states. The algorithm mentioned in the previous section is used to add each state. Since states of a new scenario are merged with existing states if possible, new paths may be established in the graph. These new paths are new scenarios that can applied to the existing network configuration.

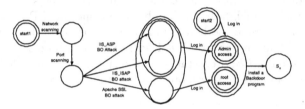

Fig. 5. Addition of a new scenario

Figure 5 shows addition of a new simple scenario to Figure 3: 1. log in to the host, and 2. install a backdoor program. One of states in the new scenario is merged with an existing state. As a result, a new path from 'Start1' to 'S_n' is possible.

3.6 Transition of a State

In STAT proposed by Vigna, et al., a transition from a state A to a state B succeeds when all the assertions of the state B are satisfied. In HSTG, in addition to assertions of a target state, assertions of parent states must be satisfied. Therefore, a transition succeeds when all the assertions of the target state and those of its parents are satisfied. Since assertions of states cannot be changed after simulation starts, it is possible for children to cache assertions of parents in advance.

4 Design of SUSH

We designed a simulator, called SUSH which stands for a Simulator Using SSF and HSTG. SSF is an acronym of Scalable Simulation Framework which is proposed in [16, 17]. SUSH has three layers: a UI layer, a Service layer, and a DB layer. The UI layer provides users tools to control and manage other layers, and tools in this layer can be run in different machine. The Service layer is a core part of SUSH. In this layer, DML files for simulations are generated and actual simulations are executed. The DB layer has DBs and their interfaces.

SUSH is currently under development. In SUSH, the UI layer and the other layers can be run in different machines. By isolating UIs from other modules, SUSH can allow more than one user to access Simulation Services simultaneously. To support multi-user modes, synchronization must be handled properly.

More detailed description about SUSH can be found in [8].

5 Summary

In this paper, we introduced a hierarchical state transition graph that has the following characteristics:

1. Inclusion and similarity can be expressed in the graph.
2. Hierarchical organization allows more efficient comparisons between states.
3. By identifying relations of states in the graph, a user can find new scenarios through finding new paths in the graph.

Contributions of this paper are:

1. HSTG introduced here allows to express multiple scenarios more efficiently than other approaches by identifying relations among states.
2. Based on HSTG, an architecture of a simulator was proposed. The proposed design of SUSH separate UI layer, service layer, and DB layer. The layered design enables easy management of systems and multi-user access.

References

1. Deraison, R.: Nessus Scanner, (http://www.nessus.org/)
2. : Internet Security Systems Internet Scanner. (http://www.iss.net/)
3. Farmer, D., Spafford, E.: The cops security checker system. In: Proceedings of the Summer Usenix Conference. (1990)
4. : N-Stealth: Vulnerability-Assessment Product. (http://www.nstalker.com/nstealth/)
5. Householder, A., Houle, K., Dougherty, C.: Computer attack trends challenge inernet security. In: Proceedings of the IEEE Symposium on Security and Privacy. (2002)
6. McDermott, J.P.: Attack net penetration testing. In: Proceedings of the 2000 workshop on New security paradigms, ACM Press (2000) 15–21

7. Steffan, J., Schumacher, M.: Collaborative attack modeling. In: Proceedings of the 17th symposium on Proceedings of the 2002 ACM symposium on applied computing, ACM Press (2002) 253–259
8. Lee, C.W., Im, E.G., Chang, B.H., Kim, D.K.: Hierarchical state transition graph for internet attack scenarios. In: Proceedings of the International Conference on Information Networking 2003. (2003)
9. Tidwell, T., Larson, R., Fitch, K., Hale, J.: Modeling Internet Attacks. In: Proceedings of the 2001 IEEE Workshop on Information Assurance and Security, West Point, NY (2001)
10. Murata, T.: Petri nets: Properties, analysis and applications. Proceedings of the IEEE **77** (1989) 541–580
11. Vigna, G., Eckmann, S., Kemmerer, R.: The STAT tool suite. In: Proceedings of the DARPA Information Survivability Conference & Exposition (DISCEX) 2000, Hilton Head, South Carolina, IEEE Computer Society Press (2000)
12. Ilgun, K.: USTAT: A real-time intrusion detection system for UNIX. In: IEEE Symposium on Security and Privacy, Oakland, CA (1993)
13. Vigna, G., Kemmerer, R.A.: NetSTAT: A network-based intrusion detection system. Journal of Computer Security **7** (1999) 37–71
14. Bordeleau, F., Corriveau, J.P., Selic, B.: A scenario-based approach to hierarchical state machine design. In: Proceedings of the Third IEEE Symposium on Object-Oriented Real-Time Distributed Computing, Newport, CA, USA (2000) 78–85
15. Eckmann, S.T., Vigna, G., Kemmerer, R.A.: STATL: An attack language for state-based intrusion detection. Journal of Computer Security (2001)
16. Cowie, J.H., Nicol, D.M., Ogielski, A.T.: Modeling the global internet. Computing in Science and Engineering (1999) 42–50
17. Lee, C.W., Im, E.G., Kim, D.K.: Design and implementation of a firewall and a packet manipulator for network simulation using SSFNet. In: Proceedings of the 7th Nordic Workshop on Secure IT Systems, Karlstad University, Sweden (2002)

A Secure Patch Distribution Architecture

Cheol-Won Lee, Eul Gyu Im, Jung-Taek Seo[1], Tae-Shik Sohn, Jong-Sub Moon[2], and Dong-Kyu Kim[3]

[1] National Security Research Institute
62-1 Hwa-am-dong, Yu-seong-gu
Daejeon, 305-718, Republic of Korea
{cheolee,imeg,seojt}@etri.re.kr
[2] CIST, Korea University
Seoul, Republic of Korea
{743zh2k,jsmoon}@korea.ac.kr
[3] Department of Computer Engineering
Ajou University, Suwon 442-749
dkkim@madang.ajou.ac.kr

Summary. Patch distribution is one of important processes to fix vulnerabilities of softwares and to ensure security of systems. Since an institute or a company has various operating systems or applications, it is not easy to update patches promptly. In this paper, we will propose a secure and consolidated patch distribution architecture with an authentication mechanism, a security assurance mechanism, a patch integrity assurance mechanism, and an automatic patch installation mechanism. We argue that the proposed architecture can allow prompt updates of patches and improve security of patch distribution processes within a domain.

Key words: Patch Distribution, Security, Patch Integrity

1 Introduction

Operating systems and application programs tend to have vulnerabilities as most of softwares do due to software bugs or design flaws, and computer incidents increases with the advance of Internet technologies because of the more widespread uses of computers. According to CERT Coordination Center, 52,658 computer incidents were reported in 2001 whereas 2134 incidents in 1997 [1]. To fix vulnerabilities and bugs of application programs and operating systems and, as a result, to reduce computer incidents, vendors of these applications and operating systems provide *patch programs* or *patches* in short to users. Prompt updates of patches would drastically reduce computer incidents. It is important for system administrators to distribute these patches securely to every host in a single domain since hosts in the same domain become more vulnerable if a patch is updated in all hosts except one host.

Patches are defined as programs that make up for weaknesses of target systems, and patches must be regularly updated in a host as soon as they are announced for systems or applications in a host. But information about required patches for a system is critical security information because attackers can easily break in the system if they know which patches are missed for the system. In addition, if a worm program is installed in a target system using corrupted patches, it will cause critical security problems in this system and systems in the same domain. Therefore, patch distribution processes are very important to ensure security in an intranet [2, 3, 4, 5].

In this paper, after researching on patch distribution procedures of various vendors, we will propose a secure patch distribution architecture for an intranet of a company. The proposed architecture supports an authentication mechanism of users and servers, a security assurance mechanism of patch distribution processes, a patch integrity assurance mechanism, and an automatic patch installation mechanism.

The rest of this paper are organized as follows: Section 2 addresses patch distribution mechanisms of vendors and previous work done by others. Our approach and a design of a secure patch distribution architecture are explain in Section 3, followed by conclusions in Section 4.

2 Previous Work and Motivations

2.1 Patch Distribution Mechanisms of Vendors

Patches are announced by software vendors to fix vulnerabilities of their products. In this section, we analyzed patch distribution mechanisms of 20 operating system vendors. It is important to provide cryptographical mechanisms, such as digital signatures, for patch distribution so that it can provide integrity, security, and authentication assurance for patches. There are two types of patch distribution: active distribution and passive distribution. In active patch distribution mechanisms, vendors notify users new patches through emails or other communication channels when new patches are available, and users download the corresponding patches from web sites or FTP sites. In passive patch distribution mechanisms, users connect to web sites or FTP sites and search necessary patches and download required patches.

The following are mechanisms used by vendors to assure authentication, integrity, and privacy.

PGP(Pretty Good Privacy) [6]: PGP signatures are provided with patches to ensure integrity of patches. A single signature can be provided for the whole package or separate signatures can be provided for different modules in the package.

HTTPs [7, 8]: HTTPs is a secure HTTP protocol based on Secure Socket Layer(SSL) to provide secure channels between servers and clients to ensure secrecy for patch distribution.

SSH(Secure SHell) [9]: Patches can be distributed using SSH to ensure secrecy.

Package: Vendors provide packages of patches so that burdens of patch update processes can be reduced.

Operating Systems	Distibution Mechanisms				
	PGP	HTTPs	SSH	Package	etc
Caldera/Open Linux	O			O	
Cobalt				O	
Compaq/Tru64 Unix		O			
Conectiva Linux	O			O	
Corel	O				
Debian Linux	O			O	
FreeBSD	O				SFSRO
HP-UX	O	O			
IBM-AIX	O	O			
Mandrake Linux	O			O	
Microsoft Windows	O				
NetBSD	O		O		
OpenBSD			O		
Redhat Linux	O			O	
SGI/IRX	O				
Sun Solaris	O	O			
Suse Linux	O			O	
Trustix	O			O	
Turbo Linux	O	O		O	
WireX/Immunix	O				

Table 1. Distribution Mechanisms of OS and Application Vendors

Table 1 shows patch distribution mechanisms of 20 operating system and application vendors. FreeBSD uses SFS read-only file system for patch distribution [10]. Among several patch distribution mechanisms, PGP signing mechanism is most widely used, but PGP signing alone cannot ensure required security properties of patch distribution procedures, such as user authentication, secrecy assurance during patch transmission. Since required patch information of a system can be used to attack the system, it is important to ensure secrecy during patch distribution. The following section summarizes possible vulnerabilities of patch distribution mechanisms of vendors.

2.2 SafePatch

The SafePatch tool [11] was developed at Lawrence Livermore National Laboratory. In this tool, a central command center downloads patches and distributes them to target systems. Even though they introduce a centralized patch management tool, the proposed their tool lacks the following.

First, their design and implementation were limited to Redhat Linux and Solaris. Since patches can be distributed for various operating systems as well as application programs, a centralized patch management tool must be able to handle various OS patches and application patches.

Second, the SafePatch tool does not have any authentication mechanisms for client or server. It is essential to provide mutual authentication between a server and a client, and, furthermore, secure channels between a server and clients must be provided to ensure secrecy and integrity of patches.

2.3 Motivations

During a patch distribution process, users assume that the patches are securely and correctly distributed to them from legitimate vendors or distributors because most of patch distribution processes are done through vendor sites. This assumption causes many problems to rise through various security vulnerabilities in patch distribution processes. The following are a list of these kinds of vulnerabilities and problems:

- Even though the key that was used by a vendor to sign patches is revoked, users may not know that the key was revoked.
- It may be difficult for users to find out that the signing key of patches is a legitimate key of vendors.
- In some cases, digital signatures are used to authenticate a part of patch instead of the whole patch.
- Harmful patches are signed with legitimate signatures.
- Signing softwares, such as PGP, or components of an operating system have trojan horse programs.

A patch distribution process can cause many problems. If a patch is announced and corresponding systems for the patch are not updated, attackers can easily find vulnerabilities of the system using the patch. To prevent attackers from taking advantages of these vulnerabilities, patches must be updated as soon as it is announced. In addition, it is important for system administrators to distribute these patches securely to every host in the same domain since hosts in the same domain become more vulnerable even though a patch is not updated in only one host.

Therefore, a new tool that can manage patch distribution processes for a domain is in need. This tool must support fast and secure distribution of patches for systems in the same domain.

3 Our Approach

A secure patch distribution architecture proposed in this paper was designed to be used in a security domain with limited size at an institute or company.

The proposed architecture includes a user authentication mechanism, a security assurance mechanism in a patch transmission process, a patch integrity assurance mechanism, and automatic patch installation in remote systems.

3.1 Secure Patch Distribution Architecture

A secure patch distribution architecture has the following components as shown in Figure 1: a patch DB, a patch distribution server, a patch manager, and patch agents.

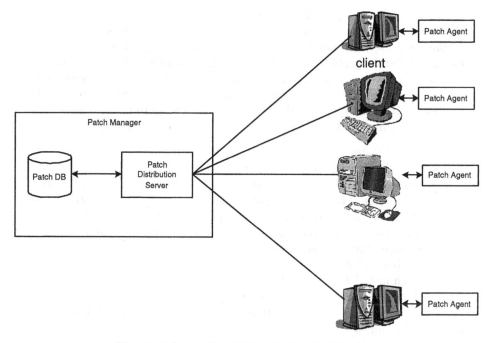

Fig. 1. A Secure Patch Distribution Architecture

Patch DB: It has profiles of client systems and users, and information about systems in the target security domain. These profiles are used to keep tracks of clients' information regarding patch updates. The patch DB has patch profiles, various patches per operating systems, patch dependency or history information, and so on. This patch DB provides a search function to users in client systems so that patch information can be looked up by users.

Patch Distribution Server: It communicates with patch agents using patch distribution protocols, fetches any required patches from the patch DB, and distributes patches to patch clients.

Patch Manager: It manages a patch DB and patch distribution servers as well as provides web-based user interfaces to administrators.

Patch Agents: A patch agent has two main parts: a patch management part and a user interface part. A patch management part is responsible for managing patches and communicating with servers. It requests required patches to the patch server, or receives patches from the patch server. Patch agents communicate with the server periodically to find new patch information. A user interface part of a patch agent provides web-based user interfaces to end users, and manages patch information through the patch management part in the client system.

Our secure patch distribution architecture has two kinds of patch distribution mechanisms: a *push mechanism* and a *pull mechanism*. In a *push mechanism*, a server sends patches to clients when new patches are read in a patch DB. Patch agents receive these new patches and install them in clients' system. In a *pull mechanism*, a client sends patch requests when its system environments are changed. The patch distribution server looks up the patch DB for new environments, and sends corresponding patches to the patch agent.

The following are detailed procedures of push and pull mechanisms.

- *Push Mechanism*
  1. Announcement of new patches by OS vendors
  2. New patches are stored in the patch DB along with their checksum information. The formats of patches are converted to a single predefined format so that patches are stored uniformly in the patch DB.
  3. Selection of clients that are required to have these new patches.
  4. Authentication of target clients
  5. Patch distribution
  6. Patch installation in the target client
  7. The client sends new patch profiles to the server
  8. The server updates patch profiles of the client in the patch DB.
- *Pull Mechanism*
  1. Change of client system environments
  2. Change of client patch profiles
  3. The client sends new patch profiles to a server
  4. The patch server searches the patch DB for new patch profiles.
  5. After mutual authentication between the patch server and the client, new patches are transmitted.
  6. Automatic patch installation in the client.
  7. The client sends new patch profiles to the patch server.
  8. The patch server updates patch profiles in the patch DB.

3.2 Patch Profile

Each client has its own patch profile, and this profile is also stored in the patch manager. A patch profile has an operating system name and a version number, application program names and version numbers, and updated patches. Patch profiles are used to find out which clients need which patches. Patch agents check the clients regularly and update their profiles.

3.3 Patch Distribution Mechanism

When distributed patches, the server needs to authenticate clients first before patches are actually distributed so that each client will get exact patches that it needs to be updated. As well as an authentication issue, there are several issues to be addressed to distribute patches securely.

Authentication Protocol

Patches that are provided to a client must be kept secret from eavesdroppers and the integrity of the provided patches must be maintained. The modified Needham-Schroeder protocol [12, 13] is used so that a client and a server can authenticate each other based on certificates [14]. This protocol enables a client and a server have an agreed session key. Secure communication channels are established based on this agreed key of the Needham-Schroeder protocol.

Client and server authentication: It is assumed that certificates of a client and the patch server are exchanged during a user registration process through an offline method.

The modified Needham-Schroeder protocol is used as follows. A represents a server and B represents a client in this example.

```
A -> B : [k1, A, r1]PB
A <- B : [k2, r1, r2]PA
A -> B : r2
```

where A and B are entities, PA and PB are public keys of A and B, r1 and r2 are random numbers, and k1 and k2 are keys to be used to generate a session key.

The server sends the client a key and a random number encrypted with the client's public key. The client replies with another key, the random number received from the server, and a random number encrypted with the server's public key. The server sends the client the received random number. Upon completion of this protocol, a client and a server are mutually authenticated.

A session key is computed as $f(k1, k2)$ using an appropriate publicly known non-reversible function f.

Once the authentication is finished, the session key is used to encrypt data between the server and the client for a certain period of time. The expiration time can be added to the session key information.

In this mechanism, a new session key is generated in each patch distribution, but this will not decrease system performance drastically because patch distribution processes are not frequent operations. In addition, it is more important to provide a secure mechanism than to provide good performance.

Secrecy and Integrity of Patches

To preserve the secrecy and the integrity of patches, the server sends patches securely using the session key after the authentication process succeeds.

A -> B : [patch, [checksum]SA]SessionKey

where SA is a secret key of A. Therefore, checksums that are encrypted with SA can be decrypted with the public key of A. Encrypted checksums may be precomputed and stored in a patch DB along with corresponding patches to improve system performance during the patch distribution process.

Integrity of patches is ensured by providing a checksum for each patch. MD5 checksums [15] are used for patches in our architecture. Secrecy of patches is obtained by encrypting a patch and its checksum with the session key. If a patch is not encrypted during transmission from a server to a client, an attacker can intercept packets and can use the patch information to attack the client.

Patch Distribution Mechanism

The patch distribution mechanism is divided into two parts: the *push* mechanism and the *pull* mechanism.

Push mechanism New patches are updated in the patch DB, and the server sends patches to clients.

```
1. SP_Server              : Selects clients for new patches
2. SP_Server -> SP_Client : [Patch Information] encrypted with the
                            public key of SP_Client
3. SP_Server <- SP_Client : Accept/Deny for the patch information
4. SP_Server <-> SP_Client : Mutual authentication
5. SP_Server -> SP_Client : Patch Transmission (encrypted with the
                            session key of step 4)
6. SP_Client              : Patch Installation
7. SP_Server <- SP_Client : [Patch Profile Information] encrypted
                            with the public key of SP_Server
8. SP_Server              : Update patch profiles
```

Pull mechanism Due to changes of system environments in a client, new patches are required.

```
1. SP_Client                      : Update patch profiles of the client
2. SP_Server <- SP_Client         : Send patch profiles
3. SP_Server -> SP_Client         : Send Accept/Deny message
4. SP_Server <-> SP_Client        : Mutual authentication
5. SP_Server -> SP_Client         : Patch Transmission (encrypted with
                                    the session key obtained from step 4)
6. SP_Client                      : Patch Installation
7. SP_Server <- SP_Client         : [Patch Profile Information] encrypted
                                    with the public key of SP_Server
8. SP_Server                      : Update patch profiles
```

3.4 Patch Automatic Installation & Management Mechanism

To install patches automatically in client systems from a server, the server uses profiles of client systems. The profile has information about OS, services, and currently installed patches, as explained before. The client composes this profile and sends it to the patch server, and the server uses these profiles to select clients when new patches are announced. Since the server has profiles of clients, it can select clients for new patches and invoke appropriate processes to install new patches. A patch agent in each client receives patches from the server and installs them. If errors occur during patch installation, the patch agent notifies users of installation errors.

4 Conclusions

In this paper, we proposed a secure and consolidated patch distribution architecture to reduce vulnerabilities in patch distribution processes. This architecture provides mutual authentication based on certificates, secure channels between a server and a client, secrecy and integrity of patches using a session key and MD5 checksums, and remote patch installation using patch profiles. This architecture can ensure more security of a secure domain by preventing leaks of patch information during patch distribution. In addition, centralized controls of the proposed architecture enable prompt updates of patches and, as a result, improve security of the domain.

The main contribution of this paper is the proposition of a secrue and consolidated patch distribution architecture that enables centralized controls of patch distribution processes. This architecture will enable prompt updates of patches and improve security of patch distribution processes within a domain.

We are currently working on the implementation of this architecture.

References

1. CERT/Coordination-Center: CERT/CC statistics,
   (http://www.cert.org/stats/cert_stats)

2. Bashar, M.A., Krishnan, G., Kuhn, M.G.: Low-threat security patches and tools. In: Proceedings of the International Conference on Software Maintenance. (1997) 306–313
3. Liu, C., Richardson, D.J.: Automated security checking and patching using TestTalk. In: Proceedings of the Fifteenth IEEE International Conference on Automated Software Engineering. (2000) 261–164
4. Michener, J.R., Mohan, S.D.: How to shrink holes in corporate data dikes. IT Professional (2002) 41–48
5. Lee, C.W., Im, E.G., Seo, J.T., Sohn, T.S., Moon, J.S., Kim, D.K.: Design of a secure and consolidated patch distribution architecture. In: Proceedings of the International Conference on Information Networking 2003. (2003)
6. Garfinkel, S.: PGP: Pretty Good Privacy. O'Reilly & Associates (1994)
7. Fielding, R., Gettys, J., Mogul, J.C., Nielsen, H.F., Masinter, L., Leach, P., Berners-Lee, T.: HyperText Transfer Protocol - HTTP/1.1. IETF Network Working Group RFC 2616 (1999)
8. Rescorla, E.: HTTP over TLS. IETF Network Working Group RFC 2818 (2000)
9. Ylonen, T., Kivinen, T., Saarinen, M., Rinne, T., Lehtinen, S.: SSH protocol architecture. IETF Network Working Group Internet-Draft draft-ietf-secsh-architecture-13.txt (2002)
10. Fu, K., Kaashoek, M.F., Mazières, D.: Fast and secure distributed read-only file system. In: Proceedings of the 4th USENIX Symposium on Operating Systems Design and Implementation (OSDI 2000), San Diego, California (2000) 181–196
11. Lawrence Livermore National Laboratory http://ciac.llnl.gov/cstc/safepatch/SafePatch_White_Paper.pdf: SafePatch. (2000)
12. Needhan, R.M., Schroeder, M.D.: Using encryption for authentication in large networks of computers. Communications of the ACM **21** (1978) 993–999
13. Menezes, A.J., van Oorschot, P.C., Vanstone, S.A.: Handbook of Applied Cryptography. CRC press (1996)
14. Rivest, R., Shamir, A., Adleman, L.: A method for obtaining digital signature and public-key cryptosystems. Communications of the ACM **21** (1978) 120–126
15. Rivest, R.: The MD5 message-digest algorithm. In: Networking Working Group RFC1321. (1992)

Intrusion Detection Using Ensemble of Soft Computing Paradigms

Srinivas Mukkamala[1], Andrew H. Sung[1,2] and Ajith Abraham[3]

{srinivas|sung}@cs.nmt.edu, ajith.abraham@ieee.org
[1]Department of Computer Science, [2]Institute for Complex Additive Systems Analysis, New Mexico Tech, Socorro, New Mexico 87801
[3]Department of Computer Science, Oklahoma State University, Tulsa, OK 74106

Abstract:

Soft computing techniques are increasingly being used for problem solving. This paper addresses using ensemble approach of different soft computing techniques for intrusion detection. Due to increasing incidents of cyber attacks, building effective intrusion detection systems (IDSs) are essential for protecting information systems security, and yet it remains an elusive goal and a great challenge. Two classes of soft computing techniques are studied: Artificial Neural Networks (ANNs) and Support Vector Machines (SVMs). We show that ensemble of ANN and SVM is superior to individual approaches for intrusion detection in terms of classification accuracy.

Keywords: Information system security, Intrusion detection, neural network, support vector machines, ensemble of soft computing paradigms

1 Introduction

This paper concerns intrusion detection and the related issue of identifying a good detection mechanism. Intrusion detection is a problem of great significance to critical infrastructure protection owing to the fact that computer networks are at the core of the nation's operational control. This paper summarizes our current work to build Intrusion Detection Systems (IDSs) using Artificial Neural Networks or ANNs [1], Support Vector Machines or SVMs [2] and the ensemble of different artificial intelligent techniques. Since the ability of a good detection technique gives more accurate results, it is critical for intrusion detection in order for the IDS to achieve maximal performance. Therefore, we study different soft computing techniques and also their ensemble for building models based on experimental data.

Since most of the intrusions can be uncovered by examining patterns of user activities, many IDSs have been built by utilizing the recognized attack and misuse patterns to develop learning machines [3,4,5,6,7,8,9]. In our recent work, SVMs are found to be superior to ANNs in many important respects of intrusion detection [9]; In this paper we will concentrate on using the ensemble of support vector machines and neural networks with different training functions to achieve better classification accuracies.

The data we used in our experiments originated from MIT's Lincoln Lab. It was developed for intrusion detection system evaluations by DARPA and is considered a benchmark for intrusion detection evaluations [10].

We performed experiments to classify each of the five classes (normal, probe, denial of service, user to super-user, and remote to local) of patterns in the DARPA data. It is shown that using the ensemble of different artificial intelligent techniques for classification gives good accuracies.

In the rest of the paper, a brief introduction to the data we used is given in section 2. In section 3 we describe the theoretical aspects of neural networks, support vector machines and the ensemble of artificial intelligent techniques. In section 4 we present the experimental results of neural networks, support vector machines and their ensemble. In section 5 we summarize our results and give a brief description of our proposed IDS architecture.

2 Intrusion Dataset

In the 1998 DARPA intrusion detection evaluation program, an environment was set up to acquire raw TCP/IP dump data for a network by simulating a typical U.S. Air Force LAN. The LAN was operated like a real environment, but being blasted with multiple attacks. For each TCP/IP connection, 41 various quantitative and qualitative features were extracted (11). Of this database a subset of 494021 data were used, of which 20% represent normal patterns.

The four different categories of attack patterns are:

a. Denial of Service (DOS) Attacks: A denial of service attack is a class of attacks in which an attacker makes some computing or memory resource too busy or too full to handle legitimate requests, or denies legitimate users access to a machine. Examples are Apache2, Back, Land, Mail bomb, SYN Flood, Ping of death, Process table, Smurf, Syslogd, Teardrop, Udpstorm.
b. User to Superuser or Root Attacks (U2Su): User to root exploits are a class of attacks in which an attacker starts out with access to a normal user account on the system and is able to exploit vulnerability to gain root access to the system. Examples are Eject, Ffbconfig, Fdformat, Loadmodule, Perl, Ps, Xterm.
c. Remote to User Attacks (R2L): A remote to user attack is a class of attacks in which an attacker sends packets to a machine over a network–but who does

not have an account on that machine; exploits some vulnerability to gain local access as a user of that machine. Examples are Dictionary, Ftp_write, Guest, Imap, Named, Phf, Sendmail, Xlock, Xsnoop.

d.   Probing (Probe): Probing is a class of attacks in which an attacker scans a network of computers to gather information or find known vulnerabilities. An attacker with a map of machines and services that are available on a network can use this information to look for exploits. Examples are Ipsweep, Mscan, Nmap, Saint, Satan.

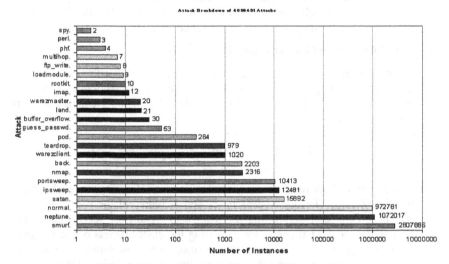

Figure 1: Intrusion detection data distribution

3 Connectionist Paradigms

Connectionist models "learn" by adjusting the interconnections between layers. When the network is adequately trained, it is able to generalize relevant output for a set of input data.

3.1. Artificial Neural Networks (ANNs)

The artificial neural network (ANN) methodology enables us to design useful nonlinear systems accepting large numbers of inputs, with the design based solely on instances of input-output relationships.

a.   **Resilient Back propagation (RP)**

The purpose of the resilient back propagation training algorithm is to eliminate the harmful effects of the magnitudes of the partial derivatives. Only the sign of the derivative is used to determine the direction of the weight update; the magnitude of the derivative has no effect on the weight update. The size of the weight change is determined by a separate update value. The update value for each weight and bias is increased by a factor whenever the derivative of the performance function with respect to that weight has the same sign for two successive iterations. The update value is decreased by a factor whenever the derivative with respect that weight changes sign from the previous iteration. If the derivative is zero, then the update value remains the same. Whenever the weights are oscillating the weight change will be reduced. If the weight continues to change in the same direction for several iterations, then the magnitude of the weight change will be increased [12].

b. **Scaled Conjugate Gradient Algorithm (SCG)**

Moller [13] introduced the scaled conjugate gradient algorithm as a way of avoiding the complicated line search procedure of conventional conjugate gradient algorithm (CGA). According to the SCGA, the Hessian matrix is approximated by

$$E''(w_k)p_k = \frac{E'(w_k + \sigma_k p_k) - E'(w_k)}{\sigma_k} + \lambda_k p_k \tag{1}$$

where E' and E'' are the first and second derivative information of global error function $E(w_k)$. The other terms p_k, σ_k and λ_k represent the weights, search direction, parameter controlling the change in weight for second derivative approximation and parameter for regulating the indefiniteness of the Hessian. In order to get a good quadratic approximation of E, a mechanism to raise and lower λ_k is needed when the Hessian is positive definite. Detailed step-by-step description can be found in [13].

c. **One-Step-Secant Algorithm (OSS)**

Quasi-Newton method involves generating a sequence of matrices $G^{(k)}$ that represents increasingly accurate approximations to the inverse Hessian (H^{-1}). Using only the first derivative information of E [14], the updated expression is as follows:

$$G^{(k+1)} = G^{(k)} + \frac{pp^T}{p^T v} - \frac{(G^{(k)}v)v^T G^{(k)}}{v^T G^{(k)}v} + (v^T G^{(k)}v)uu^T \tag{2}$$

where

$$p = w^{(k+1)} - w^{(k)}, v = g^{(k+1)} - g^{(k)}, u = \frac{p}{p^T v} - \frac{G^{(k)}v}{v^T G^{(k)}v} \tag{3}$$

and T represents transpose of a matrix. The problem with this approach is the requirement of computation and storage of the approximate Hessian matrix for every iteration. The One-Step-Secant (OSS) is an approach to bridge the gap between the conjugate gradient algorithm and the quasi-Newton (secant) approach. The OSS approach doesn't store the complete Hessian matrix; it assumes that at each iteration the previous Hessian was the identity matrix. This also has the advantage that the new search direction can be calculated without computing a matrix inverse [14].

3.2 Support Vector Machines (SVMs)

The SVM approach transforms data into a feature space F that usually has a huge dimension. It is interesting to note that SVM generalization depends on the geometrical characteristics of the training data, not on the dimensions of the input space [15]. Training a support vector machine (SVM) leads to a quadratic optimization problem with bound constraints and one linear equality constraint. Vapnik shows how training a SVM for the pattern recognition problem leads to the following quadratic optimization problem [16].

$$\text{Minimize: } W(\alpha) = -\sum_{i=1}^{l}\alpha_i + \frac{1}{2}\sum_{i=1}^{l}\sum_{j=1}^{l}y_iy_j\alpha_i\alpha_jk(x_i,x_j) \qquad (4)$$

$$\text{Subject to } \sum_{i=1}^{l}y_i\alpha_i \qquad (5)$$
$$\forall i : 0 \leq \alpha_i \leq C$$

Where l is the number of training examples α is a vector of l variables and each component α_i corresponds to a training example (x_i, y_i). The solution of (4) is the vector α^* for which (4) is minimized and (5) is fulfilled.

3.2 Ensemble of Soft Computing Paradigms

Optimal linear combination of neural networks has been investigated and has found to be very useful **Error! Reference source not found.**. The optimal weights were decided based on the ordinary least squares regression coefficients in an attempt to minimize the mean squared error.   The problem becomes more complicated when we have to optimize several other error measures. In the case of intrusion detection, our task is to design a classifier, which could give the best accuracy for each category of attack patterns. The first step is to carefully construct the different connectional models to achieve the best generalization performance for classifiers. Test data is then passed through these individual models and the corresponding outputs are recorded. Suppose the classification performance given by SVM, ANN (RP), ANN (SCG) and ANN (OSS) are a_n, b_n, c_n and d_n

respectively and the corresponding desired value is x_n. Our task is to combine a_n, b_n, c_n and d_n so as to get the best output value that maximizes the classification accuracy. The following ensemble approach was used. Determine the individual absolute error differences (example, $\left|x_n - a_n\right|$) and use the output value corresponding to the lowest absolute difference $\min\left|a_n - x_n\right|, \left|b_n - x_n\right|, \left|c_n - x_n\right|, \left|d_n - x_n\right|$. The approach is depicted in Figure 2.

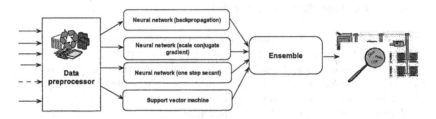

Figure 2. Ensemble approach to combine intelligent paradigms for IDS

4 Experiments

In our experiments, we perform 5-class classification. The (training and testing) data set contains 11982 randomly generated points from the five classes, with the number of data from each class proportional to its size, except that the smallest class is completely included. The normal data belongs to class1, probe belongs to class 2, denial of service belongs to class 3, user to super user belongs to class 4, remote to local belongs to class 5. A different randomly selected set of 6890 points of the total data set (11982) is used for testing different soft computing techniques.

4.1 Experiments using Neural Networks

The same data set describe in section 2 is being used for training and testing different neural network algorithms. The set of 5092 training data is divided in to five classes: normal, probe, denial of service attacks, user to super user and remote to local attacks. Where the attack is a collection of 22 different types of instances that belong to the four classes described in section 2, and the other is the normal data. In our study we used two hidden layers with 20 and 30 neurons each and the networks were trained using RP, SCG and OSS algorithms.

The network was set to train until the desired mean square error of 0.001 was met. During the training process the goal was met at 303 epochs for SCG, 66 epochs for RP and 638 epochs for OSS.

As multi-layer feed forward networks are capable of multi-class classifications, we partition the data into 5 classes (Normal, Probe, Denial of Service, and User to Root and Remote to Local).We used the same testing data (6890), same network architecture and same activations functions to identify the best training function that plays a vital role for in classifying intrusions. Table 1 summarizes the results of three different networks: network using SCG performed with an accuracy of 95.25%; network using RP achieved an accuracy of 97.04%; network using OSS performed with an accuracy of 93.60%.

Table 1: Performance of Different Neural Network Training Functions

Training algorithm	No of Epochs Trial 1	No of Epochs Trial 2	Accuracy (%) Trail 1	Accuracy (%) Trail 1
RP	67	66	97.04	95.44
SCG	351	303	80.87	95.25
OSS	638	638	93.60	93.60

Table 2: Performance of the Best Neural Network Training Algorithm (RP)

	Normal	Probe	DoS	U2Su	R2L	%
Normal	1394	5	1	0	0	99.6
Probe	49	649	2	0	0	92.7
DoS	3	101	4096	2	0	97.5
U2Su	0	1	8	12	4	48.0
R2L	0	1	6	21	535	95.0
%	96.4	85.7	99.6	34.3	99.3	

The top-left entry of Table 2 shows that 1394 of the actual "normal" test set were detected to be normal; the last column indicates that 99.6 % of the actual "normal" data points were detected correctly. In the same way, for "Probe" 649 of the actual "attack" test set were correctly detected; the last column indicates that 92.7% of the actual "Probe" data points were detected correctly. The bottom row shows that 96.4% of the test set said to be "normal" indeed were "normal" and 85.7% of the test set classified, as "probe" indeed belongs to Probe. The overall accuracy of the classification is 97.04 with a false positive rate of 2.76% and false negative rate of 0.20 %.

4.2 Experiments using Support Vector Machines

The data set described in section 4 is being used to test the performance of support vector machines. Note the same training test (5092) used for training the neural networks and the same testing test (6890) used for testing the neural networks are being used to validate the performance.

Because SVMs are only capable of binary classifications, we will need to employ five SVMs, for the 5-clas classification problem in intrusion detection, respectively. We partition the data into the two classes of "Normal" and "Rest" (Probe, DoS, U2Su, R2L) patterns, where the Rest is the collection of four classes of attack instances in the data set. The objective is to separate normal and attack patterns. We repeat this process for all classes. Training is done using the RBF (radial bias function) kernel option; an important point of the kernel function is that it defines the feature space in which the training set examples will be classified. Table 3 summarizes the results of the experiments:

Table 3: Performance of SVMs for 5 class Classifications

Class	Training Time (sec)	Testing Time (sec)	Accuracy (%)
Normal	7.66	1.26	99.55
Probe	49.13	2.10	99.70
DOS	22.87	1.92	99.25
U2Su	3.38	1.05	99.87
R2L	11.54	1.02	99.78

4.3 Experiments using Ensemble of Soft Computing paradigms

Different soft computing paradigms are carefully constructed to achieve the best generalization performance for classifiers. Test data is then passed through these individual models and the corresponding outputs are recorded. Table 4 summarizes the test results achieved for the five-class classification using 3 different neural networks, support vector machines and the ensemble of all the four.

Table 4: Performance Comparison of Testing for 5 class Classifications

Class	SVMs Accuracy (%)	RP Accuracy (%)	SCG Accuracy (%)	OSS Accuracy (%)	Ensemble Accuracy (%)
Normal	98.42	99.57	99.57	99.64	99.71
Probe	98.57	92.71	85.57	92.71	99.86
DoS	99.11	97.47	72.01	91.76	99.95
U2Su	64	48	0	16	76
R2L	97.33	95.02	98.22	96.80	99.64

5 Summary and Conclusions

Our research has clearly shown the importance of using ensemble approach for modeling intrusion detection systems. An ensemble helps to indirectly combine the synergistic and complementary features of the different learning paradigms without any complex hybridization. Since all the considered performance measures could be optimized such systems could be helpful in several real world applications.

- A number of observations and conclusions are drawn from the results reported:
- The ensemble approach out performs both SVMs and ANNs in the important respect of classification accuracies for all the five classes.
- If proper soft computing paradigms are chosen, their ensemble might help in gaining 100% classification accuracies.
- SVMs outperform ANNs in the important respects of scalability (SVMs can train with a larger number of patterns, while would ANNs take a long time to train or fail to converge at all when the number of patterns gets large); training time and running time (SVMs run an order of magnitude faster); and prediction accuracy.
- Resilient back propagation achieved the best performance among the neural networks in terms of accuracy (97.04 %) and training (67 epochs).

We note, however, that the difference in accuracy figures tend to be very small and may not be statistically significant, especially in view of the fact that the 5 classes of patterns differ in their sizes tremendously. More definitive conclusions can only be made after analyzing more comprehensive sets of network traffic data.

Acknowledgements

Support for this research received from ICASA (Institute for Complex Additive Systems Analysis, a division of New Mexico Tech) and a U.S. Department of Defense IASP capacity building grant is gratefully acknowledged. We would also like to acknowledge many insightful conversations with Dr. Jean-Louis Lassez and David Duggan that helped clarify some of our ideas.

References

[1]   Hertz J., Krogh A., Palmer, R. G. (1991) "Introduction to the Theory of Neural Computation, " Addison –Wesley.
[2]   Joachims T. (1998) "Making Large-Scale SVM Learning Practical," LS8-Report, University of Dortmund, LS VIII-Report.

[3]   Denning D. (Feb. 1987) "An Intrusion-Detection Model," *IEEE Transactions on Software Engineering*, Vol.SE-13, No 2.

[4]   Kumar S., Spafford E. H. (1994) "An Application of Pattern Matching in Intrusion Detection," *Technical Report CSD-TR-94-013*. Purdue University.

[5]   Ghosh A. K. (1999). "Learning Program Behavior Profiles for Intrusion Detection," *USENIX*.

[6]   Cannady J. (1998) "Artificial Neural Networks for Misuse Detection," *National Information Systems Security Conference*.

[7]   Ryan J., Lin M-J., Miikkulainen R. (1998) "Intrusion Detection with Neural Networks," *Advances in Neural Information Processing Systems* 10, Cambridge, MA: MIT Press.

[8]   Debar H., Dorizzi. B. (1992) "An Application of a Recurrent Network to an Intrusion Detection System," *Proceedings of the International Joint Conference on Neural Networks*, pp.78-83.

[9]   Mukkamala S., Janoski G., Sung A. H. (2002) "Intrusion Detection Using Neural Networks and Support Vector Machines," *Proceedings of IEEE International Joint Conference on Neural Networks*, pp.1702-1707.

[10]   http://kdd.ics.uci.edu/databases/kddcup99/task.htm.

[11]   J. Stolfo, Wei Fan, Wenke Lee, Andreas Prodromidis, and Philip K. Chan "Cost-based Modeling and Evaluation for Data Mining With Application to Fraud and Intrusion Detection," *Results from the JAM Project by Salvatore*.

[12]   Riedmiller, M., and H. Braun, "A direct adaptive method for faster back propagation learning: The RPROP algorithm," *Proceedings of the IEEE International Conference on Neural Networks*, San Francisco, 1993.

[13]   Moller A F, A Scaled Conjugate Gradient Algorithm for Fast Supervised Learning, Neural Networks, Volume (6), pp. 525-533, 1993.

[14]   Bishop C. M, *Neural Networks for pattern recognition*, Oxford Press, 1995.

[15]   Joachims T. (2000) "SVMlight is an Implementation of Support Vector Machines (SVMs) in C," http://ais.gmd.de/~thorsten/svm_light. University of Dortmund. Collaborative *Research Center on Complexity Reduction in Multivariate Data (SFB475)*.

[16]   Vapnik V. The Nature of Statistical Learning Theory. Springer-Verlag, New York, 1995.

[17]   Hashem, S., Optimal Linear Combination of Neural Networks, Neural Network, Volume 10, No. 3. pp. 792-994, 1995.

Part VI

Data mining, Knowledge Management and Information Analysis

NETMARK: Adding Hierarchical Object to Relational Databases with "Schema-less" Extensions

David A. Maluf
NASA Ames Research Center, Mail Stop 269-4
Moffett Field, California, USA
maluf@ptolemy.arc.nasa.gov

and

Peter B. Tran
QSS Group, Inc.
NASA Ames Research Center, Mail Stop 269-4
Moffett Field, California, USA
pbtran@mail.arc.nasa.gov

Abstract. Object-Relational database management system is an integrated hybrid cooperative approach to combine the best practices of both the relational model utilizing SQL queries and the object-oriented, semantic paradigm for supporting complex data creation. In this paper, a highly scalable, information on demand database framework, called NETMARK, is introduced. NETMARK takes advantages of the Oracle 8i object-relational database using physical addresses data types for very efficient keyword search of records spanning across both context and content. NETMARK was originally developed in early 2000 as a research and development prototype to solve the vast amounts of unstructured and semi-structured documents existing within NASA enterprises. Today, NETMARK is a flexible, high-throughput open database framework for managing, storing, and searching unstructured or semi-structured arbitrary hierarchal models, such as XML and HTML.

1 Introduction

During the early years of database technology, there were two opposing research and development directions, namely the relational model originally formalized by Codd [1] in 1970 and the object-oriented, semantic database model [2][3]. The traditional relational model revolutionized the field by separating logical data representation from physical implementation. The relational model has been developed into a mature and proven database technology holding a majority stake of the commercial database market along with the official standardization of the Structured Query Language

(SQL)[1] by ISO and ANSI committees for a user-friendly data definition language (DDL) and data manipulation language (DML).

The semantic model leveraged off from the object-oriented paradigm of programming languages, such as the availability of convenient data abstraction mechanisms, and the realization of the *impedance mismatch* [4] dilemma faced between the popular object-oriented programming languages and the underlining relational database management systems (RDBMS). Impedance mismatch here refers to the problem faced by both database programmers and application developers, in which the way the developers structure data is not the same as the way the database structures it. Therefore, the developers are required to write large and complex amounts of object-to-relational mapping code to convert data, which is being inserted into a tabular format the database can understand. Likewise, the developers must convert the relational information returned from the database into the object format developers require for their programs. Today, in order to solve the impedance mismatch problem and take advantage of these two popular database models, commercial enterprise database management systems (DBMS), such as Oracle, IBM, Microsoft, and Sybase, have an integrated hybrid cooperative approach of an *object-relational model* [5].

In order to take advantage of the object-relational (OR) model defined within an *object-relational database management system* (ORDBMS) [5][6], a standard for common data representation and exchange is needed. Today, the emerging standard is the *eXtensible Markup Language* (XML) [7][8] known as the next generation of HTML for placing structure within documents. Within any large organizations and enterprises, there are vast amounts of heterogeneous documents existing in HTML web pages, word processing, presentation, and spreadsheet formats. The traditional document management system does not provide an easy and efficient mechanism to store, manage, and query the relevant information from these heterogeneous and complex data types.

To solve the vast quantities of heterogeneous and complex documents existing within NASA enterprises, NASA at Ames Research Center initially designed and developed an innovative schema-less, object-relational database integration technique and framework referred to hereby as *NETMARK*. Developed in early 2000 as a rapid, proof-of-concept prototype, NETMARK, today, is a highly scalable, open enterprise database framework (architecture) for dynamically transforming and generating arbitrary schema representations from unstructured and/or semi-structured data sources. NETMARK provides automatic data management, storage, retrieval, and *discovery* [17] in transforming large quantities of highly complex and constantly changing heterogeneous data formats into a well-structured, common standard.

This paper describes the NETMARK schema-less database integration technique and architecture for managing, storing, and searching unstructured and/or semi-structured documents from standardized and interchangeable formats, such as XML in relational database systems. The unique features of NETMARK take advantages from the object-relational model and the XML standard described above, along with an

---

open, extensible database framework in order to dynamically generate arbitrary schema stored within relational databases, object-relational database management system.

2   BACKGROUND

2.1 Object-Relational DBMS

The object-relational model takes the best practices of both relational and object-oriented, semantic views to decouple the complexity of handling massively rich data representations and their complex interrelationships. ORDBMS employs a data model that attempts to incorporate object-oriented features into traditional relational database systems. All database information is still stored within relations (tables), but some of the tabular attributes may have richer data structures. It was developed to solve some of the inadequacies associated with storing large and complex multimedia objects, such as audio, video, and image files, within traditional RDBMS. As an intermediate hybrid cooperative model, the ORDBMS combined the flexibility, scalability, and security of using existing relational systems along with extensible object-oriented features, such as data abstraction, encapsulation, inheritance, and polymorphism.

In order to understand the benefits of ORDBMS, a comparison of the other models need to be taken into consideration. The 3x3 database application classification matrix [6] shown in Table 1 displays the four categories of general DBMS applications— simple data without queries (file systems), simple data with queries (RDBMS), complex data without queries (OODBMS), and complex data with queries (ORDBMS). For the upper left-handed corner of the matrix, traditional business data processing, such as storing and managing employee information, with simple normalized attributes, such as numbers (integers or floats) and character strings, usually needs to utilize SQL queries to retrieve relevant information. Thus, RDBMS is well suited for traditional business processing; but this model cannot store complex data, such as word processing documents or geographical information. The lower right-handed corner describes the use of persistent object-oriented languages to store complex data objects and their relationships. The lower right-handed corner represents OODBMS, which either have very little SQL-like queries support or none at all. The upper right-handed corner with the dark gray colored cell is well suited for complex and flexible database applications that need complex data creation, such as large objects to store word processing documents, and SQL queries to retrieve relevant information from within these documents. Therefore, the obvious choice for NETMARK is the upper right-handed corner with the dark gray colored cell as indicated in Table 1.

Query	RDBMS (Traditional Business Data Processing)	ORDBMS (NETMARK)
No Query	File Systems (Simple Text Editors)	OODBMS (Persistent OO Languages)
	Simple Data	Complex Data

Adapted from M. Stonebraker, "Object-Relational DBMS - The Next Wave",
Informix Software (now part of the IBM Corp. family), Menlo Park, CA

Table 1: Database Application Classification Matrix

The main advantages of ORDBMS, are scalability, performance, and widely supported by vendors. ORDBMS have been proven to handle very large and complex applications, such as the NASDAQ stock exchange, which contains hundreds of gigabytes of richly complex data for analyst and traders to query stock data trends. In terms of performance, ORDBMS supports query optimization, which are comparable to RDBMS and out performs most OODBMS. Therefore, there is a very large market and future for ORDBMS. This was another determining factor for using ORDBMS for the NETMARK project. Most ORDBMS supports the SQL3 [9] specifications or its extended form. The two basic characteristics of SQL3 are crudely separated into its "relational features" and its "object-oriented features". The relational features for SQL3 consist of new data types, such as large objects or LOB and its variants. The object-oriented features of SQL3 include structured user-defined types called *abstract data types* (ADT) [10][11] which can be hierarchical defined (inheritance feature), invocation routines called methods, and REF types that provides reference values for unique row objects defined by *object identifier* (OID) [11] which is a focus of this paper.

2.2 Large Objects

ORDBMS was developed to solve some inadequacies associated with storing large and complex data. The storage solution within ORDBMS is the large object data types called LOBs [9][11]. There are several LOB variants, namely binary data (BLOB), single-byte character data set (CLOB), multi-byte character data (NCLOB), and binary files (BFILE). BLOBs, CLOBs, and NCLOBs are usually termed *internal LOBs* [11], because they are stored internally within the database to provide efficient, random, and piece-wise access to the data. Therefore, the data integrity and concurrency of external BFILEs are usually not guaranteed by the underlining ORDBMS. Each LOB contains both the data value and a pointer to the data called the *LOB locator* [9]. The LOB locator points to the data location that the database creates to hold the LOB data. NETMARK uses LOBs to store large documents, such as word processing, presentation, and spreadsheet files, for later retrieval of the document and its contents for rendering and viewing.

2.3 Structuring Documents with XML

XML is known as the next generation of HTML and a simplified subset of the Standard Generalized Markup Language (SGML)[2]. XML is both a semantic and structured markup language [7]. The basic principle behind XML is simple. A set of meaningful, user-defined tags surrounding the data elements describes a document's structure as well as its meaning without describing how the document should be formatted [12]. This enables XML to be a well-suitable meta-markup language for handling loosely structured or *semi-structured data*, because the standard does not place any restrictions on the tags or the nesting relationships. Loosely structured or semi-structured data here refers to data that may be irregular or incomplete, and its structure is rapidly changing and unpredictable [12]. Good examples of semi-structured data are web pages and constantly changing word processing documents being modified on a weekly or monthly basis.

XML encoding, although more verbose, provides the information in a more convenient and usable format from a data management perspective. In addition, the XML data can be transformed and rendered using simple *eXtensible Stylesheet Language* (XSL) specifications [8]. It can be validated against a set of grammar rules and logical definitions defined within the *Document Type Definitions* (DTDs) or *XML Schema* [14] much the same functionality as a traditional database schema.

2.4 Oracle ROWIDs

ROWID is an Oracle data type that stores either physical or logical addresses (row identifiers) to every row within the Oracle database. Physical ROWIDs store the addresses of ordinary table records (excluding indexed-organized tables), clustered tables, indexes, table partitions and sub-partitions, index partitions and sub-partitions, while logical ROWIDs store the row addresses within indexed-organized tables for building secondary indexes. Each Oracle table has an implicit pseudo-column called ROWID, which can be retrieved by a simple SELECT query on the particular table. Physical ROWIDs provide the fastest access to any record within an Oracle table with a single read block access, while logical ROWIDs provide fast access for highly volatile tables. A ROWID is guaranteed to not change unless the rows it references is deleted from the database.

The physical ROWIDs have two different formats, namely the legacy *restricted* and the new *extended* ROWID formats. The restricted ROWID format is for backward compatibility to legacy Oracle databases, such as Oracle 7 and/or earlier releases. The extended format is for Oracle 8 and later object-relational releases. This

---

[2] The Standard Generalized Markup Language (SGML) is the official International Standard (ISO 8879) adopted by the world's largest producers of documents, but is very complex. Both XML and HTML are subsets of SGML.

paper will only concentrate on extended ROWID format, since NETMARK was developed using Oracle 8i (release 8.1.6). For example, the following displays a subset of the extended ROWIDs from a NETMARK generated schema. It is a generalized 18-character format with 64 possibilities each:

AAAAAA | BBB | CCCCCC | DDD

The extended ROWIDs could be used to show how an Oracle table is organized and structured; but more importantly, extended ROWIDs make very efficient and stable unique keys for information retrievals, which will be addressed in the following subsequent section below (3.3).

3   THE NETMARK APPROACH

Since XML is a document and not a data model per se, the ability to map XML-encoded information into a true data model is needed. The NETMARK approach allows this to occur by employing a customizable data type definition structure defined by the NETMARK SGML parser to model the hierarchical structure of XML data regardless of any particular XML document schema representation. The customizable NETMARK data types simulate the Document Object Model (DOM) Level 1 specifications [15] on parsing and decomposition of element nodes. The SGML parser is more efficient on decomposition than most commercial DOM parsers, since it is much more simpler as defined by node types contained within configuration files. The node data type format is based on a simplified variant of the *Object Exchange Model* (OEM) [13] researched at Stanford University, which is very similar to XML tags. The node data type contains an object identifier (node identifier) and the corresponding data type. Traditional object-relational mapping from XML to relational database schema models the data within the XML documents as a tree of objects that are specific to the data in the document [14]. In this model, element type with attributes, content, or complex element types are generally modeled as classes. Element types with parsed character data (PCDATA) and attributes are modeled as scalar types. This model is then mapped to the relational database using traditional object-relational mapping techniques or via SQL3 object views. Therefore, classes are mapped to tables, scalar types are mapped to columns, and object-valued properties are mapped to key pairs (both primary and foreign). This traditional mapping model is limited since the object tree structure is different for each set of XML documents. On the other hand, the NETMARK SGML parser models the document itself (similar to the DOM), and its object tree structure is the same for all XML documents. Thus, NETMARK is designed to be independent of any particular XML document schemas and is termed to be schema-less.

NETMARK is even flexible to handle more than just XML. It is also a SGML-enabled, open enterprise database framework. The term SGML-enabled means NETMARK supports both HTML and XML sets of tags through a set of customizable configuration files utilized by the NETMARK SGML parser for dynamically generating arbitrary database schema as shown in section (3.2) and in Figure 2. The

NETMARK SGML parser decomposes either the HTML or XML document into its constituent nodes and inserted the nodes as individual records within Oracle tables. This dynamic schema representation and generation without requiring to write tedious and cumbersome SQL scripts or having to depend on experienced database administrators (DBAs) saves both time and valuable resources. Thus, this makes the storage model of NETMARK a general-purpose HTML or XML storage system.

3.1 Architecture

The NETMARK architecture comprises of the distributed, *information on demand* model, which refers to the "plug and play" capabilities to meet high-throughput and constantly changing information management environment. Each NETMARK modules are extensible and adaptable to different data sources. NETMARK consists of (1) a set of interfaces to support various communication protocols (such as HTTP, FTP, RMI-IIOP and their secure variants), (2) an information bus to communicate between the client interfaces and the NETMARK core components, (3) the daemon process for automatic processing of inputs, (4) the NETMARK keyword search on both document context and content, (5) a set of extensible application programming interfaces (APIs), (6) and the Oracle backend ORDBMS.

The three core components of NETMARK consist of the high-throughput information bus, the asynchronous daemon process, and the set of customizable and extensible APIs built on Java enterprise technology (J2EE) and Oracle PL/SQL stored procedures and packages [11]. The NETMARK information bus allows virtually three major communication protocols heavily used today—namely HTTP web-based protocol and its secure variant, the File Transfer Protocol (FTP) and its secure variant, and the new Remote Method Invocation (RMI) over Internet Inter-Orb Protocol from the Object Management Group (OMG) Java-CORBA standards—to meet the information on demand model. The NETMARK daemon is a unidirectional asynchronous process to increase performance and scalability compared to traditional synchronous models, such as Remote Procedure Call (RPC) or Java RMI mechanisms. The NETMARK set of extensible Java and PL/SQL APIs are used to enhance database access and data manipulation, such as a robust Singleton database connection pool for managing check-ins and checkouts of pre-allocated connection objects.

3.2 Universal Process Flow

The NETMARK closed-loop universal process flow is shown in Figure 1. The information bus comprises of an Apache HTTP web server integrated with Tomcat Java-based JSP/Servlet container engine. It waits for incoming requests from the various clients, such as an uploaded word processing document from a web browser. The bus performs a series of conversion and transformation routines from one specific format to another using customized scripts. For instance, the NETMARK information bus will automatically convert a semi-structured Microsoft Word document into an inter-lingual HTML or XML format. A copy of the original word document, the converted HTML or XML file, and a series of dynamically generated configuration files will be handed to the NETMARK daemon process.

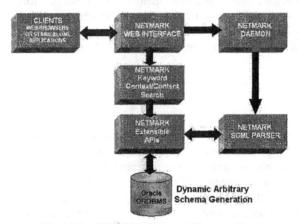

Figure 1: NETMARK Universal Process Flow

The daemon process checks for configuration files, the original processed files, and notifies the NETMARK SGML parser for decomposition of document nodes and data insertion. The daemon has an automatic logger that outputs both successful and event errors by date and time stamps with periodical archival and cleanup of log files. The daemon accepts three types of configuration files—(1) the request file, (2) the HTML/XML configuration file, and (3) the metadata configuration file. The request file is required by the daemon to proceed to process the correct information, whereas the HTML/XML configuration file and the metadata file are optional. If there is no HTML/XML configuration file provided to the daemon, a default configuration file located on the server is used. If there is no request file, the daemon issues an appropriate error message, logs the message to the log files for future reference, performs cleanup of configuration files, and waits for the next incoming request.

If the daemon can read the request file from the incoming request directory, it locks the file and extracts the name-value pairs from the request file for further processing. After extraction of the relevant attribute values, the request file is unlocked and a child process is spawned to process the incoming files. The child process locks the request file again to prevent the parent process from reprocessing the same request file and calling the SGML parser twice to decompose and insert the same document. The child process then calls the NETMARK SGML parser with the appropriate flag options to decompose the HTML or XML document into its constituent nodes and insert the nodes into the specified database schema. After the parsing and insertion completes, the source, result, and metadata files along with its corresponding configuration files will be cleanup and deleted by the daemon.

The NETMARK SGML parser decomposes the HTML or XML documents into its constituent nodes and dynamically inserts them into two primary database tables—namely, XML and DOC—within a NETMARK generated schema. The descriptions of the XML and DOC tables along with their respective relationships are listed in Figure 2. The SGML parser is governed by five different node data types, which are specified in the HTML or XML configuration files passed by the daemon. The five NETMARK node data types and their corresponding node type identifier as designated in the NODETYPE column of the XML table are as follows: (1) ELEMENT, (2) TEXT, (3) CONTEXT, (4) INTENSE, and (5) SIMULATION.

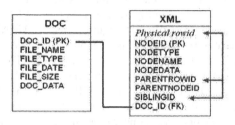

Figure 2: NETMARK Generated Schema

The node type identifier is a single character data type inserted by the SGML parser to the XML table for each decomposed XML or HTML nodes. The node type identifiers will be used in the keyword-based context and content search.

The XML table contains the node tree structure as specified by the rules governed by the HTML or XML configuration files being used by the SGML parser to decompose the original HTML or XML documents into its constituent nodes. The DOC table holds the source document metadata, such as FILE_NAME, FILE_TYPE, FILE_DATE, and FILE_SIZE. Each NETMARK generated schema contains these two primary tables for efficient information retrievals as explained in the subsequent section (3.3). In order to store, manipulate, and later on retrieve unstructured or semi-structured documents, such as word processing files, presentations, flat text files, and spreadsheets, NETMARK utilizes the LOB data types as described in section (2.2) to store a copy of each processed document. In Figure 2 both the XML and the DOC table utilize CLOB and BLOB data types, respectively within the NODEDATA attribute for the XML table and the DOC_DATA column for the DOC table.

3.3 NETMARK Keyword Search

There are two ways that the Oracle database performs queries—either by a costly full table scan (with or without indexes) or by ROWIDs. Since a ROWID gives the exact physical location of the record to be retrieved by a single read block access, this is much more efficient as the database table size increases. As implied in the earlier section (2.4), ROWIDs can be utilized to very efficiently retrieve records by using them as unique keys. The NETMARK keyword search takes advantage of the unique extended ROWIDs for optimizing record retrievals based on both context and content. The keyword-based search here refers to finding all objects (elements or attributes) whose tag, name, or value contains the specified search string.

The NETMARK keyword search is built on top of Oracle 8i *interMedia Text* index [11][17] for retrieving the search key, and it is based on the Object Exchange Model [13] researched at Stanford University as mentioned earlier in section (3). Oracle interMedia is also known as Oracle Text in later releases of Oracle 9i and formerly known as the ConText [16] data cartridge. Oracle interMedia text index creates a series of index tables within the NETMARK generated schema to support the keyword text queries. The interMedia text index is created on the NODEDATA column of the XML table as shown in Figure 2. The NODEDATA column is a CLOB data type (character data). As described in Figure 2, the NETMARK XML table is consisted of eight attributes (columns) plus one physical ROWID pseudo-column. Each row in the XML table describes a complete XML or HTML node. The main attributes being utilizing by the search are DOC_ID, NODENAME, NODETYPE, NODEDATA, PARENTROWID, and SIBLINGID from the XML table. The DOC_ID column is used to refer back to the original document file. As the name implies, NODENAME contains the name of the node; whereas NODETYPE, as described earlier in section (3.2), identifies the type of node it is and informs NETMARK how to process this particular node. Reiterating the five specialized node data types: (1) ELEMENT, (2) TEXT, (3) CONTEXT, (4) INTENSE, and (5) SIMULATION. TEXT is a node whose data are free text or blocks of text describing a specific content. An ELEMENT, similar to a HTML or XML element, can contain multiple TEXT nodes and/or other nested ELEMENT nodes. Within NETMARK search, CONTEXT is a parent ELEMENT whose children elements contain data describing the contents of the following sibling nodes. INTENSE is another CONTEXT, which itself contains meaningful data. SIMULATION is a node data type reserved for special purposes and future implementation. NODEDATA is an Oracle CLOB data type used to store TEXT data. PARENTROWID and SIBLINGID are used as references for identifying the parent node and sibling node, respectively, and are of data type ROWID.

The NETMARK keyword-based context and content search is performed by first querying text index for the search key. Each node returned from the index search is then processed based on its designated unique ROWID. The processing of the node involves traversing up the tree structure via its parent or sibling node until the first context is found. The context is identified via its corresponding NODETYPE. The context refers to here as a heading for a subsection within a HTML or XML document, similar to the <H1> and <H2> header tags commonly found within HTML pages. Thus, the context and content search returns a subsection of the document where the keyword being searched for occurs. Once a particular CONTEXT is found, traversing back down the tree structure via the sibling node retrieves the corresponding content text. The search result is then rendered and displayed appropriately.

4   CONCLUSION

NETMARK provides an extensible, schema-less, information on demand framework for managing, storing, and retrieving unstructured and/or semi-structured data.

NETMARK was initially designed and developed as a rapid, proof-of-concept prototype using a proven and mature Oracle backend object-relational database to solve the vast amounts of heterogeneous documents existing within NASA enterprises. NETMARK is currently a scalable, high-throughput open database framework for transforming unstructured or semi-structured documents into well-structured and standardized XML and/or HTML formats.

5   ACKNOWLEDGEMENT

The authors of this paper would like to acknowledge NASA Information Technology-Based program and NASA Computing, Information, and Communication Technologies Program.

References

1. Codd, E. F.: A Relational Model of Data for Large Shared Data Banks, Communications of the ACM, 13(6), (1970) 377-387
2. Hull, R. and King, R.: "Semantic Database Modeling: Survey, Applications, and Research Issues", ACM Computing Surveys, 19(3), (1987) 201-260
3. Cardenas, A. F., and McLeod, D. (eds), Research Foundations in Object-Oriented and Semantic Database Systems, New York: Prentice-Hall, (1990) 32-35
4. Chen, J. and Huang, Q.: Eliminating the Impedance Mismatch Between Relational Systems and Object-Oriented Programming Languages, Australia: Monash University (1995)
5. Devarakonda, R. S.: Object-Relational Database Systems – The Road Ahead, ACM Crossroads Student Magazine, (February 2001)
6. Stonebraker M.: Object-Relational DBMS: The Next Wave, Informix Software (now part of the IBM Corp. family), Menlo Park, CA
7. Harold, E. R.: XML: Extensible Markup Language, New York: IDG Books Worldwide, (1998) 23-55
8. Extensible Markup Language (XML) World Wide Web Consortium (W3C) Recommendation, (October 2000)
9. Eisenberg, A., and Melton, J.: SQL:1999, formerly known as SQL3, (1999)
10. ISO/IEC 9075:1999, "Information Technology—Database Language—SQL—Part 1: Framework (SQL/Framework)", (1999)
11. Loney, K. and Koch, G.: Oracle 8i: The Complete Reference, 10th edition, Berkeley, CA: Oracle Press Osborne/McGraw-Hill, (2000) 69-85; 574-580; 616-644; 646-663
12. Widom, J.: Data Management for XML Research Directions, Stanford University, (June 1999) http://www-db.stanford.edu/lore/pubs/index.html
13. Lore XML DBMS project, Stanford University (1998) http://www-db.stanford.edu/lore/research/
14. Bourret, R.: Mapping DTD to Databases, New York: O'Reilly & Associates, (2000)
15. Wood, L. et al.: "Document Object Model (DOM) Level 1 Specification", W3C Recommendation, (October 1998)
16. Oracle8 ConText Cartridge Application Developer's Guide (Release 2.4), Princeton University Oracle Reference (1998)
17. Maluf, D. A. and Tran, P. B.: "Articulation Management for Intelligent Integration of Information", IEEE Transactions on Systems, Man, and Cybernetics Part C: Applications and Reviews, 31(4) (2001) 485-496

Academic KDD Project LISp-Miner

Milan Šimůnek, simunek@vse.cz

Faculty of Informatics and Statistics, University of Economics Prague
Institute of Computer Sciences, Czech Academy of Sciences, Czech Republic

Abstract. An academic KDD system is introduced. This paper describes its architecture and project management for purpose of those interested to join the project team either as developer or user. Some possible directions of future research are also included.

1 Introduction

In this paper we are going to introduce the LISp-Miner system as a tool for KDD. It is aimed at solution of some of current problems in the area of data mining – developing new procedures and tools with stress on clear interpretation of results. This system is developed under the LISp-Miner project that adds necessary framework for work coordination and for maintaining compatibility among modules. The main features of introduced system are:

- LISp-Miner is a free and open system developed by teachers and students of the University of Economics Prague. The system can be downloaded at http://lispminer.vse.cz where the user's documentation can also be found.
- The system is developed under project management with clear project procedures and rules for analysis, implementation and writing documentation. When implementing a new module, appropriate documentation of interfaces to other modules is available for developer.
- The LISp-Miner system is suitable for teaching and research in the area of KDD. It can be used for mid-size KDD projects, pilot studies in data-mining activities and students' projects.
- Analyzed data are represented by suitable bit-strings. This approach results in a very fast mining process. Several already realized and developed modules are based on strings of bits.
- The modular architecture of the system allows other modules to be easily developed. It is also possible to create a "tailor-made" interfaces customized for domain experts in medicine, sociology, finance etc. We invite other people to participate in further development of the system.
- Particular modules of the system can be embedded in larger systems.

The LISp-Miner system is result of work of several people that contribute in developing particular modules. It is based on long time experiences in the area of data mining. Theoretical foundations were published in articles and books dated back to 1960 – see e.g. [2], [5], [4], [6], [3], [1], [15]. Older versions of some procedures were first implemented as standalone applications by their authors. The LISp-Miner system further improves functionality of included data mining procedures, new mining procedures are being added and the system benefits from synergy effect of combining all the modules together.

The goal of this paper is to describe the LISp-Miner project and its product – the LISp-Miner system – as a whole, to explain its architecture and how can it be expanded and to stress project management and development of the system.

The paper is organized as follows. The main features of the LISp-Miner system are explained in section 2. The project management is described in section 3. Implemented modules are sketched in section 4. Examples of current research topics are mentioned in section 5.

2 LISp-Miner System

The LISp-Miner system is developed at Faculty of Informatics and Statistics at University of Economics Prague from 1996. It can be freely downloaded at http://lispminer.vse.cz. It is used for teaching and research in KDD. Development of the LISp-Miner system is also topic of diploma and PhD thesis.

One of its goals is to introduce the KDD process to students and allow them to participate not only in data-mining analysis but also in developing and implementing new data-mining algorithms.

2.1 System Description

According to its academic origin a complex theory is incorporated in the system. This theory can be found in e.g. [2], [1], [7], [8], [9] and is available on-line at homepage.

One of the main advantages is system modularity. It is easy to add a new module that can use outputs from some existing modules and to provide inputs for others (see sections 2.2 and 2.3 below). Defined project management and project documentation allows others to participate in development of the system.

A special area is creation of so-called "tailor-made interfaces" fully customized for particular expert domains – e.g. medicine, sociology and finance. These interfaces can be build ad-hoc for one data-mining task or generally for tasks from a given area. There are used terms from the expert domain only (i.e. blood pressure, inflation rate) and no terms from KDD, therefore better understandable for users.

We are undertaking necessary steps to build a set of web-services around the LISp-Miner system to provide even better support for integration into large systems.

2.2 System Architecture

The basic system architecture is in Figure 1.

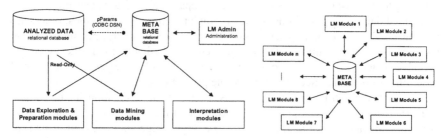

Fig. 1. The system architecture and the metabase as a central point in the architecture

Analyzed data are the input of data mining procedure. They are treated as read-only because no changes are usually possible there (e.g. capacity constraints). All KDD specific data are stored in special database called *metabase* (see section 2.3). Modules are divided into three basic groups (see Figure 1 on the left):

- Data Exploration & Preparation
- Data Mining
- Interpretation

These groups correspond to important phases of KDD process. Groups also allow maintaining complexity of each module at user-accepted level.

2.3 Metabase

The concept of modularity is based on a central *metabase* that stores all the meta-data (description of tables and columns in analyzed data, defined categorizations, data mining task description and task results etc.) the modules can use and share among themselves (see Figure 1 on the right).

All modules use metabase for storing its data and outputs. Data are shared through the whole system and metabase serves as a communication place for all modules. Structure of metabase is described in the project documents (see section 3.3) and is available for all developers.

3 Academic KDD Project LISp-Miner

There are a lot of subjects using LISp-Miner and new requests for changes and implementing new functions are received. In the same time the number of programmers developing the system is growing. It was necessary to adopt some formal principles and standards in project management and project documentation.

3.1 Project Characteristics

The main characteristic of this project is a relative large number of people contributing to its development – some of them prepare theoretical background of implemented data mining methods, some are implementing these methods, others are writing user's documentation and building web pages. Finally, people are using the system for data mining tasks and are sending proposals for further development. A difficult problem is geographical distribution of these people with no real possibility of personal meeting of the whole project team.

The second important characteristic is a permanent fluctuation in the project team. Students come and go as they graduate and they have to be replaced with younger students. It is crucial not to lose any knowledge about system or its proposed modifications after some member of team leaves or some work is finished.

3.2 Project Home Page

We have chosen the WWW as a best platform for project management and sharing information among project team members. The home page of LISp-Miner project can be found at URL http://lispminer.vse.cz. It is divided into two parts:
1. General information about the LISp-Miner system (description of used terms and data-mining methods; user's guide for all the LISP-Miner modules; the latest version of the LISp-Miner system available free to download).
2. Project development pages (list of all project documents; project procedures; project documents available for download).

The general information part is publicly available and everybody can freely download the latest version of the system. The development pages are for members of the project team only with exception of the metabase structure, which is also available for everybody.

3.3 Project Documentation

The project documentation is available for members of project after successful login into Project Development Pages. Documents are divided into categories:
- Project management documents (project organization, contacts, conventions, standards and rules for project management; document templates and other documents – list of all documents, list of released versions etc.)
- Research thesis and proposals (for further research or implementation).
- Programmer's documentation (global architecture, detail documentation for the all modules and for the meta-base).

Document templates for every document category support a uniform look of all the project documents.

4 System Modules

The current version of the LISp-Miner system consists of several modules – see Figure 2. Selected modules are described below, for others see e.g. [13] or [14].

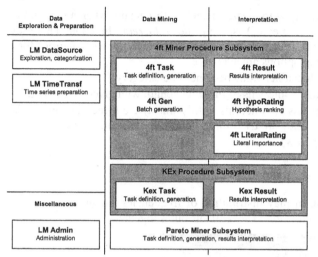

Fig. 2. Overview of existing modules

4.1 LM DataSouce

LM DataSource is designed for data exploration and mainly for data preparation.

Data preparation is the crucial phase of the whole data mining process. The most import part of data preparation is suitable categorization of values. It is possible to group together extreme values, to join two or more smaller districts into larger one or to divide non-discrete values into intervals (see section 4.1.1).

LM DataSource allows defining so called derived values from existing table columns. User can use SQL functions of the underlying database system to compute e.g. age of person from his/her date of birth, the total amount of salary as sum of base salary and bonus or to extract day of week from the date of telephone call. Using derived values of this kind can bring significant improvements to the result of data mining analysis. The other modules of the LISp-Miner system use results of data preparation through the metabase.

4.1.1 Categorization of Values

Categorization is the most import part of data preparation, because the result of the whole data mining is highly dependant on the good or bad categorization of values into categories. No interesting relationships can be found in case of non-suitable grouping of values.

Categories can be created manually or automatically. We select either one or more values for each category during manual creation of categories or some interval of values at once. It is possible to edit categories, to change order of categories, to join two or more categories into one or to divide one category into subcategories.

The most efficient way how to create categories (especially when a large number of categories has to be created) is using the *Autocreate* function. There are several possibilities how LM DataSource can automatically create categories for a given attribute. Following options are available:

- **Enumeration** ... each distinct value becomes one category;
- **Equidistant intervals** ... intervals of the same length with given beginning;
- **Equifrequency intervals** ... intervals with (almost) the same number of objects in each of them. They are used for the uniform distribution of frequencies into categories (see bellow).

4.1.2 Frequency Analysis

There is a possibility to display distribution of frequencies among categories of one attribute or even conjunction of several attributes together.

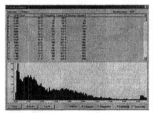 

Fig. 3. Distribution of frequencies – not sorted (on the left) and sorted (on the right)

There are both absolute frequency and relative frequency computed for each category. Cumulative frequencies from top of the list are computed too. Values from list can be exported to clipboard and paste into other Windows application (e.g. MS Word, MS Excel).

4.2 Procedure 4ft-Miner

There is a large theory of association rules related to the 4ft-Miner procedure. The presented approach was first applied in connection of development of the GUHA method of mechanized hypotheses formation [2], [5], [6]. The previous version of the 4ft-Miner procedure was described e.g. in [12].

4.2.1 Mining for Association Rules

The 4ft-Miner procedure does not use the classical A-priori algorithm. Procedure 4ft-Miner mines for association rules of the form $\varphi \approx \psi$ and for conditional association rules of the form $\varphi \approx \psi / \chi$. Here φ, ψ and χ are conjunctions of Boolean attributes automatically derived from *many-valued attributes* in various ways.

The symbol $\approx$ is called *4ft-quantifier*. The association rule $\varphi \approx \psi$ means that Boolean attributes φ and ψ are somehow associated in the sense of the 4ft-quantifier $\approx$. A conditional association rule $\varphi \approx \psi / \chi$ means that φ and ψ are associated (in the sense of $\approx$) if the condition χ is satisfied.

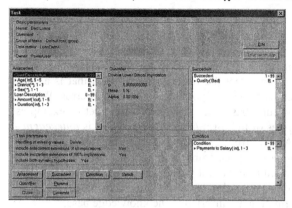

Fig. 4. 4ft-Miner task description

4.2.2 Rich Syntax

The main difference to the A-priori algorithm is ability to work with association rules with more complex syntax:

- It is possible to mine for conditional association rules of the form $A \approx B / C$. It helps to mine for association rules when A is less frequent combination.
- Rules can contain conjunction of attributes in every part of rule – i.e. $A_1 \wedge A_2 \wedge \ldots A_n \approx B_1 \wedge B_2 \wedge \ldots B_n / C_1 \wedge C_2 \wedge \ldots C_n$.
- Each attribute can refer to more than one value – e.g. A(Prague, Brno) means A can be "Prague" or "Brno".
- More than just classical implication $A \Rightarrow B / C$ can be mined for. There are 17 *4ft-quantifiers* currently available to describe desired relationship between A and B, see [8].
- Several attributes in either part of association rule can be grouped together and treated as a whole during task description (setting parameters, copying or moving to another task etc.).
- Techniques for reduction of output are implemented. Logically derived rules can be automatically removed from result, see [7].

For details of these advantages see mainly [13]. Syntactical richness leads of course to huge number of potential association rules that have to be generated and verified. The used algorithm is based on representing data by strings of bits and logical operations on that strings, for details see e.g. [5], [13]. The resulting process is then very fast and using of the 4ft-Miner procedure is interactive for data mining tasks of complexity up to 10^7 association rules to be generated.

4.3 Expert system KEX

KEX is a machine learning procedure based on association rules [1]. The version of KEX included in the LISp-Miner system goes beyond previous implementations mainly in ability to share data with other modules of the system and to use results of data preparation phase done in the LM DataSource module.

4.3.1 Building of Knowledge Base

The heart of an expert system is knowledge base that stores generalized rules about analyzed domain. So the most important task is to build proper set of rules from training data. The goal is to select the minimal set of rules that is able to classify new unseen examples with the less possible classification error. We can test expected quality of knowledge base with separated testing set of examples. Knowledge base consists of rules of the form Ant $\Rightarrow$ C, where Ant is *antecedent* – i.e. conjunction of attributes A(a) $\wedge$ B(b) $\wedge$ ... $\wedge$ X(x) and C is a classification class.

We walk through the hypotheses space from the most frequent antecedents to least frequent. For each generated rule we test whether it is significantly different from knowledge that can be at this time obtained from knowledge base. Only rules that bring important new information are inserted to the knowledge base.

Let us suppose that we have to decide whether a rule of

$$\text{Salary}(\textit{High}) \wedge \text{Age}(\textit{30ies}) \Rightarrow \text{District}(\textit{Praha})$$

should be inserted into knowledge base. Up to now we have the following rules already available (according to used algorithm):

$$\textit{<no antecedent>} \Rightarrow \text{District}(\textit{Praha})$$
$$\text{Salary}(\textit{High}) \Rightarrow \text{District}(\textit{Praha})$$
$$\text{Age}(\textit{30ies}) \Rightarrow \text{District}(\textit{Praha})$$

We combine information from these rules together into so called *compound weight* (for details see [1]). This weight is then compared to confidence of the given rule. If there is a significant difference (in the sense of χ^2 test on the given level of α), new rule is inserted into knowledge base.

After we have walked through all the rules, the knowledge base consists of set of rules that can classify new examples with desired accuracy and doesn't include redundant rules.

An important part of building of knowledge base is testing of its quality (expected classification error). There are four possible ways of test implemented (separate training set, random-split, cross-validation and full training set).

4.3.2 Consultation

After building appropriate knowledge base we can start to classify new examples. We provide values of all or some of attributes from antecedent and the most probable class is returned as result. Each consultation is accompanied with a level of uncertainty it was released with. If there is not enough information available, classification can be rejected with result "don't know".

Another way of doing consultation is provide a text file with examples for classification. Consultation is made in batch for all the included examples.

4.4 Tailor-made Interfaces

A special module named **4ftGen** is available for batch processing of 4ft-Miner procedure on a selected task. **4ftGen** can be called from any other system or even a web interface to solve a particular task.

These "tailor-made" interfaces can be built for specific task purposes – e.g. heart-disease treatment. The interface can therefore use terms from expert domain and not to confuse user by terms from data mining. Moreover it can be overall simplified to meet only needs of this particular task.

The most common way of building these interfaces is in form WWW pages. An example of tailor made interface can be found and tested at WWW page http://euromise.vse.cz/stulong-en/a-otazky/vv.

5 Further Development

The LISp-Miner system is designed as modular and therefore it is possible to add a completely new module. This new module is able to use meta-data from other modules or to provide outputs for further processing.

The project documentation contains documents that describe in detail the steps necessary to integrate a new module in the system.

There are several new mining procedures in the phase of development right now. Among others let us mention the possibility of multi-relational data mining. It is based on assumption that mining for interesting associations among bought products cannot be done on one table recording just transaction of buying each single item. It cannot be done on table describing particular products alone either. We have to combine both tables together along with its relationship *one-to-many*. Several aggregate characteristics can be computed on transaction table (e.g. average number of items bought for each product, minimal amount spend in each shop per month...) and used as additional temporally attributes of product table. Already implemented operations on strings of bits are effectively used for computing aggregate characteristics.

There are also undertaken steps towards mining for interesting relationships among many-valued attributes. It results in dealing with K×L contingency tables. Incorporating strings of bits leads, once again, to a very fast computation of frequencies from complex contingency tables.

Results of data mining have to be clearly interpreted for them having a value. It is always difficult to translate these results back for domain experts. Therefore it is always worth of thinking about the best possible (and intuitive) way of displaying results. We are trying to use several forms of 2D and 3D graphics (3D graphs, maps etc.) to find the most suitable way for communicate results to user.

References

1. Berka, P., Ivánek, J.: *Automated knowledge acquisition for PROSPECTOR-like expert systems*, In Proceedings ECML'94 (eds. Bergadano, de Raedt), Springer, 1994, pp. 339-342.
2. Hájek, P., Havránek, T.: *Mechanising Hypothesis Formation – Mathematical Foundations for a General Theory*, Springer-Verlag, 1978, pp. 396.
3. Hájek, P., Havránek, T. – Chytil, M.: *GUHA Method*, (In Czech), Praha, Academia, 1983, pp. 314.
4. Havránek, T.: The present state of the GUHA software. International Journal of Man-Machine Studies, 15, 1981, pp. 253-264.
5. Rauch, J.: *Some Remarks on Computer Realizations of GUHA Procedures*, International Journal of Man-Machine Studies, 10, 1978, pp. 23 – 28.
6. Rauch J.: *Main Problems and Further Possibilities of the Computer Realizations of GUHA Procedures*, International Journal of Man-Machine Studies, 15, 1981, pp. 283-287.
7. Rauch, J.: *Logical Calculi for Knowledge Discovery in Databases*, Principles of Data Mining and Knowledge Discovery, (eds. J. Komorowski and J. Zytkow), Springer Verlag, Berlin, 1997, pp. 47-57.
8. Rauch, J.: *Classes of Four-Fold Table Quantifiers*, Principles of Data Mining and Knowledge Discovery, (eds. J. Zytkow, M. Quafafou), Springer-Verlag, 1998, pp. 203-211.
9. Rauch, J.: *Four-fold Table Calculi and Missing Information*, JCI'S98 Association for Intelligent Machinery, Vol. II., (eds. Wang Paul), Durham, Duke University, 1998.
10. Rauch, J.: *Interesting Association Rules and Multi-relational Association Rules*, Communications of Institute of Information and Computing Machinery, Taiwan, Vol. 5, No. 2, May 2002, pp. 77-82.
11. Rauch, J., Šimůnek, M.: *Mining for 4ft Association Rules*, In Discovery Science 2000 (eds. S. Arikawa, S. Morishita), Springer Verlag, 2000, pp. 268-272.
12. Rauch, J., Šimůnek, M.: *Mining for 4ft Association Rules by 4ft-Miner*, INAP 2001, The Proceeding of the International Conference On Applications of Prolog, Prolog Association of Japan, Tokyo, October 2001, pp. 285-294.
13. Rauch, J., Šimůnek, M.: *Alternative Approach to Mining Association Rules*, In FDM02, The IEEE ICDM02 Workshop Proceedings (eds. T.Y. Lin, S. Ohsuga), IEEE, Maebashi December 2002, pp. 157-162.
14. Šlesinger, J: *Preprocessing of time series data for LISp-Miner system (In Czech)*, offered as article to Znalosti conference, Ostrava 2003.
15. Zembowicz, R., Zytkow, J.: *From Contingency Tables to Various Forms of Knowledge in Databases*, Advances in Knowledge Discovery and Data Mining (eds. Fayyad, U. M. et al.), AAAI Press/ The MIT Press, 1996, pp. 329-349.

Performance Evaluation Metrics for Link Discovery Systems

M. Afzal Upal[1]

OpalRock Technologies
42741 Center St. Chantilly, VA 20152
afzal@opalrock.com

Abstract. Recently there has been an explosion of work on the design of automated link discovery (LD) systems but little work has been done to investigate metrics to evaluate the performance of such systems. This paper states the link discovery system evaluation problem, explores the issues involved in evaluating the performance of link discovery systems by relating it to the traditional problems of evaluating classification systems, and describes metrics I derived to evaluate the LD systems being developed under DARPA's EELD program.

1   Introduction & Background

Advances in data collection technologies over the last few decades have made large amounts of data about various aspects of life available in electronic form. Various governmental and non-governmental entities collect tremendous amounts of transactional data for various reasons. Banks, for instance, collect information about deposits to and withdrawals from their customer's accounts. The banks, although primary users of this data, are by no means the only users of this information. The bank transaction data can be used by others such as law enforcement agencies (e.g., to track illegal activities such as money laundering). The problem is that the data is often not collected and structured to serve the needs of the secondary and tertiary users making it very difficult to efficiently answer their queries (e.g., which bank transactions appear suspicious enough to be investigated further for illegal activity?). This may be because of several reasons.
1. Because of differences in the mandates/objectives of the data-collectors and some data-users.
2. Because of the cost of storing data in multiple formats.
3. Because of the legal and regulatory prohibitions.
Regardless of the reasons, the secondary and tertiary users must sift through large amounts of data retrieved through queries that the databases are designed to answer and synthesize those answers to find instances that fit the patterns they are searching for. This task, known as link discovery, is different from the traditional data-mining

---

[1] This work was performed during author's tenure at Information Extraction & Transport Inc. 1901 North Fort Myer Drive, Arlington, VA 22209, under a DARPA contract.

task which involves analyzing large amounts of data to discover and represent associations based on the aggregate statistical characteristics of a sample of instances drawn from some population [4]. The link discovery task begins with such association patterns and sifts through various databases to find instances of a given pattern [2]. Recently there has been an explosion of work on the design of automated link discovery techniques primarily driven by DARPA's Evidence Extraction and Link Discovery (EELD) program but little work has been done to investigate methods to evaluate the performance of automated link discovery systems. This paper explores the issues involved in designing performance evaluation metrics for automated link discovery systems and proposes a number of metrics that can be used to measure the performance of link discovery systems.

2 The Problem

We define the link discovery problem as follows.
Given
- a set of general patterns of P different pattern types,
- a set of evidence reports such that each evidence report is a set of assertions[2], and
- a set of target pattern-types $T \subseteq P$.

Find
- sets of evidence assertions such that each set instantiates at least one target pattern $t_i \in T$. Each pattern-instance (or *case*) is described by the evidence assertions used to instantiate the general patterns of that type.

For instance, given the general money laundering pattern and evidence reports from banks, the task of an LD system is to produce the true classification, T[3].

General Patterns:
        account_number(Person, Account),
        deposit(Account, Deposit1),
        amount(Deposit1, Amount1),
        date(Deposit1, Date1),
        branch(Deposit1, Branch1),
        deposit(Account, Deposit2),
        amount(Deposit2, Amount2),
        date(Deposit2, Date2),
        branch(Deposit2, Branch2),
        greaterThan(Amount1 + Amount2, reporting_amount) → isA(Person, money_launderer);

account_number(Person, Account),

---

[2] In general an evidence report may be textual. Such reports can be translated into assertions by a wrapper.

[3] Variables start with capitals letters and constants with lower case letters in the Prolog tradition.

      deposit(Account, Deposit),
      amount(Deposit, Amount1),
      date(Deposit, Date1),
      branch(Deposit, Branch),
      withdrawal(Account, Withdrawal),
      amount(Withdrawal, Amount2),
      date(Withdrawal, Date2),
      branch(Withdrawal, Branch) $\rightarrow$ isA(Person, good_client)

Evidence:

      BankOfAmerica Report:
      account_number(client12, act123)
      deposit(act123, dep391)
      amount(dep391, 9999)
      date(dep391, july25)
      branch(dep391, 5th_ave)
      deposit(act123, dep4005)
      amount(dep4005, 9999)
      date(dep4005, july25)
      branch(dep4005, 42nd_st).

The Federal Reserve Report: equal(reporting_amount, 10000).

First Union Report:

      account_number(client45, act456).
      deposit(act456, dep678).
      amount(dep678, 9999).
      date(dep678, july25).
      branch(dep678, 5th_ave).
      withdrawal(act456, wd910).
      amount(wd910, 9999).
      date(wd910, july26).
      branch(wd910, 5th_ave).

True Classification, T:

Case T1:	*Case T2:*
isA(client12, money_launderer)	isA(client45, good_client)
account_number(client12, act123)	account_number(client45, act456)
deposit(act123, dep391)	deposit(act456, dep678).
amount(dep391, 9999)	amount(dep678, 9999).
date(dep391, july25)	date(dep678, july25).
branch(dep391, 5th_ave)	branch(dep678, 5th_ave).
deposit(act123, dep405)	withdrawal(act456, wd910).
amount(dep405, 9999)	amount(wd910, 9000).
date(dep405, july25)	date(wd910, july26).
branch(dep405, 42nd_st).	branch(wd910, 5th_ave).

The first pattern rule encodes the knowledge that if a person makes two large deposits in different branches such that the combined amount is greater than the reporting amount then the person may be a money launderer. The second pattern rule specifies that the good clients deposit and withdraw money from the same branch.

Assume that an LD System (let's call it *LDS1*) reports the following classification.

LDS1's Classification:

> *Case S11:*
>
>> isA(client12, money_launderer)
>> account_number(client12, act123)
>> deposit(act123, dep391)
>> amount(dep391, 9999)
>> date(dep391, july25)
>> branch(dep391, 5th_ave)
>> deposit(act123, dep4005)
>> amount(dep4005, 9999)
>> date(dep4005, july25)
>> branch(dep4005, 42nd_st).

The LD evaluation problem is to compare the system's classification with true classification to measure their similarity.

Given

- a set of pattern instances identified by an LD system, and
- a set of true pattern instances identified by human analysts

Find

- the amount of similarity between the pattern instances identified by the LD system and the true pattern instances.

This paper does not address the issue of how the true classification is to be derived. This could be done by the domain experts or through an automated agent [3]. For the purpose of this work, we simply assume that the true classification is available for use in the evaluation. Similarly, we also assume that the LD system has access to the ontology used by the builders of the true classification and that there is an agreement between the builders of the true classification and the LD system as to the kinds of assertions that are to be included in the system's classification.

2.1 Measuring classifier performance

The LDS evaluation problem appears to be similar to the problem of evaluating classification systems which has been previously studied [1, 8, 9]. Classification problems are usually studied as unsupervised or supervised. Unsupervised classification is defined as the process of organizing a set of data items into classes. Upal [7] reviews a number of performance evaluation metrics that can be used to measure the performance of an unsupervised classification system. Given a true classification $C_{true} = \{T_1, T_2, ..., T_i\}$ and the System's classification $C_{System} = \{S_1, S_2, ..., S_i\}$, let $f(i, j)$ and $g(i, j)$ be two similarity functions defined as follows.

$$f(i, j) = \begin{cases} 1 & \text{if } i \in T_x \text{ and } j \in T_x, 1 \geq x \geq 1 \\ 0 & \text{otherwise} \end{cases}$$

$$g(i, j) = \begin{cases} 1 & \text{if } i \in S_x \text{ and } j \in S_x, m \geq x \geq 1 \\ 0 & \text{otherwise} \end{cases}$$

Let a be the number of pairs of data items that are in the same class in both classifications, d be the number of pairs in different classifications, and let b and c be the number of the pairs of data items placed in the same class by one classification and different classes by the other (as shown in Table 1) then the most common metrics for determining similarity between two classifications are defined as:

Rand = (a+d) / (a + b +c + d)
Fowlkes and Mallows = a / sqrt((a+b) * (a + c)), and
Jaccard = a/(a+b+c).

Table 1: Contingency table for two classification.

		F	
		1	0
G	1	a	B
	0	c	D

The LD system evaluation problem can be mapped to the unsupervised classification problem by mapping assertions to be data items and cases to classes. The problem with this approach, however, is that it ignores the crucial pattern-type information available in the LD-evaluation problems.   Determining whether a pattern of a given type (such as money-laundering) exists in a set of data, and identifying instances of the patterns is often the most important part of link discovery. An alternative approach is to explore the similarities between LD and supervised classification where similar to LD systems, the task is to find out if a given data item belongs to one of the previously known classes. Precision and recall are two of the most popular metrics used to measure the performance of supervised classification systems [9]. Precision is usually defined as the proportion of test items correctly classified by the system out of the total number of items actually classified. Recall is typically defined as the proportion of test items correctly classified out of the total number of items that had to be classified.

Precision = number of items correctly classified/size of the system's classification.

(Equation 1)

Recall = number of items correctly classified/size of the true classification.

(Equation 2)

Link discovery system evaluation task can be mapped to the supervised classification task by mapping cases to data items and pattern-types to class labels. Using this mapping, we can measure the precision and recall of the example LDS1 classification to be:

Precision = 1/1 * 100 = 100 %   Recall = ½ * 100 = 50% .

While mapping an LD case onto an unsupervised classification data item allows us to measure precision and recall for classifications such as that returned by LDS1, it does not work for all LD outputs such as LDS2's classification that split information from one true case into multiple cases or merge information from multiple true cases into one system case.

LDS2's Classification

Case S22:

      isA(client12, good_client)
      account_number(client12, act123)
      deposit(act123, dep391)
      amount(dep391, 9999)
      date(dep391, july25)
      branch(dep391, 5th_ave)
      withdrawal(act456, wd910).
      amount(wd910, 9000).
      date(wd910, july26).
      branch(wd910, 5th_ave).

Case S22 contains information from two true cases; Case T1 and Case T2. Traditional supervised classifiers do not divide a data item into its components because a data item is considered to be the most elementary unit. This is not the case in LDS evaluation where evidence assertions are the atomic units. In LDS evaluation, true classification and the system's classification do not even agree on the data items.

Given the above discussion, we can redefine the LDS evaluation problem as a more general version of the supervised classification problem

Given

- System's classification of system's data items $\{(sd1, S1), (sd2, S2), ..., (sdm, Sm)\}$
- True classification of true data items $\{(td1, S1), (td2, S2), ..., (tdn, Sn)\}$

Find

- Similarity between the two classifications.

Where sdi denotes a system's data item and tdj denotes a true data item and sdi $\neq$ sdj, tdi $\neq$ tdj $\forall$ i $\neq$ j. However, Si may be identical to Sj and Ti may be the same as Tj $\forall$ i $\neq$ j. Precision and recall can then be redefined as follows. Let *similarity(i, j): {SD * TD}* $\rightarrow$ *[0, 1]* be a function that measures the similarity between two data items, let *most-similar-true-item(x)* be a function that takes a system's data item and returns the true item that is most similar to it, and let *most-similar-systems-item(x)* be a function that takes a true item and return the system's item most similar to it.

Precision = $\sum_i$ similarity(sdi, tdj) * equal(Si, Tj) / m
Where tdj = most-similar-true-item(sdi)
Recall = $\sum_i$ similarity(tdi, sdj) * equal(Ti, Sj) / n
Where sdj = most-similar-systems-item(tdi)

Note that when system's data items are identical to the true data items, as in the case of traditional supervised classification problems, the most similar true item for a given data item i is i itself and hence completely similar to it i.e., similarity(sdi, tdj) = 1. Thus precision measure reduces to Equation 1. Similarly, the recall measure reduces to Equation 2 when the true classification and the system's classification agree on the data items.

In a number of real-world classification situations, not all classification errors have the same cost [1, 7]. For instance, it may be more costly to award credit to a bad credit holder than to deny credit to a good credit holder. However, failing to identify a terrorist threat may be more costly to a terrorism identification system than to falsely identify a non-threat event as a threat case.   Given a true classification $C_{true} = \{T_1, T_2, ..., T_i\}$ and the System's classification $C_{system} = \{S_1, S_2, ..., S_j\}$ a cost matrix such as the one shown in Table 2 can be used to specify the cost of correctly/incorrectly identifying a case.

Table 2: Cost Specification Matrix. C_{SjTi} denotes the penalty for assigning a data item belonging to the true class Ti to the system's class Sj. C_{NullTi} denotes the penalty for the classification system's inability to determine a classification for a data item belonging to Class Ti.

	T_1	T_2	...	T_i
S_1	C_{S1T1}	C_{S1T2}	...	C_{S1Ti}
S_2	C_{S2T1}	C_{S2T2}	...	C_{S2Ti}
...	...	...	...	...
S_m	C_{SmT1}	C_{SmT2}	...	C_{SmTi}
NULL	C_{NullT1}	C_{NullT2}	...	C_{NullTi}

The precision and recall metrics can then be redefined to take costs into account. For instance, assuming the cost matrix shown in Table 1, the cost of LDS1's classification is 60 out of the worse possible cost of 160 hence the redefined recall is 62.5%.

Table 3: Cost specification for the money laundering problem.

	good_client	money_launderer
Good_client	−40	50
money_launderer	50	−30
NULL	50	60

Incorporating the cost, precision and recall for LD can be redefined as follows.
Precision = $\sum_i$ similarity(sdi, tdj) * equal(Si, Tj) * cost(Si, Tj) / m
Recall = $\sum_i$ similarity(tdi, sdj) * equal(Ti, Sj) * cost(Ti, Sj) / n
So far we have not addressed the problem of measuring similarity between two LD cases.   We need to address this issue before we can apply the precision and recall metrics to measure the performance of LD systems.

2.2   Measuring similarity between cases

Traditionally, the LD community has employed a number of different pattern/case-structure specific metrics for measuring case similarity and few general metrics (an exception is the Cosine co-efficient [6]) have been defined. IET has defined two similarity measures that will be used to evaluate the performance of LD systems being built under DARPA's EELD program; the minimum graph edit distance and the minimum object-edit distance.

Graph Edit Distance: The graph edit distance translates a case into a graph such that each n-arity assertion is translated into n−1 edges connecting the n-nodes (each node corresponding to an argument) using the following algorithm.

add-subgraph(Asertion, Graph)
for all assertion argument pairs (ai, aj) and assertion name n **do**
        add-edge(Graph, ai, aj, n)

add-edge(Graph, ai, aj, label)
if ai ∉ nodes(Graph) **then**
        Nodes(Graph) ← Nodes(Graph) ∪ {new node(ai)}
if aj ∉ nodes(Graph) **then**
        Nodes(Graph) ← Nodes(Graph) ∪ {new node(aj)}
Edges(Graph) ← Edges(Graph) ∪ {new edge(ai, aj, label)}

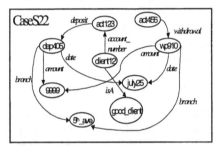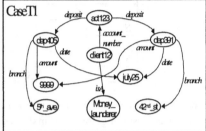

Figure 1: Graph representation of CaseT1 and CaseS22.

Figure 1 shows the graphs constructed by translating the true Case T1 and the system Case S22 into graphs using this algorithm.

The minimum edit distance between two graphs is defined as the minimum number of edit operations required to transform one graph into the other. IET's measure admits four edit operations; adding/deleting a node, adding/deleting an edge. Using this measure, we can see that the edit distance between Case-s111 and Case-T1 (shown in Figure 1) is 6 because only six edit operations need to be performed to make the two graphs isomorphic.

The object edit distance assumes that all case assertions are binary and that all important entities in a case can be identified and transformed into objects. IET's object edit distance measure uses the heuristic that each "isa"[4] assertion identifies a critical entity that should be transformed into an object.

Once all the objects has been initialized, the case assertions are parsed to populate the object features. This means finding all the assertions *about* the entity of interest and translating them into features. An assertion is about an entity if it includes that entity as an argument. IET's measure restricts this to only those assertions that mention the entity as the first argument. The assertion name becomes the feature

---

[4] IsA is a CycL assertion [5]. CycL is the common interface language currently being used
the EELD program.

name and the second argument of the assertion becomes the feature value. Figure 2 shows the objects constructed from true case T1 and the system's case S22 assuming that the critical objects in these cases include the client, the account, and the deposit/withdrawal. When value of a feature is an object itself then a link from the feature value to that object is made as is done in the case of *account* and *deposit* features as shown in Figure 2.

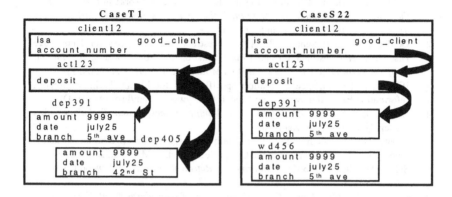

Figure 2: Object representation of Case T1 and Case S22.

The object edit distance between two cases is defined as the sum of the minimum object edit distances between the optimal object pairings between the two cases. An object pairing is defined as the one-to-one pairing of the objects contained in a case with the objects contained in the case it is being compared with. Optimal object pairing is the object pairing that minimizes the case object edit distance.[5] IET uses 1 minus normalized edit distance between two cases as the case similarity measure. The normalization constant is the size of objects in the two cases where size of an object is defined as the number of its features. IET has designed and implemented an algorithm that given a true classification and an LD system's classification measures the precision and recall of the LD system's classification.

3   Conclusion

Recently there has been an explosion of work on the design of automated link discovery (LD) systems but little work has been done to investigate methods to evaluate the performance of automated link discovery systems. This paper states the link discovery system evaluation problem, explores the issues involved in evaluating the performance of link discovery systems by relating it the traditional problems of

---

[5] If two cases contain different number of objects, null objects are added to the smaller case to make it the same size as the larger case.   Minimum edit distance between a case and a null case is the number of features in the non-null case.

evaluating classification systems, and describes the solutions that we designed to evaluate the LD systems being developed under DARPA's EELD project.

References

1. Feng, C., Sutherland, A. King, R., Muggleton, S. and Henry, R., 1993. Comparison of Machine Learning Classifiers to statistics and neural networks in *Proceedings of the Fourth International Workshop on Artificial Intelligence and Statistics*, pages 41-52, San Mateo, CA: Morgan Kaufmann.
2. Goldberg, H., and Senator, T. (1995). Restructuring Databases for Knowledge Discovery by Consolidation and Link Formation in *Proceedings of the 1st Int. Conf. on Knowledge Discovery and Data Mining* Menlo Park, CA: AAAI Press.
3. IET, 2002. Performance Evaluation Specifications for EELD, Technical Report, Information Extraction & Transport Inc. (http://www.iet.com/Projects/EELD)
4. Han, J. and Kamber, M., 2000. *Data Mining: Concepts and Techniques*, San Francisco, CA: Morgan Kaufmann.
5. Lenat, D. and Guha, R. 1990 Building large knowledge-based systems: representations and interface in the Cyc project, Reading MA: Addison-Wesley.
6. Salton, G., 1971. *The SMART Information Retreival System.* Englewood Cliffs, NJ : Prentice-Hall.
7. Upal, M. A. 1995. *Monte Carlo Comparison of Non-hierarchical Unsupervised Classifiers*, Masters Dissertation, Department of Computer Science, University of Saskatchewan.
8. Upal, M. A. and Neufeld, E. 1996 Comparison of non-hierarchical unsupervised classifiers in *Proceedings of the International Conference on Information, Statistics and Induction in Science*, pages 342-353, Singapore: World Scientific.
9. Weiss, S. and Kulikowski, C., 1991. *Computer Systems That Learn: Classification and Prediction Methods From Statistics, Neural Networks, Machine Learning, and Expert Systems.* San Mateo, CA: Morgan Kaufmann.

Generalizing Association Rules: Theoretical Framework and Implementation

Antonio Badia[1] and Mehmed Kantardzic[1]

Computer Engineering and Computer Science department, Speed Scientific School, University of Louisville Louisville, KY 40292 abadia@louisville.edu

Summary. We present a generalization of the concept of association rule. To give our approach a sound formal basis, we use the idea of *generalized quantification*, developed in logic theory. The generalization is based on allowing more complex relationships among itemsets, including *negative associations*. Negative associations are defined as associations among large itemsets which denote a strong degree of disjointness. We also propose stricter measures of relevance than confidence and support, in order to make sure that extracted rules (positive and negative) truly represent a pattern in the data. Finally, we show how to extend the A-priori algorithm to mine for the extended associations introduced with high efficiency and little overhead.

1 Introduction

Mining for association rules is one of the best known tasks in data mining ([3]). Research in the subject has produced efficient algorithms and some generalization of the original concept. In this paper, we propose a generalization of the concept of association rule based on the framework of generalized quantification. This framework allows a formal analysis of the meaning of association rules. It also makes possible to capture several relationships between itemsets, including *negative* relationships. The framework is declarative, high-level, and extendible; therefore it can be used to define new kinds of associations and to study the properties of these associations. We also discuss possible implementations of the generalized rules by modifying existing, well-known algorithms.

The paper is organized as follows. In the next section we provide some background on generalized quantification. In section 3, we define our generalization of association rules, giving a formal semantics to the concept. In section 4 we discuss an efficient implementation of the generalized rule. Finally we close with some issues that deserve further study.

2 Background

In this section we present the basic concepts of generalized quantifiers, in order to introduce some definitions and to fix the vocabulary for the rest of the paper. *Generalized Quantifiers* (henceforth, GQs) were first introduced in logical studies ([14], [12]). The concept has attracted attention lately for its uses, among others, in query languages ([6],[15],[11]) and other languages like description logics ([5]).

Given a set M, a *Generalized Quantifier (GQ)* on M is a relation among subsets of relations on M. The following definition formalizes this idea:

Definition 2.1 *Let a type be a finite sequence of positive numbers $[k_1, \ldots, k_n]$. Then a generalized quantifier Q of type $[k_1, \ldots, k_n]$ on M is an n-ary relation between subsets of $M^{k_1}, \ldots, M^{k_n}$ (i.e. between elements of $\mathcal{P}(M^{k_1}) \times \ldots \times \mathcal{P}(M^{k_n})$) that is closed under permutations (that is, if f is a permutation on M, then $Q(A_1, \ldots, A_n)$ iff $Q(f[A_1], \ldots, f[A_n])$).*

The following are examples of GQs. A universe M is fixed. We use Q as a variable over GQs, $A, B, \ldots$ as variables over sets, and write $Q(A, B)$ to indicate that A, B belong to the extension of Q, i.e. that they are in the relation denoted by Q.

all $= \{A, B \subseteq M | A \subseteq B\}$
some $= \{A, B \subseteq M | A \cap B \neq \emptyset\}$
no $= \{A, B \subseteq M | A \cap B = \emptyset\}$
at least n $= \{A, B \subseteq M | \, | A \cap B | \geq n\}$
at most n $= \{A, B \subseteq M | \, | A \cap B | \leq n\}$
H $= \{A, B \subseteq M | \, | A | = | B |\}$ (Hartig's quantifier)

All the above quantifiers are of type $[1, 1]$. Such quantifiers are called *binary* (for having two arguments) and *monadic* (each argument being just a set, i.e. a subset of M^1). Although other types exist, binary monadic have been found to be the most common and useful type in query languages ([5], [11]); they are called *standard*. Note that the first four quantifiers are first-order definable; the last one is not.

Other conditions on the definition of quantifier have been proposed. Some of them imply that the behavior of a quantifier is independent of the context, as is the case for the usual logic constants. There is an axiom that formalizes context independence:

Definition 2.2 *(EXT) Quantifier Q follows EXT if for all M, M', all A, B such that $A, B \subseteq M \subseteq M'$, $Q_M(A, B)$ iff $Q_{M'}(A, B)$[1].*

Standard (binary monadic) GQs have a very important property: all such quantifiers can be defined by certain cardinalities, as follows:

---

[1] EXT stands for *extensionality*.

Lemma 1. If Q is a standard GQ, then $Q_M(A, B)$ iff $Q_{M'}(A', B')$ holds whenever $\mid A - B \mid = \mid A' - B' \mid$, $\mid B - A \mid = \mid B' - A' \mid$, $\mid A \cap B \mid = \mid A' \cap B' \mid$, and $\mid M - (A \cup B) \mid = \mid M' - (A' \cup B') \mid$.

A proof of the lemma can be found in [19]. Let $x_1 = \mid A \cap B \mid$, $x_2 = \mid A - B \mid$, $x_3 = \mid B - A \mid$, and $x_4 = \mid M - (A \cup B) \mid$. When quantifiers follow EXT, the lemma can be simplified to ignore x_4; that is, for quantifier Q obeying EXT, the behavior of Q is determined by x_1, x_2 and x_3. Intuitively, EXT allows us to ignore the context (anything outside) of A and B. Most logical constants are supposed to be *insensitive to the context* and therefore obey EXT. For instance, all first order logic quantifiers do: **all** (defined by formula $x_2 = 0$), **some** (defined by formula $x_1 > 0$), **no** (defined by formula $x_1 = 0$),... as well as other quantifiers which are not first-order definable (**H**, defined by $x_2 = x_3$, is not first-order definable). However, with ARs, context is very important, as it is part of the way we measure support and confidence[2]. It is important to note that many logical quantifiers can be expressed by formulas on numbers. For instance, all first order quantifiers can be captured with very simple languages on natural numbers ([19]) (we showed in the previous paragraph how to define some of the first order quantifiers). Finally, we note a few simple facts that will be used in the following sections, where ARs will be modeled as special cases of generalized quantification: given the definitions above, it is easy to see that $\mid A \mid = x_1 + x_2$; $\mid B \mid = x_1 + x_3$; $\mid M \mid = x_1 + x_2 + x_3 + x_4$; and $\mid A \cup B \mid = x_2 + x_3$.

3 Generalized Association Rules

We propose to generalize association rules by defining *a family of associations between itemsets*. Such associations are defined by generalized quantifiers operating on the base sets of the itemsets. Because of the generality of the framework, we can define a large number of different associations. Because of the fact that many GQs can be defined using simple number properties, we can define such associations by simple formulas which are easy to understand and, more importantly, can be computed efficiently. Moreover, by imposing appropriate conditions on x_1, x_2, x_3 and x_4 we can design quantifiers with associated measures of relevance, thus ruling out non interesting associations.

The intuitive idea is to model a rule $X \Rightarrow Y$ as the quantifier $Q(T_X, T_Y)$. In this view, the confidence of the rule can be expressed as $\frac{x_1}{x_1 + x_2}$ (since the confidence is $\frac{|T_X \cap T_Y|}{|T_X|}$, and $T_X = (T_X - T_Y) \cup (T_X \cap T_Y)$). Also, the support can be expressed as $\frac{x_1}{x_1 + x_2 + x_3 + x_4}$, as the support is simply $\frac{|T_X \cap T_Y|}{|T|}$, and $T = (T - (T_X \cup T_Y)) \cup (T_X \cap T_Y) \cup (T_X - T_Y) \cup (T_Y - T_X)$ (i.e. the whole domain). Clearly, a problem with this approach is that $x_1 = \mid T_X \cap T_Y \mid$ is contrasted to the size of T_X (by confidence), but not to the size of T_Y; this is

---

[2] This may explain why it is so difficult to give a precise logical status to ARs.

the origin of the negative correlation problem. Such a check could be defined as an additional measure, requiring (besides support and confidence), that $x_1 + x_3$ be a parameter of the comparison. In general, we are able to define customized GQs that attempt to capture different kinds of associations that we are interested in, as well as to define different measures of the goodness of said associations. The importance of this idea is that x_1, x_2, x_3 and x_4 can be used for both purposes:

- In order to define more relationships. One particular case that we develop in this paper is that of *negative* associations. A negative association is not a negative correlation; rather, it is a strong association of negative character between large itemsets. Thus, it is not a *weak* association either, as it has its own thresholds that must be met for an association to qualify as negative.
- In order to define different measures of support and confidence (and hence avoid some of the problems commonly associated with these two measures, like *negative correlations*). We can increase the constraints put on the supporting sets and the relation between them, and the relation of the support sets to the whole.

There have been many proposals to generalize association rules, which we do not discuss for lack of space ([2, 13, 10, 17, 7, 18, 9, 1, 8, 16]). Our proposal is different from previous approaches in its use of generalized quantification (which provides a formal, declarative, high-level framework), and in the kinds of extensions proposed.

3.1 Classes of association

Current association rules cover only one type of positive association, which can be expressed through the GQ **at least n out of m**(A, B), with the following definition: $\{A, B \subseteq M \mid\mid A \cap B \mid \geq n \wedge \mid A \mid = m\}$. Here, the support for $A \Rightarrow B$ can be expressed as $\frac{x_1}{m+x_3+x_4}$, and the confidence as $\frac{n}{m}$. However, with different requirements over x_1, x_2, x_3, x_4, the GQ **at least n out of m** can be transformed into other quantifiers and may be interpreted as other types of associations. We can classify all these associations as *positive* and *negative*. Thus, we look not only at positive associations, but also at negative associations. In other words, **no** or **a few** or **all but one** are important correlations too, which can be captured easily in a generalized framework, of which the traditional notion of association rule is one (trivial) case.

We now define traditional association rules and then define the general (positive and negative) case.

Definition 3.1 A traditional association rule between itemsets A, B with minimum confidence T_{conf} and minimum support s exists iff

- $\mid A \mid > s$ and $\mid B \mid > s$ (i.e both A and B are large itemsets);
- $\mid A \cap B \mid > s$ (i.e. $A \cap B$ is also a large itemset);

- $\frac{x_1}{x_1+x_2} > T_{conf}.$

This definition has two serious deficiencies. First, both the support and the confidence are fixed in advance. Such measures are bound to be somewhat arbitrary and may not adequately capture a wide range of possible associations. Therefore, we consider general ways of expressing degrees of confidence. For now, we will consider a certain level of support s fixed, and we will refer to any itemset A simply as being large (that is, $\mid A \mid > s$). The reason to treat support and confidence in different ways is discussed in section 5. Second, this very general definition allows for rules that suffer from the problem of negative correlation. In order to avoid negative correlations, a necessary condition for rules $X \Rightarrow Y$ is that what T_X and T_Y have in common (i.e. $x_1 = T_X \cap T_Y$) is large enough *compared to the sizes of T_X, T_Y and T*. To capture this intuition, we add the following rule to the definition: $\frac{x_1}{x_1+x_2} > \frac{x_1+x_3}{x_1+x_2+x_3+x_4}$ (that is, $\frac{|T_X \cap T_Y|}{|T_X|} > \frac{|T_Y|}{|T|}$). Note that this added condition implies that $\frac{|T_X \cap T_Y|}{|T|} - \frac{|T_X|}{|T|} \times \frac{|T_Y|}{|T|} > 0$. This demands that the support for the two itemsets combined is greater than the product of the support of each itemset separately, and thus indicates *independence* (i.e. is equivalent to $P(Y|X) > P(Y)$). In our example (the rule involving milk and eggs), this would come down to $\frac{2}{3} > \frac{4}{5}$, which is obviously not met, and therefore the rule would be rejected.

Definition 3.2 A *strong positive association* between itemsets A, B with minimum confidence T_{conf} exists iff

- A and B are both large itemsets;
- $A \cap B$ is also a large itemset;
- $\frac{x_1}{x_1+x_2} > \frac{x_1+x_3}{x_1+x_2+x_3+x_4}$
- $\frac{x_1}{x_1+x_2} > T_{conf}.$

where T_{conf} is a formula with parameters x_1, x_2, x_3, x_4.

Strong positive associations represent new classes of association rules that can be mined from transactional databases. Note that not any positive association is an interesting one: some of these new classes are very useful because they represent new knowledge about the database, while others are only extensions of the theory and do not have much practical implication. As an example, **all** requires that $x_2 = 0$, and therefore its confidence must be 1. Also, **some** requires that $x_1 > 0$, so it only requires that confidence be greater than 0. Finally, **most** requires that $x_1 > x_2$, which in turn requires that confidence be greater than 0.5. Interesting associations constraint their confidence measures in more restrictive ways.

The above framework should be used by focusing on associations which put a non trivial constraint on the confidence measures; we next define a measure to show the intended use of the framework. Define $most - often(A, B)$ as $\{A, B \subseteq M \mid \forall C \subseteq M \mid T_{A,B} \mid \geq \mid T_{A,C} \mid\}$. Then this requires that

$$\frac{x_1(\text{on } A, B)}{x_1(\text{on } A, B) + x_2(\text{on } A, B)} > \frac{x_1(\text{on } A, C)}{x_1(\text{on } A, C) + x_2(\text{on } A, C)}$$ (in other words, it requires that the confidence of $A \Rightarrow B$ be greater than or equal to that of $A \Rightarrow C$, for any itemset C). We are setting $T_{conf} = \frac{x_1'}{x_1' + x_2'}$ for any itemset Z with $| T_Z |= x_1' + x_2'$ and $| T_A \cap T_Z |= x_1'$. Such association makes a non-trivial requirement of the confidence measure, but it can interpret intuitions expressed with words like *typical, popular, very often,..*

3.2 Negative Associations

Negative associations can be considered symmetrical to positive ones. Thus, in general form they can be expressed with the generalized quantifier **at most n out of m**$(A, B) = \{A, B \subseteq M \mid\mid A \cap B \mid\leq n \wedge \mid A \mid= m\}$. As before, the support for $A \not\Rightarrow B$ can be expressed as $\frac{m + x_3}{m + x_3 + x_4}$, and the confidence as $\frac{n}{m}$. For negative associations, the necessary condition for rules $X \not\Rightarrow Y$ is that T_X and T_Y have very little in common, where very little is defined in relation to the sizes of T_X and T_Y. In this case, a condition like $\frac{x_1}{x_1 + x_2} < \frac{x_1 + x_3}{x_1 + x_2 + x_3 + x_4}$ captures this intuition.

Definition 3.3 A *strong negative association* between itemsets A, B with maximum overlap T_{conf-} exists iff

- A and B are both large itemsets;
- $A \cap B$ is *not* a large itemset;
- $\frac{x_1}{x_1 + x_2} < \frac{x_1 + x_3}{x_1 + x_2 + x_3 + x_4}$
- $\frac{x_1}{x_1 + x_2} < T_{conf-}$.

where T_{conf-} is a formula with parameters x_1, x_2, x_3 and x_4.

The intuition behind the definition is to look for strong correlations of a negative nature. In particular, we are still looking at large itemsets, and there are still strict measures that must be passed for an association to be considered as negative. This attacks the problem that arises when the concept of negative association is not carefully formulated, namely that there may be many negative associations (in the sense that many more items are not related than related). Intuitively, we expect that most pairs of itemsets will not display strong associations of any kind (positive or negative); a few will show strong positive associations, and a few will show strong negative associations. These negative associations are worth considering as they still characterize the behavior of some large itemsets and may indicate a meaningful relation between the itemsets. For instance, in some domains (like diagnosis in medicine) knowing that something is not the case (a symptom is not present or not associated with a given diagnostic) is as important as knowing that something is the case.

What was said of positive associations is also true of negative associations; some of them are hardly interesting: **no** requires that $x_1 = 0$, and therefore its confidence must be 0. Also, **most not** requires that $x_1 < x_2$, so it only requires that confidence be smaller than 0.5. Finally, **some not** requires that $x_2 > 0$, which in turn requires that confidence be less than 1. Thus, as before, particular strong negative associations must be defined by giving a formula for T_{conf^-}.

4 Implementation

We turn now to the issue of finding an efficient implementation for the extended association rules just defined. We deal separately with support for generalized positive associations and generalized negative associations.

In the case of positive associations, it is obvious that the standard *a priori* algorithm can be extended to support mining such associations. The intuitive idea is that we are still looking for large itemsets, and only when it comes time to formulate the rules we have additional constraints to run, in addition to standard confidence and support. However, our additional constraints are all based on the cardinalities x_1, x_2, x_3 and x_4, which can be easily obtained with little overhead. To see why, observe that for any large itemset Z, we consider rules of the form $Z_1 \Rightarrow Z_2$, where Z_1 and Z_2 constitute a partition of Z (that is, $Z_1 \subset Z$, $Z_2 \subset Z$, $Z_1 \cup Z_2 = Z$ and $Z_1 \cap Z_2 = \emptyset$). Since Z_1 and Z_2 are also large itemsets whose cardinality was computed in some pass of the *a priori* algorithm, we know $\mid T_{Z_1} \mid$ (which is equal to $x_1 + x_2$), $\mid T_{Z_2} \mid$ (which is equal to $x_1 + x_3$), and $\mid T_{Z_1 \cap Z_2} \mid$ (x_1) -since $T_{Z_1} \cup T_{Z_2} = T_Z$, and *a priori* has computed $\mid T_Z \mid$. Thus, we can easily find out x_2 and x_3. Since the size of the whole data set, T, is easily obtainable, and $\mid T \mid = x_1 + x_2 + x_3 + x_4$, we can also deduce x_4. Thus, any additional measure introduced above can be computed with no additional passes on the database.

Most authors believe that mining negative associations is impossible with a standard Apriori algorithm, since we may find a very large number of negative associations, most of them uninteresting and insignificant ([16]). However, we show that it is possible to support efficiently the discovery of many *strong* negative association rules by modifying existing algorithms. We assume that the reader is familiar with the A-priori algorithm in its standard form ([3]). This algorithm generate candidate large k-itemsets in two steps: a *join step* in which all possible candidates are generated, and a *prune step* in which some candidates are deleted. In the prune step, it is checked whether all subsets of the newly generated itemset are large, using the crucial observation of [4]: any subset of a large itemset must be large. In our version of the algorithm, the second phase is modified as follows: when a candidate itemset is found to have weak support, it is a candidate for generation of a negative association. Such itemset is left in the hash tree but marked with a (-). Itemsets with strong support are also left in the hash tree as before, but are marked with a (+). Generation of candidates in the next step then proceeds using only the itemsets marked with a (+). Thus, this phase is the same as before and generates no more candidates than before. Note that, even though we keep more itemsets than the standard algorithm at each level (those marked with a (-)), such itemsets do not participate in the generation of further itemsets. Thus, our hash-tree is roughly of the same size as the standard hash-tree. Note also that we only consider itemsets such that all their subsets have high support (are large itemsets). Itemsets with low support are never the base of further consideration in the generation of candidates (they are always leafs in the hashing tree structure).

Finally, we introduce two extra steps in the standard algorithm. Because we consider elements in C_k with a low support which have all its subsets with high support, rules based on these sets may have very low confidence. An additional step in the algorithm will be to check the confidence of a potential negative associations. The computation is very simple because in the hash-tree structure we already counted and stored support for the given itemset and all its subsets. If the confidence value is below the given threshold value T_c^-, the association is a *strong negative association rule*. For example, if threshold for support is 40%, $T_c^- = 0.2$, and if the itemsets in the hash-tree, with corresponding supports, are

Itemset	Support (%)
A B	45
A C	40
B C	42
A B C	5

then itemset (A B C) is a candidate for generation of a strong negative association. The minimum confidence will be for the rule $\{AB\} \nRightarrow \{C\}$, and the value is $C((AB) \nRightarrow C) = \frac{S(ABC)}{S(AB)} = \frac{5}{45} = 0.11 < T_c^-$. Therefore, the rule $\{AB\} \nRightarrow \{C\}$ will be included into the set of strong negative associations discovered by the modified Apriori algorithm.

The other extra step consists of an extra pass over the hash tree which is done after computation of all large sets is finished. To explain why the additional pass is needed, consider the following example. Assume 1-itemsets A, B, C, D and 2-itemsets AB and CD are the only ones found large. This means that, in pass $k = 2$, itemsets $AC, AD, \ldots$ were found not large and will be considered as candidates for negative correlations. Also, in pass $k = 3$ sets ABC, ACD were considered but discarded in the prune step. These, too, will then be considered as candidates for negative associations. However, in this scenario the algorithm finished with $k = 3$ and never goes to the next iteration as no 3-itemset is found large. Thus, in this example no rule relating maximal large sets AB and CD (i.e. rules of the form $\{AB\} \Rightarrow \{CD\}$) is considered, even though such a rule would be covered under definition 3.3. The problem is that the union of itemsets AB and CD is a 4-itemset $(ABCD)$, which is never considered by the algorithm -for positive or negative associations. While we know that $ABCD$ cannot be large, we still have to consider negative rules based on it, since AB and CD are large. Therefore, an additional pass over the hash tree will consider such rules by computing the confidence among any two large itemsets X and Y which fulfill the following condition. Let max be the highest value of k in the execution of the Apriori algorithm (in our example, $max = 3$). If the union of X and Y is an n-itemset, with $n > max$, then $X \Rightarrow Y$ must be considered for a negative association. Note, though, that such a computation is very inexpensive, as it comes down to computing the confidence of such a rule.

We are currently experimenting with an implementation of our modified A-priori algorithm to assess the performance of the modifications proposed. Preliminary results support our hypothesis that, under our definition of *strong* negative rules, the number of such rules is small and the overhead caused by its computations is manageable.

5 Conclusion and Further Research

Generalized quantification provides a theoretical framework which allows us to define and study several definitions of relationship among itemsets, as well as new measures of relevance for them. This approach is high-level and declarative, and emphasizes a clear definition of the semantics of the relationships. In this paper, we have introduced the framework and showed how it can be used to define several confidence measures which strengthen the usual definition. We have also sketched and implementation that shows that it is possible to support efficiently the mining of this new relationships. The paper reports work in progress; the framework open up interesting avenues of research that we are currently pursuing. An important issue is to develop and study experimentally the implementations sketched here. Another issue is to extend the approach to other types of quantifiers. It must be pointed out that the concept of generalized quantifier is extremely general and powerful; we have only scratched the surface of what is possible to do in the framework. Finally, most of our efforts have concentrated on the definition of new confidence measures, while support have been left aside. However, strong correlations among small itemsets may be of interest, and hence they should be studied. Such correlations, though, require completely new approaches for their implementation, since current approaches all exploit the properties of large itemsets.

References

1. A. Abdulghany. *Cubegrades: Generalization of Association Rules to Mine Large Datasets*. PhD thesis, Rutgers University, 2001.
2. J. M. Adamo. *Data Mining for Association Rules and Sequential Patterns*. Springer, 2001.
3. R. Agrawal, T. Imielinsky, and A. Swami. Mining association rules between sets of items in large databases. In *Proceedings of ACM SIGMOD*, 1993.

4. R. Agrawal, H. Manila, R. Srikant, H. Toivonen, and A. Verkamo. *Advances in Knowledge Discovery and Data Mining*, chapter Fast Discovery of Association Rules. MIT Press, 1996.
5. A. Badia. *A Family of Query Languages with Generalized Quantifiers: its Definition, Properties and Expressiveness*. PhD thesis, Indiana University, 1997.
6. A. Badia, M. Gyssens, and D. Van Gucht. *Application of Logic Databases*, chapter Query Languages with Generalized Quantifiers, pages 235–258. Kluwer Academic Publishers, 1995.
7. R. V. V. Bollmann-Sdorra P., Hafez A. M. A theoretical framework for association mining based on the boolean retrieval model. In e. a. Kambayashi Y., editor, *Data Warehousing and Knowledge Discovery*. Springer-Verlag, 2001.
8. S. C. Brin S., Motwani R. , beyond market basket: Generalizing association rules to correlations. In *Proceedings of the 1997 ACM SIGMOD Conference on Management of Data*, pages 265–276, 1997.
9. L. Feng, Q. Li, and A. Wong. Mining inter-transactional association rules: Generalization and empirical evaluation. In *Proceedings of the Third International DAWAK Conference*, pages 31–39, 2001.
10. V. Ganti, J. Gehrke, and R. Ramakrishnan. Mining very large databases. *Computer*, 32(8):38–45, August 1999.
11. P. Y. Hsu and D. S. Parker. Improving SQL with generalized quantifiers. In *Proceedings of the Tenth International Conference on Data Engineering*, 1995.
12. P. Lindstrom. First order predicate logic with generalized quantifiers. *Theoria*, 32, 1966.
13. B. Liu and et al. Analyzing the subjective interestingness of association rules. *IEEE Intelligent Systems*, pages 47–55, September/October 2000.
14. A. Mostowski. On a generalization of quantifiers. *Fundamenta Mathematica*, 44, 1957.
15. S. Rao, D. Van Gucht, and A. Badia. Providing better support for a class of decision support queries. In *Proceedings of ACM SIGMOD*, 1996.
16. A. Savasere, E. Omiecinski, and S. Navathe. Mining for strong negative associations in a large database of customer transactions. In *Proceedings of ICDE*, 1998.
17. L. W. Tseng M. Mining generalized association rules with multiple minimum support. In e. a. Kambayashi Y., editor, *Data Warehousing and Knowledge Discovery*. Springer-Verlag, 2001.
18. D. Tsur and et al. Query flocks: A generalization 0f association-rule mining. In *Proceedings of the 1998 ACM SIGMOD Conference on Management of Data*, pages 121–142, 1998.
19. D. Westerstahl. *Handbook of Philosophical Logic*, volume IV, chapter Quantifiers in Formal and Natural Languages. Reidel Publishing Company, 1989.

New Geometric ICA Approach for Blind Source Separation

M. Rodríguez-Álvarez [1], F. Rojas[1], C.G. Puntonet[1], F.Theis[2], E.Lang[2],
R.M.Clemente[3]

[1] Departament of Architecture and Computer Technology. University of Granada (Spain)
{mrodriguez, frojas, carlos}@atc.ugr.es
[2] Institute of Biophysics, University of Regensburg (Germany)
{fabian.theis, elmar.lang}@biologie.uni-regensburg.de
[3] Area de Teoría de la Señal. Universidad de Sevilla (Spain)
ruben@cica.es

Abstract. This work explains a new method for blind separation of a linear mixture of sources, based on geometrical considerations concerning the observation space. This new method is applied to a mixture of several sources and it obtains the estimated coefficients of the unknown mixture matrix A and separates the unknown sources. In this work, the principles of the new method and a description of the algorithm are shown.

1 Introduction

The separation of independent source signals from mixed observed data is a fundamental and challenging signal processing problem. In many practical situations, one or more desired signals need to be recovered blindly knowing only the observed sensor signals. When p different source signals propagating through a real medium have to be captured by sensors, these sensors are sensitive to all sources $s_i(t)$ and thus the signal $x_k(t)$, observed at the output of sensor k, is a mixture of source signals. With a linear and stationary mixing medium the sensor signals can be described by:

$$\vec{x}(t) = A\,\vec{s}(t) \tag{1}$$

where $\vec{x}(t) = (x_1(t), \ ..., \ x_n(t))^T$ is an experimentally observable $(n \times 1)$-sensor signal vector s(t), with $\vec{s}(t) = (s_1(t), \ ..., \ s_p(t))^T$ is a $(p \times 1)$- unknown source signal vector having stochastic independent and zero-mean non-Gaussian elements $s_i(t)$, and A is a $(n \times p)$ unknown full-rank and non-singular mixing matrix. The solution of the blind signal separation (BSS) problem consists of retrieving the unknown sources $s_i(t)$ from just the observations. To achieve this it is necessary to apply the hypotheses that the sources $s_i(t)$ and the mixture matrix $A = (\vec{a}_1, ..., \vec{a}_n)^T$ are unknown, that the number n of sensors is at least equal to the number p of sources, i.e. $n \geq p$, and that the components of the source vector are statistically independent yielding:

$$p(\vec{s}) = \prod_{i=1}^{n} p(s_i) \tag{2}$$

In order to solve the BSS problem a separating matrix W is computed whose output is an estimate of the vector $\vec{s}(t)$ of the source signals such that:

$$\vec{y}(t) = W^{-1} \vec{x}(t) \tag{3}$$

Any BSS algorithm can only obtain W subject to:

$$W^{-1}A = DP \tag{4}$$

with a diagonal scaling matrix D modified by a permutation matrix P. Recently, BSS and ICA (Independent Component Analysis) have received much attention because of its potential applications in signal processing. A great diversity of estimation methods have been proposed based on some kind of statistical analysis, neural networks [7], the entropy concept [3], the geometric structure of the signal spaces [1], [6], the fixed-point algorithm FastICA [5], the maximum likelihood stochastic gradient algorithm [2], the Jade algorithm [4], among others. Several geometric procedures have been used to separate either multivalued or analog signals, by analyzing the observed sensor signals in the resulting p-dim space of observations. In the following we will present a new geometric ICA algorithm which is based on rough density estimation.

2 Principles of the new method

For $p = 2$ and with bounded values in a uniform distribution, the observed signals $(x_1(t),\ x_2(t))$ form a parallelogram in the $(\vec{x}_1, \vec{x}_2)$ space, as shown in Figure 1. We have demonstrated [8] that, through a matrix transformation, the coefficients of the matrix coincide with the slopes of the parallelogram. It can be seen that for random uniform sources, the parallelogram representing the space of observations $(\vec{x}_1, \vec{x}_2)$ is geometrically bounded within the segments between the points P_1 to P_4. The slopes of these segments give the coefficients of the estimated mixture matrix W. In order to obtain these segments, it is necessary to estimate the coordinates of those points $P_i,\ _{i\ =\ 1,\ 2,\ 3,\ 4}$. Assuming non-uniformly distributed signal as the sources, for example speech signals with an underlying super-Gaussian distribution; the form of the sensor signal distribution in the space of observations is highly non-uniform too, as can be seen in Figure 2. In this case it is not sufficient to estimate the borders of the bounded space of observations. Rather, it is necessary to detect the directions of high density in the space of observations. These directions are called ICA axes (ICA-1 ; ICA-2).

2.1 Description of the algorithm

First of all, the algorithm computes the kurtosis of each component of the sensor signals and also the correlation coefficients between all observations. This is to detect whether the underlying source signal distributions correspond to sub- or super-Gaussian distributions. According to the Central Limit Theorem, mixtures will tend to be closer to Gaussian than the original ones. Consequently, kurtoses of the mixtures will be closer to zero (Gaussian distribution) than the sources:

$$\left| Kurt(x_i) \right| \leq \max \left\{ \left| Kurt(s_j) \right| \right\} \; ; \; i, j \in [1, ..., n] \tag{5}$$

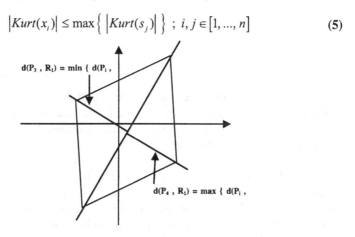

Fig. 1. Space of observations: Representative points and straight lines.

In any case, for mixtures of two signals, they will tend to preserve the sub- or super-Gaussian nature of the original signals, assuming that both sources have the same sign in the kurtosis. If the kurtoses of all observations are positive, the algorithm searches for high density regions of the sensor signal distribution. With sub-Gaussian signals, the algorithm estimates the bounding box of the parallelogram representing the space of observations. The algorithm subdivides the space of observations $(\bar{x}_1, \bar{x}_2)$ into a regular lattice of cells with N-rows and M-columns as shown in Figure 2. Then, the algorithm computes the number of cells in the lattice in which the number of points inside it is greater than a given threshold TH. The distribution of sensor signals within each of these cells then is replaced by a prototype sensor signal vector. The prototype vector mostly does not point towards the centre of the cell because its position is weighted by the density of points (x_{1i}, x_{2i}) in this cell. The next step of the algorithm finds those points which either form the border of the hyperparallelepiped or mark the high density regions of the sensor signal distribution in the space, by looking for cells that have an empty neighborhood (such cells have fewer points than the threshold TH).

Then these cells without a complete neighborhood form the border of the distribution encompassing NR data points in the space of observations. The algorithm then computes the coordinates of $P_1 = (p_{11}, p_{12})$ and $P_2 = (p_{21}, p_{22})$. The space of observations has been reduced to NR data points which, in two dimensions, represent pairs of coordinates (x_{1i}, x_{2i}). In this reduced set of NR data points, there exist data points P_1 and P_2 with largest Euclidean distance between them in the space of observations :

$$d(P_1, P_2) = \max_{i, j \in (1,2,....NR)} d(P_i, P_j) \qquad (6)$$

Once points P_1 and P_2 have been identified, the algorithm calculates the equation of the straight line R_1 which passes through these points P_1 and P_2 :

$$Ax_1 + Bx_2 + C = 0 \qquad (7)$$

being

$$A = (p_{22} - p_{12}), \quad B = (p_{11} - p_{21}), \quad C = (p_{21} - p_{12}) - (p_{22} - p_{11}) \qquad (8)$$

Next, the algorithm estimates the coordinates of the points $P_3 = (p_{31}, p_{32})$ and $P_4 = (p_{41}, p_{42})$ as follows: the straight line R_1 divides the space of observations $(\bar{x}_1, \bar{x}_2)$ into two subspaces, being R_1 the border between them. Data points which lie within one of these subspaces yield a nonzero result in Eq. (7). For example, data points lying above the straight line R_1 yield a negative result in Eq. (7). There is then one data point $P_3 = (p_{31}, p_{32})$ which provides the most negative value of all possible outcomes of Eq. (7), hence which also represents the point with the greatest Euclidean distance from the straight line R_1 in the subspace above R_1. In the same way, points in the other subspace, below the straight line R_1, yield a positive result in Eq. (7). Again, there is one point $P_4 = (p_{41}, p_{42})$ that provides the most positive value of all possible results from Eq. (7), and which is also the point with greatest Euclidean distance from the straight line R_1 in the subspace below R_1.

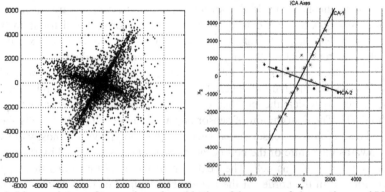

Fig. 2. Linear mixture of two real words and lattice of the space of observations and ICA axes.

Once the characteristical points of the parallelogram have been obtained, the algorithm computes either the slopes of the segments ($\overline{P_1 P_3}$ and $\overline{P_1 P_4}$ or, equivalently $\overline{P_2 P_4}$ and $\overline{P_3 P_2}$) in case of sub-Gaussian densities or the slopes of the diagonals ($\overline{P_1 P_2}$ and $\overline{P_3 P_4}$) in case of super-Gaussian densities in order to obtain the slopes of the ICA axes and the coefficients of the matrix W as in Eq. (9) (see Figure 2):

$$\left(\frac{a_{12}}{a_{22}}\right)^{-1} = \frac{p_{32} - p_{12}}{p_{31} - p_{11}} \quad ; \quad \left(\frac{a_{21}}{a_{11}}\right) = \frac{p_{42} - p_{12}}{p_{41} - p_{11}} \tag{9}$$

Using the coefficients of matrix W, the algorithm computes the inverse matrix W^{-1} and reconstructs the unknown source signals $\vec{s}(t)$ (see Eq. (3)).

2.2 Further enhancements

The computational order of the algorithm is polynomial:

$$Comput - Order = (DataPoints^2 \cdot XColumns \cdot YRows) \tag{10}$$

As a further improvement, we propose the reduction of the number of points at the beginning of the algorithm with a random elimination through all the space of the joint distribution of the mixtures as long as enough data points are kept to correctly estimate the sources. A more elaborated proposal is eliminating those points of the joint distribution of the mixtures which lay within a calculated radius near the center of the joint distribution, because they are useless for the algorithm, due to its nature of computing contours using points whose Euclidean distances are the highest. From experimental results, we have derived equations (11) and (12) for the calculation of the radius based on the kurtosis and correlation of the mixture signals.

For sub-Gaussian mixtures, the algorithm will try to find the contour of the sensor signal distribution. In this case we determine the exclusion radius as follows :

$$R = \frac{\alpha}{\rho(x)^2 + 0.1} \cdot \overline{x} \tag{11}$$

where α is a constant (experimentally, a value of $\alpha=7.5$ was applied), $\rho(x)$ is the correlation of the mixtures and

$$\overline{x} = \sqrt{\sum_{j=1}^{N} x(1, j)^2 + x(2, j)^2} \tag{12}$$

For super-Gaussian mixtures (positive kurtosis), the algorithm will search for high density regions of the joint distribution of the mixtures. Thus, the exclusion radius was calculated as:

$$R = 1.5 \cdot \overline{x} \tag{13}$$

3   Simulations and Results

The new algorithm, named as "LatticeICA", has been tested on various ensembles of artificial sensor signals with an arbitrary number of samples drawn at random from sub- and super-Gaussian distributions like uniform, Gamma, Laplacian and Delta distributions, as well as with real world speech signals. To quantify the performance achieved we calculate both a crosstalking error of the original and recovered source signals as proposed by Amari et al. [2] as well as a component wise crosstalk:

$$E(P) = \sum_{i=1}^{n}(\sum_{j=1}^{n}\frac{|P_{ij}|}{\max_k |P_{ik}|}-1) + \sum_{j=1}^{n}(\sum_{i=1}^{n}\frac{|P_{ij}|}{\max_k |P_{kj}|}-1) \tag{14}$$

where $P=(p_{ij})= W^{-1}\cdot A$. The parameter MSE (Mean Square Error) measures the similarity of the signals $s_i(t)$ and $y_i(t)$.

3.1   Speech signals.

In this simulation the algorithm separate two super-Gaussian speech voice signals with 10000 samples each. The lattice was automatically computed to be 11 rows and 11 columns, using TH = 10. The original and estimated matrices were:

$$A = \begin{bmatrix} 1 & 0.70 \\ -0.30 & 1 \end{bmatrix}; W = \begin{bmatrix} 1 & 0.65 \\ -0.28 & 1 \end{bmatrix} \tag{15}$$

The joint distribution of the mixtures points out the super-Gaussian nature of the sources (see Figure 3). The matrix performance index for this simulation was $E(W, A) = 0.113$, with Crosstalk1 (E_{s1}) = -29 dB and Crosstalk2 (E_{s2}) = -34 dB. In Figure 3 it is shown how the algorithm searches for the lines of higher density instead of the contour plot.

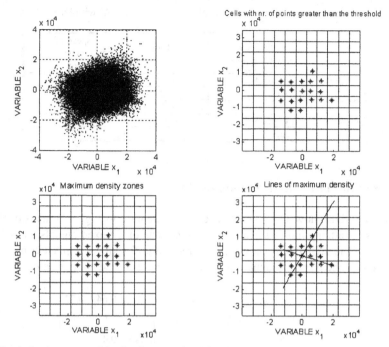

Fig. 3. Performance of the LatticeICA algorithm for a two real voice signals mixture.

3.2 Comparison with other algorithms.

In this simulation we started a more systematic exploration of the algorithm and compared the results to those obtained with two other algorithms, the FastICA [5] and Jade algorithms [4]. We tried random mixture matrixes over uniform and Laplacian mixtures of 10000 samples, running 100 simulations each time, with automatic parameters. With FastICA the number of bins has been choosen in all cases to be 180. The NRMS (normalized root mean squared error) in each case and the corresponding average convergence times (Pentium IV 1.5 GHz., 512 MB RAM, under Matlab environment) are summarized in Table 1. Although, both FastICA and Jade algorithms globally get better results than LatticeICA in most of the simulations, LatticeICA shows a great performance especially for super-Gaussian mixtures (speech signals) and it outperforms previous geometric algorithms. As a particular advantage of LatticeICA when compared with FastICA and Jade it remains its easy hardware implementation, due to the fact that it only computes simple arithmetic operations. Future enhancements in fine tuning the radius of exclusion and adjusting the final separation lines will certainly lead to a better performance.

Table 1. Comparison of performance of the algorithm (LatticeICA) with FastICA and Jade.

Source Type	Procedure	NRMS	Speed of convergence (ms.)
Uniform	Lattice ICA	0.054	808
	FastICA	0.021	501
	Jade	0.028	584
Laplacian	Lattice ICA	0.034	703
	FastICA	0.087	406
	Jade	0.009	273

3.3 Extension to higher dimensionality.

Finally, we show how this algorithm can be extended to higher dimensionality situations by attempting to separate the projections of p mixed signals from $\mathbb{R}^p$ onto $\mathbb{R}^2$. The signals are shown in figures 4 and 5 (with a Laplacian noise, a music source and a speech signal). The original and obtained matrices are:

$$A = \begin{bmatrix} 1 & 0.5 & 0.5 \\ 0.5 & 1 & 0.5 \\ 0.5 & 0.5 & 1 \end{bmatrix} \qquad W = \begin{bmatrix} 1 & 0.64 & 0.73 \\ 0.45 & 1 & 0.63 \\ 0.46 & 0.47 & 1 \end{bmatrix} \tag{16}$$

In Figure 4 can be seen the 3-dimensional mixture and the projections in each of the planes which will be the inputs to the algorithm. Figure 5 depicts the separated signals of the proposed LatticeICA algorithm.

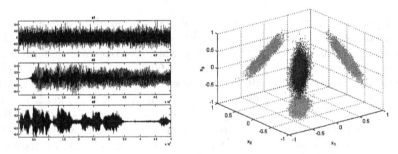

Fig. 4. Left: laplacian noise, music and speech source signals. Right: three-dimensional mixture and projections.

4   Conclusions

We have developed a new geometry-based method for blind separation of sources which greatly reduces the complexity and computational load inherent in the standard

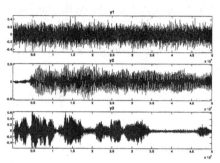

Fig. 5 Estimated signals for Simulation 3.3.

geometric ICA algorithms. This new algorithm is based on a tessellation of the input space where in each cell a code book vector is determined to represent the center of gravity of the local distribution of sample vectors. Depending on the type of distribution, either sub- or super-Gaussian, the slopes of the border lines or the diagonals are determined to obtain the coefficients of the estimated mixing matrix W. The method lends itself for an easy hardware implementation and is also very intuitive in terms of computer applications. Furthermore, this method could be used to detect the perimeter or outlines in simple two-dimensional figures. In the future we will intend to implement this method for more than two signals without using projections but working in the p-dimensional space.

Acknowledgement

This work has been supported by the Spanish CICYT Project TIC2001-2845 "PROBIOCOM - Procedures for Biomedical and Communications for BSS".

References

1.   Álvarez, M. R., Puntonet, C. G., Rojas, I.: Separation of Sources based on the Partitioning of the Space of Observations. Lecture Notes on Computer Science, Vol. 2085. Springer-Verlag, Berlin-Heidelberg-New York (2001) 762 – 769.
2.   Amari, S. I., Cichocki, A., Yang, H. H.: A new learning algorithm for blind signal separation, Proceedings of NIPS'96, (1996) 757-763.
3.   Bell A. J., Sejnowski, T. J.: An information-maximisation approach to blind separation and blind deconvolution, Neural Computation, Vol. 7 (1995) 1129-1159.

4. Cardoso, J. F.: High-order contrasts for independent component analysis. Neural Computation, Vol. 11, n° 1 (1999)157-192.
5. Hyvärinen, A., Karhunen, J., Oja, E.: Independent Component Analysis. Wiley & Sons, New York (2001).
6. Jung, A., Theis, F. J., Puntonet, C. G., Lang, E. W.: FASTGEO: A histogram based approach to linear geometric ICA. Proceedings of ICA'01 (2001) 349-354.
7. Jutten, C., Hérault, J., Comon, P., Sorouchiary, E.: Blind separation of sources, Parts I, II, III. Signal Processing, Vol. 24, n° 1 (1991) 1-29.
8. Puntonet, C. G., Prieto, A.: Neural net approach for blind separation of sources based on geometric properties. Neurocomputing, Vol. 18 (1998) 141-164.

A Taxonomy of Data Mining Applications Supporting Software Reuse

S. Tangsripairoj and M. H. Samadzadeh

Department of Computer Science, Oklahoma State University
Stillwater, OK 74078, USA
tangsri@a.cs.okstate.edu and samad@a.cs.okstate.edu

Abstract. A taxonomy is a classification of items in a systematic way based on their inherent properties and relationships. In addition to serving as a descriptive facility to distinguish among existing items, a taxonomy typically contains provisions for not only predicting items not among its baseline set, but also the ability to prescribe new items. Data mining is an advanced data analysis technique whose primary function is to extract likely useful knowledge or hidden patterns from large databases. Software reuse, the development of software systems from previously constructed software components rather than from scratch, is considered a means of improving software productivity and quality. For efficient and effective reuse, a software library or repository can be built to store and organize a collection of software components. It is indispensable that a software library provide tools for software developers to locate, compare, and retrieve reusable software components that meet their requirements. Data mining tools, techniques, and approaches can be employed to acquire useful information about software components in a software library. Such information can be beneficial to software developers in searching for desired reusable software components. In this paper, we catalog the major characteristics of several existing data mining applications supporting software reuse, and propose a taxonomy based on these characteristics. The taxonomy provides a predictive framework to help identify possible new data mining applications.

Keywords: data mining, taxonomy, software library, software reuse

1. Introduction

Data mining presents a practical solution to various problems in real world applications such as credit card fraud detection, credit approval, customer purchase behavior analysis, stock market prediction, medical diagnosis, weather forecasts, and sky object classification [1, 6, 20]. The primary function of data mining is to extract likely useful knowledge or hidden patterns from large databases [6]. This knowledge is meaningful for analysts to improve decision making, predictions, and planning [26].

Software reuse is the development of software systems from previously constructed software components [19] or existing software knowledge [8] rather than from scratch. The most outstanding benefits of software reuse over conventional software development are improving software productivity and quality, achieving saving in terms of cost, time, and effort to implement new software systems, as well as reducing maintenance costs [34].

For years, software reuse has been utilized in the software development practices of many organizations [8, 27]. In practice, the amount of software components grows steadily such that it may become difficult for software developers to keep up with all the potentially reusable software components constructed. Consequently, various organizations have built software libraries to store, organize, and catalog reusable software components. It is necessary that a software library provide the means for software developers to locate, compare, and retrieve candidates for potential reuse [12, 17].

An analogy is drawn between mining for useful knowledge in large databases and searching for reusable software components in a software library. This observation suggests that data mining tools, techniques, and approaches can be employed to acquire useful knowledge about software components in a software library. This knowledge can be beneficial to software developers in finding the closest software components, i.e., the optimum "fits", for their needs.

The main thrust of this paper was to catalog the major characteristics of several existing data mining applications supporting software reuse and to propose a taxonomy based on these characteristics. The taxonomy resulting provides a predictive framework to help identify possible new data mining applications.

2. Software Library

To establish a successful software reuse, an important prerequisite step is the construction of a software library to store, organize, and catalog collections of available software components and provide tools for developers to find and understand the most suitable software components for the task at hand [12, 17].

Building a software library involves three major tasks: defining types of reusable software components, defining classification methods for describing software components, and defining search and retrieval mechanisms for software developers to locate candidates for potential reuse.

1) Defining types of reusable software components

The types of software components that can be reused are not confined to fragments of source code. A reusable component can be "any information which a developer may need in the process of creating software [7]". In other words, products generated during the software development life cycle (SDLC) are all candidates for reuse. The following are examples of software components categorized according to the phases of the SDLC in which they are produced [31].

- *Requirement analysis and specification*: feasibility study documents, requirement documents, specification documents, etc.
- *System and software design*: design cases, design templates, design patterns, application frameworks, software architectures, user interface designs, etc.
- *Implementation and unit testing*: program/subprogram code fragments, library functions, object classes, macros, third-party software packages, etc.
- *Integration and system testing*: test plans/cases/reports, etc.
- *Operation and maintenance*: programmer's guide, user's manual, etc.

2) Defining classification methods

A classification method is a way for defining a representation or description of the software components stored in a software library [9]. By using a classification method, the software components are systematically organized into meaningful structures that enable software developers to understand software components they need without frustration and delay.

There are four primary classification methods that most existing software libraries use: enumerated classification, faceted classification, attribute-value classification, and free text keyword classification [9]. In *enumerated classification*, a software component is assigned to one of the predefined categories listed in a hierarchical structure. In *faceted classification*, a software component is represented by a set of facets and facet values. In *attribute-value classification*, a software component is described by a set of attributes and their values. In *free text keyword classification*, a software component is associated with terms that are automatically extracted from software documentation (such as manual pages and code comments) by using classic retrieval techniques.

In addition to these methods, many effective classification methods for describing software components have been proposed, e.g., an Extensible Description Formalism classification scheme [32], a combination of classification techniques [29], a multi-tiered classification scheme [30], and a Reuse Description Formalism [15].

3) Defining search and retrieval mechanisms

A search and retrieval mechanism is a means for software developers to locate candidates for potential reuse. Two classic search and retrieval mechanisms are browsing and keyword searching. *Browsing* provides a natural search method for exploring a software library. Software developers can understand the relationships among indexing terms and find the desired software components by moving up and down the hierarchy structure. *Keyword searching* enables software developers to confine their attention to a specific group of software components by formulating a query to express their domain of interest. A query usually consists of a set of keywords and operators (e.g., AND, OR, NOT, or double quotes). These operators are used to create complex queries and assist query refinements.

A large number of search and retrieval mechanisms have been developed. The following are but a few examples: a generalized behavior-based retrieval [13], an incremental query refinement [14], profile/signature matching approaches [22], retrieval with different levels of accuracy (i.e., exact match, match, and similar) [3], and using a learning agent to assist the browsing of software libraries [5].

3. Data Mining Tasks and Techniques

For decades, the advanced technology in data storage, database management systems, and data warehousing has enabled organizations to accumulate a great deal of data in very large databases. Unfortunately, the traditional data analysis mechanisms, like statistical analysis or querying systems, offer only informative summary reports, but cannot in general help extract useful information. Moreover, as the quantity of data grows progressively, these mechanisms are very expensive and time-consuming to exploit. Hence, data mining technology has emerged to alleviate some of this difficulty [6].

Data mining is a relatively new and sophisticated way of analyzing data, whose goal is to extract useful information from a large collection of data [6]. In this context, useful information encompasses hidden patterns, possibly unknown relationships among data, trends or behaviors, a summarization or generalization of the original data. This information helps analysts not only to understand data more deeply but also to make decisions more effectively.

Data mining is also known as Knowledge Discovery in Databases (KDD). In fact, strictly speaking, data mining is one of the fundamental steps of the overall KDD process, which refers to the process of transforming low-level data to high level information [6]. Typically, KDD consists of many steps including understanding the application domain, defining the data mining goals, selecting a target data set, preparing data for analysis, choosing appropriate data mining algorithms, performing data mining, evaluating the results, interpreting the derived patterns, and employing the discovered knowledge [6]. The KDD process is not linear; rather it is interactive and iterative in nature. The result of any one step may cause changes in the preceding or succeeding steps.

Data mining can carry out a wide spectrum of tasks having different objectives such as classification, clustering, associations, regression, summarization and generalization, dependency modeling, change and deviation detection, model visualization, and exploratory data analysis [2, 6, 11]. To cope with different kinds of tasks, various data mining techniques and their efficient algorithms have been developed and deployed by researchers over the last few years. Here are some examples: decision trees, mining association rules, clustering, neural networks, genetic algorithms, case-based reasoning, statistical methods, Bayesian belief networks, fuzzy sets, and rough sets [2, 6, 11]. A brief description of three of the best-known data mining tasks as well as their prominent techniques and algorithms are given below.

1) Classification

The task is to classify data items into one of several predefined classes based on their values in certain attributes [2]. For example, in massive customer databases it can be used to classify customers according to their preference for magazines. A decision tree constructed from a training set of data items is the most commonly used technique for classification. A decision tree consists of non-leaf nodes and leaf nodes, where a non-leaf node denotes a test on a single attribute value and a leaf node denotes a class. The tree is subsequently used to classify new data items, whose classes are unknown, by testing the attribute values of the new data item beginning at the root node and ending at a leaf node. Well-known examples of decision tree algorithms are ID3, CART, XAID/CHAID, and C4.5 [18].

2) Clustering

The task is to divide data items into classes or clusters according to similarity, which is evaluated by a numerical measure such as the Euclidean distance, the squared Mahalanobis distance, or the Hausdorff distance [16]. Unlike classification, these classes are not predefined but determined from the data itself. For example, it can be used to find subgroups of customers having similar purchase behaviors. Important clustering techniques are hierarchical clustering algorithms, partition algorithm, nearest neighbor clustering, fuzzy clustering, artificial neural networks for clustering (e.g., Kohonen's learning vector quantization (LVQ) and self-organizing map (SOM)), and evolutionary approaches for clustering (e.g., genetic algorithms, evolution strategies, and evolutionary programming) [16].

3) Associations

The task is to derive a set of association rules based on statistical significance. For example, it can be used for market-basket analysis, which is the process of determining which products a customer typically purchases at the same time. This kind of information helps retailers to understand customers' purchase behaviors and lead to improved decisions on product location and promotion. The Apriori algorithm is the pioneer algorithm for mining association rules. A large number of successor algorithms have been proposed to enhance the performance of the Apriori algorithm such as partition-based algorithm, hash-based algorithm, sampling-based algorithm, Dynamic Itemset Counting algorithm, mining generalized and multi-level association rules, and mining sequential patterns [2, 10].

4. Existing Applications

In the literature, numerous data mining applications supporting software reuse have been proposed by several researchers. The underlying idea behind these applications is the extraction of useful knowledge about software components in a software library. We surveyed and cataloged several existing applications. For each application, we examined its major characteristics. The descriptions of these applications appear below. The summary tables are shown in Tables 1 and 2.

- *Classifying software components*

Esteva [4] proposed *the Inductive Classification (IC) system* to determine whether or not a software module has potential for reusability. The IC system applies inductive learning techniques to produce a decision tree to classify modules into two classes: reusable and non-reusable. Each module is described by a set of attributes that are measured in terms of software complexity metrics associated with various aspects of program structure including modularity, cohesion, coupling, size, data structure, control structure, and documentation.

- *Clustering software components*

Maarek and her colleagues [23] invented *the GURU system* to construct a software library from a collection of software components. The system uses an indexing scheme based on the notions of lexical affinities and quantity of information extracted from documentation to represent a component. The system applies hierarchical agglomerative clustering methods to generate a browse hierarchy, which guides the search for appropriate software components. This technology has been applied to construct a library of 1100 AIX utilities.

Merkl and his colleagues [24] implemented *the Self-Organizing Feature Map (SOFM) system* to organize a software library according to the semantic similarity or functional similarity of the software components. The system used the self-organization map (SOM) technology, an unsupervised learning paradigm of neural networks, to create a two-dimensional map that helps visualize the structure of the software library, where software components having similar behavior are mapped onto geographically closer regions of the map. Each component is represented by a feature vector or a set of keywords extracted from the documentation. A set of 36 MS-DOS commands is contained in the experimental library.

Ye and Lo [33] also applied the SOM technique as prescribed in *the Software Self-Organizing Map (SSOM) system.* The design goals of the SSOM system were similar to those of the SOFM system. However, SSOM improved on the way to identify keywords associated with software components based on automatic indexing rather than manual indexing as used in the SOFM system. The method has been applied to a collection of 97 UNIX commands.

Pedrycz and his colleagues [28] employed the SOM technique in a new dimension, especially to analyze software measure data. This system is called *the SOM Clustering Analysis (SOMCA) system.* Each software component is characterized by a set of software metrics, e.g., lines of code, number of methods, and depth of inheritance tree. A sample of 643 JAVA classes was used as experimental data.

Lee and his colleagues [21] used genetic algorithms in *the Reusable Class Library (RCL) system* with the goal of finding optimized clusters into which software components are classified, and finding an optimal query which retrieves clusters containing software components similar to a given query. The system characterizes software components with the faceted classification method.

- *Mining reuse patterns*

Michail [25] developed *the CodeWeb system* to discover reuse patterns of library classes and member functions that are normally reused in combination by application classes. The system uses "generalized association rules", which improve upon the standard association rules mining technique by taking into account the inheritance hierarchy. Each application class, serving as a component, is associated with a set of items that indicate reuse relationships involving library classes or member functions. Five reuse relations were considered: class inheritance, class instantiation, function invocation, function overriding, and implicit invocation. The demonstration was conducted with 76 real-life C++ applications in order to mine reuse patterns for the KDE 1.1.2 core libraries.

Table 1. Characteristics of the existing applications

Tool	Data Mining Task	Data Mining Technique
The IC system	Classification	Decision Trees
The GURU system	Clustering	Hierarchical Clustering
The SOFM system	Clustering	Neural Networks (Self-Organizing Map)
The SSOM system	Clustering	Neural Networks (Self-Organizing Map)
The SOMCA system	Clustering	Neural Networks (Self-Organizing Map)
The RCL system	Clustering	Genetic Algorithms
The CodeWeb system	Associations	Mining Generalized Association Rules

Table 2. Characteristics of the existing applications

Tool	Software Component	Software Classification Method
The IC system	Pascal programs	Attribute-valued (Software metrics)
The GURU system	AIX utilities	Free text keyword
The SOFM system	MS-DOS commands	Free text keyword
The SSOM system	UNIX commands	Free text keyword
The SOMCA system	JAVA classes	Attribute-valued (Software metrics)
The RCL system	Components generated	Faceted
The CodeWeb system	C++ applications	Free text keyword (Items indicating reuse relationships)

5. Taxonomy

In this section, we propose a taxonomy that can be used to categorize data mining applications supporting software reuse. The taxonomy is based on two major characteristics of the applications: data mining task and data mining technique.

- *Data Mining Task:* Possible data mining tasks are classification, clustering, associations, regression, summarization and generalization, dependency modeling, change and deviation detection, model visualization, exploratory data analysis, etc.
- *Data Mining Technique*: Possible data mining techniques are decision trees, mining association rules, clustering, neural networks, genetic algorithms, case-based reasoning, statistical methods, Bayesian belief networks, fuzzy sets, rough sets, etc.

The data mining technique to be applied is related to the data mining task. For example, decision trees are usually for the classification task and not for the clustering task. Neural networks can be applied for both classification task (with predefined classes) and clustering task (without predefined classes). The mining association rules are exceptionally for the association task.

The taxonomy of data mining applications supporting software reuse takes the form shown in Fig. 1. Although not exhaustive due to space limitations, we believe that the taxonomy provides a predictive framework to help identify possible new data mining applications. As illustrated in Fig. 2, we apply this taxonomy to review the existing applications mentioned in Section 4. First, the applications are categorized into three groups based on the data mining task: I – classifying software components, II – clustering software components, and III – mining reuse patterns. At the second level, within Group I, there is only one group using the decision trees technique. Three subgroups using hierarchical clustering, neural networks, and genetic algorithms belong to Group II. Group III has one group using the mining generalized association rules technique. At the third level, a leaf node is labeled with the application name and indicates the class of an application. For example, the IC system is in the class of classifying software components with decision trees. The SOFM system belongs to the class of clustering software components with neural networks. The CodeWeb system is in the class of mining reuse patterns with mining generalized association rules.

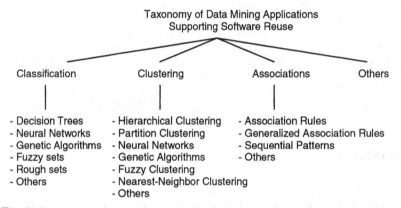

Fig. 1. Taxonomy of data mining applications supporting software reuse

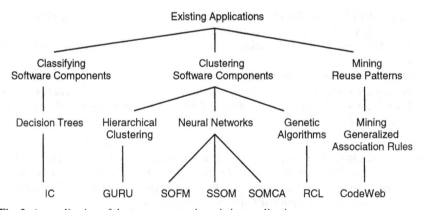

Fig. 2. An application of the taxonomy to the existing applications

6. Conclusion

In this paper, we addressed the motivation and major tasks of the construction of a software library. We introduced three of the best-known data mining tasks as well as their prominent techniques and algorithms. We observed and cataloged the major characteristics of several existing data mining applications supporting software reuse and proposed a taxonomy based on these characteristics. The taxonomy was applied to the existing applications. As seen from the observation, data mining has already been successful in classifying components, clustering components, and mining for reuse patterns. We believe that data mining is not limited to these kinds of applications. Different perspectives of applying the data mining technology to support software reuse are possible. For example, other analysis tasks of data mining may lead to different kinds of interesting knowledge. As the application of the taxonomy indicates, other data mining techniques also appear to have a promising potential.

References

[1] R. J. Brachman, T. Khabaza, W. Kloesgen, G. Piatetsky-Shapiro, and E. Simoudis. "Mining Business Databases", *CACM*, Vol. 39, No. 11, pp. 42-48, Nov 1996.

[2] M. Chen, J. Han, and P. S. Yu, "Data Mining: An Overview from a Database Perspective", *IEEE TKDE*, Vol. 8, No. 6, pp. 866-883, Dec 1996.

[3] M. M. El-Khouly, B. H. Far, and Z. Koono, "A New Multi-Level Information Retrieval Technique for Reuse Software Components", *Proc. of the IEEE Int. Conf. on Systems, Man, and Cybernetics*, pp. 773-777, Tokyo, Japan, Oct 1999.

[4] J. C. Esteva, "Learning to Recognize Reusable Software Modules Using an Inductive Classification System", *Proc. of the 5th Jerusalem Conf. on Information Technology*, pp. 278-285, Jerusalem, Israel, Oct 1990.

[5] C. G. Drummond, D. Ionescu, and R. C. Holte, "A Learning Agent that Assists the Browsing of Software Libraries", *IEEE TSE*, Vol.26, No.12, pp.1179-1196, Dec 2000.

[6] U. M. Fayyad, G. Piatetsky-Shapiro, and P. Smyth, "From Data Mining to Knowledge Discovery: An Overview", *Advances in Knowledge Discovery and Data Mining*, AAAI/MIT Press, Cambridge, MA, 1996.

[7] P. Freeman, "Reusable Software Engineering: Concepts and Research Directions", *Tutorial: Software Reusability*, the IEEE Computer Society, Washington, D.C., 1987.

[8] W. B. Frakes and C.J. Fox, "Sixteen Questions about Software Reuse", *CACM*, Vol. 38, No. 6, pp. 75-87, Jun 1995.

[9] W. B. Frakes and T. P. Pole, "An Empirical Study of Representation Methods for Reusable Software Components", *IEEE TSE*, Vol. 20, No. 8, pp. 617-630, Aug 1994.

[10] V. Ganti, J. Gehrke, and R. Ramakrishnan, "Mining Very Large Databases", *IEEE Computer*, Vol. 32, pp. 38-45, Aug 1999.

[11] M. Goebel and L. Gruenwald, "A Survey of Data Mining and Knowledge Discovery Software Tools", *ACM SIGKDD*, Vol. 1, No. 1, pp. 20-33, Jun 1999.

[12] J. Guo and Luqi, "A Survey of Software Reuse Repositories", *Proc. of the 7th IEEE Int. Conf. and Workshop on the Engineering of Computer Based Systems*, pp. 92-100, Edinburgh, UK, Apr 2000.

[13] R. J. Hall, "Generalized Behavior-based Retrieval", *Proc. of ICSE-15*, pp. 371-380, Baltimore, MD, May 1993.

[14] S. Henninger, "Using Iterative Refinement to Find Reusable Software", *IEEE Software*, Vol. 11, pp. 48-59, Sep 1994.

[15] Z. Houhamdi and S. Ghoul, "A Reuse Description Formalism", *ACS/IEEE Int. Conf. on Computer Systems and Applications*, pp. 395-401, Beirut, Lebanon, Jun 2001.

[16] A. K. Jain, M. N. Murty, and P.J. Flynn, "Data Clustering: A Review", *ACM Computing Surveys*, Vol. 31, No. 3, pp. 264-323, Sep 1999.

[17] G. Jones and R. Prieto-Diaz, "Building and Managing Software Libraries", *Proc. of the 12th IEEE Int. COMPSAC*, pp. 228-236, Chicago, IL, Oct 1988.

[18] C. Kleissner, "Data Mining for the Enterprise", *Proc. of the 31st Annual Hawaii Int. Conf. on System Science*, pp. 295-304, Kohala Coast, HI, Jan 1998.

[19] C. W. Krueger, "Software Reuse", *ACM Computing Surveys*, Vol. 24, No. 2, pp. 131-183, Jun 1992.

[20] P. Langley and H. A. Simon, "Applications of Machine Learning and Rule Induction", *CACM*, Vol. 38, No. 11, pp.55-64, Nov 1995.

[21] B. Lee, B. Moon, and C. Wu, "Optimization of Multi-way Clustering and Retrieval using Genetic Algorithms in Reusable Class Library", *Proc. of the Asia Pacific Software Engineering Conf.*, pp. 4-11, Taipei, Taiwan, Dec 1998.

[22] Luqi and J. Guo, "Toward Automated Retrieval for a Software Component Repository", *Proc. of the IEEE Conf. and Workshop on Engineering of Computer-Based Systems*, pp. 99-105, Nashville, TN, Mar 1999.

[23] Y. S. Maarek, D. M. Berry, and G. E. Kaiser, "An Information Retrieval Approach For Automatically Constructing Software Libraries", *IEEE TSE*, Vol. 17, No. 8, pp. 800-813, Aug 1991.

[24] D. Merkl, A. M. Tjoa, and G. Kappel, "Learning the Semantic Similarity of Reusable Software Components", *Proc. of ICSR-3*, pp. 33-41, Rio de Janeiro, Brazil, Nov 1994.

[25] A. Michail, "Data Mining Library Reuse Patterns Using Generalized Association Rules," *Proc. of ICSE*, pp. 167-176, Limerick, Ireland, Jun 2000.

[26] T. M. Mitchell, "Machine Learning and Data Mining", *CACM*, Vol. 42, No. 11, pp. 30-36, Nov 1999.

[27] M. Morisio, M. Ezran, and C. Tully, "Success and Failure Factors in Software Reuse", *IEEE TSE*, Vol. 28, pp. 340-357, Apr 2002.

[28] W. Pedrycz, G. Succi, M. Reformat, P. Musilek, and X. Bai, "Self Organizing Maps as a Tool for Software Analysis", *Proc. of the Canadian Conf. on Electrical and Computer Engineering*, pp. 93-97, Toronto, Ontario, Canada, May 2001.

[29] J. S. Poulin and K. P. Yglesias, "Experiences with a Faceted Classification Scheme in a Large Reusable Software Library (RSL)", *Proc. of the 17th IEEE Int. COMPSAC*, pp. 90-99, Phoenix, AZ, Nov 1993.

[30] E. Smith, A. Al-Yasiri, and M. Merabti, "A Multi-Tiered Classification Scheme for Component Retrieval", *Proc. of the 24th Euromicro Conf.*, pp. 882-889, Vesteras, Sweden, Aug 1998.

[31] I. Sommerville, *Software Engineering*, 6th Edition, Addison-Wesley Publishing Company, Reading, MA, 2001.

[32] P. A. Straub and E. J. Ostertag, "EDF: A Formalism for Describing and Reusing Software Experience", *Proc. of the IEEE Int. Symposium on Software Reliability Engineering*, pp. 106-113, Austin, TX, May 1991.

[33] H. Ye and B. W. N. Lo, "Towards a Self-Structuring Software Library", *IEE Proc.-Software*, Vol. 148, No. 2, pp. 45-55, Apr 2001.

[34] M. K. Zand and M. H. Samadzadeh, "Software Reuse Issues and Perspectives", *IEEE Potentials*, Vol. 13, Part 3, pp.15-19, Aug - Sept 1994.

Codifying the "Know How" Using CyKNIT Knowledge Integration Tools

Suliman Al-Hawamdeh

School of Communication and Information
Nanyang Technological University,
Singapore, 637718
Email: suliman@hawamdeh.net

Abstract. The increased interest in knowledge management and in particular the "Know How" has created the need for interactive and collaborative knowledge management tools. Integrated knowledge management tools such as knowledge portals provide access to wide range of facilities that encourage collaboration and promote organizational learning. Enhanced portal facilities such as personalization, summarization and ask the expert can be used to help users locate relevant knowledge sources and assist them in finding answers to queries rather than references to documents. In this paper, we discuss cyKNIT knowledge management integration tools that can be used to facilitate human interaction and help capture the "know how". The system also provides a set of tools for knowledge categorization, taxonomies, document management, discussion forums, content management, ask the expert, and advance searching and extraction facilities.

Keywords: cyKNIT, Know How, Knowledge Management, Portals, Tacit Knowledge

1 Introduction

Knowledge portals have emerged as key tools for supporting knowledge activities with the organization. They are single-point-access software systems intended to provide easy and timely access to information and people across the organization. Knowledge portals attempt to integrate corporate internal and external information sources and use them to support the organization business process.

The early portals focused on information aggregation only (Radding, 2000). They focused on specific types of contents □ such as business intelligence reports used by people in a certain workgroup. This generation of portals acts like a central intelligence warehouse for users to find central information to be shared by all employees. Some organizations still retain this form of portals today as a basic and functional tool for employees. Today, knowledge portals are still evolving and a desirable trait that researchers like to do is to incorporate artificial intelligence into the evolutionary process. With this feature in place, the portal can support decision-making process. By facilitating and improving key business decisions, it reduces operating costs, increases corporate revenues by integrating business information across the organization and employing collaborative processing to monitor decisions and actions (Szuprowicz, 2000). Knowledge portals are designed to provide easier access to corporate resources and aggregated business information. They help users to organize and locate information within the repository systems. When specific information is needed, users are able to pull it out regardless of the format it was stored in.

Organizing internal and external resources in an effective and efficient way is another area of concern when it comes to the development and implementations of knowledge portals. For many organisations, the thought of organising the vast growing digital information manually is horrifying. If the system can automatically determine the important topics contained within these resources and classify them that will cut the need for manual classification process and reduce cost. But while many vendors claim to be able to automatically generate concepts and build taxonomies, in reality, this process still remains very complex and problematic. This may be due to the complex nature of the natural language processing and the limitations of the techniques used.

Given the limitations of the statistical based classification techniques and the complexity associated with using natural language processing, many researchers and industrial experts are turning to semiautomatic environment using taxonomies. Taxonomy, such as a hierarchical tree of terms, is a structure for providing guidance over a classification system (Bailey, 1994; Mohan, 2000; Sacco, 2000). It shows users, groupings and relationships that can emerge from the information in various patterns. It is also possible to include the organization structure, expertise and processes as part of the organization taxonomy. This enables people to refer to the experts in the event they are not able to find answers in documents and databases.

2   Knowledge Integration Tools

Davenport and Prusak (1998) stated that, "providing access to people with tacit knowledge is more efficient than trying to capture and codify that knowledge electronically or on paper.... the richest tacit knowledge in organisations is generally limited to locating someone with the knowledge, pointing the seeker to it, and encouraging them to interact". While this is true, we believe that efforts should still be made to capture and codify tacit knowledge. An integrated approach in which access to explicit and implicit knowledge is provided for will enhance the knowledge sharing and creation environment.

cyKNIT provides access to physical and digital resources similar to that of many digital libraries. It is an integration of different applications and digital resources. The primary function of cyKNIT is to assist users in locating relevant information from diversified digital resources. But due to the scalability and adaptability of the system, cyKNIT can be used as a knowledge portal to capture, process and maintain organisational memory. cyKNIT Knowledge portal consists of the following modules:

1. Document & Record Management
2. Collaboration and discussion forums
3. Content Management and e-Learning
4. Ask the expert and cyKNIT database
5. Advance Searching and Extraction
6. Personalization & UI.

Ask the Expert module in cyKNIT contains a directory of subject experts. This is similar to a "Yellow Pages" directory consisting of subject experts' profiles, their availability and contact details. Since experts are not expected to be available online all the time, queries can be posted and routed by emails and the experts can attend to the queries when they are available. Depending on the mode of operation, some subject experts such as librarians can be available online to answer questions. The important part of this module is the ability to capture the interaction between the users and subject experts.

2.1 Personalisation and User Interface Module

This module provides a seamless integration between the users and cyKNIT modules. One of the tricky areas in designing cyKNIT interface is the interaction with the subject experts due to the fact that the experts are not always available online to answer queries. Questions can be posted to the experts if users indicated to the system that they are not satisfied with the information retrieved from both the knowledge database and the digital resources. When users submit their queries to the cyKNIT, the query-processing module will first consult the knowledge database to check if such queries have been handled before. If the request is new, the system will proceed to search the digital resources available. Some of the digital resources may possess different searching interfaces. Hence, cyKNIT interface is able to synchronise these searches and allow users to get information from consult these resources. If after consulting the various digital resources, users are still not satisfied, cyKNIT will suggest or recommend a subject expert from the subject expert directory depending on the nature of the requests, and users have the option to accept or reject this offer and conduct another search. If subject experts are available online, requests will be routed to them and with whom users can interact directly with. If the experts are off-line, cyKNIT will route an email containing users' requests and users contact details for further action.

2.2   Advance Searching and Extraction Module

Two types of search facilities are available in cyKNIT. The first search facility is the Query Processing Module within cyKNIT databases including questions and answers captured by Ask the Expert module. The second searching facility is the meta search engines for external resources. Due to the large number of digital resources available in the system, some of these digital resources have their own proprietary search interfaces. In this case automating the search function will be difficult. Different digital resources possess different indexing structures. Some could use free text or keyword indexing, others might use categories or hierarchical structure with navigation capabilities and limited keywords search functionality. Query formulation may vary from one digital resource to another. The search results depend to large extent on the query formulation and the user familiarity with the database. Some resources might rely heavily on navigation, thus making automated query processing difficult. There are different types of tools that can help users to search information from different digital resources.

- Bibliographic tools. They provide access to references and bibliographic information rather than to the documents themselves. In this case indexing and searching are applied to document surrogates, such as titles or abstracts. Examples include OPACS, OCLC, and DIALOG.
- Full-text access tools. These tools attempt to collect and index resources based on the document content. Automatic indexing reduces the amount of human effort and allows searches through massive amounts of information. Examples include Web Search engines, full text databases such as ACM and IEEE Digital libraries
- Meta-search engines. Meta-search engines do not have there own databases or indexing services. Instead, they send queries simultaneously to multiple search engines. Meta-search engines integrate search results, eliminate duplications, and rank the results through their own criteria. Examples of these search engines include Copernic and SavvySearch.
- Intelligent Agents. Intelligent agents are autonomous software designed to perform specific tasks. Intelligent agents have the capability to coordinate and collaborate with other agents, adapt to a new environment and learn from experience using machine learning and knowledge discovery.

Given the different types of tools that can be used to locate information from digital resources, it is desirable that the intelligent search interfaces being used is capable of hiding this complex searching environment from the user. By making all these access tools and the technology behind them transparent to the user, it is more likely that the system will meet users need for an effective and a user-friendly system. An intelligent agent operating in the query-processing module should be able to take the user's query, submit it to the various digital resources, consolidate the results and present them to the user. While this is the desirable way of locating relevant information from the large number of digital resources, automating query is difficult due to the various indexing structures and search strategies. Though some of the resources will still require manual browsing or navigation, most of the indexing services on the Web including the digital libraries can be interfaced using meta search engines and intelligent agents.

2.3   Ask the Expert

This is an important module that captures queries and answers from the subject experts. The answers given to the users by the expert are captured, formatted and stored in the database. The database also contains other information such as tips on how to perform searches on different digital resources, tutorials and learning materials that may help users to conduct better searches. It also contains best practices and successful search strategies. It captures user feedback and suggestions such as how access to digital resources can be further improved. The most important function of the database is the recording of the interaction that takes place between users and the subject experts. The database is fully searchable and dynamic as answers to queries are automatically updated each time they are created. Answers from the experts are integrated via the Tacit Knowledge Capture Module, which consists of two parts. The first is a directory containing details of subject experts such as areas of consultations, contact details and availability. The directory is some sort of 'Yellow Pages' that creates knowledge profiles of library and subject experts. It helps locate expertise anywhere within and outside the organisation using knowledge structure such as taxonomies and knowledge trees. Library and subject experts can update their records in the 'Yellow Pages'. They can also update and change their expertise profiles as and when they become more knowledgeable in other areas.

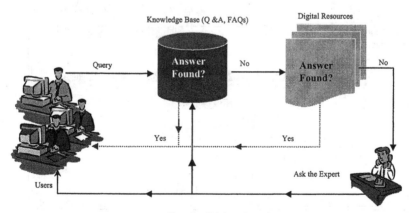

Figure 1: cyKnit Query Processing

The other part of the Tacit Knowledge Capture Module is the extraction of answers and tips from the interaction that takes place between users and experts. The Module captures and codifies answers that are representations of an expert's tacit knowledge. Users' queries are forwarded to the experts and answers submitted by the experts are sent to the users directly. The answers are also captured in the Knowledge Capture module and subjected to value-adding processes such as labelling, indexing, sorting, abstracting, standardisation and then integration into the Knowledge Database.

Queries and answers along with other useful information such as tips, best practices, frequently asked questions and users experience are indexed into cyKNIT database. When the user submits a new query, cyKNIT will first try to determine if the new query matches any of the previously processed queries by the subject experts. If the answer to the new query is found in the database, cyKNIT will use that to update the searches and send results to the user. If the new query does not match those in the database, cyKNIT will proceed to search the digital resources, integrate the results to the database and present it to the user. In the event that users are still not satisfied with the results, they have the option to refer to the subject experts or conduct another search. If the subject expert attends to the query, the answer is sent to the user and concurrently captured into cyKNIT database.

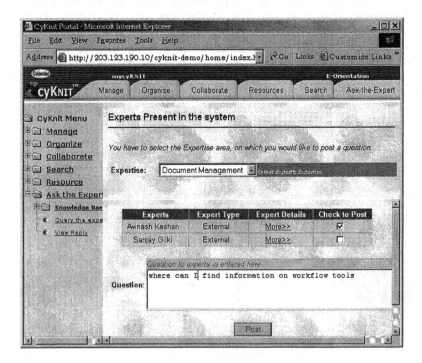

Figure 2: Ask the expert Interface in cyKNIT

3   Conclusion

The increased interest in knowledge management and the need to manage the "know how" has impacted the design and development of intelligent information systems including knowledge portals. Knowledge portals need to go beyond the simple concept of digitisation and provide more sophisticated search and collaboration tools. The integration of tacit knowledge in the form of "Ask the Expert" services where experts provide answers to queries and direct the users to relevant information is essential. cyKNIT as a knowledge integration tool helps users to capture, organize and process explicit knowledge as well as provides a set of tools that help users to obtain answers from the experts. The system allows the users to consult wide range of digital resources, aggregate the answers and send them back to the user. If the users are still not satisfied with the answers, the system will then direct the questions to subject experts using the expert directory or yellow pages. The ability of the system to capture the answers from the experts and store them for future use reduces the need for consulting the experts again for similar queries. For many users, looking for answers rather than documents, this module is extremely useful. This not only saves time, but also provides the missing human touch needed in many information management systems.

References

Bailey, K. D. (1994). Typologies and taxonomies: An introduction to classification techniques. Thousand Oaks, California: Sage Publications.

Davenport, T.H. and Prusak, L. (1998). *Information Ecology: Mastering the Information and Knowledge Environment.* New York: Oxford University Press.

Mahon, B. (2000). Taxonomies - tools and developments. *In A. Gilchrist & P. Kibby (Eds.), Taxonomies for business: Access and connectivity in a wired world,* pp.162-170. London: TFPL.

Radding, A. (2000). Internet Portals Get Personal. *Informationweek.com.* December 11: pp. 101-108.

Sacco, G.M. (2000). Dynamic taxonomies: A model for large information bases. *IEEE Transactions on Knowledge and Data Engineering,* 12(3), 468-479.

Szuprowicz, B. (2000). Implementing Enterprise Portals: Integration Strategies for Intranet, Extranet, and Internet Resources. South Carolina: Computer Technology Reseach Corporation.

Data Mining Techniques in Materialised Projection View

Ying Wah Teh[1] and Abu Bakar Zaitun[1]

[1]Faculty of Computer Science and Information Technology, University of Malaya,
50603, Kuala Lumpur, Malaysia
{tehyw, zab}@um.edu.my
http://www.fsktm.um.edu.my/

Abstract. This paper investigates one of the important factors like the attributes specified in the criteria of a query and their influence on the response time of decision support query processing. The study of the attributes is based on materialised projection view to select relevant attributes to form a new table or relation in order to minimise accessing irrelevant attributes. As it is very costly to form a new table for every query, top priority users like production managers and frequently accessed attributes are the two major parameters used to build a new table. We introduce the data mining techniques in our study such as by using decision tree to perform a materialised projection view on a relation/table.

1 Introduction

Relational databases support views which are also called virtual tables, defined in terms of queries [10]. A view needs not be explicitly stored, and is not physically materialised. Ramakrishnan and Gehrke [7] state that *a view is just a relation, but a view stores a definition, rather than a set of tuples.*

The database stores all of an organisation's information, some of which are shared among departments. Views are created to control the users' access to information. The situation becomes exponentially difficult when organisations' need to release their production databases to customers and suppliers.

One of the major strengths of views is that it helps the organisation to enforce security access rules based on virtual tables. One of the weaknesses, however, is that it only stores a definition and deals with the security issues of database rather than the response time of a query.

A materialised view can be materialised by physically storing the results of the view in the database. The major strength of using materialised view is speed -- it is often quicker to access a materialised view than to recompute the corresponding query from scratch [5]. The ability to use materialised views yields significant performance benefits, not only to novel applications built on a data warehouse, but also to existing business applications. The importance of materialised views stems from the constant decrease in cost of data storage [5]. The main drawback of this approach is that it is very costly to create materialised views for every query. The algorithm of materialised view technique will be discussed in Section 4.   An intelligent way of handling the materialised view is one of the main issues of this research.

To determine whether to materialise the relation in order to speed up the performance of a query is by knowing how frequently tables or relations are accessed. The materialised techniques were used in speeding up data communication. Decision support query processing environments and large website databases usually consist of many individual queries requesting or manipulating relatively small amounts of data. As the system grows, and as users issue more queries against the database, the database administrators usually try to improve the response time by scaling up, which means increasing the server's power. CPUs can be added; CPUs can be replaced with faster ones, more memory can be added, network can be expanded, or faster disk drives with smarter controllers can be added, but at some point, the available resources for scaling up will be exhausted, i.e., the machine's—or budget's—limit will be met [9]. SQL Server 2000 introduces a solution for the growing need for more processing power—scaling out [9]. When scaling out, a huge table will be split into smaller tables, each of which is a subset, or partition, of the original table. Each partition will be placed on a separate server. To access the data on any of the partitions, define a view with the same name on each of the servers, making transparent the fact that the data is distributed among several nodes.

2 Motivation

Tackling a large amount of information [9], query by partitioning the underlying data [6], allowing new indexes [10] and one materialised repository [11] are the major improved query processing techniques currently proposed to improve the response time of a query in the database environment for high priority users like managers to access in the Web environment [3]. These topics motivate us to discuss the issues about the benefit of having materialised techniques in a decision support environment

The response time of a particular query is affected by – irrelevant attributes, irrelevant tuples, as well as irrelevant attributes and tuples. Among the problems identified include

- The number of irrelevant attributes affects the response time of a query
- The number of irrelevant tuples affects the response time of a query
- The number of irrelevant attributes and tuples affects the response time of a query

3 Research Objectives

Based on the problems described in the previous section, the study attempts to address the pertinent issues with regards to the factors of irrelevant attributes. Relevant access is especially important for high priority users. The reason is that the cost of high priority users accessing computers is higher than normal users. Recognising the important role of the right selected attributes in query processing, the question in the following is posed:

- What are the important criteria or quality attributes that must be materialised in the relation/table for high priority users

It is necessary to have a set of guidelines for a database administrator to manage the high priority users for accessing the database activities. The existing access log files are comprehensive in terms of access activities for all user levels. Indeed, establishing simple high priority users access database activities could have positive results in improving query response time. A set of guidelines has been established to help in keeping database access of high priority users. The following research questions will bring into focus the objective of the study of the attribute factor in the query response time.

- Has the adoption of the high priority user   access's log file contributed to good query response time?
- Are the high priority user's access log files useful and practical?

The main objective of this study is to improve query response time by using a simple high priority user access's log file.

Decision support queries typically involve several joins, a grouping with aggregation, and/or sorting of the resulting tuples. Claussen et al. [2] proposed two classes of query evaluation of such queries. The algorithms are based on (1) early sorting and (2) early partitioning – or a combination of both. The idea is to push the sorting and /or the partitioning to the leaves, i.e., the base relations of the query evaluation plans (QEP) and thereby avoids sorting or partitioning large intermediate results generated by the joins.   Ross and Li [8] described two new join algorithms that are based on join indices. One of the algorithms is sorted-based and the other is partitioned-based. Both algorithms benefit from not completely materialising the join result, rather they store the temporary result on disk in two ordered files to be merged to obtain the join result.

From their work [8], [2], it is known that it is important to start optimisation from the base relation to reduce large, immediate results.

One of the techniques employed in improving the response time of a query is the creation of a set of materialised views [4]. An intelligent way of handling materialised view are one of the main issues dealt with

4   Data Mining Techniques

To explain how data mining techniques are applied to generate the materialised projection view, it is assumed that a relation has 14 attributes involved in a query processing. Given a relation has 182 bytes, how many instruction cycles do CPUs need to fetch the tuples from the hard disk and place them in the main memory?

If the server has a 386-based processor and the data bus speed is 32 bits per instruction cycle (32 bits means 4 bytes) then 182 / 4 = 46 instruction cycles are needed.

If the server has a Pentium-based processor and the data bus speed is 64 bits per instruction cycle (64 bits mean 8 bytes) then 182 / 8 = 23 instruction cycles are needed

In the Pentium-based server, one disk page can only take less than 8 bytes. Emphasis is placed on the characteristics of hardware to optimise the main memory usage to select candidate attributes to form a new relation. A high priority user's log file keeps track of a high priority user's access, such as a manager issuing the decision support queries from time t_1 to t_{35}, as shown in Fig. 1.

Time	Queries
t_1	select a_1, a_2, b_2 from A, B where A.a_2 = B.b_1;
t_2	select a_1, a_6, a_8 , b_2, b_3 c5 from A, B, C where A.a_2 = B.b_1 and B.b_2 = C.c_1;
t_3	select a_1, a_{10}, a_{11}, b_2, b_3 c_2, c_5 from A, B, C where A.a_2=B.b_1 and
	B.b_2 = C.c_1;
t_4	select a_3 from A;
t_5	select a_4 from A;
t_6	select a_5 from A;
	:
t_{35}	select a_1, a_c, b_1, b_2 c5 from A, B, C where A.a_1 = B.b_1 and B.b_2 =C.c_1;

Fig. 1. A high priority user's access log file

The algorithms in Fig. 2 demonstrate how the log file is processed and generate the frequently accessed table.

```
Open the access log file as input
Open a frequently accessed attributes file as output
Do
    Read a record from the log file
    Keep count for respective attributes and store  them into respective array
    of attributes
While (not end of file)
Do
    Read data from the array of attributes
    Store it into frequently accessed attributes file
While (not end of array)
```

Fig. 2. Algorithm for Generating Frequently Accessed Attributes

Table 1 is the output generated by the algorithm in Fig. 2.

Table 1. Frequently Accessed Attributes Table

Attribute Names	Frequently Access F(x)
a_1	7
a_2	9
:	:
a_{12}	4
a_{13}	4
a_{14}	5

4.1 Multiple-Attribute Access

The concept of multiple-attribute access can be explained by selecting t_1 and t_4 as examples (see Fig. 1).

t_1 select a_1, a_2, b_2 from A, B where $A.a_2 = B.b_1$;
t_4 select a_3 from A;

The attributes a_1 and a_2 in t_1 are identified as *True* status for multiple-attribute access, whereas the attribute a_3 in t_4 are identified as *False* status for multiple-attribute access.

Fig. 3 demonstrates the algorithm to process the access log file and generate the table, which involves multiple- attributes access.

Table 2 is the output generated by the algorithm in Fig. 3.

```
Open access log file as input
Open the file of multiple-attribute access as output
Do
      Read a record from the access log file
      If  multiple-attribute access at a time
        Keep count for respective attribute and
        Store it into respective array of attributes
      End if
While (not end of file)
Do
   Read data from the array of attributes
   Store it into multiple-attribute access file
   While (not end of array)
```

Fig. 3. Algorithm for Generating a Table for Multiple-Attribute Access

Table 2. A Query Involving Multiple-Attribute Access

Attribute Names	F(>x) ?
a_1	True
a_2	True
:	:
a_{12}	False
a_{13}	False
a_{14}	False

4.2 Decision Tree Based on the Size of Attribute, Multiple-Attribute Access and Mean Sequence

The choice of an attribute is normally based on the size of an attribute. If the number of bytes in an attribute is less than 8 bytes, it will be qualified to be materialised. Why is such an attribute qualified? The main idea comes from previous section regarding the data bus of the Pentium-based processor. The data bus of the Pentium-based processor is 64 bits per instruction cycle (64 bits mean 8 bytes). The main purpose is to make use of the internal computer architecture fetch the data from the harddisk to main memory. Otherwise, it checks the size of an attribute that fall between 9 and 16 bytes and the status of multiple-attribute access is *True*, it will be qualified to be partitioned. This is followed by checking the size of the attributes greater than 16 bytes and the frequently accessed attribute having the number of frequently access $F(x)$ that is more than the mean of the total frequently access attributes. It will be qualified to be materialised if this is verified. The detailed explanation of the decision tree process is described by the algorithm given in Fig. 5.

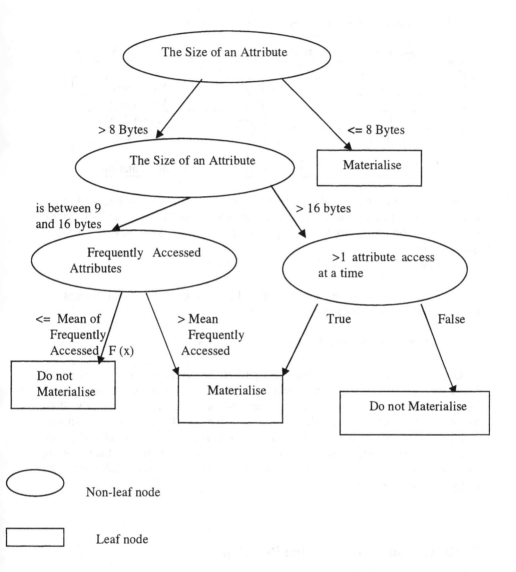

Fig. 4. Decision Tree Based on the Size of Attributes, Multiple-Attribute Access at a Time and Mean Oriented

4.3   Algorithm for Generate Materialised Projection View (MPV)

Table 3 is the output generated by the algorithm given in Fig. 5.

Table 3. Materialised Project View With a Small Training Data Set

X	Len(x)	F(x)	F(>x)?	Class
a_1	30	7	True	Materialise
a_2	45	9	True	Materialise
a_3	18	4	False	Materialise
a_4	24	6	False	Materialise
:	:	:	:	:
a_{14}	16	5	False	Do not Materialise

Based on the results in Table 3, a_1, a_2, a_3, a_4, a_6, a_7, a_8, a_9, a_{10} and a_{11} will form a new materialised projection view A_1 after the time t_{35}. As shown in the high priority user's access log file, whenever the high priority user is in t_{36}, he/she issues the same query as in t_1.

t_{36} select a_1, a_2, b_2 from A, B where $A.a_2 = B.b_1$;

Since the attributes a_1 and a_2 can be found in A_1, the database management system will rewrite the query above as follows:

t_{36} select a_1, a_2, b_2 from A_1, B where $A_1.a_2 = B.b_1$;

One of main drawbacks of using a log file to select attributes is that it is only based on a certain period of user accessing behaviour. To make this materialised view more dynamic, we need to keep track of the usage of A_1 continually. If the usage of A1 is below certain threshold value, that is, 50% of usage, the database management system will reselect the attributes to form a new materialised projection view automatically. The materialised process needs to start from the access log file again from t_1 until current time. This incurs overhead of materialisation, especially if it has to be done frequently.

5   Evaluation of Data Mining Prototypes

To test the query response time of our data mining prototype and to explore how well the different query processing techniques perform when accessing large tables, we used a standard SQL benchmark supplied by the Transaction Processing Performance Council (TPC) (See [11]).

The Table 5 shows the performance TPC-H sample queries

Open frequently accessed attributes file as input
Open the file of multiple-attribute access as input
Open a Materialised Projection View file as output
Calculate the Mean of frequently accessed from frequently accessed attributes file
Do
Read a record from the frequently accessed attributes file
   Read a record from the file of multiple-attributes access
  If the number of bytes for an attribute is between 1 and 8
     Store it with 'Materialise' information into MPV file
   Else if the number of bytes for an attribute is between 9 and 16 and Multiple
     attributes access is 'True'
     Store it with 'Materialise' information into MPV file
   Else if number of bytes for an attribute > 16 and frequently accessed
     attributes > Mean of Frequently Accessed F (x)
     Store it with 'Materialise' information into MPV file
  Else
     Store it with 'Do not Materialise' information into MPV file
  End if
While (not end of file)

Fig. 5. Algorithm for Generating Materialised Projection View

Table 5. The Performance TPC-H Sample Queries

Query Name	Without MPV	With MPV
Query 1	1:17 min	46 sec
Query 2	17:09 min	15:04 min
Query 3	2:47 min	56 sec
Query 4	50 sec	35 sec
Query 5	20 sec	11 sec
Query 6	3:25 min	1:28 min
Query 7	18 sec	8 sec
Query 8	14 sec	11 sec
Query 9	37 sec	11 sec
Query 10	3:08 min	40 sec
Query 11	7 sec	4 sec
Query 12	11 sec	10 sec
Query 13	46 sec	9 sec
Query 14	4:51 min	2:05 min
Query 16	6 sec	3 sec
Query 17	3:54 min	2:16 min
Query 18	2:06 min	43 sec
Query 19	1:39 min	19 sec

5   Conclusions

The results of this work are expected to encourage wider use of data mining techniques in the physical design of a data warehouse. The current tools do not allow
data warehouse administrators or decision makers to optimise the performance of
their data warehouse. The main contribution of this work is both conceptual and
practical.   It successfully illustrated how the Pre-processed Data Mining and data
mining technique can help to generate the redundant data structures. Pre-processed
data mining in high priority user's access log file is one of most significant challenges in the research.

References

1. Bruno, C., Kilmartin, G., Tolly, K., : Fast relief for slow Web sites Network World,
   Available from World Wide Web:
   http://www.nwfusion.com/news/1999/1101tollyfeat.html.
2. Claussen et. al.: Exploiting early sorting and early partitioning for decision support
   query processing. The VLDB Journal (2000)190-213.
3. Clement, T. Y., Meng, W.: Principles of Database Query Processing for Advanced
   Applications The Morgan Kaufmann Series in Data Management Systems, Jim Gray, Series Editor (1997).
4. Agrawal, S., Chaudhuri, S., Narasayya, V. :Automated Selection of Materialized Views
   and Indexes for SQL Databases. In, Proceedings of the 26th International Conference on
   Very Large Databases, San Francisco, Morgan Kaufmann (2000).
5. Kawaguchi, A.: Implementation Techniques for Materialized Views, Doctoral Dissertation, Columbia University (1998).
6. Kenneth , A. R.: Query Processing and Optimization. Available from World Wide Web:
   http://www.cs.columbia.edu/info/research-guide/html/node38.html.
7. Ramakrishnan R., Gehrke, J.: Database Management Systems. McGraw-Hill (1999).
8. Ross, K.A., Li, Z.: Fast Joins Using Join Indices, VLDB Journal (1999)1-24.
9. Distributed Partitioned Views. Available from World Wide Web:
   http://www.sqlmag.com/Articles/Index.cfm?ArticleID=9086
10. Ullman, J.: Principles of Database, Second Edition, Computer Science Press (1982).
11. Teh, Y. W., Zaitun, A. B.: Data Mining Techniques in Index Techniques. In, Third
    International Conference on Intelligent Systems Design and Applications, Tulsa, USA,
    August 2003.

Data Mining Techniques in Index Techniques

Ying Wah Teh[1] and Abu Bakar Zaitun[1]

[1]Faculty of Computer Science and Information Technology, University of Malaya,
50603, Kuala Lumpur, Malaysia
{tehyw, zab}@um.edu.my
http://www.fsktm.um.edu.my/

Abstract. Redundant data structures such as indexes have emerged as some of the improved query processing techniques for dealing with very large data volumes and fast response time requirements of a data warehouse. This paper investigates factors like the use of tuples specified in the criteria of a structured query language (SQL) query and their influence on the response time of a query in a data warehouse environment. Experience of the pioneers in the data warehouse industry in handling queries by using redundant data structures has already been well established. Since we are given very large data storage nowadays, redundant data structures are no longer a big issue. However, an intelligent way of managing storage or redundant data structures that can lead to fast access of data is the vital issue to be dealt with in this paper.

1 Introduction

As hardware becomes increasingly powerful in recent years, how do the current business Information Technology (IT) infrastructures exploit this very large storage of hard disks, powerful PCs and networks available today? Decision support systems form the analytical systems of business IT infrastructures. From the hardware part, decision support systems are making use of the extremely powerful processors, high-capacity memory, very large storage and high-speed networks. A decision support system is one of the business intelligence technologies that help companies translate a wealth of business information into tangible and profitable outcome.

According to Chaudhuri [5], before developing a decision support system, there are three fundamental issues given below to be addressed and resolved.

- What data to gather and how to conceptually model the data and manage its storage.
- How to analyse the data, and
- How to efficiently load data from several independent sources

The first issue is to concentrate on the logical database design and physical database design aspects of data warehouses. A data warehouse provides place store data and supply information for decision makers. As we are given very large data storage nowadays, redundant data structures are no longer a big issue. However, the intelligent way of managing storage that can lead to fast access to data is the vital issue to be dealt with. A redundant data structure (e.g. an index or a materialised view) also determines which query processing technique to be applied. The second issue is to address analysing the data using data mining tools such as DB Miner [8] or RuleQuest [18]. The third issue correlates to the process of extracting legacy data from several independent sources, defining and normalising it for acceptance into the data warehouse. Some of the back-end tools for doing this task are IBM's Visual Warehouse [11] or Acta Technologies's ActaWorks [1].

Redundant data structures (e.g., indexes and materialised views) are introduced to process complex queries more efficiently. Selecting the right elements to build redundant data structures can greatly improve the quality of service for decision makers. On the other hand, unnecessary data structures may result in substantial overhead due to the creation and update costs for redundant data storage [6].

Despite its importance, the problem of automatically determining the right set of redundant data structures to build and maintains from a data warehouse has received little attention. The task of deciding which redundant data structure creates and maintains is a complex function. Indeed, the choice of redundant data structure changes as the workload and data distribution evolves. Today, this challenging task falls on the data warehouse administrators. Only a few data warehouse administrators can do justice to the task of picking the right redundant data structure for a data warehouse [6]. If the data warehouse administrators can't justify his/her selection, decision makers need a simple tool to pick the right redundant data structure for a data warehouse.

2   Query Processing Techniques

The historical perspectives of the query processing techniques are discussed in Section 2.1. In this section, file processing, simple index, B-Tree index and views are discussed. The present scenarios of the query processing techniques are discussed in Section 2.2. In this section, BitMap index and single-column indexes are discussed.

2.1   Historical Perspectives of Query Processing Techniques

The background of query processing is strongly related to the history of database development. The beginnings of query processing starting from file system, the file can be in text file or binary file format. A database is an organised collection of related file. The advantages of databases over file processing are reduced redundancy, integrated data, and integrity. The way the database organises data depends on model of database. One of database model is relational model. In 1969, Dr. Edgar F. Codd published his work on the relational model of file. This model affects heavily in current and future work of database.

File processing [4] refers to the use of computer files to store data in persistent memory (i.e. the data is stored even after the computer has been turned off and re-started). A programmer needs to know at least one-third generation language like C, Pascal, or COBOL for writing a data retrieval program to access the relevant information from a file system. Data retrieval programs are in the context of query processing. The programmer needs to perform a sequential scan or full scan of all the records in the file in order to locate data of interest. File processing needs the programmer or user to implement files for each application [4]. The strength of file processing is that it is faster than database approach in accessing data because it only involves in the departmental basis.   The main drawback of this approach is that the programmer needs to write the data retrieval program to access the data for respective departments. Most of file processing approach use sequential scan or full scan that will scan a file F from the top until the end of the File for accessing data as shown in Table 1. A file F consist of fix number of fields $(a_1, a_2, ..., a_n)$ and variant number of records $(i_{11}, i_{12}, ..., i_{14}, .., i_{1n}, ..., i_{m1}, i_{m2}, ..., i_{m4}, .., i_{mn})$

Table 1. File F

a_1	a_2	...	a_4	...	a_n
i_{11}	i_{12}	..	i_{14}	...	i_{1n}
:	:	:	:	:	:
:	:	:	:	:	:
i_{m1}	i_{m2}	:	i_{m4}	:	i_{mn}

Sequential scan or full scan is more suitable for the small data volume environ-ment (e.g. department basis). Sequential scan or full scan is not only applied in the file processing, but also it is used in the images database as well [20]. In the images database environment, a user doesn't have enough information about images; he/she needs to sequential scan through all images in the images database. In the file proc-essing environment, there are two separate file systems built for two departments even though they access to the same information. It is difficult to maintain the con-sistency between files. Database systems were developed to prevent problem of data consistency.

A database management system (DBMS) is a system that provides all the func-tionality needed to create, modify and maintain a database or collection of data-bases. Since the database is collection of files which make up the large amount of data. There are a number of techniques developed to speed up in accessing informa-tion such as simple index, B-Tree index and View.

Database Management Systems (DBMSs) were developed in the late 1960s [21] that included simple indexes called hashed key.   Hashed key [21] allows users to access information very quickly by a unique value such as a Customer Number, Sup-plier Number or Order Number.

This was a great improvement over sequential reads of large flat files. Instead of scanning the File F as shown in Table 1 fully from top to bottom, a hashed key as shown in Table 2 was used to reduce the number of readings. The hashed key creates a list of record identification which acts as pointers to records in the File F. The strength of simple index is to access the relevant records through the simple index. The number of bytes in simple index is smaller than the File F. Then it will save memory spaces. The main drawback of this approach is that the users need to know the exact key value to access the record [19]. This technique is more suitable for transaction processing systems since it needs to have a unique value to create a hashed key based index.

Table 2. Customer Number Hashed Key

Customer Number	Record Identification Number
1001	1
1002	2
...	...

Hashed key based index has a limit application due to the requirement of exactly key value to access data. Users seldom know all the illogical numbers that database applications require, so they need to access data in some other way. B-tree indexes also developed along the way [21]. B-tree indexes allow users to carry out partial key lookups and view the data in sorted order as shown in Table 3. B-tree indexes offered a vast improvement over hashed keys in terms of flexibility such as partial key lookup and exactly key lookup [19]. The main strength of this approach is that it is allowed users to create non-unique key index to access data as shown in Table 2. The non-unique key index is an important for reporting purposes. The main drawback of this approach is a very costly to create for every query. The intelligent way of handling the B-tree index is the vital issue to be dealt in this research.

Table 3. Gender B-Tree Index

Gender	Record Identification Number
Female	3
Female	4
...	...

2.2 Present Scenario

In most of the transaction processing systems use simple indexes, B-tree indexes or views, the primary key is selected as an index in the relation to improve the response time of transaction processing requirements (e.g. insert, delete and modify tuples). However, in decision support systems, a user normally issues a query that only requires a small portion of the result of relations and the predicate is non-primary key. To meet this query requirement, a secondary key of index must be generated in order to process the query quickly. In most relational databases, only one RID index can be used at a time. So, even if there are multiple indexes on one table, they cannot be used in combination [14].

In following section, we introduce present scenarios of query processing techniques, namely: BitMap index techniques and single-column indexes. BitMap index techniques and single-column indexes are built in the most of decision-support systems. The feature, strength and weakness of each technique will be discussed in the following section.

BitMap indexing has become a promising method in decision support systems. By using bit-vector approach, it is found to be more space and CPU-efficient than RID index [16]. Traditionally, primary and secondary indexes are treated as RID that are used to speed up the query's response time [21]. A RID occupies at least 8 bits, while a BitMap index occupies only 1-bit pointer to a tuple of the relation. Therefore, BitMap indexes are more space saving than RID. In [3], present a general framework to study the design space of BitMap indexes for queries in a data warehouse. They found that the increase of space requirement for a BitMap index is directly proportional to the response of a query. Moreover, the space requirement of a BitMap index is based on the cardinality of the index attribute of the base relation. Chan and Ioannidis present the design criteria for bit-sliced indexes with respect to continuous range selections. Edelstein [9] describes two problems (such as high-cardinality data and update performance) in BitMap indexes. As a rule of thumb, somewhere between 100 and 1,000 distinct values (depending on the size of an organisation data warehouse) would be considered high-cardinality data, and thus would be inappropriate for a pure BitMap index. For example, a range retrieval using simple BitMaps on high-cardinality data, such as "Find all cars bought between Feb. 5, 1994 and July 8, 1995," would require evaluating more than 500 BitMaps - one for each day in the range. Unless this problem is solved, query performance will suffer because the database system will still need to scan data pages to find the records that meet the selection criteria [9]. The second problem is that updating a BitMap to reflect changes in data can take substantially longer than updating a plain B-tree. Modifying bitmaps takes a fair amount of overhead in finding the bits to switch on or off, as well as the need to add BitMaps when new values are added to the database [9]. The strength of the BitMap index is found to be more space and CPU-efficient than the RID index. The weakness of the BitMap index will work well only with low-cardinality data. The intelligent way of handling the BitMap index is the vital issue to be dealt in this research.

Index intersection offers greater flexibility for database user environments in which it is complicated to decide all of the queries that will be run against the database. A good strategy would be to define single-column indexes on all columns that will be frequently queried and let index intersection handle situations [13].

SQL Server 7.0's new query processor index intersection which lets query processor to consider multiple indexes from a given relation/table, construct a hash relation/table based on those multiple indexes, and employ the relation/hash table to reduce I/O for a given query. The intelligent way of handling the single-column indexes is the vital issue to be dealt in this research.

3 Our Research Perspective: Data Mining in the Physical Level of Data Warehouse

Most of researchers apply data mining in the application level of data warehouse. We propose to apply data mining the physical design of data warehouses. The physical design highlights how the data is physically stored, and what data structures are used to speed up the response of a query. An architectural overview of our approach to redundant data structures (such as indexes) is shown in Fig. 1. The workload of this architecture only caters for Select statements; therefore, our architecture generates redundant data structures that are more suitable for a data warehouse.

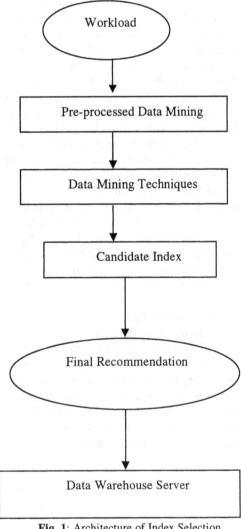

Fig. 1: Architecture of Index Selection

4 Evaluation of Data Mining Prototypes

To test the query response time of our data mining prototype and to explore how well   the different query processing techniques perform when accessing large tables, we used a standard SQL benchmark supplied by the Transaction Processing Performance Council (TPC). This benchmark permits large tables to be quickly generated, and provides a set of sample queries that can be pertained to the resulting relations. We loaded the resulting relational tables into Microsoft SQL Server 7.

The generated TPC-H simulates a decision-support system with large volumes of data, is loaded into tables of having the following Table name, number of tuples and number of attributes as shown Table 4a.

Table 4a. TPC-H Database

Table Name	Number of tuples	Number of Attributes
Part	200,000	9
Supplier	10,000	7
Partsupp	800,000	5
Customer	150,000	8
Nation	25	4
Lineitem	6,000,000	16
Region	5	3
Orders	1,500,000	9

The Table 4b shows the performance TPC-H sample queries.

Table 4b. The Performance TPC-H Sample Queries

Query Name	Without Indexes	With Index
Query 1	1:17 min	45 sec
Query 2	17:09 min	2 sec
Query 3	2:47 min	13 sec
Query 4	50 sec	9 sec
Query 5	20 sec	9 sec
Query 6	3:25 min	1:42 min
Query 7	18 sec	9 sec
Query 8	14 sec	3 sec
Query 9	37 sec	15 sec
Query 10	3:08 min	16 sec
Query 11	7 sec	4 sec
Query 12	11 sec	11 sec
Query 13	46 sec	4 sec
Query 14	4:51 min	26 sec
Query 16	6 sec	5 sec
Query 17	3:54 min	1 sec
Query 18	2:06 min	28 sec
Query 19	1:39 min	19 sec

Not all samples queries could be performed without alteration on Microsoft SQL v7 because the language accepted by Microsoft SQL Server v7 does not match to the SQL2 standard. Some SQL2 functions (such as extract) required to be reworked using comparable Microsoft SQL v7 constructs.

5   Conclusions

The anticipation of future growth of database capacity for a data warehouse has led to this research. The present scenario of query processing techniques has a strong desire for the intelligent way of handling the various query processing techniques. Data warehouse performance tuning, Microsoft's AutoAdmin, Microsoft SQL 2000's tuning wizard were the intelligent way of handling the various query processing techniques. Data warehouse performance tuning are too many parameters required for selection which is not suitable for inexperience data warehouse administrators or decision makers. Microsoft's AutoAdmin and Microsoft SQL 2000's tuning wizard uses the optimiser estimated cost for all the SQL statement (Insert, Delete, Update and Select) which might not be suitable for a data warehouse (read-only statements). Moreover, Microsoft SQL 2000's tuning wizard is not open-source software, it is impossible for changing the existing codes. Therefore, we propose data mining techniques as an intelligent way for handling the query processing techniques in this research. Data mining techniques in the physical design of a data warehouse to generate redundant data structures such as indexes and materialised view have emerged as some of the improved query processing techniques for dealing with very large data volumes.

References

1. Acta, *ActaWorks*. Available from World Wide Web:
   http://www.acta.com/products/actaworks/index.htm. Last modified: February 26, 2002.
2. Bellatreche, L., Karlapalem, K., Mohania, M.: Some Issues in Design of Data Warehouse Systems , Developing Quality Complex Database Systems: Practices, Techniques, and Technologies, Dr. Shirley A. Becker (Eds.), Ideas Group Publishing, (2001) 125-172.
3. Chan, C. Y., Ioannidis, Y. E., 1998. Bitmap index Design and Evaluation. In: Proceeding of SIGMOD. (1998) 355 – 366.
4. CERN: What is file processing? Available from World Wide Web:
   http://wwwinfo.cern.ch/db/aboutdbs/faqs/what_is_file_processing.html. Last Modified : March 29, 2002.
5. Chaudhuri, S., Dayal U., Ganti, Venkatesh: Database Technology for Decision Support Systems, Computer, IEEE Computer Society (2001).
6. Chaudhuri, S., Narasayya, V.: An Efficient Cost-Driven Index Selection Tool for Microsoft SQL Server. In,: Proceedings of the 23rd International Conference on Very Large Databases , Athens, Greece, (1997) 146-155.

7. Dataspace: Business Solutions. Available from World Wide Web: http://www.dataspace.com/biz_define.htm. Last modified: March 12, 2002.

8. DBMiner: DB Miner. Available from World Wide Web: http://www.dbminer.com/index.html. Last modified: March 20, 2002.

9. Edelstein, H.: Faster Data Warehouses. Information Week, December 4, 1995.

10. GeneCards: Knowledge Discovery In Biology and Medicine. Available from World Wide Web: http://bioinfo.weizmann.ac.il/cards/knowledge.html. Last modified: February 26 2002.

11. IBM: Visual Warehouse. Available from World Wide Web: http://www-3.ibm.com/software/data/vw/. Last modified: February 26 2002.

12. Inmon, W.: Building the Data Warehouse, 2nd Edition. John Wiley & Sons, Inc (1996).

13. Lau, H.: Microsoft SQL Server 7.0 Performance Tuning Guide. Available from World Wide Web: http://msdn.microsoft.com/library/default.asp?url=/library/en- us/dnsql7/html /msdn_sql7perftune.asp. Last modified: October 16, 2002.

14. Mattison, R.: Data Warehousing and Data Management, Second Edition, McGraw-Hill Series (1998).

15. NCR: EXAMIND AG Selects Teradata Data Warehouse Solution. Available from World Wide Web: http://www.ncr.com/media_information/2001/dec/pr120301.htm. Last modified: December 3, 2001.

16. Neil, P. O'., and Quass, D.: Improved Query Performance with Variant Indexes. In: Proceeding of SIGMOD, Tucson, Arizona (1997).

17. PAKDD: Toward the Foundation of Data Mining. Available from World Wide Web: http://www.mathcs.sjsu.edu/faculty/tylin/pakdd_workshop.html. Last modified: February 20 2002.

18. RuleQuest: RuleQuest Research Data Mining Tools. Available from World Wide Web: http://www.rulequest.com/index.html. Last modified: 20 March, 2002.

19. Ramakrishnan R. and Gehrke, J.: Database Management Systems. McGraw- Hill (1999).

20. Stehling, R.O., Nascimento, M.A. and Falcão, A.X.: Techniques for Color-Based_ Image Retrieval. Book of Intelligent Multimedia Document, Kluwer (2002).

21. Ullman J.: Principles of Database, Second Edition, Computer Science Press (1982).

Decision Tree Induction from Distributed Heterogeneous Autonomous Data Sources

Doina Caragea, Adrian Silvescu, and Vasant Honavar

Artificial Intelligence Research Laboratory,
Computer Science Department, Iowa State University,
226 Atanasoff Hall, Ames, IA 50011-1040, USA
{dcaragea,silvescu,honavar}@cs.iastate.edu

Summary. With the growing use of distributed information networks, there is an increasing need for algorithmic and system solutions for data-driven knowledge acquisition using distributed, heterogeneous and autonomous data repositories. In many applications, practical constraints require such systems to provide support for data analysis where the data and the computational resources are available. This presents us with *distributed learning* problems. We precisely formulate a class of distributed learning problems; present a general strategy for transforming traditional machine learning algorithms into distributed learning algorithms; and demonstrate the application of this strategy to devise algorithms for decision tree induction (using a variety of splitting criteria) from distributed data. The resulting algorithms are *provably exact* in that the decision tree constructed from distributed data is identical to that obtained by the corresponding algorithm when in the batch setting. The distributed decision tree induction algorithms have been implemented as part of INDUS, an agent-based system for data-driven knowledge acquisition from heterogeneous, distributed, autonomous data sources.

1 Introduction

Recent advances in computing, communications, and digital storage technologies, together with development of high throughput data acquisition technologies have made it possible to gather and store large volumes of data in digital form. For example, advances in high throughput sequencing and other data acquisition technologies have resulted in gigabytes of DNA and protein sequence data, and gene expression data being gathered at steadily increasing rates in biological sciences. These developments have resulted in unprecedented opportunities for large-scale data-driven knowledge acquisition with the potential for fundamental gains in scientific understanding (e.g., characterization of macromolecular structure-function relationships in biology) in many data-rich domains. In such applications, the data sources of interest are typically physically distributed, heterogenuous, and are often autonomous.

Given the large size of these data sets, gathering all of the data in a centralized location is generally neither desirable nor feasible because of bandwidth and storage requirements. In such domains, there is a need for knowledge acquisition systems that can perform the necessary analysis of data at the locations where the data and the computational resources are available and transmit the results of analysis (knowledge acquired from the data) to the locations where they are needed. In other domains, the ability of autonomous organizations to share raw data may be limited due to a variety of reasons (e.g., privacy considerations). In such cases, there is a need for knowledge acquisition algorithms that can learn from statistical summaries of data (e.g., counts of instances that match certain criteria) that are made available as needed from the distributed data sources in the absence of access to raw data. Furthermore, data sources may be heterogeneous in structure (e.g., relational databases, flat files) and/or content (e.g., names and types of attributes and relations among attributes used to represent the data). Data-driven learning occurs within a context, or under certain ontological commitments on the part of the learner. In most applications of machine learning, the ontological commitments are implicit in the design of the data set. However, when several independently generated and managed data repositories are to be used as sources of data in a learning task, the ontological commitments that are implicit in the design of the data repositories may or may not correspond to those of the user in a given context. Hence, methods for context-dependent dynamic information extraction and integration from distributed data based on user-specified ontologies are needed to support knowledge aquisition from heterogeneous distributed data [1, 2].

Against this background, this paper presents an approach to the design of systems for learning from distributed, autonomous and heterogeneous data sources. We precisely formulate a class of distributed learning problems; present a general strategy for transforming a large class of traditional machine learning algorithms into distributed learning algorithms; and demonstrate the application of this strategy to devise algorithms for decision tree induction (using a variety of splitting criteria) from distributed data. The resulting algorithms are *provably exact* in that the decision tree constructed from distributed data is identical to that obtained by the corresponding algorithm in the batch setting (i.e., when all of the data is available in a central location).

Agent-oriented software engineering [3, 1] offers an attractive approach to implementing modular and extensible distributed computing systems. For the purpose of this discussion, an agent is an encapsulated information processing system that is situated in some environment and is capable of flexible, autonomous action within the constraints of the environment so as to achieve its design objectives. The distributed decision tree induction algorithms described in this paper have been implemented as part of INDUS (Intelligent Data Understanding System), an agent-based system for data-driven knowledge acquisition from heterogeneous, distributed, autonomous data sources.

2 Distributed Learning

To keep things simple, in what follows, we will assume that regardless of the structure of the individual data repositories (relational databases, flat files, etc.) the effective data set for learning algorithm can be thought of as a table whose rows correspond to instances and whose columns correspond to attributes. We will later discuss how heterogeneous data can be integrated and put in this form. In this setting, the problem of learning from distributed data sets can be summarized as follows: The data is distributed across multiple autonomous sites and the learner's task is to acquire useful knowledge from this data. For instance, such knowledge might take the form of a decision tree or a set of rules for pattern classification. In such a setting learning can be accomplished by an agent that visits the different sites to gather the information needed to generate a suitable model (e.g., a decision tree) from the data (*serial distributed learning*). Alternatively, the different sites can transmit the information necessary for constructing the decision tree to the learning agent situated at a central location (*parallel distributed learning*). We assume that it is not feasible to transmit raw data between sites. Consequently, the learner has to rely on information (e.g., statistical summaries such as counts of data tuples that satisfy particular criteria) extracted from the sites. Our approach to learning from distributed data sets involves identifying the information requirements of existing learning algorithms, and designing efficient means of providing the necessary information to the learner while avoiding the neeed to transmit large quantities of data.

2.1 Exact Distributed Learning

We say that a *distributed learning* algorithm L_d (e.g., for decision tree induction from distributed data sets) is *exact* with respect to the hypothesis inferred by a batch learning algorithm L (e.g., for decision tree induction from a centralized data set) if the hypothesis produced by L_d using distributed data sets $D_1, \cdots, D_n$ stored at sites $1, \cdots, n$ (respectively), is the same as that obtained by L from the complete data set D obtained by appropriately combining the data sets $D_1, \cdots, D_n$. Similarly, we can define exact distributed learning with respect to other criteria of interest (e.g., expected accuracy of the learned hypothesis). More generally, it might be useful to consider *approximate* distributed learning in similar settings. However, the discussion that follows is focused on exact distributed learning.

2.2 Horizontal and Vertical Data Fragmentation

In many applications, the data set consists of a set of tuples where each tuple stores the values of relevant attributes. The distributed nature of such a data set can lead to at least two common types of data fragmentation: *horizontal fragmentation* wherein subsets of data tuples are stored at different sites; and *vertical fragmentation* wherein subtuples of data tuples are stored at different sites. Assume that a data set D is distributed among the sites $1, \cdots, n$ containing data set fragments $D_1, \cdots, D_n$. We assume that the individual data sets $D_1, \cdots, D_n$ collectively contain enough information to generate the complete dataset D.

2.3 Transforming Batch Learning Algorithms into Exact Distributed Learning Algorithms

Our general strategy for transforming a batch learning algorithm (e.g., a traditional decision tree induction algorithm) into an exact distributed learning algorithm involves identifying the information requirements of the algorithm and designing efficient means for providing the needed information to the learning agent while avoiding the need to transmit large amounts of data. Thus, we decompose the distributed learning task into *distributed information extraction* and *hypothesis generation* components.

The feasibility of this approach depends on the information requirements of the batch algorithm L under consideration and the (time, memory, and communication) complexity of the corresponding distributed information extraction operations.

In this approach to distributed learning, only the information extraction component has to effectively cope with the distributed nature of data in order to guarantee provably exact learning in the distributed setting in the sense discussed above. Suppose we decompose a batch learning algorithm L in terms of an information extraction operator I that extracts the necessary information from data set and a hypothesis generation operator H that uses the extracted information to produce the output of the learning algorithm L. That is, $L(D) = H(I(D))$. Suppose we define a distributed information extraction operator I_d that generates from each data set D_i, the corresponding information $I_d(D_i)$, and an operator C that combines this information to produce $I(D)$. That is, the information extracted from the distributed data sets is the same as that used by L to infer a hypothesis from the complete dataset D. That is, $C[I_d(D_1), I_d(D_2), \cdots, I_d(D_n)] = I(D)$. Thus, we can guarantee that $L_d(D_1, \cdots, D_n) = H(C[I_d(D_1), \cdots, I_d(D_n)])$ will be exact with respect to $L(D) = H(I(D))$.

3 Decision Tree Induction from Distributed Data

Decision tree algorithms [4, 5, 6] represent a widely used family of machine learning algorithms for building pattern classifiers from labeled training data. They can also be used to learn associations among different attributes of the data. Some of their advantages over other machine learning techniques include their ability to: select from all attributes used to describe the data, a subset of attributes that are relevant for classification; identify complex predictive relations among attributes; and produce classifiers that can be translated in a straightforward manner, into rules that are easily understood by humans. A variety of decision tree algorithms have been proposed in the literature. However, most of them select recursively, in a greedy fashion, the attribute that is used to partition the data set under consideration into subsets until each leaf node in the tree has uniform class membership.

The ID3 (Iterative Dichotomizer 3) algorithm proposed by Quinlan (Quinlan, 1986) and its more recent variants represent a widely used family of decision tree learning algorithms. The ID3 algorithm searches in a greedy fashion, for attributes that yield the maximum amount of information for determining the class membership of instances in a training set S of labeled instances. The result is a decision tree that correctly assigns each instance in S to its respective class. The construction of the decision tree is accomplished by recursively partitioning S into subsets based on values of the chosen attribute until each resulting subset has instances that belong to exactly one of the M classes. The selection of attribute at each stage of construction of the decision tree maximizes the estimated expected information gained from knowing the value of the attribute in question.

Different algorithms for decision tree induction differ from each other in terms of the criterion that is used to evaluate the splits that correspond to tests on different candidate attributes. The choice of the attribute at each node of the decision tree greedily maximizes (or minimizes) the chosen splitting criterion. To keep things simple, we assume that all the attributes are discrete or categorical. However, all the discussion below can be easily generalized to continuous attributes.

Often, decision tree algorithms also include a pruning phase to alleviate the problem of overfitting the training data. For the sake of simplicity of exposition, we limit our discussion to decision tree construction without pruning. However, it is relatively straightforward to modify the proposed algorithms to incorporate a variety of pruning methods.

3.1 Splitting Criteria

Some of the popular splitting criteria are based on entropy [4] which is used by Quinlan's ID3 algorithm and its variants, and the Gini Index [5] which is used by Breiman's CART algorithm, among others. More recently, additional splitting criteria that are useful for exploratory data analysis have been proposed [6].

Consider a set of instances S which is partitioned into M disjoint subsets (classes) $C_1, C_2, ..., C_M$ such that $S = \bigcup_{i=1}^{M} C_i$ and $C_i \cap C_j = \emptyset \ \forall i \neq j$. The estimated probability that a randomly chosen instance $s \in S$ belongs to the class C_j is $p_j = \frac{|C_j|}{|S|}$, where $|X|$ denotes the cardinality of the set X. The estimated *entropy* of a set S measures the expected information needed to identify the class membership of instances in S, and is defined as follows:

$entropy(S) = \sum_j \frac{|C_j|}{|S|} \cdot \log_2\left(\frac{|C_j|}{|S|}\right)$. The estimated *Gini index* for the set S containing examples from M classes is defined as follows: $gini(S) = 1 - \sum_j (\frac{|C_j|}{|S|})^2$. Given some impurity measure (either the entropy or Gini index, or any other measure that can be defined based on the probabilities p_j) we can define the estimated *information gain* for an attribute a, relative to a collection of instances S as follows: $IGain(S,a) = I(S) - \sum_{v \in Values(a)} \frac{|S_v|}{|S|} I(S_v)$ where $Values(A)$ is the set of all possible values for attribute a, S_v is the subset of S for which attribute a has value v, and $I(S)$ can be $entropy(S)$, $gini(S)$ or any other suitable measure.

It follows that the information requirements of decision tree learning algorithms are the same for both these splitting criteria; in both cases, we need the relative frequencies computed from the relevant instances. In fact, additional splitting criteria that correspond to other impurity measures can be used instead, provided that these measures can be computed based on the statistics that can be obtained from the data sets. Examples of such splitting criteria include misclassification rate, one-sided purity, one-sided extremes [6]. This turns out to be quite useful in practice since different criteria often provide different insights about data. Furthermore, as we show below, the information necessary for decision tree construction can be efficiently obtained from distributed data sets. This results in provably exact algorithms for decision tree induction from horizontally or vertically fragmented distributed data sets.

3.2 Distributed Information Extraction

Assume that given a partially constructed decision tree, we want to choose the best attribute for the next split. Let $a_j(\pi)$ denote the attribute at the jth node along a path π starting from the attribute $a_1(\pi)$ that corresponds to the root of the decision tree, leading up to the node in question $a_l(\pi)$ at depth l. Let $v(a_j(\pi))$ denote the value of the attribute $a_j(\pi)$, corresponding to the jth node along the path π. For adding a node below $a_l(\pi)$, the set of examples being considered satisfy the following constraints on values of attributes: $L(\pi) = [a_1(\pi) = v(a_1(\pi))] \wedge [a_2(\pi) = v(a_2(\pi))] \cdots [a_l(\pi) = v(a_l(\pi))]$ where $[a_j(\pi) = v(a_j(\pi))]$ denotes the fact that the value of the jth attribute along the path π is $v(a_j(\pi))$.

It follows from the preceding discussion that the information required for constructing decision trees are the counts of examples that satisfy specified constraints on the values of particular attributes. These counts have to be obtained once for each node that is added to the tree starting with the root node. If we can devise distributed information extraction operators for obtaining the necessary counts from distributed data sets, we can obtain exact distributed decision tree learning algorithms. Thus, the decision tree constructed from a given data set in the distributed setting is exactly the same as that obtained in the batch setting when using the same splitting criterion in both cases.

Horizontally Distributed Data

When the data is horizontally distributed, examples corresponding to a particular value of a particular attribute are scattered at different locations. In order to identify the best split of a particular node in a partially constructed tree, all the sites are visited and the counts corresponding to candidate splits of that node are accumulated. The learner uses these counts to find the attribute that yields the best split to further partition the set of examples at that node. Thus, given $L(\pi)$, in order to split the node corresponding to $a_l(\pi) = v(a_l(\pi))$, the information extraction component has to obtain the counts of examples that belong to each class for each possible value of each candidate attribute.

Let $|D|$ be the total number of examples in the distributed data set; $|A|$, the number of attributes; V the maximum number of possible values per attribute; n the number of sites; M the number of classes; and $size(T)$ the number of nodes in the decision tree. For each node in the decision tree T, the information extraction component has to scan the data at each site to calculate the corresponding counts. We have: $\sum_{i=1}^{n} |D_i| = |D|$. Therefore, in the case of serial distributed learning, the time complexity of the resulting algorithm is $|D||A| \cdot size(T)$. This can be further improved in the case of parallel distributed learning since each site can perform information extraction in parallel. For each node in the decission tree T, each site has to transmit the counts based on its local data. These counts form a matrix of size $M|A|V$. Hence, the communication complexity (the total amount of information that is transmitted between sites) is given by $M|A||V|n \cdot size(T)$. It is worth noting that some of the bounds presented here can be further improved so that they depend on the height of the tree instead of the number of nodes in the tree by taking advantage of the sort of techniques that are introduced in [7, 8].

Vertically Distributed Data

In vertically distributed datasets, we assume that each example has a unique index associated with it. Subtuples of an example are distributed across different sites. However, correspondence between subtuples of a tuple can be established using the unique index. As before, given $L(\pi)$, in order to split the node corresponding to $a_l(\pi) = v(a_l(\pi))$, the information extraction component has to obtain the counts of examples that belong to each class for each possible value of each candidate attribute. Since each site has only a subset of the attributes, the set of indices corresponding to the examples that match the constraint $L(\pi)$ have to be transmitted to the sites. Using this information, each site can compute the relevant counts that correspond to the attributes that are stored at the site. The hypothesis generation component uses the counts from all the sites to select the attribute to further split the node corresponding to $a_l(\pi) = v(a_l(\pi))$. For each node in the decision tree T, each site has to compute the relevant counts of examples that satisfy $L(\pi)$ for the attributes stored at that site. The number of subtuples stored at each site is $|D|$ and the number of attributes at each site is bounded by the total number of attributes $|A|$. In the case of serial distributed learning, time complexity is given by $|D||A|n \cdot size(T)$. This can be further improved in the case of parallel distributed learning since the various sites can perform information extraction in parallel. For each node in the tree T, we need to transmit to each site, the set of indices for the examples that satisfy corresponding constraint $L(\pi)$ and get back the relevant counts for the attributes that are stored at that site. The number of indices is bounded by $|D|$ and the number of counts is bounded by $M|A|V$. Hence, the communication complexity is given by $(|D| + M|A|V)n \cdot size(T)$. Again, it is possible to further improve some of these bounds so that they depend on the height of the tree instead of the number of nodes in the tree using techniques similar to those introduced in [7, 8].

Distributed versus Centralized Learning

Our approach to learning decision trees from distributed data based on a decomposition of the learning task into a distributed information extraction component and a hypothesis generation component sites provides an effective way to deal with scenarios in which the sites provide only statistical summaries of the data on demand and prohibit access to raw data. Even when it is possible to access the raw data, the distributed algorithm compares favorably with the corresponding centralized algorithm which needs access to the entire data set whenever its communication cost is less than the cost of collecting all of the data in a central location. It follows from the preceding analysis that in the case of horizontally fragmented data, the distributed algorithm has an advantage when $MVn \cdot size(T) \leq |D|$ since the cost of shipping the data is given by its actual size, which is given by $|D||A|$. In the case of vertically fragmented data, the corresponding conditions are given by $size(T) \leq |A|$ since the cost of shipping the data is given by its actual size, which has a lower bound of $|D||A|$. These conditions are often met in the case of large, high-dimensional data sets.

4 Data Integration in INDUS

The discussion of distributed learning in the preceding section assumed that it is possible to extract the information needed by the learning algorithm (e.g., counts of instances that satisfy specific constraints on the values of the attributes) from the distributed data sources. This is rather straightforward in the case of data sources that have a homogeneous structure and semantics. However the heterogeneity in the structure and semantics of data can make this task significantly more complex in real-world knowledge discovery tasks [2].

INDUS is designed to provide a unified query interface over a set of distributed, heterogeneous and autonomous data sources which enables us to view each data source *as if* it were a table. For the purpose of decision tree learning, we formulate queries whose executions return tables containing data of interest (e.g., counts). They are driven by the learning algoritm which used them whenever a new node needs to be added to the tree.

5 Summary and Discussion

Efficient learning algorithms with provable performance guarantees for knowledge acquisition from distributed data sets constitute a key element of any attempt to translate recent advances in our ability to gather and store large volumes of data into an ability to effectively use the data to advance our understanding of the respective domains (e.g., biological sciences, atmospheric sciences) and decision support tools. In this paper, we have precisely formulated a class of distributed learning problems and briefly presented a general strategy for transforming a class of traditional machine learning algorithms into distributed learning algorithms. We have demonstrated the application of this strategy to devise algorithms for decision tree induction (using a variety of splitting criteria) from distributed data. The resulting algorithms are *provably exact* in that the decision tree constructed from distributed data is identical to that obtained by the corresponding algorithm when it is used in the batch setting. This ensures that the entire body of theoretical (e.g., sample complexity, error bounds) and empirical results obtained in the batch setting carry over to the distributed setting. The proposed distributed decision tree induction algorithms have been implemented as part of INDUS, an agent-based system for data-driven knowledge acquisition from heterogeneous, distributed, autonomous data sources.

The distributed learning problem has begun to receive considerable attention in recent years. However, many of the algorithms proposed in the literature [9, 10, 11] do not guarantee generalization accuracies that are provably close to those obtainable in the centralized setting. Typically, they deal with only horizontally fragmented data. Furthermore, several of them are motivated by the desire to for scale up batch learning algorithms to work with large data sets by partitioning the data and parallelizing the algorithm. In this case, the algorithm typically starts with the entire data set in a central location; the data set is then distributed across multiple processors to take advantage of parallel processing. In contrast, in the distributed scenario discussed in this paper, the algorithm may be prohibited from accessing the raw data; even when it is possible to access the raw data, it may be infeasible to gather all of the data at a central location (because of the bandwidth and storage costs involved).

The algorithm proposed in [12], is closely related to our algorithm for learning decision trees from vertical fragmented data using entropy or information gain as the splitting criterion. It provides a mechanism for obtaining counts from *implicit tuples* in the absence of a unique index for each tuple in the data set by simulating the effect of *join* operation on the sites without enumerating the tuples. In contrast, our algorithms assume the existence of a unique index, but are more general in other respects (ability to deal with both horizontal and vertical fragmentation, incorporation of multiple splitting criteria). Our approach can be modified using an approach similar to that used in [12] in the absence of unique indices.

Our work on the algorithms and software for data-driven scientific discovery has been driven, at least in part, by the needs of emerging data-rich disciplines such as biological sciences where there is an explosive growth in number, diversity, and volume of data being archived in publicly accessible, distributed, autonomous data repositories [2].

References

1. V. Honavar, L. Miller, J. Wong, Distributed knowledge networks, in: Proc. of the IEEE Conf. on IT, Syracuse, NY, 1998.
2. V. Honavar, A. Silvescu, J. Reinoso-Castillo, D. Caragea, C. Andorf, D. Dobbs, Ontology-driven information extraction and knowledge acquisition from heterogeneous, distributed biological data sources, in: Proceedings of the IJCAI-2001 Workshop on Knowledge Discovery from Heterogeneous, Distributed, Autonomous, Dynamic Data and Knowledge Sources, 2001.
3. N. Jennings, M. Wooldridge, Agent-oriented software engineering, in: J. Bradshaw (Ed.), Handbook of Agent Technology, AAAI/MIT Press, 2001.
4. R. Quinlan, Induction of decision trees, Machine Learning 1 (1986) 81–106.
5. L. Breiman, J. Friedman, R. Olshen, C. Stone, Classification and Regression Trees, Wadsworth, Pacific Grove, CA, 1984.
6. A. Buja, Y. Lee, Data Mining Criteria for Tree-Based Regression and Classification, 2000.
7. J. Shafer, R. Agrawal, M. Mehta, Sprint: A scalable parallel classifier for data mining, in: VLDB'96, Proc. of 22th Int. Conf. on VLDB, September 3-6, 1996, Mumbai (Bombay), India, Morgan Kaufmann, 1996.
8. J. Gehrke, V. Ganti, R. Ramakrishnan, W. Loh, Boat – optimistic decision tree construction, in: Proc. SIGMOD Conf., Philadelphia, Pennsylvania, 1999.
9. W. Davies, P. Edwards, Dagger: A new approach to combining multiple models learned from disjoint subsets, in: ML99, 1999.
10. P. Domingos, Knowledge acquisition from examples via multiple models, in: Proc. of the 14th ICML, Nashville, TN, 1997.
11. A. Prodromidis, P. Chan, S. Stolfo, Meta-learning in distributed data mining systems: Issues and approaches, in: H. Kargupta, P. Chan (Eds.), Advances of Distributed Data Mining, AAAI Press, 2000.
12. R. Bhatnagar, S. Srinivasan, Pattern discovery in distributed databases, in: Proc. of the AAAI-97 Conf., Providence, RI, 1997.

Part VII

**Computational Intelligence
in Management**

Using IT To Assure a Culture For Success

Raj Kumar

Director, Aim Knowledge Management Systems Pvt. Ltd., Mumbai - 400005, India

Abstract. It is 40 years since the ARPANET conceived interactive computing but Information Technology (IT) has barely raised the capability of teams. The reason is IT's inability to manage the heavy and unpredictable workflows of daily knowledge exchange. The intelligence presented harnesses IT to organize and anticipate the decision-maker. Prototype operations establish its thesis that all unstructured collaboration on a Document is synthesized by known repeatable actions; universal norms determine the actions possible next. Replication technology, controlled by the norms, extends the marvelous communication offline and across servers. The give and take culture induced increases the enterprise's capability for success.

Keywords. Business Communication, give and take, interactive processes, virtual teamwork, Knowledge Interaction Management, unstructured collaboration

1. Lessons from Babel and the babble thereafter

Sun Tzu's war treatise blueprinted success 2500 years ago: *"If you know yourself and know the enemy, you need not fear the result of a hundred battles. If you know yourself but not the enemy, for every victory gained you will also suffer a defeat. If you know neither the enemy nor yourself, you will succumb in every battle."* [29].

Structures evolved around the written word to organize for the *'know'*. IT's speed, storage and dissemination swamped them, with no viable alternative. Information overload followed [14]. The ability of teams to organize for success [24 30], viz., evaluate facts[1], define problems, strategize, maintain order, manage culture, innovate, be audacious, etc., has stagnated. The rapid change underway requires teams improve their capability to manage the future with prudence [7 17 25]. Drucker observed [7] in 1999: *"The most important contribution management needs to make in the 21st century is ...... to increase the productivity of knowledge work and the knowledge worker."* IT is making efforts but in the words [21] of Ray Ozzie, creator of major group-ware products, Lotus Notes and Groove: *"We have not even begun to explore how technology can help us to augment our own feeble memories and to effectively work with each other across time and space"*.

---

[1] Human nature tends to see through tinted glasses [23].

Drucker's declaration in 1988 [6] that the Colonial Administration in India was the '*epitome of successful Knowledge Management*' shows the way. Its success lay in organized collaboration to evolve opinion, without same-time contact or process definition. This give and take spontaneously forwards an issue with context, an opinion and an expectation. Conducted at ones own convenience but with accountability, it supports debate for swift clarity [27], involves experts, and provides time for contemplation. Its collective responsibility aids learning, strengthens risk taking ability and motivates its flow. It is a major component of Ikujiro Nonaka's seminal definition [20]: "*knowledge management is not to do with processing `objective' information but with concentrating on tapping the tacit and often highly subjective insights, intuitions and hunches of individual employees and making these insights available for testing and use by the company as a whole*".

The times demand faith in and growth of the enterprise's wisdom [7 2 6] and the capability to apply it to strategy, reach conclusions, innovate and take action [15 24 30]. It is concluded that focused and free give and take supported by access to experience and unobtrusive follow-up, hereafter called G&T, can achieve this.

2. The attempt to re-build higher

Spurred by the indispensability of IT for personal communication, Knowledge Management (KM) has sought to establish structures for the interactive processes or daily Business Communication. However, Information Week's review [1] of the fierce competition in 2000 to capture the KM market reflects IT's low rel evance for the G&T: "*As with all knowledge-management efforts, the technology is far less important than the business processes and human cultural issues that influence success. The right company can succeed with practically any technology*".

A statement of Andy Grove, Chairman of Intel, to Business Week in 2000 [18] explains IT's impact on knowledge exchange: "*When it comes down to the bulk of knowledge work, the 21st century works the same as the 20th century. You can reach people around the clock, but they won't think any better or any faster just because you've reached them faster. The give and take remains a limiting factor*".

IT interprets the interactive need as more expressive means of communication to speed creation [22]. Ozzie bypasses the needs of G&T capability like managing expectations and chaos for systematic knowledge capture with: "*People work best when they can 'self-organize' cooperating spontaneously in free-form ways*". A Conventional Wisdom[2] has developed that IT cannot organize enterprise-wide G&T since it is spontaneous and unpredictable. Email has emerged as the only means for coordination of unstructured collaboration across processes and departments. It is entrenched as the intranet backbone but serves G&T poorly:

- Users must[3] self-organize to harness process IT like collaboration, and content IT like Document Management for their daily Business Communication.

---

[2] A powerful term coined by the economist John K Galbraith [8].

[3] Isolated knowledge is dangerous [9]. Strategy development needs teamwork.

- The chaos of unstructured work is amplified by the build up of issues and spam;
- Users must categorize mails for knowledge capture. It is unlikely to be uniform;
- Emails have no logical life. Its interface needs housekeeping to avoid clutter;
- Routing to professional colleagues and communities absorbs energy and time;
- Emails do not capture the expectation that initiates interaction. Outstanding expectations on issues in circulation are a key source of anxiety;
- Email technology does not distinguish between the needs of business and personal communication. It provides poor tracking, knowledge and control of downstream thinking, action and response times. The mail is not the message.

Collaboration technology improves upon email for assimilation and application of knowledge. However, it lacks privacy and tracking for control and follow-up. Tedious check in/out, coordination of process software, chaos of multiple issues, etc., like for email, impair the work experience. Effective teamwork is affected.

Same-time IT like video-conferencing, meeting on-line, etc., has its benefits but cannot replace asynchronous and unstructured collaboration and capture of the exchange, viz. G&T, to raise the response capability. To compensate, KM seeks to individually organize the interactive processes of success like innovation, strategy, etc. Where this is not possible [11], it motivates personnel to self-organize for systematic capture and sharing of their knowledge. The results are poor. The KPMG 2000 survey revealed just 1% of the respondents had succeeded at implementing KM strategies though 81% had invested in IT for KM [14]. Cynicism has followed to undermine KM: *"Like many a business concept, KM has evolved from a hot buzzword to a phrase that now evokes more skepticism than enthusiasm"* [2].

3. Re-laying the foundations

This paper presents the science of Knowledge Interaction Management to establish a way of working that harnesses IT for fostering enterprise-wide G&T. As a by-product the way records the knowledge flows in context, consistently codified.

The science draws inspiration from the bicycle's evolution. It was still a fad in 1891, 74 years after invention [31] and 2 years after the air tire. Edouard Michelin had the vision [19] that the 'marvelous sensation' of air supported by a replaceable tire would lead to a revolution. He invented the replaceable tire in 1891. Chicago mailmen adopted the bicycle by 1895 and the price of a quality healthy horse fell.

The science seeks to provide IT the intelligence for satisfying the needs of virtual workers [12 16] and creating a marvelous work experience (See Appendix). This is essential as *"These virtual work-horses don't want their time wasted"* [12]. They demand [5] an intuitive way of working, a supportive organization structure, transparent technology, creative autonomy and time and energy conservation. Preferences like turf, politics, priority and reliability abort adoption where the system falls short. The intelligence harnesses IT, not create it, to satisfy the needs and overcome IT's failings for Business Communication, an amalgam of Communication, Collaboration, Coordination, Document and Performance Management.

4. The science of Knowledge Interaction Management

Darwin's theory of Natural Selection [3] implies relationships evolve to govern operations when many alternative forms of conduct are possible. Violation of the relationships leads to anarchy. Knowledge Workers have long relied upon interactions with peers in the context of Documents to function as a team and progress their work. It stands to reason that a science governs the work of teams and the conduct of their interactions. The primary tenets of the science are as follows:

- The unstructured daily work of an organization is distributed amongst groups. Groups report to a pyramid. The hierarchy posts are also group members.
- The number of hierarchy levels may vary. The levels contribute experience and are delegated authority to resolve issues. All assignments are flexible.
- Personnel can hold multiple posts across hierarchies and so belong to different groups. On a personal level this sets up a honeycomb structure.
- The organization, its groups and members may be distributed over a wide area. A Resource Map assists swift routing of work by responsibility.
- Work groups engage in defined Pursuits, e.g., Legal & Estate matters, Customer Service, Complaints, Payments, etc. Their Pursuits gives them their profile. Activities or stages of completion may be associated with each Pursuit.
- The Reasons why issues or Cases arise for a work group are linked to its Pursuits. A Case can have multiple Reasons. The linkages between Group, Pursuit, Reason and Case create the Knowledge Maps for the organization.

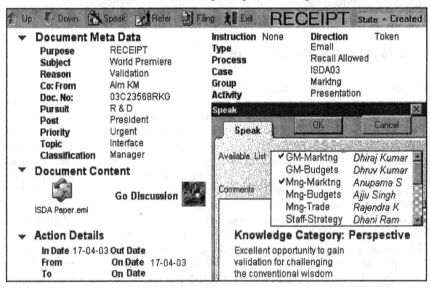

Fig. 1. Document Workspace. Exists for each Document entry in the smart interface, see Fig. 4, defined for each person. Screenshot shows the possible Action buttons applicable to the Document. The *Speak* Action has been selected. The personnel who qualify are shown for multiple selections. On *OK*, the *Speak* Action and opinion recorded shall be updated

- The events that take place on a Case may be captured by Documents that have a structured part to capture meta-data such as Customer, Group, Pursuit, etc., and an unstructured part to capture content. The structured part relates the Document to the organization structure. Typical Document data is shown in Fig.1.
- The Knowledge Maps assist comprehensive and swift capture of a Document's meta-data, e.g., the Group, Pursuit, etc., given the Reason.
- All Maps are integral to the daily work. This assures their regular maintenance.
- Document's have a life cycle: creation followed by circulation at the discretion of personnel and an end in the archives. The Document Purpose, e.g., Receipt or Response, decides how a Document shall be created and enter the archives.
- The Document State records the work progressed in relation to the life cycle.
- There is a cause and effect relationship between the Documents that accumulate on a Case. History is the chain of Documents and the discussion that accumulates. The discussions share the meta data of their parent Document, see Fig.2.

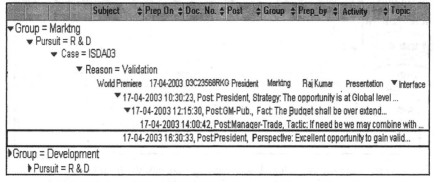

Fig. 2. Discussion. Recorded for each Document and linked to its workspace, see Fig.1

- 18 repeatable Actions account for at least 90% of all Document processes. The balance 10% may be modeled using the same parameters. Action on a Document may lead to an outstanding expectation. A Token controls right of Action.
- The Document Pursuit, its Purpose, its State and any outstanding expectation are the primary determinants of the Actions possible on it. The evolution of teamwork has established the norms that govern the possible Actions.
- The conversion of knowledge to commitment follows an identifiable process. The flow in Fig. 3, for example, defines the knowledge elements of daily decision-making on an event. The tacit knowledge heads may be used for capture.

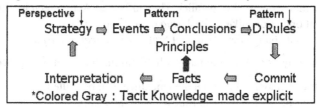

Fig. 3. Converting Knowledge to Commitment

5. Intelligence from the Science Of Interactions

The science covers the means for work and interaction. The interactive processes viz., evaluation, innovation, strategizing, etc., or their mix, are treated as collaboration on content, synthesized by a set of repeatable Actions. Norms and personnel discretion decide the Actions assembled and their order of assembly. Knowledge exchange is thus perceived as independent of the interactive process. The science's norms and relationships apply across departments to help anticipate:

- The Document meta-data using the Knowledge Map and any leading data;
- The next Action that may be assembled to the Document's processing; and
- The personnel who can satisfy the Action selected for the Document.

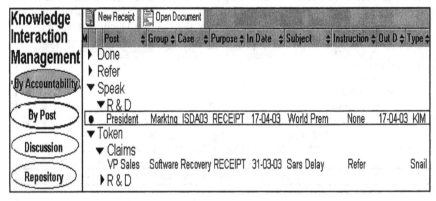

Fig. 4. Smart Interface. The categorization follows automatically from Document capture and Action. Views provide for meaningful drill down. The interface gives the message

Fig. 4 shows the private interface. It presents all the Documents a person is involved with, automatically categorized for maximum focus. The outstanding expectation like Refer, Speak, etc., identifies the Documents-in-process. Following selection of a Document, personnel select one of the Actions offered per the norms in the Document workspace shown in Fig.1. Actions initiate:

- Transfer of accountability like move *Up* or *Down*. These Actions transfer the *Token* for Action and update *Done* to sender, viz., no expectation;
- Interactions with expectations like please *Speak* or send *Information*. The *Token* is transferred and sender updated with the expectation. These interactions may be either sequential or parallel. They influence Action recipients can take;
- Document operations like make *Memo* or add content; and
- Processes that change the Document State like *Cancel* or *Dispatch*.

An Action effects coordination and collaboration. On selection it processes:

- The posts or personnel to whom the Document can be routed;
- The posts or personnel that can follow up on the Action;
- Creation of entry in recipient/s interface/s or change of the Document's state;

- Reorganization of the User's interface per the Action taken; and
- Categorization of the opinion registered (same as the Document meta-data) and its relationship with the main discussion – it could be a sub-string.

Each Action has its own set of norms. The norms are tenets of the science. They were "discovered" by hypothe sizing possible combinations of Actions until all circumstances were covered and the Actions functioned in harmony. The assembly of Actions can execute any valid Document workflow as shown in Fig. 5.

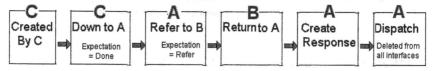

Fig. 5. Typical Spontaneous, Unstructured Workflow On A Document

The following illustrate the common sense and reality that founds the norms:

- The Document forwarder transfers the right of Action by awarding the Token.
- If an Action like 'send Information' is outstanding, the Document ca nnot be concluded downstream, viz., its State has to be maintained.
- A Document that has been received must be linked to the archives and re-corded. A Response Document has to be created for dispatch.

The norms also define logic for conflict-free replication (could be PtoP), e.g., cancellation of any upstream Action deletes the Actions assembled downstream.

The norms form the core of the science. By way of the Actions they integrate the expectations with the organization structure and the Document taxonomy. They provide the intelligence to anticipate and conduct the daily interactions from the interface. The global similarity of teamwork makes the science, its tenets and the intelligence derived universally valid[4]. The interface harnesses technology for:

- Effortless step by step coordination of Business Communication, viz., organiz-ing contribution and the conduct of spontaneous and free form (unpredictable) interactions on each Document leading to a unique polyglot process;
- Eliminating all anxiety that logically is an undesirable sequence of Actions;
- Smart capture of the Document metadata and content, with historical links;
- Control over the chaotic progress of work on each Document;
- Smart codification of Actions and the knowledge exchanged for continuous co-ordination, contribution, capture, corporate storage, access, and sharing;
- A natural way to develop and manage content in context, with accountability;
- Conducting pre-defined workflows where no expectations are outstanding;
- A flexible organization structure that performs re-routing for re-assignments, smart security and smart formation of professional communities;

---

[4] Supported by KPMG 2000 KM Survey [14], Item 1.9: "*.... there were no significant - differences between respondents' views on a sector or geographic b asis.*"

- Privacy and classification of the Documents to limit access;
- Dynamic x-categorization of the interface (particularly by outstanding expectation) for swift review of Action and collaboration on each pending Document;
- A two level categorized Knowledge repository: Level 1 provides event history and content, and Level 2 the flow of opinion on Level 1 as shown in Fig.2; and
- Intelligent replication between networked servers, and with the User's mobile computer, for comprehensive communication. Norms eliminate file sharing conflicts for anytime and anywhere working to promote the User's convenience (essential for the mobile user. Also enhances contemplation time).

6. Delivering a foundation for success

Licklider's path -breaking conception of the ARPANET [13] as *"spirit of community"* and *"computer as a communication device"* is forty years old, but knowledge exploitation by teams has barely changed since the phone's debut in 1876. The reason is the poor work experience of email's interface retained in the intranet infrastructure by the conventional wisdom for all coordination. The adoption of IT for enterprise-wide G&T becomes dependent on motivation and culture [11].

The intelligence developed powers a smart interface to organize and anticipate the decision-maker. This fills the lacunae in IT's intranet infrastructure for a practical and flexible means to conduct the daily interactive processes. The step by step approach evolves teamwork and applies across departments. Its marvelous work experience assures adoption. The way of working transparently harnesses IT to rewrite the conventional wisdom and make the daily Business Communication effortless. Email is confined to personal communication. Systematic capture of all knowledge flows is a by-product. Personnel self-interest [27], like involvement and reduced anxiety, is released from the constraints of self-organization to drive the flow and reuse of knowledge. A knowledge driven work culture is induced.

The smart interface may not accelerate the G&T but it does fulfill the critical need [4]: *".. knowledge management has to be "baked into" the job. It's got to be part of the fabric of the work to import knowledge when it's needed and export it to the rest of the organization when it's created or acquired"*. The enhanced capability assists rapidly formed teams manage the chaos of multiple issues with multiple peers, significantly improve judgment, and achieve the flexibility, focus, and unity to move fast with clarity and purpose. The architecture sets the organization's managers free from the mechanics of Knowledge Management to engage in leadership [7 24 26] and respond to credible warnings like: *"Adapt or die...shape a flexible organization that is capable of responding to unpredictable events"* [10].

The intelligence has proven itself for coordination and management of expectations in challenging prototype operations on a department scale. The superior exploitation of technology for better coordination alone improved performance by over 25%. The Appendix lists the major deliverables of the smart interface developed for enterprise operations. The likely superior management direction fostered by the leveraging of wisdom and unity achieved shall enhance the value delivered.

Appendix: Smart Interface – Deliverables For A Marvelous Work Experience

The intelligence vested in the interface:

- Anticipates and conducts all spontaneous free form interactions on unstructured documents, without imposing a learning burden.
- Effortlessly coordinates all the ad-hoc interaction on a Document. Linked with Collaboration, this captures all opinions expressed in context, and automatically categorizes the knowledge for corporate wide access and reuse.
- Establishes a reliable procedure for decision making in the daily operations, with free flow of knowledge and reuse of experience.
- Manages the chaotic and complex interactions [28] common in departmental and x-departmental handling of issues.
- Delivers a personalized way to manage multiple role assignments.
- Easily incorporates structured workflow sub-systems or integrates with them.
- Creates means for distributed team members to interact at their convenience.
- Provides each knowledge worker a clear identity. Prompt, accurate and unobtrusive performance reports, with downstream Document recall where relevant assist exercise of responsibility.
- Performs all housekeeping and categorization to manage multiple issues with multiple people. The interface facilitates drill down per criteria of interest.
- Intelligently implements Document access security and collaboration norms with easy means to modify the settings.
- Provides effortless and systematic Document capture with links to the history.
- Supports a flexible organization structure with automatic rerouting of outstanding Documents in the event of a reassignment.
- Provides prompt means to dispel the Knowledge Worker's anxiety:
  - Do procedures protect me from my judgment being clouded by my fears, hopes and ambition?
  - Who is accountable for Action on the Document?
  - Who are the group members working on the Document?
  - Is my issue handling performance or that of my group satisfactory?
  - Am I protected from oversights?
  - What is the progress on my expectations?
  - Is my responsibility on this shared? Etc.

Performance of all work on the system enables high quality knowledge capture as a by-product. Extraordinary productivity benefits follow, including centralization of expensive overhead expenditures, access to competencies, etc.

All the interaction and work benefits are made available offline and to team members distributed across servers with the use of replication technology (in - theory, PtoP technology like Groove may serve the purpose also). The interface intelligence resolves all conflict problems arising from the sharing of files.

References

1. Angus J (August 7, 2000) Harnessing Corporate Knowledge. InformationWeek.com, http://www.informationweek.com/798/knowledge.htm
2. Berkman Eric (April, 2001) When Bad Things Happen to Good Ideas. - DarwinMag.com, http://www.darwinmag.com/read/040101/badthings.html
3. Darwinist Evolution. http://rivendell.fortunecity.com/lair/316/bio/darwin.htm
4. Davenport TH (Nov, 1999) Knowledge Management - Round Two. CIO Magazine
5. Davenport TH, Cantrell S (January 2002) The Art of Work: Facilitating the Effectiveness of High-End Knowledge Workers. Outlook Journal, http://www.accenture.com/xd/xd.asp?it=enweb&xd=ideas%5Coutlook%5C1.2002%5Cart.xml
6. Drucker Peter F (1988, Jan/Feb) Coming of the New Organization. Harvard Bus. Rev.
7. Drucker Peter F (1999) Management Challenges for the 21st Century. Harper Business
8. Galbraith JK (1958) The Affluent Society. Review, http://www.oakridger.com/stories/121301/opE_1213010048.html
9. Goleman D (1997) Vital Lies, Simple Truths: The Psychology of Self-Deception. - London: Bloomsbury (1st published 1985), P6
10. Grove Andrew (September 18, 1995). Fortune, pp 229-230
11. Harigopal U, Satyadas A (2001) Cognizant Enterprise Maturity Model. IEEE Transactions On Systems, Man, And Cybernetics – Part C: Applications And Reviews, Vol 31, No. 4
12. Harvard Management Commn Letter (2000) Creating Successful Virtual Organizations. Harvard Business School Newsletter, http://www.4cmg.com/InsideC/VirtOrg.htm
13. Hauben M, History Of The ARPANET. http://www.dei.isep.ipp.pt/docs/arpa.html
14. KPMG (Dec 1999) Knowledge Management Research Report 2000, Item 1.6, p 3
15. Jennings J (2002) Less is More. Penguin Group, p 173
16. McDonald T (Spring 2000) Real Results:A Whole New Ball Game. Executive-Forums, http://www.executiveforums.com/newsletters/spring00.pdf
17. McEwen (30 June 2001) Human Nature and Investments Don' t Mix www.nzse.co.nz/exchange/publications/investor-column-2001/investor_column_30jun01.html
18. McGovern G (June 2001) Is Speed God Or The Devil. New Thinking, http://www.gerrymcgovern.com/nt/2001/ nt_2001_08_06_speed_god_devil.htm
19. Michelin Company. http://www.michelin.com/corporate/en/innovation/how1891.jsp
20. Nonaka I (1991) The Knowledge Creating Company. Harvard Business Review
21. Ozzie R (August 13, 2001) Q&A with Ray Ozzie. http://www.technologyreview.com/articles/wo_mcdonald081301.asp
22. Ozzie R interviewed by Breen Bill (May 2001) Jazzed About Work. FastCompany, - issue 46, p192, http://www.fastcompany.com/magazine/46/ideazone.html
23. Pinker S (September 2002) The Blank Slate: The Modern Denial of Human Nature. Viking (Penguin Putnam)
24. Porras IJ, Collins JC (1999) Built to Last: Successful Habits of Visionary Companies. Harper Business
25. Prahalad CK, Hamel G (April 1996) Competing for the Future. HBS Press, Ch 1 & 3
26. Semler Ricardo (1993) Maverick: The Success Story ....Unusual Workplace. Arrow
27. Semler Ricardo (2003) The Seven-Day Weekend. Random House
28. Snowden D (May 2002) Complex acts of knowing – paradox and descriptive self-awareness. Special issue of the Journal of Knowledge Management – Vol 6, No. 2
29. Sun Tzu (500 BC) The Art Of War. http://www.cheraglibrary.org/taoist/artwar.htm, Chap.3
30. Stern CW (Ed), Stalk G (Ed) (1998) Perspectives on Strategy from The Boston - Consulting Group. John Wiley & Sons, Inc.
31. Timeline. http://www.exploratorium.edu/cycling/timeline.html

Gender Differences in Performance Feedback Utilizing an Expert System: A Replication and Extension

Tim O. Peterson[1], David D. Vanfleet[2], Peggy C. Smith[3] and Jon W. Beard[4]

[1]Department of Management, Oklahoma State University, 700 N Greenwood Avenue, Tulsa, OK 74106, USA, email: top@okstate.edu

[2]Arizona State University West, School of Management, P.O. Box 37100, Phoenix, AZ 85069, email: David_D_Van_Fleet@asuwest.online.west.asu.edu

[3]University of Tulsa, College of Business, 600 South College Avenue, Tulsa, OK 74104, email: pcsmith@utulsa.edu

[4]Southern Illinois University, Department of Computer Management and Information Systems, Box 1106 Edwardsville, IL 62026, email: jbeard@siue.edu

Abstract. Expert system proponents claim that expert systems can provide managers with the knowledge they need to perform managerial tasks. One domain for expert system support is in giving performance feedback. This study replicates an earlier study that used only male managers. The result of the current laboratory study using female managers provides interesting empirical findings. The implications of these findings, limitations to the study, and future research are discussed.

1. Introduction

Recently numerous articles on business applications of intelligent systems including expert systems have appeared. Many of these focus on the technical aspects of building an expert system or on functional applications (e.g. Chuleeporn, Holsapple, & Madden 2001; Warkentin, 1991). Human resource management functions that are procedural in nature and well-structured processes have also been reported on by Warkentin and Griffeth (1992). However, far less has been published on expert systems that assist managers in performing their human resource tasks such as providing performance feedback.

Effective performance is critical to organizations, and feedback is important to improving performance (Geddes, 1993). Managers may fail to provide feedback, especially negative feedback. Managers describe giving negative feedback as one of the most difficult and unpleasant tasks they face (Plax, Beatty, & Feingold, 1991). Yet, when performance feedback is given in a consistent and clear manner performance does improve (Hawkins, Penley, & Peterson, 1983). One reason advanced for the delay in development of expert systems in this area is that little research has been conducted on the effectiveness of such systems (Cabrera & Bonache, 1999; Yoon, Guimaraes, & O'Neal, 1995). In 1986, Lewin said, "Whether these new systems help managers is still a question to be answered." Not much has changed since that time.

One empirical study (Peterson & Van Fleet, 1990) found that male managers using an expert system significantly improved their ability in identifying appropriate behavioral responses during a negative performance feedback session. The expert system also assisted the managers in avoiding incorrect behaviors. While promising, there are limitations to this study. To prevent ratee-rater interaction and organizational effects, all subjects used in the Peterson and Van Fleet (1990) study were white male managers obtained from a single organization located in a southwestern city. Therefore, this study cannot be generalized to women or other organizations. Our study replicates that one but uses women managers from many different organizations. It also extends their examination of the system's effect on experienced and inexperienced managers.

Research Question

If expert systems can provide all managers with the cognitive information processing skills to perform better on the job, these systems will afford bright options for organizations. Thus, the research question under study is: Can the use of an expert system provide women managers with the knowledge and skills they need to function better on a performance feedback task?

Methods

To test this research question, a laboratory study was conducted replicating the earlier work by Peterson and Van Fleet. A laboratory study was deemed the most appropriate method for evaluating the potency of the new intervention to control for possible confounding variables. The expert system selected for this study was Performance Mentor (For a more complete description see Peterson & Van Fleet, 1990). This system has been tested for content validity by comparing human experts' advice to its advice for a variety of feedback scenarios (Rufflo, 1987). In addition, external reviews report face validity (deJaager, 1986; PC BusinessSoftwareview, 1986). Performance Mentor uses material from over 100 sources dealing with feedback, human relations, interpersonal communications, and personality theory to develop over 350 rules upon which advice is based (Rufflo, 1987; Schlitz, 1986).

A check on the appropriateness of the system's advice was conducted. Three human resource management scholars, knowledgeable of the performance feedback process, independently reviewed the case files used in the study. Since only the poor performer would be used in the study, the three scholars were asked to focus on the poor performer's case file and review the videotape. Finally, without the assistance of the system, they were asked to select ten behavioral intentions that they felt were most appropriate for a manager to use in the situation. The ten behavioral intentions were selected from the same worksheet provided to the subjects. The management scholars agreed with the system's advice 100%, 90%, and 100% respectively.

Subjects. All subjects were white women managers obtained from multiple organizations located in a southwestern city. A total of fifty subjects were used. Finding women managers proved to be a little more difficult since there are fewer women in managerial positions than men.

Procedure. Subjects role-played a division manager of an organization. Each manager (subject) initially reviewed the performance records and background information on each of three fictitious purchasing agents. Upon completing the review of the records and background information, the manager completed a performance evaluation form on each purchasing agent. Next, the manager prepared to provide performance feedback to one of her purchasing agents. At this point, to further assure realism, the manager watched a five-minute videotape of the purchasing agent at work. By reproducing these salient characteristics of the work environment, Weick (1967) suggests that a valid laboratory organization can be created and realism enhanced.

At the point where managers (subjects) were informed that they would need to provide one of the purchasing agents with performance feedback, the experimental manipulation was introduced. While the subjects were told that the purchasing agent would be selected at random, in actuality, all subjects were to give feedback to the poor performer. Half were instructed that they would be using a computer program that would assist them in preparing for the performance feedback session and they were furnished a six-page handout of instructions but no other training on its use. The other half were instructed to plan what they would do during the performance feedback session with no mention made of an expert system.

Measure. Subjects were asked to identify the ten behaviors that they intended to use from a list of fifty provided. The fifty behavioral intentions were developed from the performance feedback literature. The behavioral intentions list contained: 1) ten appropriate intentions for negative feedback situations (i.e., "Acknowledge his experience and tell him you know he can do the job if he wants to"); 2) ten intentions appropriate for positive feedback situations but not for negative feedback situations (i.e., "Tell him he is doing an excellent job and with a little more work you will be able to give him a merit pay raise"); 3) twenty intentions inappropriate in any situation (i.e., "Tell the individual that he should be embarrassed at his performance"); and 4) ten phatic intentions not intended to share information (Verderber, 1981) (i.e., "Call the individual in immediately"). Three independent raters who had taught communication and performance feedback were asked to identify the category to which these behavior intentions belonged. These raters correctly classified the intentions 97%, 90%, and 95% of the time respectively.

Since this study emphasized both "doing the right things," and "avoiding the inappropriate things" the behavioral intention score was derived from the identification of both the appropriate and inappropriate intentions.   Each appropriate intention identified by a subject was scored as +10.   Thus, a subject's score could range from zero (no appropriate intentions identified) to 100 points (all ten appropriate intentions identified).   Each inappropriate intention identified by a subject was scored as -10.   Thus, a subject's score could range from zero (no inappropriate intentions identified) to -100 points (all ten inappropriate intentions identified).

2.   Experimentation Results

Counter to the earlier findings on male managers, these results do not support the hypothesis that an expert system can help women managers do their jobs better. Figure 1 shows that women managers using the expert system were only able to identify only slightly more appropriate intentions than those not using the system (63% verses 58%).   The t-test performed on these two independent samples confirmed that there is no difference between the two samples.

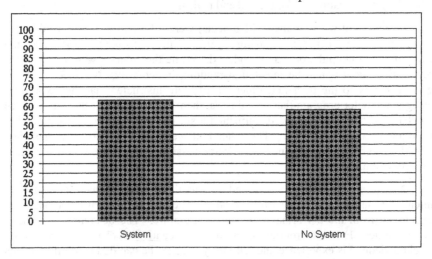

Figure 1. Percent Identified Correct Behavioral Intentions

Figure 2 shows the same pattern of results for the inappropriate behavioral intentions.   Women managers using the expert system identified only slightly less inappropriate behavioral intentions than the women managers not using the expert system.   Again the t-test showed no significant difference between these two values.

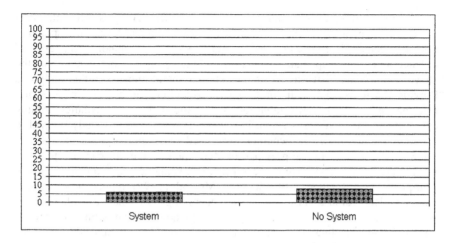

Figure 2. Percent Identified Inappropriate Behavioral Intentions

We also examined the impact of experience. This was determined by asking how often the subjects had provided performance feedback during their professional career. It was decided that inexperienced would be defined as a response of three times or less and experienced would be determined by a respond of six times or more. A t-test ($t = 4.076$, $p < .001$) revealed that these two groups are significantly different from each other. The inexperienced managers had provided performance feedback less than one time on the average while the experienced managers had performed the task over twenty times on the average. We combined the experience-level of the manager with whether they had used the expert system or not to create a new variable with four values (1 = inexperienced with a system, 2 = inexperienced without a system, 3 = experienced with a system, and 4 = experienced without a system). One-way ANOVA results confirmed that there was no significant difference between the four groups on either the appropriate behavioral intention score or the inappropriate behavioral intention score.

3.   Conclusions and Discussions

The literature on managers' ability to provide negative performance feedback to subordinates is replete with examples of how managers struggle with this task (Alvero, Bucklin, & Austin 2001). Our findings would indicate that this is not necessarily true from women managers. It is interesting to note that the women managers using the expert system identified exactly the same percentage of appropriate behavioral intentions as the men using the expert system in the earlier study (63% women verses 63% men). However, the women who did not use the expert system scored much better than the men who did not use the expert system (58% women verses 39% men). What might explain these findings?

A recent study of both women managers and men managers (Moskal, 1997) reports that women managers scored significantly better than men on responding to others' needs. The report said that women provide more support in the terms of guidance and follow up than do men. It is interesting to note that one of the behavioral intentions that was selected most often by the subjects in this study was "Provide the individual with some general guidelines on how to improve her performance." This specific intention is a phatic item. It contributes nothing substantive to the feedback session; however, the women in this study seemed to be compelled to mark this whether they used the expert system or not. In reviewing the printed advice provided by the expert system for each subject, not once did the expert system recommend providing some general guidelines on how to improve performance. In addition, Moskal (1997) reports that women managers also scored significantly better on giving performance feedback than their male colleagues. Does this mean that the expert system is unnecessary for women? The anecdotal information received from the women who used the expert system would disagree with this assertion. Repeatedly during the debriefing, the women who used the expert system referred to it as useful, informative, and relevant. One subject said, "That software was really useful. It reinforced what I thought I needed to do. That gave me more confidence as I prepared for the feedback session." The expert system seems to have still provided an important function by increasing the efficacy of this subject.

This study addresses one of the limitations of the earlier research. We now have some results that tell us how women managers perform on a performance feedback task using an expert system. However, there is still more work to be done. One of the limitations concerns the behavioral intentions scores themselves. While everything was done to convince the subjects that they would actually give face-to-face feedback, there is nonetheless a difference between indicating what one intends to do and actually doing it. They were repeatedly told that they would provide feedback, that the feedback session would be videotaped, and that they would be evaluated on how well they were able to "translate" their intentions as indicated on the checklist to behaviors during the actual feedback session. What they checked may or may not have corresponded to what they would have done. The results of this study support the continuation of this research stream. Furthermore, we believe that this research will provide practitioners with knowledge needed to perform managerial tasks in a confident, professional, and humane manner. Replications of this study with a mixture of managers and subordinates are essential in order to confirm the current findings and to extend the validity of these findings. The examination of how managers would deal with an average performer should also be explored. In addition, future research must include actually having the subject provide the feedback to the subordinate (confederate), perhaps with and without performance feedback training before using the expert system. Also, the Feedback Intervention Theory (FIT) (Kluger & DeNisi, 1996) provides additional research questions that should be examined using an expert system. One is: "Toward which process(es) are the feedback intervention messages of the expert system directed - meta-task, task motivational, or task learning?" Another is: "How do the personalities of the manager and the performer (subordinate) interact to change the advice provided by the expert system?" Performance Mentor has personality components for both the manager and performer although they were held constant in this study to prevent contamination to the variables under examination. Answers to such research questions will either add to the validity of the expert system or identify necessary changes to the system's rule-base. In either case, the research will extend the application of intelligentCounter to the earlier findings on male managers, these results do not support the hypothesis that an expert system can help women managers do their jobs better.

Acknowledgements

The authors appreciate the helpful comments made by Claudette C. Peterson, on earlier drafts of this paper.

References

Alvero, A., Bucklin, B., & Austin, J. 2001. An objective review of the effectiveness and essential characteristics of performance feedback in organizational settings (1985-1998). *Journal of Organizational Behavior Management,* 21(1):3-29.

Cabrera E.F. & Bonache, J. 1999. An expert HR system for aligning organizational culture and strategy. *Human Resource Planning,* 22(1):51-60.

Chuleeporn, C., Holsapple, C.W., Madden, D.L. 2001. Supporting managers' internal control evaluations: An expert system and experimental results. *Decision Supports Systems,* 30: 437-449.

deJaager, J. 1986. Software evaluation of performance mentor. *HRD Review,* 4(5):23 (Reprint).

Geddes, D. 1993. Examining the dimensionality of performance feedback messages: Source and recipient perceptions of these attempts to influence. *Communication Studies,* 44, 200-215.

Hawkins, B.L., Penley, L.E., & Peterson, T.O. (1983). An examination of performance, communication, and the communication-performance relationship as a function of supervisor and subordinate ethnicity. In K.H. Chung (Ed.) *Proceedings of the 43rd Annual Convention of the Academy of Management:* 55-59.

Kluger, A.N. & DeNisi, A. 1996. The effects of feedback interventions on performance; A historical review, a meta-analysis and a preliminary feedback intervention theory. *Psychological Bulletin,* 119(2):254-284.

Lewin, A.Y. 1986, 14 August. Presentation at the National Academy of Management Conference, Chicago, IL.

Moskal, B.S. 1997. Women make better managers. *Industry Week*, 246(3): 17-19.

PC BusinessSoftwareview 1986. Your PC as personnel counselor, September: 1-2.

Peterson, T.O. & Van Fleet, D.D. 1990. Casting managerial skills into a knowledge based system. In M. Masuch (Ed.), *Organization, management, and expert systems*: 81-103. New York, NY: Walter de Gruyter.

Plax, T.G., Beatty, M.J., & Feingold, P.C. 1991. Predicting verbal plan complexity from decision rule orientation among business students and corporate executives. *Journal of Applied Communication Research*, 19: 242-262.

Rufflo, M.F. 1987, August 28. [Interview with Mr. Michael F. Rufflo, cofounder of Human Edge Software Corporation and AI Mentor Incorporated].

Schlitz, J. (1986). *US Government and AI Mentor Inc.* Sign Site License Agreement, Press Release, AI Mentor, Inc..

Verderber, R.F. 1981. *Communicate!* (3rd Edition). Belmont, CA: Wadsworth Publishing Company.

Warkentin, M.E. 1991. AI in business and management: An overview. *PC AI*, 5(6): 26-28.

Warkentin, M.E. & Griffeth, R.W. 1992. AI in business and management. *PC AI*, 6(4): 56-58.

Weick, K.E. 1967. Organizations in the laboratory. In V. Vroom (Ed.) *Methods of organizational research*: 1-56. Pittsburgh, PA: University of Pittsburgh Press.

Yoon, Y., Guimaraes, T., & O'Neal, Q. 1995. Exploring the factors associated with expert systems success. *MIS Quarterly*, 19(1): 83-106.

Part VIII

Image Processing and Retrieval

Image Database Query Using Shape-Based Boundary Descriptors

Nikolay M. Sirakov[1], James W. Swift[1], and Phillip A. Mlsna[2]

[1] Dept. of Mathematics and Statistics, P.O. Box 5717, Northern Arizona University,
Flagstaff, AZ 86011 Nikolay.Sirakov@nau.edu Jim.Swift@nau.edu
[2] Dept. of Electrical Engineering, P.O. Box 15600, Northern Arizona University, Flagstaff,
AZ 86011 Phillip.Mlsna@nau.edu

1 Introduction

Interest in the potential of digital images has increased enormously over the last few years. The internet collection of images has been estimated well in excess of 30×10^6. Examples of large subject-specific image collections include medical and environmental databases. The latter might contain images of earth surface or subterranean objects obtained via remote sensing or from drill and borehole sensors.

Internet users and environmental specialists are exploiting the opportunities offered by the ability to access and manipulate remotely stored images [8]. However, the process of locating a desired image in a large and dynamic collection emerges as a challenging problem [9]. Hence, the image database query problem is becoming widely recognized and the search for solutions is an increasingly active area.

Problems with traditional methods of image indexing [6] have led to the rise of interest in techniques for retrieving images on the basis of automatically derived features such as color, texture, and shape — *Content-Based Image Retrieval* (CBIR).

One of the main problems in CBIR is to locate a desired image or object among a rich and extensive image database. While it is perfectly feasible for a small collection to be searched simply by browsing, more effective querying techniques are needed with collections containing thousands of items. The query types are naturally classified into three levels of increasing complexity:

Level 1 – query by *primitive* features: color, texture, shape, or location.
Level 2 – query by *derived logical* features.
Level 3 – query by *abstract* attributes.

This paper describes an approach and system designed to address queries of level 1.

A survey of the methods designed for retrieving images on the basis of color similarity shows that most of them are based on color histograms [21], cumulative color histograms [20], or region-based color querying [2]. Each image added to the collection is analyzed to compute a color, which is then stored in the database. The matching process retrieves those images whose color histograms most closely match those of the query.

Texture similarity can often be useful in distinguishing between areas of similar color (such as sky and sea, or leaves and grass). A variety of techniques have been used for measuring texture similarity. The best established methods rely on second-order statistics calculated from both query and stored images, or periodicity, directionality, and randomness [13]. Alternative methods of texture analysis for retrieval include the use of Gabor filters [14] and fractals [11].

Two main types of shape features are commonly used for query purposes:

global features such as aspect ratio, circularity, and moment invariants;
local features such as sets of consecutive boundary segments.

Alternative methods proposed for shape matching have included elastic deformation of templates [16, 3]; comparison of directional histograms [10, 1]; shocks [12]; and skeletal representations of object shape that can be compared using graph matching techniques [22].

In our approach, we represent a 2D region as the center of an N-D ellipsoid in N-D vector feature space (VFS) [17]. Each vector component represents a single feature of potential user interest. To each feature is assigned a weight of importance that allows the user to skip those features he considers unimportant for a certain query. Our system extracts features based on gray level and geometric information present in each image. Searching is accomplished by measuring the similarity of feature vectors. Each feature vector is used as an index to its image.

With respect to the above classification, we employ the following global image features: the number of regions in an image and the distances between corresponding regions in different images. The local shape features used in the present approach are: essential (critical) boundary points, the boundary segments and their number [18, 19], and B-spline control points that yield an approximation to the region's boundary [4, 5].

Based on the theoretical concepts, we have developed an intelligent system capable of finding regions similar to a desired region within a dynamic image database. Queries are formulated by computing the above feature characteristics for the query region(s). Queries are then answered by computing the same set of features for every region present within each stored image. The system retrieves stored images whose region(s) features most closely match those of the query. To validate the capability of this system, we performed experiments using an image database that contains images of 2D sections obtained from subsurface objects.

A discussion about advantages and disadvantages of our approach is given at the end of the paper.

2 Image Description and Querying

An important question in CBIR is how to construct the query in order to optimize the search. Assume a given image database (IDB): $IDB = (I^1, \ldots, I^n)$.

Denote by $F = (f_1, \ldots, f_m)$ the set of features sought by all regions O_i situated in the IDB. By $D = (D_1, \ldots, D_m)$ we denote the vector formed by the numerical intervals, where each feature takes a value. If f_i takes all the values from D_i, for $i = 1, \ldots, m$ the obtained vectors represent m-dimensional Vector Feature Space (VFS). A basis of this space can be constructed by using the set of m-D vectors $e_i = (0, 0, \ldots, u_i, \ldots, 0)$, where u_i is the relative unit element of D_i for $i = 1, \ldots, m$. It follows that every $O_i \in IDB$ can be represented by a vector V_i.

Definition 1 *We say that the region O_j is similar to the region O_i if $O_j \in E_i$, where E_i is the ellipsoid*

$$\sum_{n=1}^{m} \frac{(f_n - f_{ni})^2}{\Delta_{ni}^2} w_n = 1 \qquad \text{with center } O_i.$$

One can denote similarity by "$\approx$". The length of the interval is $2\Delta_{ni}$, where f_{ni}, the nth feature of the ith region, takes values. The parameter $w_n = 0$ or 1 and allows the user to skip those features considered unessential for the particular query.

Statement 1 *The region O_j is similar to the region O_i iff*

$$\sum_{n=1}^{m} \frac{(f_{nj} - f_{ni})^2}{\Delta_{ni}^2} w_n \leq 1 \qquad (1)$$

The number of features and their intervals depend on the regions' nature and the information that the IDB contains about them. If an IDB is a dynamic one, some previous knowledge about the new region's nature is necessary in order to determine the corresponding intervals Δ_{ni}.

Remember that each region in the IDB and the query region could be represented as m-dimensional vectors. A query vector can be derived from a query image. The vector derived from the database image is used as an index to that image. We apply equation 1 to measure the similarity between the query vector and a member of the IDB.

The next items present the set of features we use as components of the query vector.

3 2D Boundary Description

Consider a closed polygon situated in a finite plane O_{xy}, where point O is attached to the lower left corner, the axis O_x^+ is oriented toward the lower right, and O_y^+ toward the upper left corner. We define the notion "regularity" as a direction of motion of a point along the closed polygon such that each regularity S^i concludes the following angle φ^i with the axis O_x^+:

$$S^1 : \varphi^1 = \frac{\pi}{2}, \ S^8 : \varphi^8 = \frac{3\pi}{2}, \ S^2 : \varphi^2 = 0, \ S^7 : \varphi^7 = \pi, \ S^3 : \varphi^3 = (0, \frac{\pi}{2}),$$

$$S^6 : \varphi^6 = (\pi, \frac{3\pi}{2}), \ S^4 : \varphi^4 = (\frac{3\pi}{2}, 2\pi), \ S^5 : \varphi^5 = (\frac{\pi}{2}, \pi) \qquad (2)$$

We say that S^i and S^j are: different if $i \neq j$; similar if $i = j$, but $|\varphi^i - \varphi^j| \leq d^1$; almost equal if $d^1 = 0$ but $|l^i - l^j| \leq d^2$; equal if $d^1 = 0$, $d^2 = 0$. We denote the notions as follows: similarity by "$\approx$", almost equality by "$\cong$", and equality by "$\equiv$".

Note the defined relations are reflexive, symmetric, and transitive. Also, the regularities are invariant with respect to plane translation and scaling but not to rotation.

Since we consider closed polygons, any vertex A could be the First Point listed in the polygon Description (FPD). In our work, we use (x_{min}, y_{min}) as the FPD.

The notion "consistency" we define as consecutive connected regularities. This way, exactly one consistency describes each closed polygon with respect to the FPD. For example, the closed polygon shown in Fig. 1 is described by the following consistency:

$$S^5_i \xrightarrow{i=1,2} S^3_i \xrightarrow{i=3,5} S^2_6 \rightarrow S^1_7 \rightarrow S^2_8 \rightarrow S^4_i \xrightarrow{i=9,10} S^6_i \xrightarrow{i=11,12} S^7_{13} \rightarrow S^6_{14} \rightarrow Cl \qquad (3)$$

Fig. 1. A closed polygon whose vertices are A_i and its regularities are S_i.

It follows from the regularities definition that each consistency expands into four finite numerical sequences: R_g - the sequence of different regularities; R_p - the sequence of repetition indicators, which represent the successive appearance of regularity in the consistency; A_n - contains the angles of the regularities; L_e - contains the lengths of the regularities. It follows that $R_p = 1$ for S^1, S^2, S^7, and S^8.

For example, the consistency shown in (3) expands into the following numerical sequences: $R_g = 5, 3, 2, 1, 2, 4, 6, 7, 6$; $R_p = 2, 3, 1, 1, 1, 2, 2, 1, 1$; $A_n = \varphi_1, \ldots, \varphi_{14}$; $L_e = l_1, \ldots, l_{14}$.

Definition 2 *A boundary point A is an essential point if $A = S^i \cap S^j$ for $i \neq j$.*

The FPD is an essential point.

It follows from definition 2 that the essential boundary points are critical points. Also, the number of the essential points equals the number of different arc segments that form the boundary and the number of quadratic B-splines to be used for boundary approximation. Therefore, the essential points of a closed polygon represent a shape distinguishing property.

The above concepts show that polygon description by consistency is invariant with respect to plane translation and scaling but not to rotation. To overcome the latter disadvantage, we employ a boundary support.

4 Boundary Support

Suppose $Sh \subseteq R^2$ is a compact shape. Assume Sh is equal to the closure of its interior. Let us consider an angle θ and denote the tangent line to the Sh at the point P by: $x \cos \theta + y \sin \theta = d$. Then the tangent line at the point Q is $x \cos(\theta + \pi) + y \sin(\theta + \pi) = -d$ (See Fig 2).

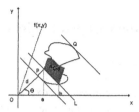

Fig. 2. A compact shape Sh, two tangent and one secant line.

Denote by $B = \{(\theta, d) \in [0, 2\pi) \times R | l(\theta, d)\}$. Since the set Sh is compact, the set B is bounded. In general, B can retrieve the convex hull of Sh because each point of B corresponds to a tangent line of the set Sh. Assuming Sh is connected, the boundary of B is made up of two curves $g(\theta) = d$ and $g(\theta + \pi) = -d$, where g is a 2π periodic function as in Fig. 3 (right). Now assume further that g is differentiable at θ^*. The point of tangency of the line $x \cos \theta^* + y \sin \theta^* = d^*$ can then be computed as follows. Find the intersection of:

$$L_1 : x \cos \theta^* + y \sin \theta^* = h(\theta^*) \quad and \quad L_2 : x \cos(\theta^* + \varepsilon) + y \sin(\theta^* + \varepsilon) = h(\theta^* + \varepsilon)$$

and take the limit as $\varepsilon \to 0$. Then an approximation for the line L_2 is:

$$x(\cos \theta^* - \varepsilon \sin \theta^*) + y(\sin \theta^* + \varepsilon \cos \theta^*) = d^* + \varepsilon h'(\theta^*).$$

Therefore, the set $L_1 \cup L_2$ can be replaced by $L_1 \cup L_2'$, where $L_2' : -x \sin \theta^* + y \cos \theta^* = h'(\theta^*)$. Therefore:

$$\begin{bmatrix} \cos \theta^* & \sin \theta^* \\ -\sin \theta^* & \cos \theta^* \end{bmatrix} \begin{bmatrix} x \\ y \end{bmatrix} = \begin{bmatrix} h(\theta^*) \\ h'(\theta^*) \end{bmatrix} \tag{4}$$

Equation (4) presents a rotation and applying it to the region given in Fig. 3 (left) we found the set B given in Fig. 3 (right). Since the matrix in (4) is a matrix of rotation, it has a unique inverse used in the following transformation:

$$\begin{bmatrix} \cos \theta^* & -\sin \theta^* \\ \sin \theta^* & \cos \theta^* \end{bmatrix} \begin{bmatrix} h(\theta^*) \\ h'(\theta^*) \end{bmatrix} = \begin{bmatrix} x \\ y \end{bmatrix} \tag{5}$$

Since (5) is the inverse of (4), it may be employed to retrieve the initial shape Sh if the set B is given. Also, it follows from the formulated concepts that description of boundary support by consistency is invariant with respect to plane rotation.

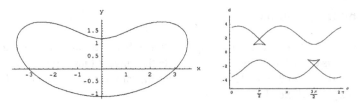

Fig. 3. Left: a concave 2D region. Right: the boundary support of this region.

5 B-Splines and Boundary Control Points

We stated in section 3 that the number of essential boundary points equals the number of different arc segments that form the boundary of a 2D region. In general, a boundary segment of a region can be approximated by a second-order B-spline, which represents a smooth parametric curve defined by three control points. Hence, a set of pixels forming a closed boundary can be approximated by a periodic sequence of B-spline segments. An important property of B-splines is that any given control point exerts only local influence on the shape.

The basic approach of boundary approximation by B-splines is described in [4, 5], which has recently improved upon earlier work [15]. Essentially, the process involves some initial smoothing of the original boundary, an identification of certain geometrically important points (essential points) along the boundary, and the formation of a circulant Toeplitz matrix that combines this information with the basic second-order B-spline equation. The control points are obtained by solving this system using standard matrix techniques.

For such a closed, periodic second-order B-spline curve, the number of B-spline segments matches the number of essential points and the number of control points produced. The set of control points can be considered an efficient encoding of the shape, size, and position of the original boundary. Thus, shape-based comparisons and queries can be accomplished by means of control point set similarity. By comparing shapes using only the control points instead of the boundary pixels, this provides the capability of saving significant amounts of both memory and execution time.

6 Experimental Results

On the basis of the theoretical concepts is developed an intelligent system capable of determining regions in a large and dynamic image database. The system opens the images from the IDB one by one and can search for one or multiple desired regions given in a query image. For each region, the system extracts and measures the following features: the number of regions in the query image; the number of segments into which falls the boundary of each region; the distance between one and the same region located in two different images; the sequences R_g and R_p for each region; and the minimum and maximum gray levels of each region. If (1) holds, the considered IDB region(s) are similar to the query region(s).

Queries are formulated by computing the values of the above shape features for each query region. Queries are then answered by computing the values for the same set of features for every region within each stored image. The system retrieves those stored images whose regions' features most closely match those of the query.

The user defines the intervals Δ_n for each feature $n = 1, \ldots, m$.

As an output, our system provides a record of the titles of all images where are found regions similar to the query set of regions. Also, there is a record of the images into which are found different regions. For both, similar and dissimilar regions, is given their location in the corresponding image.

The system runs under Windows 98/2000/NT. That part of the software that extracts the features, computes their values, and performs the matching runs under DOS in order to decrease the run time. The latter is very important in case of a large number of images to be processed.

To validate the capability of the developed intelligent system, we performed experiments using a set of images from IDB that contains 2D sections obtained from geological objects such as gravel deposits, ore bodies, and groundwater units. The regions have different sizes, gray levels, and locations in the different images. The background of each image is uniform, but could vary from image to image.

Fig. 4. The query image.

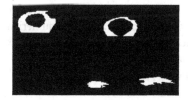

Fig. 5. First image – two sections of gravel deposit and two of ore body.

The query image is shown in Fig. 4. The size of the image in Fig. 5 is 2184×3244 pixels, in Fig. 6 (left) is 413×120, and in Fig. 6 (right) it is 250×140.

The process of searching (matching the desired region with the IDB regions), after running the system, is shown in Fig. 7. The final result is provided as a list of images in table 1. In column 1 is the image title, while columns 2-5 contain the coordinates of the upper left and lower right corners of the rectangle that envelops

Fig. 6. Second and third images representing sections of permeable and impermeable ground-water units, respectively.

the corresponding region. This way the regions most similar to the query one are localized by their images and positions in the images.

Table 1. Example of query results.

Image	x_1	y_1	x_2	y_2	Comments
200L	43	61	163	126	
200L	275	83	396	148	
200L	228	386	311	437	
200L	379	388	490	425	Results: 100.00% OK!
005R	70	21	285	85	
053L	47	39	238	133	

7 Advantages, Conclusions, and Future Directions

The described system is designed for image indexing and region searching in an un-organized and dynamic IDB. The system does not use any information about the IDB organization for searching purposes. Employing regularities and essential boundary points provides the opportunity for using finite numerical sequences [19] to describe the boundary. Hence, the matching process is restricted to comparing finite numerical sequences, which is faster than the graph matching techniques reported in [12, 22]. Results typically require 1 to 1.5 seconds on an 850MHz Pentium processor.

The use of B-splines, control points, and finite numerical sequences to model 2D region boundaries significantly reduces the memory and execution time required to match regions in a large and dynamic IDB.

Another advantage of our approach is that it combines the use of local and global shape features as well as gray level. Most of the methods separately use color [2], texture [11], or shape features [1] for the goal of region matching.

Comparing our approach to that of Gadi [7], ours provides the opportunity for searching multiple desired regions in images that contain multiple regions.

The main disadvantage of our approach is its sensitivity with respect to the boundary discontinuity. It follows that a single boundary might be approximated by more than one polygon. Ongoing research is addressing this problem.

Work is currently underway to develop appropriate similarity metrics for CBIR using vectors of B-spline control points and other gray-level information as well as to expand the use of feature vectors to image indexing.

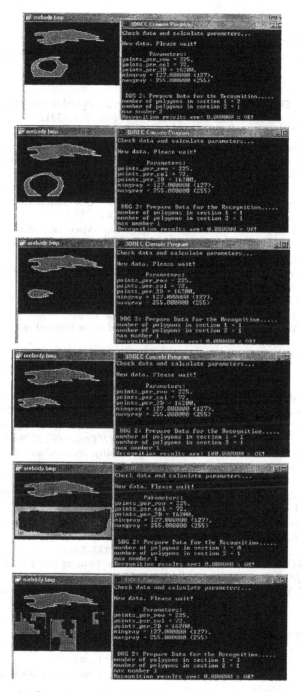

Fig. 7. Query results. In each case, the right window describes the current comparison. The left window contains the query region above the derived region from the current image.

Acknowledgements

Thanks to IM-BAS and Asst. Prof. R. Mironov for valuable software help provided. Mr. Tarek El Doker is acknowledged for his B-spline approximation assistance.

References

1. Androutsas D, et al (1998) "Image retrieval using directional detail histograms" *Storage and Retrieval for Image and Video Databases VI* Proc. SPIE **3312**:129–137

2. Carson C S, et al (1997) "Region-based image querying" *Proc. of IEEE Workshop on Content-Based Access of Image and Video Libraries*, San Juan, Puerto Rico, pp. 42–49

3. del Bimbo A, et al (1996) "Image retrieval by elastic matching of shapes and image patterns" *Proc. of Multimedia '96*:215–218

4. El Doker T A, Mlsna P A (2002) "Efficient region boundary approximation using adaptive smoothing and second-order B-splines" *Proc. IEEE Southwest Symposium on Image Analysis and Interpretation*, Santa Fe, NM, 139–143

5. El Doker T A, Mlsna P A (2002) "Efficient unsupervised estimation of second-order B-spline contour descriptors" invited paper *Proc. IEEE Midwest Symposium on Circuits and Systems*, Tulsa, OK

6. Enser P G B (1995) "Pictorial information retrieval", *Journal of Documentation* **51**(2):126–170

7. Gadi T, et al (1997) "Fuzzy similarity measure for shape retrieval", *Proc. Intl. Conf. Vision Interface '99*, Trois Rivieres, Canada, 386–389.

8. Gudivada V N, Raghavan V (1995) "Content-based image retrieval systems" *IEEE Computer* **28**(9):18–22

9. Jain R (1995) "World-wide maze" *IEEE Multimedia* **2**(3), 3

10. Jain A K, Vailaya A (1996) "Image retrieval using color and shape" *Pattern Recognition* **29**(8):1233–1244

11. Kaplan L M, et al (1998) "Fast texture database retrieval using extended fractal features" *Storage and Retrieval for Image and Video Databases VI* Proc. SPIE **3229**:162–173

12. Kimia B B, et al (1997) "A shock-based approach for indexing of image databases using shape" *Multimedia Storage and Archiving Systems II* Proc. SPIE **3229**:288–302

13. Liu F, Picard R W (1996) "Periodicity, directionality, and randomness: World features for image modelling and retrieval" *IEEE Trans. on PAMI* **18**(7):722–733

14. Manjunath B S, Ma W Y (1996) "Texture features for browsing and retrieval of large image data" *IEEE Trans. on Pattern Analysis and Machine Intelligence* **18**:837–842

15. Meier F M, et al (1997) "An efficient boundary encoding scheme which is optimal in the rate distortion sense" *Proc. IEEE Intl. Conf. on Image Processing* **2**:9–12

16. Pentland A, et al (1996) "Photobook: tools for content-based manipulation of image databases" *Intl. Journal of Computer Vision* **18**(3):233–254

17. Sirakov N (1997) "Several 3D objects reconstruction by using vector feature space" *Proc. of RecPad97* Coimbra, Portugal, 107–111

18. Sirakov N, Muge F H (2001) "Comparing of 3D reconstructed objects" *Proc. Application of Computers and Operations Research in the Mineral Industries* Beijing, China, 75–80

19. Sirakov N M, Muge F (2001) "A system for reconstructing and visualizing three-dimensional objects" *Intl. Journal on Computers and Geosciences* **27**(1):59–69

20. Stricker M, Orengo M (1995) "Similarity of color images" *Storage and Retrieval for Image and Video Databases III*, Proc. SPIE **2420**:381–392

21. Swain M J, Ballard D H (1991) "Color indexing" *International Journal of Computer Vision*, **7**(1):11–32

22. Tirthapura S, et al (1998) "Indexing based on edit-distance matching of shape graphs" *Multimedia Storage and Archiving Systems III* Proc. SPIE **3527**:25–36

Image Retrieval by Auto Weight Regulation PCA Algorithm

W.H. Chang and M.C. Cheng

Department of Information Management, Fortune Institute of Technology,
Kaohsiung, Taiwan, R. O. C.
E-mail: { cwh1, MCheng}@center.fjtc.edu.tw

Abstract. A new image retrieval method by integrating color and edge features is proposed in this study. We also implement a practical interface to retrieve images from image database that are relevant to the user query. The proposed system uses auto weight regulation PCA (Principal Component Analysis) algorithm for similarity measure and image retrieval.

In our proposed system, there are three steps: features extraction, similarity measure, and image retrieval. In the first step, we extract color features and edge features of a query image and save extracted features as code word. Similarity measure is done by auto weight regulation PCA algorithm in the second step. Then retrieve the most relevant images from image database optimally by comparison the projection value in codebooks with query image. The effectiveness and practicality of the proposed method has been demonstrated by various experiments.

1   Introduction

Lots of image retrieval methods have been proposed. In [2] color features are used to retrieval images. In [3] and [4] the shape features are used. These methods although providing overall system performance but time consuming in processing. In our proposed system, the color features is based on the perceptual HSI color space and the shape features was extracted by non-maximum suppression edge detection. We integrate color and edge features for image retrievals for more flexible and better performance. The image database is summarized with identifying features saved in color and edge pattern code books respectively and computed by auto weight regulation PCA algorithm prior to retrieve. The effectiveness of the proposed method can speed up similarity measure and relevant images retrieval procedure.

This paper is organized as follows: In Section 2, we describe our proposed algorithm. System architecture is described in Section 3. The experimental results are presented in Section 4. Conclusions and future work are given in Section 5.

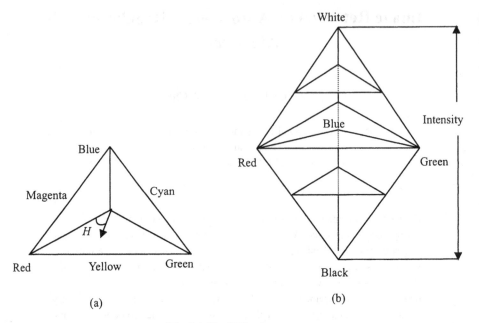

Fig. 2.1 The HSI color space.

2. The Proposed Image Retrieval Method

2.1 The Color Feature Extraction

Several widely used three-dimensional color spaces like RGB, YUV and HSV have been studied and analyzed. Among these color spaces, which one is suitable for image representation is still has no conclusion. In our proposed system, we adapt HSI color space because of its similarity and perceptibility characteristics. Fig. 2.1 (a) and (b) show the relationship of H, S, and I components of the HSI color space. In Fig. 2.1(a), the Hue component H of a color P is the angle related to Red axis.

For a given color (R, G, B) represented by the RGB color system, the values of R, G, and B are all integers in the range of [0, 255] and represent the values of red, green and blue, respectively. The values of hue(H), saturation(S), and intensity(I) can be transformed from the values of R, G, and B by the following formulas:

$$H = \cos^{-1}\left\{\frac{\frac{1}{2}[(R-G)+(R-B)]}{[(R-G)^2+(R-B)(G-B)]^{1/2}}\right\} \tag{1}$$

$$S = \frac{3}{(R+G+B)} \times [\min(R, G, B)] \tag{2}$$

$$I = \frac{1}{3} \times (R+G+B) \tag{3}$$

Quantization scheme achieves less storage and retrieval efficiently. The number of color bins used is normally small than the real color space and total number of colors used to represent image. Therefore, a number of colors have to group into one bin and this is called color clustering. Ideally, colors cluster into the same bin should be more similar than colors of different bins [1].

We convert RGB or True color type image into HSI coordinates using formulas (1), (2), and (3). Then, we cluster hues of an image into 6 bins. Since hue is represented as circle and primary hues (Red, Yellow, Green, Cyan, Blue, and Magenta) are located on the circle every 60 degrees apart in the HSI color space. In our proposed system, there are 4 bins for saturation and 4 bins for intensity. Because of the human visual system is more sensitive to hue than to saturation and intensity [1], the bin number of hues is more than intensity and saturation. There are 4 grades of intensity for gray level, in which the R, G, and B components are the same but have different intensity value. The extracted color features of an image are represented as a code word and save into feature database. The code word C of an image color is represented as follow:

$$C = \{filename, C_1, C_2, ..., C_{100}\} \tag{4}$$

2.2 The Edge Feature Extraction

In our proposed system, we employ the Canny edge detector to extract edge pattern from an image. The Canny edge detector uses two different threshold to detect strong and weak edges, and it includes the weak edge in the object only if they are connected to strong edges. This method is therefore less likely than the others to be fooled by noise, and more likely to detect true weak edges [5]. However, the matching procedure is time consuming. Our method is based on auto weight regulation PCA algorithm that depends on comparing projection value and can speed up retrieval procedure.

The Canny edge detector involves three stages:

1. Directional gradients are computed by smoothing the image and numerical differentiating the image to compute the x and y gradients.
2. Non-maximum suppression finds peaks in the image gradient.
3. Threshold and locates edge strings.

After edge detection, the edge image is segmented into 16 by 16 blocks. And then, we calculate the edge probability for each block as follow:

$$S_i = \frac{E_i}{N}, \quad i = 1, 2, ..., 256. \tag{5}$$

Where E_i is the edge pixels number of block i, N is the total edge pixels number of edge image, and S_i is the edge probability of block i.

The edge code word S of an image is computed by equation (5). Then, we represented edge code word S as follow:

$$S = \{filename, S_1, S_2, ..., S_{256}\}, \quad i = 1, 2, ..., 256. \tag{6}$$

Where S_i is the edge probability of block i and is calculated by equation (5).

2.3 The Auto Weight Regulation PCA Algorithm

The PCA is first presented by H. Hotelling and plays an important role for the application of image compression and signal processing. In this paper, we design an auto weight regulation PCA to find out the first three principal component directions D_1, D_2, and D_3 of an image features and find out their respective eigen values λ_1, λ_2, and λ_3. The percentage of variance information keep by first three principal components directions is denoted by P_1, P_2, and P_3 and compute as follows:

$$P_1 = \frac{\lambda_1}{\sum_{i=1}^{n} \lambda_i}, \quad P_2 = \frac{\lambda_2}{\sum_{i=1}^{n} \lambda_i}, \quad P_3 = \frac{\lambda_3}{\sum_{i=1}^{n} \lambda_i} \tag{7}$$

Where n is the dimension of each image feature. The auto weight regulation W_1, W_2, and W_3 is designed as follows:

$$W_1 = \frac{P_1}{P_1 + P_2 + P_3}, \quad W_2 = \frac{P_2}{P_1 + P_2 + P_3}, \quad W_3 = \frac{P_3}{P_1 + P_2 + P_3} \tag{8}$$

The projection value α for each code vector v in a code book is determined by the following equation:

$$\alpha = (D_1 \times W_1 + D_2 \times W_2 + D_3 \times W_3) \times v \tag{9}$$

Our proposed auto weight regulation PCA algorithm has two stages, namely feature database creation stage and query processing stage, and can be detailed as follows:

Feature database creation stage:

1. Extracting color code word and edge code word for each image in the image database according to equation (4) and (6), respectively.

2. Store extracted color code word and edge code word into color code book and edge code book, respectively. Both color code book and edge code book are stored in the feature database.
3. Calculating the correlation matrix R_E for edge code book and correlation matrix RC for color code book.
4. Calculating the eigenvectors and eigenvalues for correlation matrix R_E and R_C.
5. Arranging the eigenvalues in decreasing order and pick up first three eigenvalues, λ_1, λ_2, and λ_3.
6. Calculating first three principal components D_1, D_2, and D_3 according to first three eigenvalues obtained in step 5.
7. Calculating the percentage of variance information keep by first three principal component directions P_1, P_2, and P_3 according to equation (7).
8. Calculating auto weight regulation W_1, W_2, and W_3 according equation (8).
9. Calculating projection value for each color code word and edge code word according to equation (9). Collect projection values of each color code word and edge code word then save as matrices P_C and P_E, respectively.
10. Store P_C, P_E, D_1, D_2, D_3, W_1, W_2, and W_3 for query processing stage.

Query processing stage:
1. Calculating edge code word v_1 and color code word v_2 of query image according to equation (6) and (4), respectively.
2. Calculating projection value of edge code word v_1 for query image according to equation (9).
3. Compare projection value obtained in step 2 with P_E and retrieving top m images which projection values are most closed to v_1, where m can be defined by user.
4. Extracting projection values of color code word of top m images obtained in step 3 from P_C and store as matrix P_C'.
5. Calculating projection value of color code word v_2 for query image according to equation (9).
6. Compare projection value obtained in step 5 with P_C' and retrieving top n images which projection values are most closed to v_2, where n can be defined by user and $n \leq m$.
7. Show n images obtained in step 6 to the user.

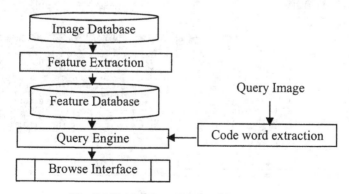

Fig. 3.1 The image retrieval architecture.

3. System architecture

According to retrieval method proposed in section 2, we develop image retrieval architecture to implement our proposed retrieval method. The image retrieval system architecture is shown in Fig. 3.1.

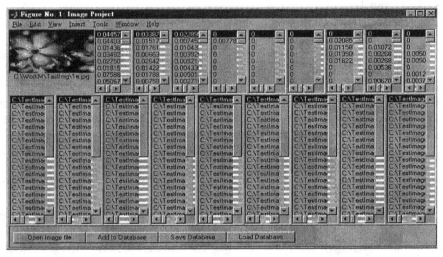

Fig. 4.1 Features extraction interface.

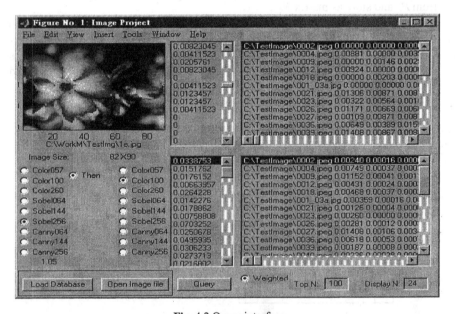

Fig. 4.2 Query interface.

4. The Prototype System and Experiment Results

4.1 The Prototype System

To provide an evaluating environment and a friendly user interface, we implemented a prototype system that contains two interfaces. Features extraction interface extracted color and edge code words in off-line by color cluster and non-compression edge detection algorithm. Users may specify a query image by query interface, through auto weight regulation PCA algorithm mechanism to retrieval the most relevant images. The interfaces of our prototype system are shown in Fig. 4.1 and Fig. 4.2.

We made experiments on a mixture image database. Images in the image database are collected from WWW and classified image database into several classes manually for precision measure. The system was implemented in MATLAB on PC with Intel Celeron 1.2 GHz CPU, 256M RAM and windows platform.

4.2 The Evaluation Model

Precision is the ratio of the number of relevant images retrieved Y to the total number of images retrieved X. Perfect precision means that all retrieval images are relevant.

$$\Pr ecision = \frac{X \cap Y}{X}$$

Another issue is processing time T. T means when user specify a query image, timing the feature extraction and retrieve relevant images procedure. Ideally we want to build a lower processing time and higher precision system.

For evaluating the effectiveness and accuracy of the image retrieval system, we design many sets of feature combinations shown in Fig. 4.3. The explanation of Fig. 4.3 is shown in Table 1.

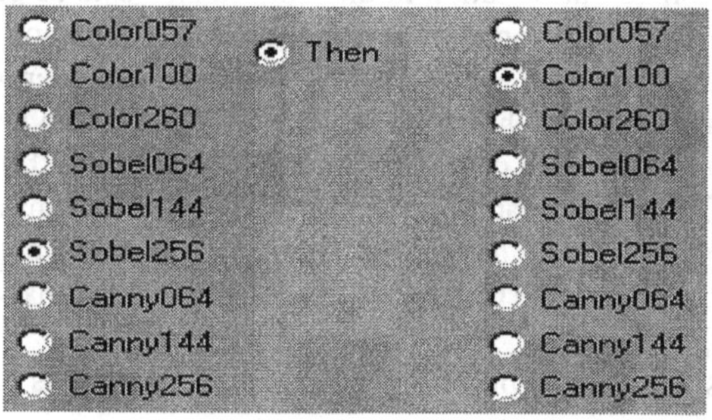

Fig. 4.3 Different feature combination.

Table 1. The explanation of Fig. 4.3

Terms	Descriptions	Goals	Methods
Color057	Color clustering algorithm	Color feature extraction	6Hx3Sx3I+3 grey levels
Color100	Color clustering algorithm	Color feature extraction	6Hx4Sx4I+4 grey levels
Color260	Color clustering algorithm	Color feature extraction	16Hx4Sx4I+4 grey levels
Sobel064	Sobel edge detection algorithm	Edge feature extraction	8x8 blocks
Sobel144	Sobel edge detection algorithm	Edge feature extraction	12x12 blocks
Sobel256	Sobel edge detection algorithm	Edge feature extraction	16x16 blocks
Canny064	Canny edge detection algorithm	Edge feature extraction	8x8 blocks
Canny144	Canny edge detection algorithm	Edge feature extraction	12x12 blocks
Canny256	Canny edge detection algorithm	Edge feature extraction	16x16 blocks

Table 2. Evolution for many sets of feature combinations

Color \ feature / Edge feature	Color057		Color100		Color260	
	T (sec)	Precision (%)	T (sec)	Precision (%)	T (sec)	Precision (%)
Sobel144	3.19	80.42	3.19	81.22	3.52	82.03
Sobel256	3.19	82.23	3.19	84.21	3.52	87.03
Canny144	3.02	80.21	3.02	81.62	3.35	82.43
Canny256	3.02	82.13	3.02	84.71	3.35	87.16

Ideally we would like to reduce the amount of data compared to its original image size and maintain as much information in the reduced size data as possible, for these reasons, we would find out a optimal set of features combination for our system, processing time T and *Precision* are two issues we should consider, the evolution for sets of feature combinations is shown in Table 2. In Table 2, the test images are all 128 by 192 pixels size and each data is obtained by average of 20 times on different test images. From Table 2, we can observe that the more color bins and edge blocks, the more processing time and higher precision, but to achieve less storage and retrieval efficiently, we adapt the Canny256 and color100 as optimal combination ,the reasons as following:

1. The precision is less than color260 2.8% but processing time is outstanding better than color260 9.8%.

2. Hue is represented as circle and primary hues are located on the circle every 60 degrees in HSI color space, color100 means 6H*4S*4I+4 grey levels HSI features.
3. Canny edge detection is less likely than the others to be fooled by noise, and more likely to detect true weak edges [5].

4.3 The Experiment Results

Experiment 1: The query image is 128x192 pixels size and shown on the top-left corner of Figure 4.4. Querying is executed on our mixed image database. Before query, there are 20 relevant images selected by manually. The query result is shown in Figure 4.4. The total features extraction time is 3.3 sec. Among 18 extracted images there are 15 extracted images are relevant, the *Precision* is 0.833. In this experiment, all extracted images are natural scene image.

Experiment 2: The query image is a sketch image and shown on the top-left corner of Figure 4.5. Querying is executed on our mixed image database. Before query, there are 22 relevant images are selected by manually. The query result is shown in Figure 4.5. The total features extraction time is 3.01 sec. Among 18 extracted images there are 17 extracted images are relevant, the *Precision* is 0.944. In this experiment, there are no natural scene image appear on the experiment result.

5. Conclusions and Future work

This paper proposes an optimal set of feature combination to achieve less storage and retrieve efficiently, also designs the image database indexing/filtering mechanism by Auto Weight Regulation PCA algorithm according to the color and edge pattern features. A new approach of image retrieval system has been proposed.

In the future, we will focus on the following topics:
1. More advanced feature descriptors like texture or different shape representation methods will be incorporated into the system.
2. User feedback mechanism will be developed to increase retrieval accuracy.
3. More experiments will be conducted to prove Auto Weight Regulation PCA algorithm can be used as a new image classification method.

References:

1. Ching-Sheng, Wang and Timothy, K. Shih, "An Intelligent Content-Based Image Retrieval System Based On Color, Shape and Spatial Relations", University of Tamkang Ph.D's. Thesis, 2001.
2. M. J. Swain and D. H. Ballard, "Color indexing", *International Journal of Computer Vision*, vol.7, pp.11-32, 1991.

3. B. M. Mehtre, M. S. Kankanhalli and W. F. Lee, "Shape measures for content based image retrieval: a comparison", *Information Processing & Management*, vol.33, pp 319-337, 1997.
4. A.K. Jain and A. Vailaya, "Image retrieval using color and shape", *Pattern Recognition*, vol.29, pp.1233-1244, 1996.
5. http://www.mathworks.com/access/helpdesk/help/toolbox/images/enhanc10.shtml

Fig. 4.4 Experiment result of experiment 1.

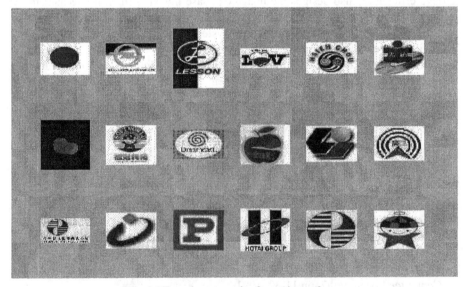

Fig. 4.5 Experiment result of experiment 2.

Improving the Initial Image Retrieval Set by Inter-Query Learning with One-Class SVMs

Iker Gondra[1], Douglas R. Heisterkamp[1], and Jing Peng[2]

[1]Department of Computer Science, Oklahoma State University
Stillwater, Oklahoma 74078, USA
{doug, gondra}@cs.okstate.edu
[2]Department of Electrical Engr. & Computer Science, Tulane University
New Orleans, Louisiana 70118, USA
jp@eecs.tulane.edu

Abstract. Relevance Feedback attempts to reduce the semantic gap between a user's perception of similarity and a feature-based representation of an image by asking the user to provide feedback regarding the relevance or non-relevance of the retrieved images. This is intra-query learning. However, in most current systems, all prior experience is lost whenever a user generates a new query thus inter-query information is not used. In this paper, we focus on the possibility of incorporating prior experience (obtained from the historical interaction of users with the system) to improve the retrieval performance on future queries. We propose learning one-class SVMs from retrieval experience to represent the set memberships of users' query concepts. Using a fuzzy classification approach, this historical knowledge is then incorporated into future queries to improve the retrieval performance. In order to learn the set membership of a user's query concept, a one-class SVM maps the relevant or training images into a nonlinearly transformed kernel-induced feature space and attempts to include most of those images into a hyper-sphere. The use of kernels allows the one-class SVM to deal with the nonlinearity of the distribution of training images in an efficient manner, while at the same time, providing good generalization. The proposed approach is evaluated against real data sets and the results obtained confirm the effectiveness of using prior experience in improving retrieval performance.

1 Introduction

The rapid development of information technologies and the advent of the World-Wide Web have resulted in a tremendous increase in the amount of available multimedia information. As a result, there is a need for effective mechanisms to search large collections of multimedia data, especially images. In traditional image retrieval, keywords are manually assigned to images and, for any particular query, images with matching keywords are retrieved [12]. However, it is usually the case that all the information contained in an image cannot be captured by a few keywords. Furthermore, a large amount of effort is needed to do keyword assignments in a large image database and, because different people may have different interpretations of image contents, there will be inconsistencies [12].

In order to alleviate some of these problems, Content-Based Image Retrieval (CBIR) was proposed. Some early systems include [5, 9]. A CBIR system extracts some features (such as color, shape, and texture) from an image. The features are then the components of a feature vector which makes the image correspond to a point in a feature space. In order to determine closeness between two images, a similarity measure is used to calculate the distance between their corresponding feature vectors. However, because of the gap between high level concepts and low level features and the subjectivity of human perception, the performance of CBIR systems is not satisfactory [12].

Relevance feedback attempts to overcome these problems by gathering semantic information from user interaction. In order to learn a user's query concept, the user labels each image returned in the previous query round as "relevant" or "not relevant". Based on the feedback, the next set of images is retrieved to the user for labeling. This process iterates until the user is satisfied with the retrieved images or stops searching. Many approaches for improving the performance of relevance feedback have been proposed [11, 13]. Recently, Support Vector Machines (SVM) have been applied to CBIR systems with relevance feedback to significantly improve retrieval performance [3]. However, in most current systems, all prior experience based on past queries is lost whenever a user generates a new query. That is, the system is adapting to the current user without using any long-term, inter-query learning.

A few approaches [6, 8, 16] attempt inter-query learning. That is, relevance feedback of past queries are used to improve the retrieval for a current query. Both [8] and [16] take the approach of complete memorization of prior history. Then the correlation between past image labeling is merged with low-level features to rank images for retrieval. In [16] the extra inter-query information is efficiently encoded as virtual features. In [6] Latent Semantic Analysis was used to provide a generalization of past experience. The initial results from the three approaches for inter-query learning show a tremendous benefit in the initial and first iteration of retrieval. Inter-query learning thus offers a great potential for reducing the amount of user interaction by reducing the number of interactions needed to satisfy a query.

In this paper, we propose using one-class SVMs to capture users' query concepts and utilize them as previous experience to be used in future queries. In order to learn the set membership of a user's query concept, a one-class SVM maps the relevant or training images into a nonlinearly transformed kernel-induced feature space and performs risk minimization by attempting to include most of those images into a hyper-sphere of minimum size. The use of kernels allows the one-class SVM to deal with the non-linearity of the distribution of training images in an efficient manner, while at the same time, providing good generalization. In addition, the geometric view of one-class SVM allows a straightforward interpretation of the density of past interaction in a local area of the feature space and thus allows the decision of exploiting past information only if enough past exploration of the local area has occurred.

The rest of this paper is organized as follows. Section 2 gives a brief introduction to SVMs and describes one-class SVMs in detail. A description of our proposed approach for improving retrieval performance by using SVMs to capture historical information and fuzzy classification to incorporate it into the relevance feedback method is presented in Section 3. In Section 4, we report experimental results which confirm the effectiveness of our approach. Concluding remarks are presented in Section 5.

2   Support Vector Machines

A Support Vector Machine (SVM) is a system for training linear learning machines in a kernel-induced feature space efficiently while at the same time, respecting the insights provided by generalization theory and exploiting optimization theory [4]. The objective of support vector classification is to create a computationally efficient method of learning "good" separating hyperplanes in a high dimensional feature space, where "good" corresponds to optimizing the generalization bounds given by generalization theory [4].

Suppose we are given training data $\{x_1, x_2, ...,x_n\}$ that are vectors in some space $X \in \mathfrak{R}^d$ and their corresponding class labels $\{y_1, y_2, ..., y_n\}$ where $y_i \in \{-1, 1\}$. The task of a learning machine would be to learn the mapping $x_i \rightarrow y_i$. The machine is defined by a set of possible mappings $x \rightarrow f(x, \alpha)$, where the functions $f(x, \alpha)$ are labeled by the adjustable parameters α [2]. If there are no restrictions on the family of functions $f(x, \alpha)$ from which we choose our trained machine f, even though f may have zero error on the training data, it may not generalize well on unseen data. This problem is known as overfitting and it drove the initial development of SVMs [2]. Statistical learning theory, or VC (Vapnik-Chervonenkis) theory, shows that the best generalization performance can be obtained when the "capacity" of the learning machine is restricted to one that is suitable to the amount of available training data [2]. Suppose we have a class of separating hyperplanes $(x \cdot w) + b = 0$, where $w \in \mathfrak{R}^n$ and $b \in \mathfrak{R}$, corresponding to decision functions $f(x) = \text{sign}((x \cdot w) + b)$. It can be shown that the optimal hyperplane (i.e., the one that minimizes the generalization error or the bound on the actual risk) corresponds to the one with maximal margin of separation between the two classes [2]. The optimal hyperplane has the smallest "capacity" (also known as the lowest "VC dimension"). In order to find the optimal separating hyperplanes, a constrained quadratic optimization problem is solved. The solution has an expansion $w = \sum_i \alpha_i x_i$. Those points for which $\alpha_i > 0$ are called "support vectors" and lie on one of the separating hyperplanes. All other points have $\alpha_i = 0$ thus the support vectors are the critical elements of the training set [2]. The final decision function is of the form $f(x) = \text{sign}(\sum_i \alpha_i (x \cdot x_i) + b)$.

In order to generalize to the case where the decision function is not linearly separable, SVMs first map the data into some other (possibly infinite dimensional) feature space F using a mapping $\Phi: \mathfrak{R}^n \to F$. Because both the quadratic optimization problem and the final decision function depend on the data through dot products in F (i.e, on functions of the form $\Phi(x_i) \bullet \Phi(x_j)$), if we are given a "kernel function" K such that $K(x_i, x_j) = \Phi(x_i) \bullet \Phi(x_j)$, we could just use K without even having to know what Φ is [2]. This is known as the "kernel trick" and it allows SVMs to implicitly project the original training data to a higher dimensional feature space.

2.1   One-Class SVM

In a one-class classification problem, data from only one of the classes (the target class) is available. For instance, when a user labels some images as "relevant" and others as "non-relevant", information about one class (i.e., the one corresponding to the user's query concept) is given by the "relevant" images. However, the "non-relevant" images do not provide any class information since they can belong to any class. Thus, in one-class classification, the task is to create a boundary around the target class such that most of the target data is included while, at the same time, minimizing the risk of accepting outliers (i.e., data that does not belong to the target class) [15].

The strategy that we will follow to capture a user's query concept (i.e., the target class) is to map the "relevant" images (i.e., the training data) to a higher dimensional feature space and then try to include most of those images into a hyper-sphere. That is, given training data $\{x_1, x_2, ...,x_n\}$ that are vectors in some space X $\in \mathfrak{R}^d$, we have to find the smallest hyper-sphere (so that the risk of including outliers is minimized) that includes most of the training data. Thus, the task is to minimize the following objective function (in primal form):

$$\min_{R \in \mathfrak{R}, \, \xi \in \mathfrak{R}^n, \, a \in F} R^2 + C \sum_i \xi_i$$

$$\text{such that } \|\Phi(x_i) - a\|^2 \leq R^2 + \xi_i, \qquad \xi_i \geq 0, \text{ for } \forall i$$

where R and a are the radius and center of the hyper-sphere, C gives the tradeoff between the radius of the hyper-sphere and the number of training data that can be included, and $\Phi: \mathfrak{R}^n \to F$. By setting partial derivatives to 0 in the corresponding Lagrangian we obtain the following expression for the center of the hyper-sphere:

$$a = \sum_i \alpha_i \Phi(x_i)$$

Replacing partial derivatives into the Lagrangian and noticing that the center a is defined as a linear combination of $\Phi(x_i)$, which allows us to use a kernel, we obtain the following objective function (in dual form):

$$\min \sum_{i,j} \alpha_i \alpha_j K(x_i, x_j) - \sum_i \alpha_i K(x_i, x_i) \quad \text{such that } 0 \leq \alpha_i \leq C, \quad \sum_i \alpha_i = 1$$

This is a quadratic programming problem and the optimal α's can be obtained using a quadratic programming method [15]. In order to determine the ranking of an image x in the database (in terms of belonging to a particular query concept), the following function can be used [15]:

$$f(x) = R^2 - \|\Phi(x) - a\|^2$$
$$= R^2 - \sum_{i,j} \alpha_i \alpha_j K(x_i, x_j) + 2 \sum_i \alpha_i K(x_i, x) - K(x, x)$$

where images with higher values are closer to the hyper-sphere and thus, are more likely to belong to the same query concept.

In [3] image retrieval performance is successfully improved by using a one-class SVM for intra-query learning. Scholkopf proposed another approach [14] in which a largest margin hyperplane is used to separate the training data from the origin. When the training data has unit norm this is identical to the approach taken by Tax [15] (explained above). In this paper, as in [3], we will follow the approach developed by Tax [15].

3   Proposed Approach

Using one-class SVMs we obtain set membership knowledge (about previous users' query concepts), which can be visualized as hyper-spheres in feature space. In order to integrate this prior experience with a user's current query, we do a fuzzy classification of the user's query concept. When a query is submitted, we determine whether it falls into one of the existing one-class SVMs. Because it is very common for an image to be ascribed into many different concepts, we expect to have queries that fall into many hyper-spheres. One possible way of selecting the images that will be presented to the user would be to perform a hard classification by retrieving the KNN images closest to the nearest center (i.e., KNN images to the closest prototype). However, this purely exploitative approach is not a very good strategy since the query is considered not to be an outlier by several one-class SVMs. It may as well be ascribed to the concept corresponding to any one of the other hyper-spheres. Furthermore, the query may be ascribed to a combination of different concepts. Therefore, instead, we use the ideas from possibilistic cluster analysis [7] and assign a degree of membership to each one of the one-class SVMs (i.e., to each cluster) according to the degree by which the query can be ascribed to its particular concept. A possibilistic cluster analysis drops the probabilistic constraint that the sum of the degrees of membership of each image to all one-class SVMs is equal to one [7]. Therefore, possibilistic cluster partitions are especially useful in the classification of images because it is often the case that an image cannot be assigned to any one of the existing clusters [7].

Let's introduce some terminology. Let $D = \Re^d$ be the *data space* of d-dimensional image vectors, $C = \{c_1, c_2, ..., c_n\}$ be the *concept space* (i.e., the set of concepts corresponding to all existing one-class SVMs), and $R = \{\{c\} \mid c \in C\}$ be the *result space*. The result of a data analysis is a mapping $\mu: X \to \{c\}$, where $X \subseteq D$ and $c \in C$. Then, $A(D, R) = \{\mu \mid \mu: X \to K, X \neq \phi, K \in R\}$ is called an *analysis space* [7]. The *fuzzy set* of X is a mapping $\mu\colon X \to [0,1]$ and the set containing all fuzzy sets of X is denoted by $F(X) = \{\mu \mid \mu\colon X \to [0,1]\}$. Then, $A_{\text{fuzzy}}(D, R) = A(D, \{F(K) \mid K \in R\})$ is the *fuzzy analysis space* for $A(D, R)$. A result of an analysis $\mu\colon X \to F(K)$ is called a *possibilistic cluster partition* if $\forall k \in K\colon \sum_{x \in X} \mu(x, k) > 0$, where $\mu(x, k)$ is interpreted as the degree of membership of $x \in X$ into the cluster $k \in K$ [7].

The fuzzy c-means algorithm carries out a data analysis by minimizing an objective function. It searches for an optimal set of spheres (i.e., clusters) of d-dimensional points in a feature space. The clusters are represented by their corresponding centers (i.e., by their prototypes) and Euclidean distance is used as the measure of distance between a point x and a prototype a_i [7]. In our case, the set of clusters (in the form of one-class SVMs) is formed by the historical interaction of users with the system. We use the following membership function to assign degrees of membership to the n hyper-spheres into which a query x falls [7]:

$$\mu(x, a_i) = 1 / \sum_{j=1}^{n} \left(\|\Phi(x) - a_i\|^2 / \|\Phi(x) - a_j\|^2 \right) \text{ for } \forall i$$

where a_i is the center of the ith hyper-sphere and $\Phi\colon \Re^n \to F$. Therefore, the degree of membership of a query into a one-class SVM is based on the relative distances between the query and the centers of all hyper-spheres into which it falls. Suppose a query x falls into n hyper-spheres. If c_i denotes the concept that is embodied by the ith hyper-sphere then the belief (or our degree of confidence) that x is delivering concept c_i is equal to $\mu(x, a_i)$.

Our approach for selecting the set of images that will be presented to the user (i.e, the retrieval set) is based on combining exploitation and exploration while maximizing the number of relevant images in the retrieval set. The results of experiments conducted in [1] for learning users' text preferences suggest that, for simple queries (i.e., queries that can be ascribed to one concept), a purely exploitative strategy delivers very good performance. However, for complex queries (i.e., queries that can be ascribed to more than one concept), there is a tradeoff between faster learning of the user's query concept and the delivery of, in our case, more relevant images. In other words, for a complex query, we may be able to maximize the number of relevant images presented to the user by selecting images that can be ascribed to the concept with the largest $\mu(x, a_i)$ (i.e., by pure exploitation). However, we may not be able to learn the user's query concept unless some exploration is also done. The approach that we take is to combine exploitation and exploration while, at the same time, attempting to maximize the number of relevant images that are presented to the user.

To fully exploit the relevance feedback information provided by the current user, we set $\mu(x, a_{\text{current}}) = w$, where a_{current} is the center of a hyper-sphere formed by using all the feedback provided by the current user (or the original query image when no relevance feedback iterations have been performed yet) and $0 \le w \le 1$ signifies our confidence that a_{current} captures the user's query concept. To form the retrieval set, sample representative images from each hyper-sphere in which the query falls are included. The number of representatives that a particular concept c_i has in the retrieval set is proportional to $\mu(x, a_i)$. Thus, if $N(c_i)$ denotes the number of images of concept c_i that appear in the retrieval set then $N(c_i) < N(c_j)$ whenever $\mu(x, a_i) < \mu(x, a_j)$. Because a query may fall into many hyper-spheres but only a fixed number of images is to be retrieved, priority is given to hyper-spheres with higher μ and, after the fixed number of images is reached, the remaining hyper-spheres with smaller μ are ignored. This allows us to perform exploitation and exploration while, at the same time, maximizing the possibility of presenting relevant images to the user. The search of the historical one-class SVMs could be done efficiently by building an M-Tree in feature space [10].

At system startup there is no historical interaction of users with the system and thus, no prior experience. A query is then considered to be the center of its own cluster, $w = 1$, and the retrieval set is formed by the same steady-state procedure outlined above. Similarly when there is prior historical information but the query does not fall into any hyper-sphere. The images marked as relevant by the user during the relevance feedback iterations are used as training data and a one-class SVM is used to learn the set membership (in the sense of a user's query concept). Figure 1 shows a block diagram of the system.

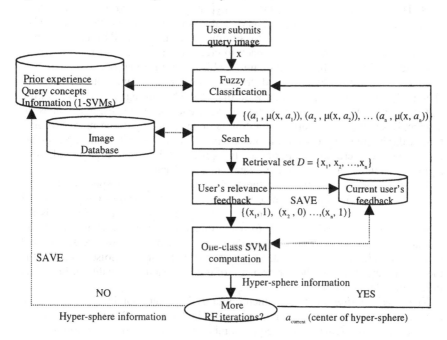

Fig. 1. System diagram

4   Experimental Results

We compare the performance of our approach against that of relevance feedback methods that do not use historical information. The response of our technique with respect to different amounts of experience (data level) and with respect to quality (noise level) of historical information is also investigated. The retrieval performance is measured by *precision*, which is defined as

$$precision = \frac{\text{number of relevant images retrieved}}{\text{number of images retrieved}}$$

The *Texture* database, from MIT Media Lab, consists of 640 images of size 128x128. There are 15 classes (different textures) and each image is represented by a 16-dimensional feature vector. To determine the free parameters, a ten-fold cross-validation was performed. Our approach was then evaluated with different amounts of experience (data level) using a Gaussian kernel with width $s = 0.1$, and a misclassification penalty $C = 1/(p*n)$, where n is the number of training images, with $p = 0.001$. The weight for exploration of historical information is $(1-w)$, where $w = 0.25$. The number of images in the retrieval set $k = 20$. Figure 2a shows the precision of the initial retrieval (i.e., with no relevance feedback iterations) with respect to different data levels. The values reported are the average of the ten tests. The level of data is the number of hyper-spheres relative to the number of images in the database. We can observe that, with low data levels (less than 1/3), precision is less than that obtained by simply retrieving the KNN images in input or feature space, which is the approach to create the initial retrieval set taken by relevance feedback methods that do not use historical information.

In order to avoid the initial decrease in performance we could adaptively change the value of the parameter w so that at the beginning, when there is little historical information, w is large and, as experience accumulates, w becomes increasingly smaller (i.e., more exploration is done). Thus by doing more exploitation at the beginning we would avoid the initial drop in performance. Nevertheless, with the increasing experience, the precision becomes higher and, with significant data levels, there is a large gain in performance.

In order to investigate the robustness of our approach with respect to poor historical information, simulated noise was added. Figure 2b shows the precision of the initial retrieval at different amounts of noise. The percentage of noise is the probability that a users' feedback for each image is flipped (i.e., the probability that an image marked as "relevant" by the user will be marked "non-relevant", and vice versa). We can observe that, with reasonable amounts of noise, data level is the dominating factor on the performance of our approach and quality of historical information has a small effect.

The *Segmentation* database, which contains 2310 outdoor images, was taken from the UCI repository at http://www.ics.uci.edu/~mlearn/MLRepository.html. There are 7 classes (each with an equal number of instances) and a 19-dimensional feature vector is used to represent each image. A one-fold cross validation was conducted to determine the free parameters. Figure 2c shows the precision of the initial retrieval at different data levels with $s = 25$, $p = 0.001$, $w = 0.25$, and $k = 20$. Similarly, we can observe that with significant data levels (more than 1/2 in this case), our approach performs better than relevance feedback methods that do not use historical information. From Figure 2d we can observe that our method outperforms an approach based on using a one-class SVM for intra-query learning only (i.e., with data level 0). These results support the efficacy of our method.

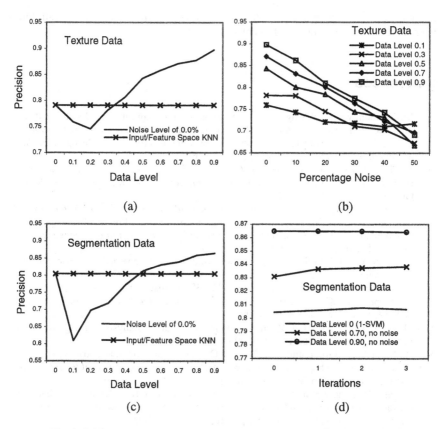

Fig. 2. Initial Retrieval Set: (a),(c) Precision vs Data Level, (b) Precision vs Noise Percentage, (d) Precision vs Iteration of Relevance Feedback

5   Conclusions

This paper presented an approah for incorporating historical information into a relevance feedback system to improve image retrieval performance. By training one-class SVMs with users' feedback, we can learn users' query concepts and accumulate retrieval experience. Using a fuzzy classification approach we exploit both current and historical information to improve retrieval performance. Initial investigation suggests that our approach improves retrieval in the initial set where a traditional intra-query approach requires an iteration of relevance feedback to provide improvement. Therefore, our method reduces user interaction by reducing the number of iterations needed to satisfy a query. Furthermore, it is robust to poor historical information. Our future research will focus on methods for combining or merging hyper-spheres (i.e., users' concepts). This may be desirable when the amount of historical information is very large. Also, we will investigate some systematic scheme for adaptively changing w so that the amount of exploration done is proportional to the amount of experience.

References

1. M. Balabanovic. Exploring versus exploiting when learning user models for text recommendation. *User Modeling and User-Adapted Interaction*, 8(1-2):71-102, 1998.
2. C. J .C. Burges. A Tutorial on Support Vector Machines for Pattern Recogntion. *Data Mining and Knowledge Discovery*, 2(2):121-167, 1998.
3. Y. Chen, X. Zhou, and T. S. Huang. One-Class SVM for Learning in Image Retrieval. In *Proceedings of IEEE International Conference on Image Processing*, Thessaloniki, Greece, pp. 34-37, October 2001.
4. N. Cristianini and J. Shawe-Taylor. *An Introduction to Support Vector Machines and other kernel-based learning methods*. Cambridge University Press, 2000.
5. C. Faloutsos, M. Flickner, and et al. *Efficient and effective querying by image content*. Technical Report, IBM Research Report, 1993.
6. D. Heisterkamp. Building a latent-semantic index of an image database from patterns of relevance feedback. In *Proceedings of 16th International Conference on Pattern Recognition*, Quebec City, Canada, pp. 132-135, August 2002.
7. F. Hoppner, F. Klawonn, R. Kruse, and T. Runkler. *Fuzzy Cluster Analysis*. John Wiley & Sons, Inc., New York, 1999.
8. M. Li, Z. Chen, and H. J. Zhang. Statistical correlation analysis in image retrieval. *Pattern Recognition*, 35(12):2687-2693, 2002.
9. W. Niblack, R. Barber, and et al. The QBIC project: querying images by content using color, texture and shape. In *Proceedings of SPIE Storage and Retrieval for Image and Video Databases*, San Jose, California, pp. 173-187, February 1993.
10. J. Peng, B. Banerjee, and D. Heisterkamp. Kernel index for relevance feedback retrieval in large image databases. In *Proceedings of 9th International Conference on Neural Information Processing*, Singapore, November 2002.
11. J. Peng, B. Bhanu, and S. Qing. Probabilistic feature relevance learning for content-based image retrieval. *Computer Vision and Image Understanding*, 75(1/2):150-164, 1999.
12. Y. Rui and T. Huang. Relevance Feedback: A Power Tool for Interactive Content-Based Image Retrieval. *IEEE Transactions on Circuits and Systems for Video Technology*, 8(5):644-655, 1998.
13. Y. Rui, T. Huang, and S. Mehrotra. Content-based image retrieval with relevance feedback in mars. In *Proceedings of IEEE International Conference on Image Processing*, Santa Barbara, California, pp. 815-818, October 1997.
14. B. Scholkopf, R. Williamson, A. Smola, and J. Shawe-Taylor. SV Estimation of a Distribution's Support. In *Proceedings of Neural Information Processing Systems*, Denver, Colorado, pp. 582-588, November 1999.
15. D. M. J. Tax. *One-class classification*. PhD thesis, Delft University of Technology. (http://www.ph.tn.tudelft.nl/~davidt/thesis.pdf), June 2001.
16. P. Yin, B. Bhanu, K. Chang, and A. Dong. Improving retrieval performance by long-term relevance information. In *Proceedings of 16th International Conference on Pattern Recognition*, Quebec City, Canada, pp. 533-536, August 2002.

Tongue Image Analysis Software

Dipti Prasad Mukherjee and Dwijesh Dutta Majumder

Electronics and Communication Sciences Unit, Indian Statistical Institute, Calcutta, India. Email: dipti@isical.ac.in

Abstract. In this paper we describe an application software for tongue image analysis. The color, shape and texture of tongue are very important information for medical diagnostic purpose. Given a tongue image, this software derives features specific to color, shape and texture of the tongue surface. Based on the feature characteristics of a normal tongue image, the possible health disorder associated with a particular patient is ascertained. A set of conventional image processing algorithms is integrated for this purpose. The treatment management for similar past cases with respect to a given patient case can also be retrieved from the image and the associated database based on image similarity measure.

1 Introduction

In this paper we are reporting a working image processing system that utilizes concepts of traditional Indian and Chinese medical practices. In the ancient Indian and Chinese medical school tongue is considered as the mirror of human body [8]. In this traditional form of treatment and in modern medicine as well, the color, shape and texture of tongue are very important indicators for disease management. The image processing system described in this paper captures the tongue image and derives necessary features representing color, shape and texture of the tongue. The system then retrieves similar tongue images with corresponding patient history and disease management details. Based on finer queries the system can assign disease class label to a particular image.

The system is an attempt to store and code a very important ancient medical practice that seems to be loosing out against the ultra modern diagnosis management system. The major goal of the ancient tongue based diagnostic management system was prevention through regular inspection. Specifically four types of disorders are mainly evident from the appearance of the tongue. They are Gastro-intestinal disorder, Respiratory disorder, Renal disorders and Anemia and Malnutrition. The aim of this software is to use it as a self-assessment tool where patient on a regular basis can identify and take care of his/her ailments to a great extent. For self-assessment as practiced in traditional Indian and Chinese medicine, the observation from tongue is correlated and coupled with the observation of pulse rate and the color of the eye. We feel that the current system can be a useful complimentary tool to modern healthcare system.

The working image processing system has both data acquisition and data analysis module. The data acquisition module consists of a digital video camera. After extensive study *Nikon Coolpix 990* digital video camera is selected as the imaging device as the color response of the digitized image of this camera is closest to the doctors' visual assessment of tongue color. Through extensive experimentation, we have fixed an imaging process where the patient is seated approximately twelve inches away from the camera stand in fronto-parallel direction. The picture is taken under normal clinic lighting condition.

The data analysis module can be used as stand-alone software that processes the captured images. The idea of separating the analysis module from the data acquisition unit is to widen its acceptability. Even without the expensive data acquisition hardware, the full functionality of self-assessment is possible with the analysis tool using a mirror image of the tongue. The analysis module at present contains 500 tongue images. The image analysis and the corresponding query processing are part of the analysis module, which is described in detail in this paper.

The first step of tongue image analysis is to extract the tongue region from the image excluding the background and the mouth. In the next section we present the tongue region extraction algorithm. In section 3, color, shape and texture features of the tongue region are derived. This is followed by the description of the analysis module in section 4. In section 5, conclusion is presented pointing towards the on going development related to data analysis module.

2 Tongue Region Segmentation

Fig. 1(a) shows a typical image captured using the Nikon Coolpix 990 camera for the tongue image analysis system. In this step we extract the tongue region, which is the largest segment central to the image matrix. Even though the dominant color is red there are instances, usually due to Anemic or Gastro-intestinal disorder, when the tongue looses its usual redness. This prohibits us to use any dominant color information for tongue region segmentation. We pose this image segmentation problem as a two-class clustering problem. The cluster to our interest is the central tongue region while the other class is the background class consisting of background, portion of mouth and any other irrelevant details. We have used fuzzy c-means to cluster the image pixels. The red, green and blue pixel values of the digitized color image are used as feature vectors for clustering.

For fuzzy c-means clustering algorithm, the familiar least-squared error criterion is applied [3]:

$$J_m(U,\mu) = \sum_{\Omega}\sum_{i=1}^{C} (u_i(x,y))^m \parallel d_i(x,y) \parallel^2 . \qquad (1)$$

Here, U is the fuzzy C-class partition. Given a feature vector representing a color pixel $\mathbf{I}(x, y)$ at location (x, y), the measure $\parallel d_i(x,y)\parallel = \parallel \mathbf{I}(x,y) - \mu_i \parallel$ is the distance between the feature vector and the ith cluster center μ_i. The distance is weighed by the fuzzy membership value of each scale space vector $u_i(x,y)$ corresponding to ith class. The fuzzy exponent m is taken as $m=2$. For every feature vector $\mathbf{I}(x, y)$, the error criterion $J_m(U,\mu)$ is minimized subject to the conditions $\sum_{i=1}^{l} u_i(x,y) = 1$,

$0 < \sum_{\Omega} u_i(x,y) < |\Omega|$ and $u_i(x,y) \geq 0$.

For feature vector at (x, y), the fuzzy membership is updated according to

$$u_i(x,y) = 1/[\sum_{j=1}^{C} \left(\frac{d_i(x,y)}{d_j(x,y)} \right)^{2/(m-1)}]. \qquad (2)$$

The initial fuzzy membership value is generated using uniformly distributed random number generator. For the clustered image, the cluster center is updated for both tongue and the background classes according to

$$\mu_i = \frac{\sum_{\Omega} (u_i(x,y))^m \mathbf{I}(x,y)}{\sum_{\Omega} (u_i(x,y))^m}. \qquad (3)$$

The algorithm is terminated for insignificant (1-2% of the current value) changes in μ_i between consecutive iterations. To reduce processing time, the final classified image is obtained by classifying the pixels based on highest cluster membership value for each pixel.

The clusters are labeled as tongue class and background class based on the fact that the largest single region spatially central to the image is the tongue class. All other segments representing background class are discarded including falsely detected tongue classes other than the one mentioned above, if any.

The software also has the provision of using scale space clustering of the tongue image following [1]. For situations where small disconnected and isolated regions of background class are detected inside the tongue class, may be due to diseased condition, scale space clustering can eliminate these isolated regions extracting the overall tongue shape without compromising the tongue area or the sharpness of the tongue edge. Note that at this stage we are interested to extract the overall shape of the tongue. Details internal to the tongue region are analyzed in the section 3.

Fig. 1(b) is the two class clustered image of **Fig. 1(a)**. **Fig 1(c)** shows the tongue portion of the image of **Fig. 1(a)** based on the tongue cluster information of **Fig. 1(b)**.

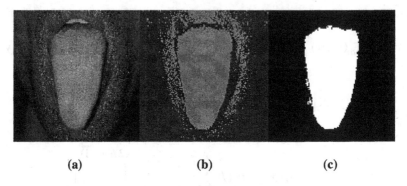

(a)   **(b)**   **(c)**

Fig. 1(a). A normal tongue image. **(b)** After two class fuzzy c-means clustering. **(c)** Extraction of the tongue cluster spatially central to the image.

In the next section, we segment the tongue region based on diagnostic interest.

2.1 Demarcation of symptomatic tongue regions

In the traditional medical diagnostic system, the tongue region is spatially divided into four different clusters: region I to IV [8]. Each of these regions is the indicator of the smooth functioning of certain human organs. Fig. 2(a) shows region I to IV along with the name of the organs whose performance is reflected in the respective tongue region. The central region (region I) mirrors functioning of Spleen and Stomach, the upper and the lower tip regions (region II and III) represent Kidney (region II) and Lung and Heart (region III) respectively. The rest part on the tongue region (region IV) on the sides of the central region indicates smooth functioning of Liver and Gall Bladder. We now specify the technique to automatically identify the tongue regions I to IV.

Region I: This is a circular region central to the overall tongue shape. The center of gravity of the tongue cluster is identified and a circular region of radius 20 pixels are selected as region I. The region I radius of 20 pixels is set as per Doctor's advice after experimentation and calibration of pixel space with respect to actual image size (200x200 pixels) and tongue region.

Region II and III: For the overall tongue region as shown in **Fig. 1(c)**, we have calculated the orientation of the shape through its principal axis passing through the center of gravity. The upper and lower tangents at the points of intersection of the principal axis with the circle representing region I, divide the tongue shape in regions II and III respectively.

Region IV: The rest part of the tongue is clubbed together for further analysis as region IV.

Examples of determination of region I to IV for **Fig. 1(c)** is shown in **Fig. 2(b)**. For clarity horizontal lines approximating the tangent lines are drawn.

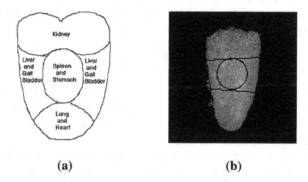

(a)   (b)

Fig. 2(a). Regions within tongue surface that mirrors the performance of certain human organs. (b) Extraction of similar regions as in (a) using proposed technique.

For the tongue regions I to IV, we need to calculate a set of feature vectors akin to medical observations. In the next section we describe this.

3 Region Feature Estimation

As noted earlier the three important parameters to assess the tongue condition are color, shape and texture of the tongue. Following are schemes to derive corresponding features for color, shape and texture parameters.

Color features: In assessing the color of a tongue, the doctors usually stress on two things. The brightness of the color or more appropriately the 'paleness' of the specific tongue region is an important indicator. To code this information, for a given region we calculate the brightness and color saturation component. In order to determine brightness and saturation, the r-g-b pixel values are converted to HSV space [5]. The mean saturation (S_m) and mean value (V_m) component of all the pixels of a particular region are the color feature vectors. Depending on the disease condition, S_m and V_m have variances and these values are also stored in color feature vector space for a particular region.

Shape features: The significant information for tongue shape are depicted through the overall symmetry of the tongue and the degree of waviness of the tongue contour. Even teeth marks including uneven development of teeth and jaws due to some secondary ailments may be assessed from the tongue contour. To encode the symmetry information, the orientation of the tongue is calculated from the principal axis passing through the centroid. To encode the regularity information of the tongue contour, we have used the Fourier shape descriptors [6]. Tongue contour is traced as a single pixel thick ordered list of coordinate points and it is subjected to Discrete Fourier Transform (DFT). The one dimensional DFT of the contour coordinate sequence $u(n) = x(n) + jy(n)$ of length N is given as,

$$a(k) = \frac{1}{N} \sum_{n=0}^{N-1} u(n) e^{\frac{-j2\pi kn}{N}}, \quad 0 \le k \le N-1. \tag{4}$$

The complex coefficients $a(k)$ represents the Fourier descriptors. We have observed that first 10 Fourier coefficients are good enough to describe the regularity of the contour. While the orientation information provides a global shape measure, the Fourier descriptors represent the local variation of contour.

Texture features: Texture feature is not always visually predominant and sometimes require minute observation by an expert. The periodic or sometimes quasi-periodic texture patterns have vital clues for diagnostic purpose. Localized opponent color feature based approaches are used to estimate texture of images of tongue surface [9]. In our approach the tongue image is approximated as an AM-FM model [2]. From the dominant component analysis of the AM-FM model, the horizontal and vertical instantaneous frequencies at pixel locations are determined.

A 2-D AM-FM function $\mu(x, y)$ takes the form $\mu(x, y) = a(x, y) \exp[j\psi(x, y)]$, where $a(x, y)$ and $\psi(x, y)$ are arbitrary real-valued functions. Without loss of generality, we assume that $a(x, y) \ge 0$. The AM and FM components of interest that are contained in $\mu(x, y)$ in (5) are the instantaneous amplitude $a(x, y)$ and the instantaneous frequency vector $\nabla \psi(x, y) = [u(x, y) \quad v(x, y)]^T$. The functions $u(x, y)$ and $v(x, y)$ are the horizontal and vertical instantaneous frequencies of $\mu(x, y)$. Given $\mu(x, y)$, the AM and FM functions may be calculated using the straightforward demodulation formulae:

$$\nabla \psi(x, y) = \text{Re}\left[\frac{\nabla \mu(x, y)}{j\mu(x, y)}\right], \tag{5}$$

and

$$a(x, y) = |\mu(x, y)|, \tag{6}$$

which yield exact solutions at all points where $\mu(x, y) \neq 0$. The AM function $a(x, y)$ may be interpreted as the image contrast function, while $\nabla \psi(x, y)$ characterizes the local texture orientation and granularity. The corresponding 2-D discrete version is detailed in [2].

We have used an appropriately designed bank of Gabor filters to determine the AM-FM components. Let $G_i(x, y)$ be the response of a particular Gabor filter with impulse response $g_i(x, y)$ and frequency response $G_i(\omega)$. The AM function $a_l(x, y)$ is then estimated using

$$a_l(x, y) \approx \left|\frac{G_i(x, y)}{G_i[\nabla \psi_l(x, y)]}\right|, \tag{7}$$

At each pixel, we define the dominant component as the one that dominates the response $G_i(x, y)$ of the filter that maximizes the selection criterion

$$\Pi_i(x, y) = \frac{|G_i(x, y)|}{\max_\omega |G_i(\omega)|}. \tag{8}$$

This criterion tends to select filters that are dominated by large amplitude components with instantaneous frequencies lying near the maximum transmission frequency of the filter.

For the tongue region, dominant component analysis is performed on the V space of HSV transformed space of the RGB color image. In the next section we present the use of the color, shape and texture feature vectors.

4 Analysis Module

The feature vectors related to the color and shape of the tongue regions are taken as the signature vector for a particular tongue image. The search and service of the queries to tongue image database are performed using the signature vector. The localized instantaneous center frequency information depicting the periodicity of the texture is utilized only in case the doctor wants to use it as diagnostic assisting aid. For a doctor, the texture information sometimes acts as a confirmatory step especially when the texture periodicity is difficult to assess using normal eye.

The core of the analysis module is to generate a set of inferences based on the derived feature vectors and the stored feature vectors of the images present in the database. Henceforth, the digitized image of a patient to be subjected to image analysis is referred as candidate image. The major tasks performed by this inference generation module are:

1. Assigning possible disease label to the tongue candidate image: About 100 normal tongue images are analyzed and their mean color and shape feature vectors represent normalcy condition and their possible variance for each of the four tongue regions as shown in **Fig. 2(a)**. At present the disease class label is assigned for four specific disorder types discussed in the introduction. For each of this disorder type, a mean signature vector is already established after experimentation with approximately 300 patients. The possible disorder group label is assigned with minimum distance between the candidate image signature vector and the signature vectors of the four disorder groups and that of the normal group. The distance measure is taken as Euclidean. The detailed statistical analysis of the result and corresponding health care management is presented elsewhere [4].

2. Supporting region specific query: This is of two types. Given a region (between I to IV), a possible disease label can be enquired. The label is assigned based on the distance of derived features from that of the five different groups (four abnormal disorders plus the normal). In case similar images for past cases need to be consulted, the similar images are retrieved, again based on the distance between region specific feature vectors. The images are displayed starting with the image having highest region specific similarity value.

3. Retrieving similar images: This is a straightforward extension of step 2 where the distances between the entire signature vector of the candidate and that of the entire database images are calculated. Matching tongue images based only on color content is given in [7]. Again, the display starts with images having minimum distance in feature space with respect to the database images. The similar images are retrieved along with corresponding peripheral symptoms and treatment management and its performance details. This helps doctors to readily streamline and formulate the treatment management in conformity with past history of related cases.

An example of similar image retrieval is shown in **Fig. 3**. **Fig. 3(a)** is the candidate image of an anemic patient. First four images similar to the candidate image based on minimum distance between overall signature vectors are shown in **Fig. 3(b)** to **(e)**. Note that in this particular case similar image recall is approximately 83%. For our database of 500 images currently grouped into four different categories (125 in each group) related to four disorder groups (Gastro-intestinal, Respiratory, Renal and Anemia and Malnutrition), similar image recall rate is between 70 to 88%. The similar image recall is defined as the number of images displayed from the same group of the candidate image (as determined by the expert clinician) with respect to the total number of images present in each category expressed in percentage.

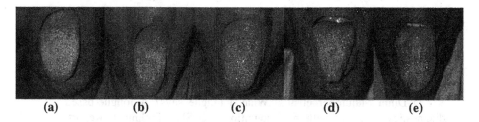

| (a) | (b) | (c) | (d) | (e) |

Fig. 3: For the candidate image (**a**), the first four similar images in descending order of similarity measure are shown from (**b**) to (**e**).

5 Conclusions

In this paper we present the description of a software tool that assists the doctors in inspecting tongue image and correlate the findings with other clinical observations. Also, given a tongue image, the software automatically clusters it to relevant disease group. The system is at present being used by the doctors. We are working on a number of aspects to improve the performance. The extraction of tongue regions can be improved matching the description given in ancient scriptures, an example of which is shown in Fig. 2(a). We are experimenting with fractal dimensions of intensities along the vertical line scan of the tongue region. Also, generating a certainty factor while assessing the disease class label will be an important task. The improvement of the image database management through efficient image matching will be attractive especially for scalable larger tongue image database. In the next phase we would like to incorporate a learning scheme that can take care of symptom variability detected in terms of color and texture features of different tongue regions.

Acknowledgement

The authors wish to thank IRIIM (Indian Research Institute of Integrated Medicine. Mourigram, WestBengal, India) for their assistance in connection to the system development, data acquisition and medical analysis of tongue images. The work is partially supported by the Ministry of Information Technology, Government of India.

References

1.   S.T. Acton and D.P. Mukherjee, "Scale space classification using area morphology," in *IEEE Trans. on Image Processing*, 9(4): 623-635, April 2000.
2.   S.T. Acton, D.P. Mukherjee, J. Havlicek, A.C. Bovik, "Oriented texture completion by AM-FM reaction-diffusion," in *IEEE Trans. on Image Processing*, 10(6): 885-896, June 2001.
3.   J.C. Bezdek, Pattern recognition with fuzzy objective function algorithms, Plenum Press, NewYork, 1981.
4.   D Dutta Majumder et al., "What a tongue says? A tongue-based diagnostics through pattern recognition", ICSIT Technical Report, 2002/3.
5.   J.D. Foley *et al*, Computer Graphics: Principles and Practice, Addison-Wesley, Reading, MA, 1990.
6.   R. Jain, R. Kasturi and H. Schunk, *Machine Vision*, McGraw-Hill, Inc., New York, 1995.
7.   C H Li and P C Yuen, "Tongue image matching using color content", *Pattern Recognition, Vol. 35, No. 2, pp. 407-419, 2002.*
8.   Giovanni Maciocia, *Tongue Diagnosis in Chinese Medicine*, Redwing Book. New York, 1987.
9.   P C Yuen, Z Y Kuang, W Wu and Y T Wu, "Tongue Texture Analysis using Opponent Colour Features for Tongue Diagnosis in Traditional Chinese Medicine", *Texture Analysis in Machine Vision*, (Ed. M. Pietikäinen). Series in Machine Perception and Artificial Intelligence - Vol. 40, pp. 179-188, World Scientific, November 2000.

2 D Object Recognition Using the Hough Transform

Venu Madhav Gummadi, Thompson Sarkodie-Gyan

Vgummadi@nmt.edu, New Mexico Tech, Socorro, NM-87801

Abstract

The objective of this paper is to identify 2D object features in object recognition in order to ensure quality assurance. In this paper we use Hough transform technique to identify the shape of the object by mapping the edge points of the image and also to identify the existing straight lines in the image. The Edge Detection Algorithm is applied to detect the edge points by the sharp or sudden change in intensity. Object recognition is usually performed using the property of shape as a discriminator. The shape of the edges, their size, orientation, and location can be extracted from the objects, which is used for object recognition.

1 Introduction

The advanced manufacturing technology demands for the assemble of components and also for a very high quality control. This calls for a good vision-based inspection systems with corresponding analytical methodologics to recognize the quality of the components during manufacture.

In recent years emphasis has been concentrated on advance manufacturing technology focusing principally on the manufacturing metrology. This is the art and science of measurement and control in the industry with the ultimate aim of precision, quality etc. Hence the monitoring of the quality of a manufacturing product is an essential prerequisite for marketability.

The goal of image recognition system often involves in identification of objects in the images and the establishment of the relationship among the objects. One of the main challenges in object recognition is to develop a flexible, adaptable and, capable system for performing complex image analysis tasks and of extracting information from varying scenes or images.

This transformation of signals (the images) to symbols (the interpretation) in the visual domain is one of the most important and most difficult tasks. The reasons are:

- The way with which humans capture the image content, group image features into meaningful objects and attach semantic descriptions to images has not been fully understood to automate the image analysis procedure.
- The needed computational power for real-time digital image analysis outruns even the best workstation.

Hough Transform, was first introduced by Paul Hough in 1962 [1] with the aim of detecting alignments in T.V lines. In the following years it became the basic technique for a number of image analysis applications. The Hough Transform is used to detect the geometric shapes like Lines, Circles, ellipse etc in the images. The Basic idea of Hough transform is that it identifies the shape by mapping edge points of the image into parameter space from image space. A complicated problem in the image space is converted to the detection of peaks in the parameter space. The method is robust to the presence of noise since noisy image points are unlikely to contribute to a peak in the parameter space. Ballard [10] described how to optimize circle detection by the use of gradient information.

In the rest of the paper, a brief introduction to our experimental setup is given in section 1. An introduction to edge detection and its applications are given in section 2. In section 3 we explain the concept of Hough transform which is a widely used technique for performing efficient binary convolution of the shape model with the edge image. In section 4 we describe the concept of object recognition and also the attributes for identifying the objects. Section 5 gives overview of the experiments performed. In section 6 we conclude our work.

2 Experimental Setup

In our experiments we explore ways to select features and corresponding techniques for implementing an object recognition vision system. The method proposed extracts the features that are used to characterize the invariant transformations in the object. Noise, which is the primary cause for the discrepancies between the objects, should be removed before edge detection is applied for better performance. The method presented in this paper is based on edge detection in computer images and the object can be recognized even if it is translated or rotated.

The experimental setup is as shown in the *figure1*. The camera and the base of the objects that have to be analyzed should be kept stable and constant. This experiment is operated under normal lighting conditions. The necessary equipment for 2D Object Recognition is:

- Sensor -CCD Camera
- Frame Grabber
- Back Lighting
- Computer – Image Preprocessing and analysis.

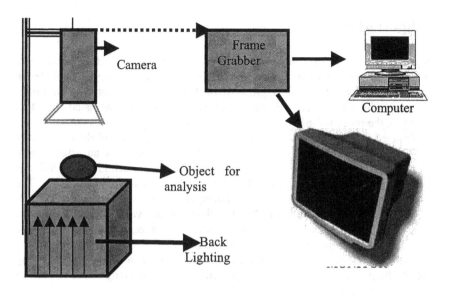

Fig. 1. Experimental Setup

2.1 CCD Camera:

The component is scanned by the CCD camera, in which the digital images are given in pixel arrays. These arrays are fed into the computer with the help of a frame grabber. The frame grabber we used can capture up to 25 frames per second.

2.2 Backlighting:

Commonly used backgrounds, are green screen, back lighting and, direct lighting. The technique, used in our experiments, is backlighting. Backlighting is a technique wherein we pass white light from a light source under the object. It thus increases the contrast and removes the noise. Using this technique we can clearly distinguish between the object boundaries.

2.3 Image Acquisition:

The images that are captured by the CCD camera are being transferred to the computer for analysis with the help of a frame grabber. The frame grabber captures about 25 frames per second. The captured images are then saved on the computer, which can be used for further analysis.

3 Edge Detection

Significant changes in an image are caused due to the edges that occur on a boundary between two different regions in an image. Edges characterize boundaries and are therefore a problem of fundamental importance in image processing. The need for edge detection is that it provides strong visual clues that can help in the recognition process.

The obtained image is first converted into a grayscale image and then the places with sharp or significant changes in intensity are defined as an edge. These grayscale values are used to calculate the gradient values. Each grayscale value is compared to its neighboring grayscale value both horizontally and vertically. The sum of the difference in the grayscale values, both horizontally and vertically is the gradient. There is a certain threshold or the tolerance value and if the gradient is below the threshold value then it is considered as a zero. If it more than the threshold value then the gradient value is noted.

Thus, the edge detection of an image can be done. This is a low level processing stage. A sharp filter has to be used to prevent the unwanted edges and to remove the noise from the images.

3.1 Edge Tracking:

The edge tracking execution starts from the top-left pixel in the image. Each pixel is evaluated with the surrounding pixels, so that a proper edge-starting node can be located. The algorithm, which we designed, will keep tracking the edges, and it connects the next edge pixel and like wise to all the detected edge points.

4 Hough Transform

The Hough transform is widely used for performing efficient binary Convolutions of the shape model with the edge image. The Hough transform is usually applied to an edge image. It is used to isolate the features of a particular shape within an image. Hough transform has many applications as most manufactured parts contain feature boundaries, which can be described by regular curves or straight lines. Hough Transform is therefore the methodology for executing the transition from a visual-stimulus oriented representation to the symbolic representation of shape necessary for pattern recognition.

4.1 Line Detection Using Hough Transform

The basic idea in Hough transform technique is that each input measurement indicates its contribution to a consistent solution. We can analytically describe a line segment in a number of forms. The convenient equation is

$$x \cos \theta + y \sin \theta = r$$

Where **r** is the length of a normal from the origin to this line and θ is the orientation of **r** with respect to the x-axis. For any point (**x, y**) on this line r and θ are constant.

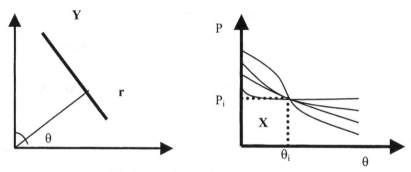

Fig. 2. Hough Transform Principle

If we plot (**r**, θ) values defined by each (x_i, y_i) points in the Cartesian image space map to the curves in the polar Hough space. This point-to-curve transformation is the Hough transformation for straight lines. The points that are collinear in the Hough parameter space become readily apparent as they yield curves which intersect at a common (**r**, θ) point.

The transform is implemented by quantizing the Hough parameter space into finite intervals or accumulator cells. Each (x_i, y_i) is transformed into a discredited (**r**, θ) curve and the accumulator cells, which lie along this curve, are incremented. Peaks in the accumulator array represent strong evidence that a corresponding straight line exists in the image. The Hough transform can be used to identify the parameters of the curve, which best fits a set of given edge points. The output of the Hough transform is to determine both what the features are and how many of them exist in an image.

4.2 Shape Analysis

The results from the Hough transform are taken and the Hough Space is clearly analyzed. The number of lines of an object is represented by the local maxima and counting the number of local maxima can be used to find out the number of lines. The Length of each line can be calculated. Using all the above results the shape of the object can be identified and the object can be identified.

5 Object Recognition

The Attributes are extracted from the objects and a knowledge base is constructed. All the attributes map to the specific pattern in a group of patterns. In the following example the object X has some attributes X1, X2....Xn, and then it specifies that it is most likely to be in Pattern A.

Ex:

Object   Attributes   Recognition

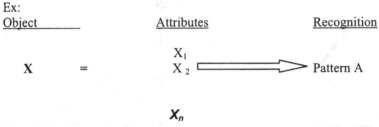

X = X_1
           X_2 $\Longrightarrow$ Pattern A

0 X_n

5.1 Object Recognition Algorithm

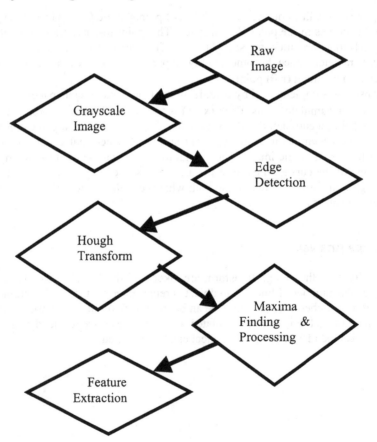

6 Experimental Results

The objects, which are to be analyzed, are taken. The CCD Camera captures the objects that are kept on the observatory surface. The Camera is positioned at a particular level and at a particular angle for all the experiments. The Object that is captured is sent to the computer for Image Processing.

The obtained object is taken and it is converted to the grayscale image. The converted grayscale image is given as the input for detecting the edges. The gradient based Edge Detection algorithm is used to detect the edges. The pixels are extracted from the images and the grayscale values are calculated for each of the pixels. The grayscale values are used to calculate the gradient values. Each grayscale value is compared to its neighboring grayscale value both horizontally and vertically. The sum of the difference in the grayscale values, both horizontally and vertically is the gradient. There is a certain threshold or the tolerance value and if the gradient is below the threshold value then it is considered as a zero. If it more than the threshold value then the gradient value is noted.

Thus, the edge detection of an image can be done. Edge Detection is used to extract the significant features and discard the background. These significant features can be extracted and they can be used to recognize and differentiate between the objects. The gradient values are evaluated from the top left pixel and the starting node is located. Then the tracking of edges and the edge points are done. After obtaining all the parameters, the necessary statistics of the object can be calculated. The Edge Detection is thus done and the various examples are shown below-

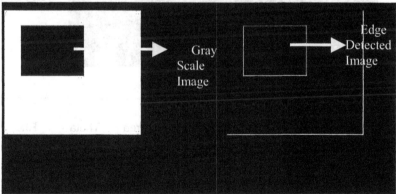

Fig. 3. Gray Scale Image is converted into Edge Detected Image

After the Edge Detection is done, the edge-detected image is given as an input to the Hough transform. The edge points from the Edge Detection are given as inputs to the Hough transform, a curve is generated in polar space for each point in Cartesian space. The curves generated by collinear points in the gradient image intersect in peaks in the Hough transform space. The intersection points characterize line segments of the straight-line segments of the original image.

Then the local maxima are calculated from the accumulator array. The examples of the Hough Transform are shown below-

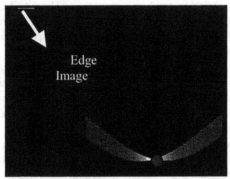

Fig.4. Example 1

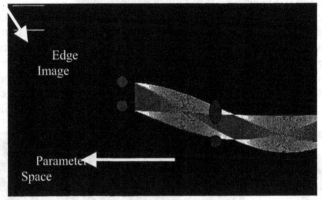

Fig.5. Example 2

Component	Parameters	Number of Lines	Area	Theta	Rho
Object 1	3,28	4	1728	47	161
	5,82				
	35,90				
	38,93				
Object 2	5,34	4	1998	43	191
	6,36				
	42,88				
	44,92				

Table 1. Experimental results for all components Results

7 Conclusions

The main objective of this research paper is to apply the technique of Hough Transform for detection and recognition of quality products in advanced manufacturing processing. Notably, the Hough Transform is that type of recognition metrology that is also capable of coping with noise, gaps in outlines. Experimental Results show that this transform is suitable for detection and recognition of various objects. Features of the objects are extracted even in the presence of noise and still the attributes are confidently determined. Hence, it may be proposed that the Hough Transform technique is a powerful tool capable of detection and recognition of industrial components.

8 Acknowledgements

I would like to thank Dr. Thompson Sarkodie-Gyan and his former students for giving me valuable guidance and valuable information. I would also thank many of my friends for providing me valuable suggestions.

9 References

1. P.V.C Hough. Method and means for recognizing complex patterns. Technical report, 1962. U.S
   Patent No. 3069654.
2. V.F Leavers. Shape Detection in Computer Vision Using the Hough transforms. Springer-Verlag, 1992.
3. V.F Leavers. Which Hough Transform? CVGIP: Image Understanding, 58(2):250-264, Sep 1993.
4. E.R. Davies. Machine Vision. Academic Press Limited, 1990.
5. Nello Zuech and Richard K.Miller, "Machine Vision", The Fairmont Press, Inc. 1987.
6. H.Ballard, "Generalizing the Hough Transform to detect arbitrary Shapes" Pattern Recognition, Vol. 13, pp 111-122, 1981.
7. Richard O. Duda and Peter E. Hart, "Use of the Hough Transformation to Detect Lines and Curves in Pictures" Communications of ACM, Vol. 15, pp 11-15, Jan.1972.
8. Lam, C. W., Campbell A. and Sarkodie-Gyan, T., "Experimental Investigation to classify Turbocharger Components", 27th ISATA, Dedicated Conference On Robotic, Motion and Machine Vision, Achen, Germany, 31st Oct- 4th Nov 1994.
9. Antony Barrett, " Computer Vision and Image Processing", Chapmen K Hall, 1991.
10. D.H. Ballard. Generalizing the Hough transform to detect arbitrary shapes. Pattern Recognition, 13(2): 111-122,1981.

Part IX

**Optimization, Scheduling
and Heuristics**

Population Size Adaptation for Differential Evolution Algorithm Using Fuzzy Logic

Junhong Liu, Jouni Lampinen

Department of Information Technology, Lappeenranta University of Technology, P.O.Box 20, FIN-53851 Lappeenranta, Finland

Abstract: The Differential Evolution algorithm is a floating-point encoded Evolutionary Algorithm for global optimization over continuous spaces. The objective of this study is to introduce a dynamically controlled adaptive population size for the Differential Evolution algorithm by the means of a fuzzy controller. The controller's inputs incorporate the changes in objective function values and individual solution vectors between the populations of two successive generations. The fuzzy controller then uses these data for dynamically adapting the population size. The obtained preliminary results suggest that the adaptive population size may result in a higher convergence rate and reduce the number of objective function evaluations required.

Key Words: Differential Evolution, Evolutionary Algorithm, Fuzzy logic, Control parameter adaptation

1. Introduction

Since the introduction of the Differential Evolution (DE) by Price and Storn [11], DE has been tested extensively against a wide variety of artificial and real-world minimization problems [3, 9, 10, 12, 13]. As the main motivations for applying DE, typically the advantages, like a simple structure, ease of use, speed of convergence, and robustness are mentioned in the literature.

The classical DE algorithm is controlled by three control parameters, which are all kept fixed throughout the entire optimization procedure. The mutation amplification, F, is a real-valued constant factor that controls the differential mutation operation. The crossover parameter, CR, controls the probability of the crossover operation. The third control parameter, NP, is the number of population members.

In previous research reports discussing DE algorithm and applications [1, 3, 4, 5, 6, 7, 8, 9, 10, 12, 13, 14, 15, 17] practical advices and recommendations were given concerning the effective settings for the DE's control parameters. These parameters' values determine whether the algorithm will find a near-optimum solution and whether it will find such a solution efficiently. However, there still exists a lack of knowledge on how to find the best values for the control parameters of DE for a given function. Unfortunately the advices currently available in the literature are still rather incomplete. Basically, they specify only useful ranges for control parameters and provide some help with the initial guess for their values. Therefore, the trial-and-error method has to be used for fine-tuning the control parameters for effective and efficient search. For that reason, choosing the right parameter values sometimes turns out to be a laborious and time-consuming task.

In order to simplify the usage of DE and avoid the problems with tuning its control parameters by trial-and-error, adaptive DE algorithms (ADEs) have been proposed to dynamically adjust the control parameters. A self-adaptive approach to control the mutation amplification and crossover parameter was proposed in [1]. In [15] the control parameters of DE were adapted by controlling the population diversity. One way to build ADEs involves the application of fuzzy logic controllers (FLCs) that use fuzzy rule bases built through the expertise, experience and knowledge of DE, which have become available as a result of empirical studies. This kind of Fuzzy Adaptive DE (FADE) has been proposed in [6, 7]. In general, there are three levels where the adaptation may take place:

- Population-level adaptation adjusts the control parameters that are to be applied to the entire population as, for example, in [6, 7].
- Individual-level adaptation is centered on the consideration of the individual members of the population rather than the ensemble as a whole.
- Component-level adaptations alter an individual component of the function parameter vector; the parameters are manipulated independently from each other.

These approaches include the application of FLCs for adapting the control parameters controlling the operation of DE.

In [5], the influence of three control parameters on the performance of DE was investigated by conducting experiments with a set of benchmark test functions. The obtained empirical results revealed that ways of improving the effectiveness, efficiency and robustness of DE were possible to be found. It was concluded that: (1) The compound effect of parameters of DE should be concerned in order to make DE perform well for a particular problem at hand; (2) The optimal control parameter settings are rather heavily problem dependent. As an outcome, there are obvious difficulties in selection of the parameter settings and in understanding their influence on the DE's performance both individually and in combination.

In [6], it was suggested that the usage of FLCs [2, 16] to control DEs can be considered for solving two problems to which a standard DE may be subjected due to inappropriate control parameters settings: low convergence velocity and high failure risk. These problems are due to: (1) Control parameters not well chosen initially for a given task; (2) Parameters are always kept fixed through the whole process and having no response to population's information (function parameter vectors and function values and their changes) even though the local environment, the local objective function landscape, in which the population of DE operates in different phases of its convergence is varying; (3) Lack of *a priori* knowledge about the search space and it's landscape.

In [7], two FLCs were controlling the mutation control parameter, F, and crossover control parameter, CR. These two parameters were adapted individually for each component of population members. Parameter vector change and function value change over the whole population during the previous two generations were applied as the inputs for both FLCs. The outputs of the FLCs were constrained into the range of [0,2] for F, and into the range of [0,1] for CR. FLCs were used for controlling the control parameters of DE in order to: (1) Allow choosing the initial control parameters freely; (2) Adjust the control parameters on-line to dynamically adapt to new situations. A set of function was tested to explore the performance of FADE, which showed that FADE results in an improved convergence velocity with higher problem dimensions in tested cases.

In this article FADE applying a single FLC to control the population size, NP, is under investigation. The two inputs of FLC are: the linearly depressed objective function value change and the corresponding parameter vector change over the whole population during the previous two generations. The Ackley's path function is used for demonstrating the investigated approach.

The paper is organized as follows: In Sect. 2, the possibility of implementing fuzzy adaptation for the population size of DE is analyzed and the basic idea of FADEs is described. In Sect. 3, a FADE model is introduced. In Sect. 4, an instance of the FADE model and the related FLC for adapting the population size are described. In Sect. 5, FADE with the adaptive population size is experimentally demonstrated, and finally in Sect. 6 some concluding remarks are offered.

2. The Differential Evolution Algorithm

2.1 Brief description

DE is a parallel direct search method that utilizes NP D-dimensional parameter vectors:

$$X_{i,G}, i = 1,2,\ldots, NP, \tag{1}$$

as a population for the generation G, where NP, the number of population members, doesn't change during the optimization process to minimize the objective function:

$$f(X): \mathfrak{R}^D \to \mathfrak{R}, \tag{2}$$

by finding an optimal parameter vector:

$$X = (x_1, x_2,\ldots, x_D), \tag{3}$$

where X denotes a vector composed of D objective function parameters.

The initial population is chosen randomly to cover the entire parameter space uniformly. Later on a uniform probability distribution function will be assumed for all random decisions, unless otherwise stated. The crucial idea behind DE is a scheme for generating trial parameter vectors. Basically, DE generates new trial parameter vectors by adding the weighted difference between two population vectors to a third vector subject to perturbation. After crossover, if the resulting trial vector yields a lower or equal objective function value than a predetermined population member, the newly generated trial vector will replace it in the following generation's population; otherwise, the old vector is retained. The basic principle, however, is extended when it comes to the practical variants of DE [10].

2.2 Some characteristics

The effect of the population size, NP, on the performance of DE is demonstrated in Fig. 1 using the Ackley's path function as an example. In this example the DE's control parameters are set to: $F = 0.9$, $CR = 0.9$. For the objective function, the dimensionality is $D = 5$, and the global optimum value is $f(X^*) = 0$, from which maximum error $e = 10^{-6}$ is allowed. The maximum number of generations was $G_{max} = 500$. For every tested value of $NP \in [D+1,100]$, 100 test runs are performed. Fig. 1-a shows how many of the performed 100 independent experiments failed, i.e., the best-found objective function value exceeded the given allowable error, e, after the maximum number of generations. Fig. 1-b shows the average of function evaluations versus NP and Fig.1-c shows the average number generations versus NP for reaching the global optimum with the specified accuracy.

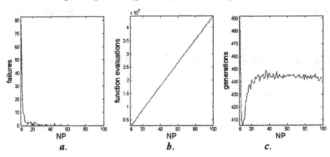

Fig. 1. Performance of DE on Ackley's path function: (*a*) Failure times – NP; (*b*) Function evaluations – NP; (*c*) Generations – NP

From Fig. 1, it can be observed that

- when NP is too small, the process will most probably fail.
- when NP is too big, the process will converge slowly.
- when NP is increased from the lower limit, the probability of failure decreases quickly at the beginning.

When NP is higher than $10 \cdot D$, the reduction of generations for convergence is very small and can be completely compromised by the linearly increasing number of function evaluations. When NP is set suitably, DE will converge both reliably and quickly. When increasing NP, from a certain point forward the probability of failure will further decrease only marginally while the computational effort will keep on increasing approximately linearly. The problem is that this particular population size is a problem dependent property and therefore not known *a priori*. Furthermore, there is no reason to believe that the optimal strategy is to keep the population size fixed throughout the optimization process.

At this stage, DE with the adaptive population size, like FADE using FLC for adaptation could be of help. FADE is based on FLC designed according to the following steps: the inputs and output are defined based on the information obtained from the DE's population during the previous two generations. Each input and output has an associated set of linguistic labels, whose meaning is specified through membership function of fuzzy sets attempting to take into consideration the uncertainties and non-linearity that may appear in optimization of an objective function. The bounded ranges of values for every input and output are determined in order to define these membership functions over it. Then the fuzzy rules describing and mapping the relationship between the inputs and output are defined by using the experience and the knowledge of DE experts.

3. Adaptation of DE's parameters

In this section, we present a general proposal for a mechanism based on FLCs for the adaptation of DE's control parameters. Its main features are the following:

- It adapts the DE's control parameters at population-level by applying FLCs. Control parameters' values are computed for the whole population by FLCs on the basis of features associated with the whole populations.
- The fuzzy rule bases of FLCs are based on the results of [5] and the basic ideas of another non-linear optimization method (Gradient method).
- After a pre-specified number of generations, FLCs evaluate and adjust the control parameters.

The idea of FADE can be expressed using the control diagram in Fig. 2. The generic version of FADE is presented in Fig. 4 based on the classical DE in Fig. 3.

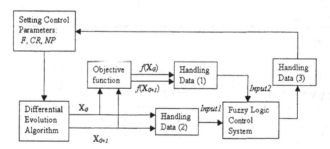

Fig. 2. Control diagrams of DE and FADE: *Input1* – information of the parameter vectors and *Input2* – information of the function values

4. FADE with Adaptive Population Size

For the model of FADE with the adaptive population size the following steps are followed:

Step 1. System Parameters and Fuzzy Sets Membership Functions: An adaptation method is obtained by applying the mean square root of differences concerning the whole population over two successive generations during the optimization process:

---

let D denote the dimension, G a generation, F the mutation amplification, CR the crossover operator, NP the population size and $X_{i,G}$ the ith individual in the population of Gth generation
input *D, NP, F, CR, and initial bounds: $x_j^{(L)}, x_j^{(U)}, j = 1,2,...,D$*
initialize the population $P^{(G=0)} = \{X_{1,0}, X_{2,0}, ..., X_{NP,0}\}$ as
for each *individual $i \in P^{(G=0)}$*

$$X_{i,0} = \{x_{j,i,0} = rand_j[0,1] \cdot (x_j^{(U)} - x_j^{(L)}) + x_j^{(L)}, j = 1,2,...,D\}$$

end for each
evaluate $P^{(G=0)}$
$G = 1$
while *the stopping criterion is not satisfied* **do**
    forall *$i <= NP$*
      randomly select $r_1, r_2, r_3 \in (1,.. NP), j \sim r_1 \sim r_2 \sim r_3$
      forall *$j <= D$*

$$u_{j,i,G} = \begin{cases} x_{j,r_3,G-1} + F \cdot (x_{j,r_1,G-1} - x_{j,r_2,G-1}) & if \;\; rand_j[0,1] \le CR \vee j = D_i \\ x_{j,i,G-1} & otherwise \end{cases}$$

      end forall
      repair $u_{j,i,G}$ if it is outside the initial boundary as $u_{j,i,G} = rand_j[0,1] \cdot (x_j^{(U)} - x_j^{(L)}) + x_j^{(L)}$

$$X_{i,G} = \begin{cases} U_{i,G} & if \;\; f(U_{i,G}) \le f(X_{i,G-1}) \\ X_{i,G-1} & otherwise \end{cases}$$

    end forall
    evaluate $P^{(G)}$
    $G = G+1$
end while
return *the best encountered solution*

Fig. 3. The DE algorithm

$$vpc = \sum_{j=1}^{D} \sqrt{\frac{1}{NP^{(n)}-1} \sum_{i=1}^{NP^{(n)}} \left(x_{j,i}^{(n)} - E(x_j^{(n)})\right)^2} - \sum_{j=1}^{D} \sqrt{\frac{1}{NP^{(n-1)}-1} \sum_{i=1}^{NP^{(n-1)}} \left(x_{j,i}^{(n-1)} - E(x_j^{(n-1)})\right)^2}$$

$$vfc = c \cdot \left(\sqrt{\frac{1}{NP^{(n)}-1} \sum_{i=1}^{NP^{(n)}} \left(f_i^{(n)} - E(f^{(n)})\right)^2} - \sqrt{\frac{1}{NP^{(n-1)}-1} \sum_{i=1}^{NP^{(n-1)}} \left(f_i^{(n-1)} - E(f^{(n-1)})\right)^2} \right) \tag{4}$$

$$\left. \begin{aligned} d_{31} &= 1 - (1+|vpc|) \cdot e^{-|vpc|} \\ d_{32} &= 1 - (1+|vfc|) \cdot e^{-|vfc|} \end{aligned} \right\} \tag{5}$$

where $f_i^{(n)}$ represents the objective function value $f \in \mathfrak{R}^{NP}$ for the ith individual of the nth generation's population, $i = 1, 2, ..., NP$; $x_{j,i}^{(n)}$ represents the component in the ith individual and j^{th} parameter to be optimized in the population matrix $X_{NP \times D}$ for the nth generation, $i = 1, 2, ..., NP, j = 1, 2, ..., D$; c is a positive constant.

vpc is called the variance change of objective function parameters in magnitude and depressed into the range of $[0,1]$ as d_{31}; *vfc* is called the variance change of objective function value and depressed into $[0,1]$ as d_{32}.

The input variables of FLC are numerical values of observed characteristics like d_{3j}s as stated in Eqs. 4 and 5, and the output variable of FLC is y_{NP}, whose value will be transformed into the control parameter value, *NP*, later on. y_{NP} is adapted based on the absolute values of *vpc* and *vfc*. The underlying principles are:

---

let D denote the dimension, G a generation, F the mutation amplification, CR the crossover operator, NP the population of size and $X_{i,G}$ the ith individual in the population of Gth generation
input *D and initial bounds:* $x_j^{(L)}, x_j^{(U)}, j = 1, 2, ..., D$
input *the initial values for NP, F, CR*
initialize the population $P^{(G=0)} = \{X_{1,0}, X_{2,0}, ..., X_{NP,0}\}$ *as*
foreach *individual* $i \in P^{(G=0)}$

$$X_{i,0} = \left\{ x_{j,i,0} = \text{rand}_j[0,1] \cdot (x_j^{(U)} - x_j^{(L)}) + x_j^{(L)}, j = 1, 2, ..., D \right\}$$

end for each
evaluate $P^{(G=0)}$
$G = 1$
while *the stopping criterion is not satisfied* **do**
    adjust values of F, CR and NP for every certain number of generations
    if $NP^{(G)} > NP^{(G-1)}$,
    Then *get new bounds for D parameters:* $x_{j,new}^{(L)}, x_{j,new}^{(U)}, j = 1, 2, ..., D$

      add $(NP^{(G)} - NP^{(G-1)})$ *more initial individuals into the population* $P^{(G-1)}$ *as*
      for each *individual* $i \in [NP^{(G-1)}+1, NP^{(G)}]$

$$X_{i,G-1} = \left\{ x_{j,i,G-1} = \text{rand}_j[0,1] \cdot (x_{j,new}^{(U)} - x_{j,new}^{(L)}) + x_{j,new}^{(L)}, j = 1, 2, ..., D \right\}$$

      end for each
    else if $NP^{(G)} < NP^{(G-1)}$,
      take only the first $NP^{(G)}$ *individuals of* $P^{(G-1)}$ *as* $P^{(G-1)} = \left\{ X_{1,G-1}, X_{2,G-1}, ..., X_{NP^{(G)},G-1} \right\}$

    end if
    forall $i <= NP^{(G)}$
      randomly select $r_1, r_2, r_3 \in (1, ... NP^{(G)}), j \neq r_1 \neq r_2 \neq r_3$
      forall $j <= D$

$$u_{j,i,G} = \begin{cases} x_{j,r_1,G-1} + F \cdot (x_{j,r_2,G-1} - x_{j,r_3,G-1}) & \text{if } \text{rand}_j[0,1] \leq CR \vee j = D_i \\ x_{j,i,G-1} & \text{otherwise} \end{cases}$$

      end forall
      repair $u_{j,i,G}$ *if it is outside the initial boundary us* $u_{j,i,G} = \text{rand}_j[0,1] \cdot (x_j^{(U)} - x_j^{(L)}) + x_j^{(L)}$

$$X_{i,G} = \begin{cases} U_{i,G} & \text{if } f(U_{i,G}) \leq f(X_{i,G-1}) \\ X_{i,G-1} & \text{otherwise} \end{cases}$$

    end forall
    evaluate $P^{(G)}$
    sort the population members by values in ascending order
    $G = G+1$
end while
return *the best encountered solution*

Fig. 4. Fuzzy Adaptive Differential Evolution algorithm

- y_{NP} should be large when the absolute value of *vpc* is large, indicating that the actual solution is away from the expected.
- y_{NP} should be small when the absolute value of *vfc* is small, showing that the actual solution is close to the expected.

The inputs and output are treated as fuzzy variables, with 3 elements for d_{31}, d_{32} and y_{NP} individually, values are then assigned as grades of membership in 3 subsets as follows:

- **S** is "small";
- **M** is "medium";
- **B** is "big".

Among a set of membership functions, Gaussian membership function, f_g, is used for each input and output (Fig. 5). The definitions are characterized by the membership functions in Table 1.

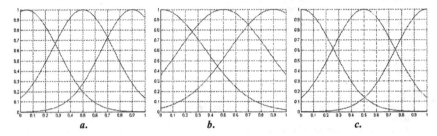

Fig. 5. Membership functions: (*a*) for input 1; (*b*) for input 2; (*c*) for output

Table 1. Membership functions

Inputs,Outputs	Membership functions
d_{31}	$\mu_S(d_{31}) = f_g(d_{31}, 0.25, 0.05)$
	$\mu_M(d_{31}) = f_g(d_{31}, 0.25, 0.5)$
	$\mu_B(d_{31}) = f_g(d_{31}, 0.25, 0.9)$
d_{32}	$\mu_S(d_{32}) = f_g(d_{32}, 0.35, 0.01)$
	$\mu_M(d_{32}) = f_g(d_{32}, 0.35, 0.5)$
	$\mu_B(d_{31}) = f_g(d_{31}, 0.25, 0.9)$
y_{NP}	$\mu_S(y_{NP}) = f_g(y_{NP}, 0.25, 0.01)$
	$\mu_M(y_{NP}) = f_g(y_{NP}, 0.25, 0.5)$
	$\mu_B(y_{NP}) = f_g(y_{NP}, 0.25, 1)$

Note: S = small; M = medium; B =big.

Table 2. The rules of functions

Rules	d_{31}	d_{32}	y_{NP}
1	S	S	S
2	S	M	M
3	S	B	B
4	M	S	M
5	M	M	M
6	M	B	B
7	B	S	B
8	B	M	B
9	B	B	B

Note: S = small; M = medium; B = big.

Step 2. Determine Fuzzy Rules: There are "9" rules in Table 2 that determine the values of y_{NP} based on d_{31} and d_{32} as in Eqs. 4 and 5. Each rule has two inputs and one output, and describes a characteristic of the system and represents a mapping from the input space to the output space.

Step 3. Select Control Strategy: The fuzzy control strategy is used to map the given inputs to an output. Mamdani's fuzzy inference method is used. The output of each rule is a fuzzy set. These fuzzy output sets are then aggregated into a single fuzzy output set.

Step 4. Determine Defuzzification Strategy: Defuzzification is mapping from a space of fuzzy output into a space of real output. The input is a fuzzy set μ and the output is a single real value y^*. The centroidal defuzzification technique (CDT) is chosen to utilize all the information in the fuzzy distribution to compute the centroid represented as

$$y^* = \sum_{j=1}^{n} \mu_z(w_j) \cdot w_j \Big/ \sum_{j=1}^{n} \mu_z(w_j), \tag{6}$$

the defuzzified value is a value in the range [0,1], and not directly the value for the DE's parameter, NP, but the value for NP is obtained as follows:

$$NP = \begin{cases} Int(y^* \cdot 15 \cdot D), & \text{if } y^* \neq 0 \\ 10 \cdot D, & \text{if } y^* = 0 \end{cases} \tag{7}$$

Fig. 6 depicts the resulting control surface of the described adaptation system.

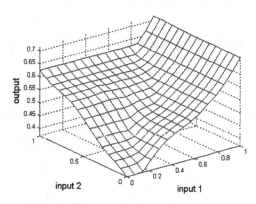

Fig. 6. The control surface for adaptive NP

5. Experimental demonstration

A test function commonly known as Ackley's path function is for demonstrating. It is a widely used multi-modal and continuous test function that is scalable to any number of dimensions. To facilitate its use for optimization and to achieve a standardization of the global optimum to an objective function value of zero, the function is formulated as follows:

$$f(x) = -a \cdot e^{\left(-b \times \sqrt{\frac{1}{n}\sum_{i=1}^{n} x(i)}\right)} - e^{\left(\frac{1}{n}\sum_{i=1}^{n} \cos(c \cdot x(i))\right)} + a + e \tag{8}$$

where $a = 20$, $b = 0.2$, $c = 2 \cdot \pi$, $i = 1:n$ (dimensionality), $x(i) \in [-32.768, 32.768]$, $\min f(0) = 0$.

In Fig. 7 the two-dimensional version of the Ackley's function and a detail of it, is illustrated.

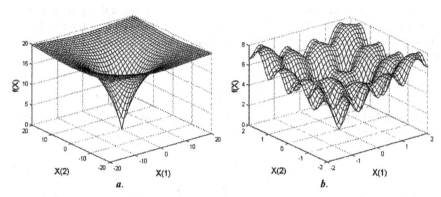

Fig. 7. A 3-D representation of Ackley's path function: (*a*) $x(i) \in [-20,20]$, $i = 1, 2$; (*b*) $x(i) \in [-2,2]$, $i = 1, 2$

The settings of DE and FADE in tests are given in Table 3. All results are averaged over 30 independent runs.

Table 3. Settings for algorithms

Method, Parameters	Settings for Test #1	Settings for Test #2
Strategy	DE/rand/1/bin	DE/rand/1/bin
Test problems	$\min f(X)$	$\min f(X)$
Dimensionality	50	50
Number of individuals	500	FLC
Crossover parameter	0.9	0.9
Mutation amplification	0.9	0.9

Note: Each setting is kept constant per run for test #1; the population size is controlled by FLC for test #2, the initial population size is identical to test #1.

Fig. 8 shows the convergence history averaged from 30 independent tests using traditional DE and FADE. The setting of FLC presents a subjective view of an expert with respect to the objective characteristics of DE. The population size setting strategy of FADE can be viewed as the control surface of DE (Fig. 6). The change in behavior (convergence) of FADE in accordance with fuzzy control action can be observed from Fig. 8. It can be observed that FADE performed better than DE for a fixed population size from the viewpoint of the best-found solution, and also considerably reduced the number of function evaluations for obtaining the best solution.

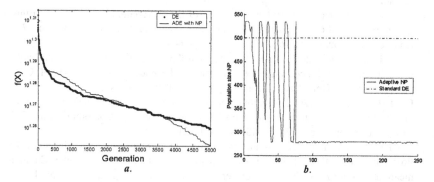

Fig. 8. Behavior: (*a*) Comparison of DE and FADE (*D* = 50); (*b*) Changes of *NP* during 5000 generation (*NP* starting from $10 \cdot D$)

6. Conclusions

In this article a fuzzy system is proposed to control dynamically the population size of DE algorithm based on the previous two generations' features: the changes of the objective function values for the individuals, and the changes of the individual parameter vectors.

By utilizing a fuzzy logic approach, FADE is capable of adapting the population size dynamically. Based on human knowledge and expertise, the fuzzy rules are designated to expedite the convergence velocity of DE through the dynamic population size adaptation.

The obtained first experimental results suggest that the proposed FADE algorithm may perform better than those using all the fixed parameters. It converged faster than the traditional DE in terms of the number of objective function evaluations and thereby it reduced the computational load.

However, due to the limited experimentation performed so far, the conclusions drawn should be considered preliminary. In the future, we intend to continue with further investigations on the performance of the proposed approach.

7. References

[1] Abbass H A (2002) The self-adaptive Pareto differential evolution algorithm. In: Proceedings of Congress on Evolutionary Computation, May 12–17, pp 831–836, Honolulu, USA.
[2] Bandemer H (1995) Fuzzy sets, fuzzy logic, fuzzy methods with applications. Wiley, Chichester (UK).

[3] Lampinen J (1999) A Bibliography of Differential Evolution Algorithm. Lappeenranta University of Technology, Department of Information Technology, Laboratory of Information Processing, Tech. Rep. October 16.

[4] Lampinen J, Zelinka I (2000) On Stagnation of the Differential Evolution Algorithm. In: Proceedings of 6th International Mendel Conference on Soft Computing, June 7–9, Brno, Czech Republic, pp 76–83. ISBN 80-214-1609-2.

[5] Liu J H, Lampinen J (2002) On Setting the control parameter of the Differential Evolution Algorithm. In: Proceedings of 8th International Mendel Conference on Soft Computing, June 5–7, Brno, Czech Republic, pp 11–18. ISBN 80-214-2135-5.

[6] Liu J H, Lampinen J (2002) Adaptive Parameter Control of Differential Evolution. In: Proceedings of 8th International Mendel Conference on Soft Computing, June 5–7, Brno, Czech Republic, pp 19–26. ISBN 80-214-2135-5.

[7] Liu J H, Lampinen J (2002) A Fuzzy Aaptive Differential Evolution Algorithm. In: Proceedings of 17th IEEE Region 10 International Conference on computer, Communications, Control and Power Engineering, Oct. 28–31, Beijing, China, pp 19–26. ISBN 0-7803-7490-8 7-900091-55-6.

[8] Lopez Cruz I L, Van Willigenburg L G, Van Straten G (2001) Parameter Control Strategy in Differential Evolution Algorithm for Optimal Control. In: Hamza M H (ed) Proceedings of the IASTED International Conference Artificial Intelligence and Soft Computing, May 21-24, Cancun, Mexico, pp 211–216. ACTA Press, Calgary (Canada). ISBN 0-88986-283-4, ISSN 1482-7913.

[9] Price K V (1999) An Introduction to Differential Evolution. In: New Ideas in Optimization, David Corne, Marco Dorigo and Fred Glover (eds). McGraw-Hill, London (UK), pp 79–108.

[10] Storn R (1996) On the Usage of Differential Evolution for Function Optimization. In: Biennial Conference of the North American Fuzzy Information Processing Society, Berkeley, pp 519–523. IEEE, New York, USA.

[11] Storn R, Price K (1995) Differential Evolution – a Simple and Efficient Adaptive Scheme for Global Optimization over Continuous Spaces. Tech. Rep. TR-95-012, ICSI, March.

[12] Storn R, Price K (1997) Differential Evolution – A simple evolution strategy for fast optimization. Dr. Dobb's Journal 22(4), pp 18–24 and 78, April.

[13] Storn R, Price K (1997) Differential Evolution – a Simple and Efficient Heuristic for Global Optimization over Continuous Spaces. Global Optimization, 11(4), pp 341–359, December. Kluwer Academic Publishers.

[14] Zaharie D (2002) Critical values for the control parameters of Differential Evolution algorithms. In: Proceedings of 8th International Mendel Conference on Soft Computing, June 5–7, Brno, Czech Republic, pp 62–67. ISBN 80-214-2135-5.

[15] Zaharie D (2002) Parameter Adaptation in Differential Evolution by Controlling the Population Diversity. In: Proceedings of 4th International Workshop on Symbolic and Numeric Algorithms for Scientific Computing, Oct. 9–12, Petcu D, Negru V, Zaharie D, Jebelean, Mirton T (eds), pp 385–397, Timisoara, Romania, ISBN 973-585-785-5.

[16] Zimmermann H J (1986) Fuzzy set theory – and its applications. Kluwer-Nijhoff, Boston (USA).

[17] Šmuc T (2002) Improving Convergence Properties of the Differential Evolution algorithms. In: Proceedings of 8th International Mendel Conference on Soft Computing, June 5–7, Brno, Czech Republic, pp 80–86. ISBN 80-214-2135-5.

Intelligent Management of QoS Requirements for Perceptual Benefit

George Ghinea[*], George D. Magoulas[*], and J.P.Thomas[+]

[*] Department of Information Systems and Computing, Brunel University,
Uxbridge, UB8 3PH, United Kingdom

[+] Department of Computer Science, Oklahoma State University,
Tulsa, OK 74106, USA

Abstract. The vision of a new generation of network communication architectures, which deliver a *Quality of Service* based on intelligent decisions about the interactions that typically take place in a multimedia scenario, encourages researchers to look at novel ways of matching user-level requirements with parameters characterising underlying network performance. In this paper, we suggest an integrated architecture that makes use of the objective-technical information provided by the designer and the subjective-perceptual information supplied by the user for intelligent decision making in the construction of communication protocols. Our approach opens the possibility for such protocols to dynamically adapt based on a changing operating environment.

1 Introduction

The concept of Quality of Service (QoS) in distributed multimedia systems is indelibly associated with the provision of an acceptable level of application performance. In turn this performance is itself dependent on both the perceived, subjective quality of the application as well as its robustness to network congestion.

The focus of our research has been the enhancement of the traditional view of QoS with a user-level defined *Quality of Perception* (QoP). This is a measure which encompasses not only a user's satisfaction with multimedia clips, but also his/her ability to perceive, synthesise and analyse the informational content of such presentations. As such, we have investigated the interaction between QoP and QoS and its implications from both a user perspective [5] as well as a networking angle [6]. Although the problem of multimedia application-level performance is closely linked to both the user perspective of the experience as well as to the service provided by the underlying network, it is rarely studied from an integrated viewpoint. Clearly, this is a very unsatisfactory state of affairs.

In this paper, we try and rectify this situation by proposing a scheme for QoP management based on an integrated architecture that takes into account both user-centric QoP and low-level QoS parameters. Such a scheme would then, by appropriate management of these QoS parameters, provide the potential of ensuring an optimum QoP in a distributed multimedia setting.

In the next section, we take a closer look at the reasons behind the current segregated approach. Section 3 presents empirical results in the measurement of human QoP given a multimedia presentation and describes the framework that shall be used as a test bed for the new approach. Section 4 shows how the Analytic Hierarchy Process (AHP) can be used to obtain a QoP-oriented ordering of QoS parameters while Section 5 provides a practical example of our approach. Lastly, conclusions are drawn and possibilities for future work are identified in Section 6.

2 Background and Motivation

The networking foundation on which current distributed multimedia applications are built either do not specify QoS parameters (also known as best effort service) or specify them in terms of traffic engineering parameters such as delay, jitter, and loss or error rates. However, these parameters do not convey application-specific needs such as the influence of clip content and informational load on the user multimedia experience. As a result the underlying network does not consider the sensitivity of the application performance to bandwidth allocation. There is thus an architectural gap between the provision of network-level QoS and application-level QoP requirements of the distributed multimedia applications. This gap causes distributed multimedia systems to inefficiently use network resources and results in poor end-to-end performance which in turn has a direct negative impact on the user experience of multimedia.

Previous work in the area of subjective quality assessment of multimedia applications has concentrated on the entertainment side of multimedia. Accordingly, work has been done to appraise the enjoyment of multimedia presentations shown at differing frame rates [1], while [14] examined the effect that random media losses have on user-perceived quality. The bounds within which lip synchronisation can fluctuate without undue annoyance on the viewer's part have also been studied, [13], as has the establishment of metrics for subjective assessment of teleconferencing applications [14].

The inherent difficulty of trying to link subjective sentiments about the quality of the presentation with the facts and figures of network parameters is reflected in the relative scarcity of work trying to bridge the application-network gap. In essence, three approaches can be identified. The first approach tries to bridge the application-network gap *implicitly*. By this it is understood that there is no explicit mapping between application-level user requirements and the QoS provided by the network. What happens rather is that the user specifies, usually through a Graphical User Interface, his/her desired presentation quality [2], [10]. This is typically provided through the means of sliders or radio buttons via which the user would specify, for example, the desired playback frame rate, spatial resolution, as well as the acceptable synchronisation delay between the audio and video streams.

In the second approach an *explicit* mapping linking application-level user requirements to network QoS is actually given. Such a mapping can either be defined on a per layer basis, such as a network to transport to session to application-layer mapping, or directly between application and network-level parameters [**12**].

The last approach is, in essence, a more restrictive version of the first. What happens here is that the user is played short-duration *probes* of differing deliverable qualities of the multimedia material in question and (s)he then specifies which of the given sample qualities is acceptable [**8**]. Apart from the obvious goal of polling user preferred multimedia quality, the probe-based approach is advantageous from the point of view that it tests current network conditions. Thus, any choice that the user might make in as far as desired multimedia quality goes is guaranteed to be delivered - at least in the initial stages - by the network.

In this paper we address the problem of bridging the application-network gap from a multi-attribute decision making perspective. We have sought to use this approach to integrate results from our work on user-level Quality of Perception with the more technical characterisation of QoS. Our ultimate aim is to provide a communications architecture which uses an adaptable communications protocol geared towards human requirements in the delivery of distributed multimedia.

3 Evaluating User-perceived Quality of Service

Our approach to evaluating user-perceived QoS (QoP) has been mainly empirical, as is dictated by the fact that its primary focus is on the human-side of multimedia computing. What has been done is that users from diverse backgrounds and ages were presented with a set of 12 short (30 - 45 seconds' duration) multimedia clips in MPEG-1 format. These were chosen to be as varied as possible, ranging from a relatively static news clip to a highly dynamic rugby football sequence. All of them depicted excerpts from real world programs and thus represent informational sources which an average user might encounter in everyday life. Each clip was shown with the same set of QoS parameters, unknown to the user. After each clip, the user was asked a series of questions based on what had just been seen and the experimenter duly noted the answers. Lastly, the user was asked to rate the quality of the clip that had just been seen on a scale of 1 - 6 (with scores of 1 and 6 representing the worst and, respectively, best-perceived qualities possible).

Because of the relative importance of the audio stream in a multimedia presentation as well as the fact that it takes up an extremely low amount of bandwidth compared to the video it was decided to transmit audio at full quality during the experiments. Parameters were, however, varied in the case of the video stream. These include both spatial parameters (such as colour depth) and temporal parameters (frame rate). Accordingly, two different colour depths were considered (8 and 24-bit), together with 3 different frame rates (5, 15 and 25 frames per second - fps). A total of 12 users have been tested for each (*frame_rate, colour_depth*) pair. In summary the results (see [5] for a more detailed coverage) obtained in the QoP experiments show that:

- A significant loss of frames (that is, reducing the frame rate) does not proportionally reduce the user's understanding and perception of the presentation. In fact, in some instances (s)he seemed to assimilate more information, thereby resulting in more correct answers to questions. This is because the user has more time to view a frame before the frame changes (at 25 fps, a frame is visible for only 0.04 sec, whereas at 5 fps a frame is visible for 0.2 sec), hence absorbing more information. This observation has implications on resource allocation.
- User assimilation of the informational content of clips is characterised by the *wys<>wyg* (what you see is not what you get) relation. What this means is that often users, whilst still absorbing information correctly, do not notice obvious cues in the clip. Instead the reasoning process by which they arrive at their conclusions is based a lot on intuition and past experience.
- Users have difficulty in absorbing audio, visual and textual information concurrently. Moreover, if the user perceives problems with the presentation (such as lip synchronisation) users will disregard them and focus on the audio message if that is considered to be contextually important. This implies that critical and important messages in a multimedia presentation should be delivered in only one type of medium, or, if delivered concurrently, should be done so with maximal possible quality.
- Highly dynamic scenes, although expensive in resources, have a negative impact on user understanding and information assimilation. Questions in this category obtained the least number of correct answers. However the entertainment value of such presentations seems to be consistent, irrespective of the frame rate at which they are shown. The link between entertainment and content understanding is therefore not direct and this is further confirmed by the second observation above.

All these results indicate that technical-oriented QoS must also be specified in terms of perception, understanding and absorption of content - Quality of Perception in short - if multimedia presentations are to be truly effective.

3.1 A Framework for QoP Adaptation

Distributed guaranteed services need to incorporate capabilities for responding to QoP and QoS changes originating from the user/applications or the system/network. To achieve these changes, the networked multimedia system will require fast renegotiation protocols and adaptive mechanisms. The renegotiation protocols will rely on dependable and simple monitoring and recognition algorithms to detect requests for QoS changes or system degradations. The envisioned adaptive mechanisms should include update mechanisms for resource allocation in response to detection of system degradation. The challenge will be to make monitoring, detection and adaptation mechanisms efficient and fast [9].

The *Dynamically Reconfigurable Stacks Project* (DRoPS) provides an infrastructure for the implementation and operation of multiple adaptable protocols [4]. The core architecture is embedded within the Linux operating system, is accessible through standard interfaces, such as *sockets* and the UNIX *ioctl* (I/O control) system calls, has direct access to network devices and benefits from a protected, multiprogramming environment. The architecture allows additional QoS maintenance techniques, such as flow shaping (to smooth out bursts in traffic), at the user or interface level, and transmission queue scheduling, at the device queue level.

DRoPS-based communication protocols are composed of fundamental mechanisms, called *microprotocols*, which perform arbitrary protocol processing operations. The complexity of processing performed by a microprotocol is not defined by DRoPS and may range from a simple protocol function, such as a checksum, to a complex layer of a protocol stack, such as TCP. In addition, protocol mechanisms encapsulated within a microprotocol may be implemented in hardware or software. If appropriate hardware is available, the microprotocol merely acts as a wrapper, calling the relevant hardware function. Microprotocols are encapsulated in loadable modules, allowing code to be dynamically loaded into a running operating system and executed without the need to recompile a new kernel. Each such microprotocol can be implemented via a number of adaptable functions. For instance, an acknowledgement protocol can be implemented either as an Idle Repeat Request or a Per Message Acknowledgement Scheme.

4 User-centred Design with Multi-criteria Constraints

In linking perceptual considerations with low-level technical parameters, the design process should take into account the subjective judgement of the end-user. The end-user may not clearly specify a desired parameter value and prefers the use of linguistic phrases to describe the objectives, e.g. *"A is equally important as B"*; *"A is slightly more important than B"*, or more complicated ones, such as *"How much more preferable is a flow control protocol when compared to a checksum algorithm with respect to obtaining a satisfactory audio quality?"*; *"How important is a specific acknowledgement scheme for user perception of video?"*. This does result in inherent imprecision. Although information about questions like the previous ones is vital in making correct design decisions, it is very difficult, if not impossible, to quantify them correctly, i.e. the main problem is how to quantify the linguistic choices selected by the decision-maker during the evaluation of the pairwise comparisons. All methods that use this approach eventually express the qualitative answers of a decision-maker into some numbers. To this end, we have applied Saaty's AHP formalism [11] to obtain a method which, from combined user-, application- and network-level requirements, ultimately results in a protocol configuration specifically tailored for the respective user-needs. Thus, within the QoP framework, each multimedia application can be characterised by the relative importance of the video (V), audio (A) and textual (T) components as conveyors of information, as well as the dynamism (D) of the presentation. This is in accordance with the experimental QoP results obtained which emphasise that multimedia QoP varies with: the number of media flows, the type of medium, the type of application, and the relative importance of each medium in the context of the application. On the other hand, 5 network level QoS parameters have been considered in our model: bit error (*BER*), segment loss (*SL*), segment order (*SO*), delay (*DEL*) and jitter (*JIT*). Together with the *V*, *A*, *T* and *D* parameters these constitute, in Saaty's terminology, the *criteria* on the basis of which an appropriate tailored communication protocol is constructed. In DRoPS, the functionality of this protocol is realised through a number of 9 microprotocols, spanning 4 broad functionality classes [6].

By applying Saaty's methodology we obtain a total of 10 matrices. To this end, the decision-maker (both the designer and the user) has to express his/her opinion about the value of a single pairwise comparison at a time. Nine of these matrices give the relative importance of the various microprotocols (*alternatives*, in Saaty's vocabulary) with respect to the criteria identified in our model, while the last of these matrices details *pairwise comparisons* between the criteria themselves.

$$ALTwrtBER = \begin{bmatrix} 1 & 1 & 1 & 1 & 1 & 1 & 1 & 1/3 & 1/9 \\ 1 & 1 & 1 & 1 & 1 & 1 & 1 & 1/3 & 1/9 \\ 1 & 1 & 1 & 1 & 1 & 1 & 1 & 1/3 & 1/9 \\ 1 & 1 & 1 & 1 & 1 & 1 & 1 & 1/3 & 1/9 \\ 1 & 1 & 1 & 1 & 1 & 1 & 1 & 1/3 & 1/9 \\ 1 & 1 & 1 & 1 & 1 & 1 & 1 & 1/3 & 1/9 \\ 1 & 1 & 1 & 1 & 1 & 1 & 1 & 1/3 & 1/9 \\ 3 & 3 & 3 & 3 & 3 & 3 & 3 & 1 & 1/5 \\ 9 & 9 & 9 & 9 & 9 & 9 & 9 & 5 & 1 \end{bmatrix} \tag{1}$$

The formulation of the 10 matrices has been based on the results of the experiments described in Section 3. An example of one of the former type of matrices, i.e. of the different alternatives with respect to one of the criteria (bit error rate in this case) is given in Relation (1). Each entry a_{ij} of the matrix *ALTwrtBER* defines the numerical judgement made by the designer and/or the user in comparing criterion to criterion. Psychological experiments have shown that individuals cannot simultaneously compare more than 7 objects (± 2) [7]. Thus, usually, pairwise comparisons are quantified by using a scale of nine grades, which describe the relative importance of the criteria [11]. If a_{ij} is a point on this nine-point scale, i.e. $a_{ij} \in \{1,2,3,...,8,9\}$, then $a_{ji} = 1/a_{ij}$ also holds [11]. For example, in Relation (1) the judgement *"microprotocol A is equally important as microprotocol B with respect to BER"* corresponds to a weighting of $a_{ij} = 1$, while the judgement *"microprotocol A is absolutely more important than microprotocol B"* would correspond to a value of $a_{ij} = 9$. Intermediate terms can also be assigned when compromise is needed between two adjacent characterisations. Note that in Relation (1), the considered microprotocols are, in order {*no sequence control, strong sequence control, no flow control, window-based flow control, IRQ, PM-ARQ, no checksum algorithm, block checking, full Cyclic Redundancy Check*}. For example, as far as bit error rate is concerned (see Relation 1), the only microprotocols that have an impact upon it are the checksum algorithms. The strongest of these methods, the full CRC, has the highest weight (a value of $a_{ij} = 9$) in comparison with all the others, while a relatively weak block checking algorithm is considered to be *moderately more important* ($\alpha_{8j}=3$, j=1,2,...,7) than microprotocols from other functionality classes.

While all the previous nine matrices considered have a constant form, the matrix of each criterion with respect to all the other criteria named *CRwrtCR*, shown below, is the only one whose values may fluctuate as a result of changes in the operating environment, as well as a consequence of changes in *user preferences* and *perceptions*. Relation (2) provides an instance of this matrix used in our model; the respective criteria are, in order, *BER, SO, SL, DEL, JIT, V, A, T* and *D*.

$$
CRwrtCR=
\begin{bmatrix}
1 & 1/2 & 1 & 1/4 & 1 & | & 2 & 4 & 4 & 3 \\
2 & 1 & 1 & 1/4 & 1/3 & | & 5 & 4 & 5 & 4 \\
1 & 1 & 1 & 1/3 & 1/2 & | & 4 & 6 & 4 & 4 \\
4 & 4 & 3 & 1 & 5 & | & 6 & 7 & 6 & 5 \\
1 & 3 & 2 & 1/5 & 1 & | & 4 & 6 & 6 & 6 \\
- & - & - & - & - & + & - & - & - & - \\
1/2 & 1/5 & 1/4 & 1/6 & 1/4 & | & 1 & 1/2 & 3 & 2 \\
1/4 & 1/4 & 1/6 & 1/7 & 1/6 & | & 2 & 1 & 3 & 2 \\
1/4 & 1/5 & 1/4 & 1/6 & 1/6 & | & 1/3 & 1/3 & 1 & 1 \\
1/3 & 1/4 & 1/4 & 1/5 & 1/6 & | & 1/2 & 1/2 & 1 & 1
\end{bmatrix}
\tag{2}
$$

An average user, though, would have difficulty in *a priori* judgement of varying technical parameters such as delay, jitter, error and loss rates on highly subjective attributes such as perception, understanding and satisfaction. Whilst this is true for QoS attributes at the level of the transport service, users are better able to quantify their requirements in terms of more abstract characteristics like the prioritisation of core multimedia components such as *V, A* and *T*. The matrix in Relation (2) reflects this situation and could conceptually be split-up into 4 sub-matrices, which are:

- A 5×5 matrix, in the upper left part of matrix (2), giving the relative importance of the *BER, SO, SL, DEL* and *JIT* criteria with respect to one another. This matrix changes dynamically during the course of the transmission of a multimedia clip. For example, as a result perhaps of a delay-intolerant audio application being subjected to a period of high network delays, the sub-matrix can reflect this situation by changing the numerical judgements to reflect a more radical bias in favour of the delay component.

- A 4×4 matrix, located in the bottom right of the matrix given by (2). Here user input can reflect personal choices of the relative importance of the video, audio and textual components in the context of the application, as well as a relative characterisation of the dynamism of the multimedia clip. Whilst, these values can be changed dynamically depending on the visualised scene, *a priori* values in this case could reflect the result of user- consultations. This indeed is the case with our QoP experiment [5], where a broad base of people were polled about their opinions for a wide range of clips with a variety of subject matter (action movie, animated movie, band, chorus, commercial, cooking, documentary, news, pop music, rugby, snooker, weather forecast). Thus, users characterised the News clip as being relatively static, i.e. dynamism is *Low*, and considered that the video and audio media components conveyed a *High* informational load, in contrast to the text component which was judged to have a *Medium* informational weight.

- A 5×4 matrix and a 4×5 (one of which is the transpose of the other) which reflect designer choices of the relative importance of the five QoS parameters considered on V, A, T and D. The elements of these matrices remain fixed at run-time, and, in our particular case, reflect the results of our previous work on QoP [5].

Following the AHP, the weights w_i, $i=1,...9$ denoting the relative importance of each criterion i among the p criteria ($p=9$) are evaluated using the formula:

$$w_i = \frac{\left(\prod_{j=1}^{p} a_{ij}\right)^{1/p}}{\sum_{i=1}^{p}\left(\prod_{j=1}^{p} a_{ij}\right)^{1/p}} \quad i = 1,2,...,p \tag{3}$$

and a higher priority setting corresponds to a greater importance.

Pairs among alternatives are also compared with respect to the ith criterion and then a weight $w_{j,i}$, which denotes how preferable is the alternative j with respect to the criterion i, is derived. As previously, there is a total of p pairwise comparisons in the matrix and weights are calculated following Relation (3). The weighted product model [3] is used to compare alternative v with alternative u. This can be done by multiplying a number of ratios, one for each criterion. Each ratio is raised to a power, which is equivalent to the relative weight of the corresponding criterion, i.e.

$$P_{v/u} = \prod_{i=1}^{p}\left(\frac{w_{v,i}}{w_{u,i}}\right)^{w_i} \tag{4}$$

If the ratio $P_{v/u}$ is greater or equal to one then the conclusion is that the alternative v is more preferable than alternative u. In the maximisation case, the best alternative is the one that possesses the highest value among all others.

5 Application Example

For the 12 multimedia clips considered in the QoP experiments, we have mapped the terms *Low, Medium* and *High* to the numeric values of 2, 4 and 8, respectively. However, these represent *absolute* values within the context of each clip. The AHP formalism, however, concerns itself with *pairwise* comparisons between components. These were provided in our case by the ratios of the numeric values corresponding to the components being pairwise compared. Thus, if a multimedia category is adjudged to have a high video informational content and a low textual one, then the pairwise comparison between the two will have a value of 8/1 = 8. Analogously, if, in the same clip, one compared the video with a medium audio informational component, then the corresponding ratio will be 8/4 = 2. These are the considerations upon which the *a priori* 4×4 matrices in the bottom right of Relation (2) were generated. Of course, during the course of visualising the respective clips, a user can tune these parameters to better suit his/her preferences (see Figure 1).

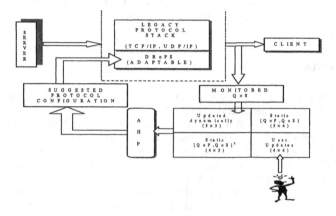

Fig. 1. AHP-based architecture for QoP management

The applicability of the proposed approach can be illustrated by means of a multimedia user scenario: a currently executing BER-sensitive application suddenly experiences deterioration in BER. Then this parameter becomes absolutely important in Relation (2) with respect to the others i.e., $a_{1j}=9$ for $j=2,...,5$. This value is dynamically updated in the top-left 5×5 matrix (see Figure 1). The result of the AHP (calculations are omitted due to space limitation) then shows that the microprotocol that should preferentially be managed under the circumstances is the full CRC - the strongest such microprotocol available to correct the quality loss in BER. This is in contrast to the scenario applicable just before the sudden surge in BER, which gave preference to the Strong Sequence Control microprotocol as a result of applying Relations (3) and (4) on the matrix shown in Relation (2).

6 Conclusions

We have presented a method of obtaining, in the context of a multimedia application, a priority order of low-level QoS parameters, which would ensure that user level QoP is maintained at an optimum level. In our case, such QoP management would be ensured through dynamic protocol adaptation, as a result of changes in user preferences or the operating environment. Our approach has distinct advantages over previous work. Firstly, since it is already based on the results of extensive QoP tests, users themselves do not necessarily need to specify preferences in the interface of an application incorporating the mapping, thereby resulting in a more streamlined viewing process. It thus permits the maintenance of high-level QoP (without user interference) as low-level QoS varies. Secondly, since the framework includes results linking a human's ability to perceive, analyse and synthesise informational results, it is a more comprehensive approach than preceding attempts. The viability itself of actually implementing this approach in practice and using it in a QoP-oriented management scheme for QoS parameters will be further investigated within the DRoPS framework.

References

1. R.T. Apteker, J.A. Fisher, V.S. Kisimov, H. Neishlos: Video Acceptability and Frame Rate, IEEE Multimedia, 2(3), (1995) 32-40.
2. S. Ardon, R. DeSilva, B. Landfeldt, A. Seneviratne: Total Management of Transmissions for the End-User, a Framework for User Control of Application Behaviour, Proc. 4th Int. Workshop on High Performance Protocol Architectures, London, June 1998.
3. S.J. Chen & C.L. Hwang: Fuzzy multiple decision making, Lecture Notes in Economics and Mathematical Sciences, Vol. 375, NY, Springer, 1992.
4. R.S. Fish & R.J. Loader: A kernel based adaptable protocol architecture, Proc.of the 24th Euromicro Conference, Västerås, Sweeden (1998) 1029-1036
5. G. Ghinea & J.P. Thomas: QoS Impact on User Perception and Understanding of Multimedia Video Clips, Proc. of ACM Multimedia '98, Bristol, UK, 1998.
6. G. Ghinea, J.P. Thomas, R.S. Fish: Quality of Perception to Quality of Service Mapping Using a Dynamically Reconfigurable Communication System, Proc. of IEEE Globecom '99, Rio de Janeiro, Brazil
7. G.A. Miller: The magical number seven plus or minus two: some limits on our capacity for processing information, Psychological Review, 63, (1956) 81-97.
8. K. Nahrstedt, K., A. Hossain, S. Kang: A Probe-based Algorithm for QoS Specification and Adaptation, Proc. 4th Int. Workshop on QoS, 115-124, Paris, France, 1996.
9. K. Nahrstedt: Challenges of Providing End-to-End QoS Guarantees in Networked Multimedia Systems, ACM Computing Surveys Journal, December 1995
10. D. Reininger, D. Raychaudhuri, M. Ott: A Dynamic Quality of Service Framework for Video in Broadband Networks, IEEE Network, Nov/Dec (1998) 22 - 34.
11. T. Saaty, "The Analytic Hierarchy Process", McGraw-Hill, New York (1980.)
12. Seneviratne, M. Fry, V. Withanan, V. Sapapramdu, A. Richards, E. Horlait: Quality of Service Management for Distributed Multimedia Applications, Proc. of the Phoenix Conf. on Computer and Communications, Phoenix, Arizona, IEEE Press, 1994.
13. R. Steinmetz: Human Perception of Jitter and Media Synchronisation, IEEE Journal on Selected Areas in Communications, 14(1), (1996) 61 - 72
14. A. Watson & M.A. Sasse: Measuring Perceived Quality of Speech and Video in Multimedia Conferencing Applications, Proc. of ACM Multimedia, Bristol, U.K., 1998
15. D. Wijesekera & J. Srivastava: Quality of Service (QoS) Metrics for Continuous Media, Multimedia Tools and Applications, 3(1) (1996) 127-166.

Integrating Random Ordering into Multi-heuristic List Scheduling Genetic Algorithm

Andy Auyeung, Iker Gondra, and H. K. Dai

Department of Computer Science, Oklahoma State University
Stillwater, Oklahoma 74078, USA
{wingha, gondra, dai}@cs.okstate.edu

Abstract. This paper presents an extension of a Multi-heuristic List Scheduling Genetic Algorithm. The problem is to find a schedule that will minimize the execution time of a program in multi-processor platform. Because this problem is known to be NP-complete, many heuristics have been developed. List Scheduling is one of the classical heuristic solutions. The idea of Multi-heuristic List Scheduling Genetic Algorithm is to find an "optimal" combination of List Scheduling heuristics, which outperforms only one heuristic. However, a very important drawback of this algorithm is that it does not take all the search space into consideration. We improved it by introducing random ordering that allows it to cover the entire search space while at the same time focusing on the heuristics region. An experimental comparison is then made to against both Multi-heuristic List Scheduling Genetic Algorithm and Combined Genetic-List Algorithm.

1 Introduction

The multi-processor scheduling problem is that of finding an assignment of tasks to processors and an execution order so that a performance criterion (usually running time) is optimized. With the exception of some special cases, it is known to be NP-complete [1]. In this paper, we devote our attention to deterministic scheduling; that is, task execution times and precedence relations between tasks are fixed and known before running time. We assume that precedence relations among tasks in a program are represented as a directed acyclic graph (DAG).

Because of the intractability of the problem, several heuristic-based methods have been proposed that can obtain optimal or near-optimal solutions under different assumptions [2]-[7]. Hellstrom and Kanal [2] mapped the multi-processor scheduling problem into a neural network model by using asymmetric mean-field networks. Hou et al. [4] first introduced a genetic algorithm (GA) for task scheduling (HGA). The main drawback of HGA is that it cannot efficiently search all valid schedules due to the limitations of its crossover operator. In HGA, a chromosome (i.e., a schedule or a solution) is represented by a set of m strings, where m is the number of processors. Each string represents the particular tasks assigned to the corresponding processor, and the execution order of the tasks is determined by the order of appearance in the string. It is easy to observe that, under this representation, randomly generated chromosomes (from the initial population) or chromosomes obtained after performing crossover may not represent valid schedules since the precedence constraints may be violated. In order to overcome this problem, the following method was proposed in HGA. Each task t in the DAG is assigned a random *height* whose value is in the range $hp(t) < height(t) < hs(t)$, where $hp(t)$ is the maximum length of a path between a task with no immediate predecessors and an immediate predecessor of t, and $hs(t)$ is the maximum length of a path between a task with no immediate predecessors and an immediate successor of t. If there is a directed edge from task t_i to task t_j in the DAG, then $height(t_i) < height(t_j)$. Therefore, ordering the tasks in a string according to their *height* guarantees the validity of a schedule.

Corrêa et al. [5] observed three main drawbacks of HGA. First, because of the scheme used to generate the initial random population, the distribution of tasks over the processors is not uniform. Second, because crossover is performed by choosing a random *height*, which is used as a threshold for assigning tasks to either a left or right part, and then exchanging the right parts of both parents, many valid schedules cannot be generated. In fact, Corrêa et al. observed that the search space might contain no optimal solution. Third, they indicated that HGA does not use any problem-specific information that could be used to improve the quality of the solutions that are generated. In fact, the knowledge that is used is a structural one and it consists of verifying the validity of solutions (i.e., ensuring that precedence constraints are satisfied). Corrêa et al. developed Full Search GA (FSG) [5] to overcome the first and the second drawback. To overcome the third drawback, they developed Combined Genetic-List Algorithm (CGL) which integrates knowledge about the multi-processor scheduling problem into FSG in the form of new crossover and mutation operator-based on List Scheduling heuristics. The results of the experiments conducted in [5] showed that CGL performed much better than either FSG or HGA in terms of the quality of the solutions (i.e., schedules with a much shorter execution time were found). However, the running time of CGL was much higher than that of HGA.

List Scheduling maintains a list of "ready" tasks ordered by relative priorities according to some optimization criteria. Whenever a processor becomes available, the task with the highest priority is fetched and assigned to the processor. In Multi-heuristic List Scheduling GA (MHG) [10], four common List Scheduling heuristics for assigning priorities to tasks are used. Briefly, it considers combinations of four List Scheduling heuristics and uses GA to find good combinations.

In this paper, we present an extension of Multi-heuristic List Scheduling GA. The original algorithm is described in detail in [10]. The main improvement consists of introducing random ordering that allows the algorithm to cover the entire search space, while at the same time focusing on a smaller subset corresponding to combinations of the four List Scheduling heuristics.

The remainder of this paper is organized as follows. First, in Section 2, we present the problem of task scheduling on a multi-processor system and introduce

List Scheduling along with four common heuristics. A brief introduction to GAs is also presented. In Section 3, the main ideas of MHG are reviewed and an example where it cannot find an optimal solution is presented. Multi-heuristic with Random Ordering List Scheduling GA (MRG) is then introduced in Section 4. Finally, the experimental comparison of MRG with both MHG and CGL is presented.

2   Preliminaries

2.1   Multi-Processor Scheduling

The multi-processor scheduling problem consists of determining an assignment of tasks to processors and the execution order of the tasks so that execution time is minimized. We assume that there are m identical processors connected by a complete interconnection network and with communication costs ignored. There is a set of n tasks with arbitrary precedence constraints and processing time. The set of tasks is represented as a DAG, $G = (V, E)$, where each node in V represents a task ($|V| = n$) and E is the set of directed edges connecting the nodes. That is, $E = \{(t_i, t_j)$ | a precedence relation exists between tasks t_i and t_j }. A simple DAG along with a valid scheduling is illustrated in Figure 1.

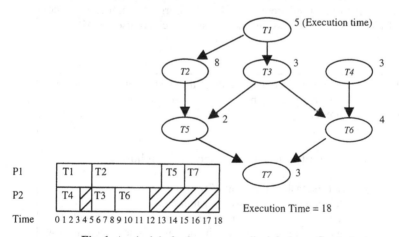

Fig. 1. A schedule for 2 processors displayed as a Gantt chart

2.2   List Scheduling

The problem of scheduling tasks on a multi-processor system is known to be NP-complete for general cases. As a result, several heuristic-based methods have been proposed in the literature [2] – [7]. One common heuristic-based method, to which many schedulers are classified, is List Scheduling. List Scheduling derives its name from the fact that a list of "ready" nodes (i.e., nodes with no unscheduled predecessors) is maintained and ordered by relative priorities based on an underlying heuristic. Whenever a processor becomes available, the highest priority node in the list is chosen and assigned to the processor. This process is then repeated until all the nodes have been scheduled. The schedulers in this class are classified according to the particular heuristic that is used to assign priorities to nodes and to select the best processor to run the task [1]. Algorithm A below shows the general List Scheduling algorithm.

The particular assignment of priorities to tasks results in different schedules because tasks are selected for execution in different orders. The length of a path in a DAG is defined to be the sum of the execution times of all nodes along the path including the initial and final nodes. The level of a node in a DAG is the length of the longest path from the node to an exit node (a node with no successors). The co-level of a node in a DAG is similar to its level except that lengths are measured from the start nodes of the DAG instead of from the exit node. Figure 2 shows a DAG along with the levels and co-levels of all the tasks.

The results of empirical performance studies [3] on List Scheduling indicate that level-based heuristics are the best at approximating the optimal schedule. Below, we introduce HLF (Highest Level First) and HCF (Highest Co-level First) as well as two commonly used heuristics, LNSF (Largest Number of Successors First) and LPTF (Largest Processing Time First), which are used in MHG.

1	Assign a priority to each node in the DAG. Initialize a priority queue for ready tasks by inserting every task that has no immediate predecessors. Sort tasks in decreasing order of task priorities.
2	While priority queue is not empty do the following:
	a) Get the task at the front of the queue.
	b) Select an idle processor to run the task.
	c) After all immediate predecessors of a task are executed, insert that successor into the priority queue since it is now ready.

Algorithm A. General List Scheduling Algorithm [1]

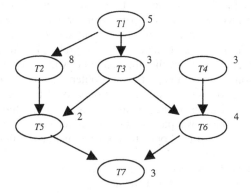

TASK	1	2	3	4	5	6	7
LEVEL	18	13	10	10	5	7	3
CO-LEVEL	5	13	8	3	15	12	18

Fig. 2. Levels and co-levels of tasks in a DAG

The path yielding the level of a node in a DAG is called the "critical path" be-cause any delay of a task on this path will increase the total execution time. There-fore, HLF attempts to minimize program execution time by avoiding delays in the critical path; HCF is based on a similar idea. The motivation behind LNSF is that, by scheduling a task with the largest number of successors, the number of "ready" nodes is maximized. Therefore, idle time will be less likely. LPTF's idea is similar to that of HLF. By getting the longest tasks done first, LPTF minimizes delays in long tasks.

2.3   Genetic Algorithms

The basic principles of GAs were established by John Holland in 1975. Holland's insight was to be able to represent the fundamental biological mechanisms that permit system adaptation into an abstract form that could be simulated on a com-puter for a wide range of problems. He introduced bit strings to represent feasible solutions (or individuals) in some problem space. GAs are analogous to the natural behavior of living organisms. Individuals in a population compete for resources. Those individuals that are better adapted survive and have a higher probability of mating and generating descendants. Therefore, the genes of stronger individuals will increase in successive generations.

A GA works with a population of individuals, each representing a feasible so-lution to a given problem. During each iteration step, called a generation, the indi-viduals in the current population are evaluated and given a fitness value, which is proportional to the "goodness" of the solution in solving the problem. Individuals are represented with strings of parameters or genes known as chromosomes.

The phenotype, the chromosome, contains the information that is required to construct an individual (a solution to the problem). The phenotype is used by the fitness function to determine the genotype, which denotes the level of adaptation of the chromosome to the particular problem. To form a new population, individuals are selected with a probability proportional to their relative fitness. This ensures that well adapted individuals (good solutions) have more chances of being reproduced. Once two parents have been selected, their chromosomes are combined and the traditional operators of crossover and mutation are applied to generate new individuals.

GAs have been successful in solving complex problems that are not easily solved through conventional methods [9] for several reasons. They start with a population of points rather than a single point. Therefore, many portions of the domain are searched simultaneously and, as a result, they are less prone to settling at local optima during the search. GAs work with an encoding of the parameter set, not the parameters themselves. Because they do not depend on domain knowledge in performing the search, inconsistent or noisy domain data are less likely to affect them as is common with hill-climbing or domain-specific heuristics [9].

3   The Multi-heuristic List Scheduling Genetic Algorithm

3.1   Overview

Given the problem of finding a schedule by using a combination of the four heuristics: level, co-level, number of successors, and processing time, MHG finds a new heuristic, which is based on a good weighting of the four heuristics. In other words, suppose that we are given a particular instance of the multi-processor scheduling problem (i.e., a DAG and a number of processors). If, for example, the weight given to HLF is greater than zero and the weight of the remaining three heuristics is zero, then the priority numbers assigned to each one of the tasks will be based on HLF alone. On the other hand, if there are at least two heuristics with weights greater than zero, then the priority number assigned to each task will be a sum of the weighted priorities given by the heuristics with nonzero weights.

3.2   Shortfalls

By only considering combinations of List Scheduling heuristics, MHG focuses on the most promising region of the solution space. However, a reduced search space may not contain the optimal solution and in some cases it may even be far away from good solutions. For instance, Adam et al [3] demonstrated experimentally that HLF heuristic is within 5% of the optimal in 90% of cases. Therefore, it would be reasonable to believe that the subset of the solution space corresponding to schedules generated by using HLF is a promising region in the sense that, most of the time, it contains the optimal or a near-optimal solution. The results of experiments conducted in [10] show that schedules found by MHG outperform those found by each one of the four List Scheduling heuristics alone. Therefore, this argument can also be applied to combinations of the four heuristics. However, by reducing the solution space we run the risk of missing good solutions (i.e., good schedules for a particular DAG). As a trivial example of this observation, let us use the DAG in Figure 3, where there are no precedence relations among the tasks. Suppose that there are two processors. Because every task has no predecessors and no successors, the values for the level and co-level are equal to the processing time (so if a task has the largest processing time, it will also have the largest level and co-level) and the value for the number of successors is the same for all the tasks (which is zero). Therefore, the weights assigned for each of the four List Scheduling heuristics can be ignored since the resulting scheduling will be decided by the processing times of the tasks alone. Figure 3 shows a schedule found by MHG and an optimal schedule. We can observe that, regardless of the values for the weights of the four List Scheduling heuristics, MHG will never find an optimal schedule for this particular DAG. The results of our experiments (presented in Section 5) suggest that the percentage of DAGs with optimal schedules outside the "region" covered by MHG is significant.

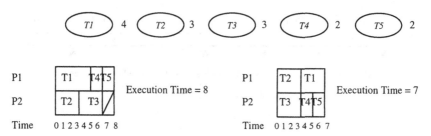

Fig. 3. Sample DAG and 2 possible schedules

In short, there is a tradeoff between exploitation (of the most promising region of the solution space) and exploration (of the entire solution space). MRG presented in the next section uses random ordering that allows for the exploration of the entire solution space while focusing on the most promising region.

4   Multi-heuristic with Random Ordering List Scheduling Genetic Algorithm

In MRG, a solution to the problem is a weighting of the four heuristics (HLF, HCF, LNSF, LPTF) and a fifth "heuristic", which we call RTF (Random Task First). The objective of the GA is to find an optimal weighting of the five heuristics. RTF assigns priorities to tasks randomly. That is, a "seed" is used to generate a sequence of "random" numbers representing RTF priorities, which are assigned to the tasks in the DAG. Because the number representing the "seed" for the pseudo-random number generator is also included as part of a chromosome, a solution (i.e., a schedule) is completely determined by and can be recovered from the corresponding chromosome. Therefore, a chromosome represents a set of six ordered numbers. With the exception of the "seed", a number and its position in the chromosome corresponds to the weight assigned to the corresponding heuristic. MRG uses a one-dimensional array of 64 bits to represent the chromosome. The solution (heuristic) is encoded by dividing the chromosome into five 8-bit cells and a 24-bit cell. The 8-bit cells correspond to the weights (a number with range 0 to 255), and the 24-bit cell corresponds to the "seed". The index of the cell indicates the heuristic to which the corresponding weight should be assigned. Figure 4 below shows the representation of MRG:

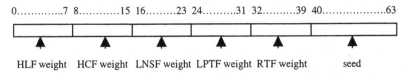

Fig. 4. Solution Encoding

The second improvement made to the algorithm is that the priority value assigned by each one of the heuristics to a task is normalized. That is, instead of assigning the magnitude of a heuristic's priority value, a normalized value in the range 1 to n is assigned. Because all five heuristics assign values in different ranges, normalizing to a common range is a more fair approach.

We use the classical one-point crossover and bit-flip mutation as the genetic operators. In one-point crossover the chromosomes of the parents are cut at some randomly chosen common point and the resulting sub-chromosomes are swapped. In bit-flip mutation each gene (bit) in a chromosome is flipped with a small probability given by the mutation rate. The fitness function is a procedure that does a Multi-heuristic List Scheduling based on the particular heuristic encoded in the input chromosome and returns the completion time of the last task being executed. Algorithm B below shows the evaluation function of MRG.

1	For each node x in the DAG compute the following: $p(x)=(p_{HLF}*w_{HLF})+(p_{HCF}*w_{HCF})+(p_{LNSF}*w_{LNSF})+(p_{LPTF}*w_{LPTF}) + (p_{RTF}*w_{RTF})$ where p(x) is the priority assigned to x $\quad\quad$ p_{HLF} is priority HLF assigns to x (level of x), normalized $\quad\quad$ p_{HCF} is priority HCF assigns to x (co-level of x), normalized $\quad\quad$ p_{LNSF} is priority LNSF assigns to x (successors of x), normalized $\quad\quad$ p_{LPTF} is priority LPTF assigns to x (processing time of x), normalized $\quad\quad$ p_{RTF} is priority RTF assigns to x (x^{th} random number), normalized $\quad\quad$ w_{HLF} is the weight assigned to HLF $\quad\quad$ w_{HCF} is the weight assigned to HCF $\quad\quad$ w_{LNSF} is the weight assigned to LNSF $\quad\quad$ w_{LPTF} is the weight assigned to LPTF $\quad\quad$ w_{RTF} is the weight assigned to RTF
2	Initialize a priority queue for ready tasks by inserting every task that has no immediate predecessors. Sort tasks in decreasing order of task priorities (use task identification as tiebreaker).
3	While priority queue is not empty do the following: $\quad$ a)$\quad$ Get the task at the front of the queue $\quad$ b)$\quad$ Select an idle processor to assign to the task $\quad$ c)$\quad$ After all immediate predecessors of a task are scheduled, insert $\quad\quad\quad$ that successor into the priority queue since it is now ready
4	Return the completion time of the last task scheduled

Algorithm B. Evaluation function of MRG

5   Experimental Results

In order to verify the effectiveness of MRG, we compared MHG, CGL and MRG on randomly generated DAGs (and random number of processors $m > 1$). The algorithms were written in C++ and the simulations were performed on a SUN E3000 with two UltraSparc 250 MHz processors. We compared the execution times of schedules found by the algorithms (Figure 4). We have also measured the times (in seconds) required for finding schedules (i.e., the time until convergence to a solution) (Figure 5). The GA control parameters for MHG, MRG, and CGL are as follows: Maximum number of generations: 100, Population size: 30, Crossover rate: 0.5, and Mutation rate: 0.2

From the figures below we can observe that both MHG and MRG performed better than CGL in terms of the time needed to find a solution and the execution time of the scheduling. Because of the high time complexity of CGL, we did not run the simulation for more than 150 tasks. However, MHG and MRG continue to have a behavior similar to that of Figures 5 and 6 for more than 150 tasks. We observed that the difference in execution time between MHG and MRG was very small (practically insignificant), even though, most of the time, MHG found slightly better schedules than MRG in terms of the execution time of the schedules. We expected this to occur since MHG focuses on the most promising region of the search space while MRG has some departures from it. However, 12.3% of the schedules found by MRG had a shorter execution time than those found by MHG. Therefore, the experimental results confirm the effectiveness of the improvements made.

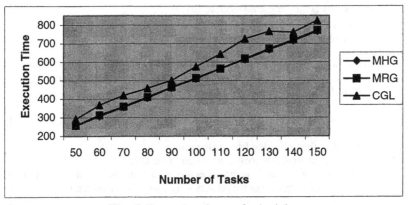

Fig. 5. Execution times of schedules

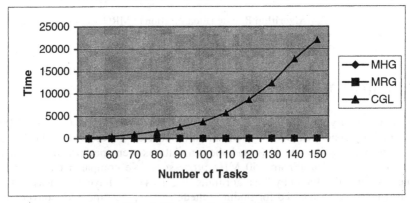

Fig. 6. Time to find schedules

6   Concluding Remarks

This paper presented an extension of Multi-heuristic List Scheduling GA proposed in [10] for solving the multi-processor scheduling problem. By focusing only on combinations of List Scheduling heuristics, MHG ran the risk of missing good solutions. We improved it by introducing random ordering that allows it to cover the entire search space while at the same time focusing on a smaller subset corresponding to combinations of List Scheduling heuristics. This allows for the exploration of the entire solution space while focusing on the most promising region. The results of our experiments showed that the MHG and MRG perform better than CGL in terms of both the times needed to find a schedule and its execution time. The reason for the comparatively low time complexity of both MHG and MRG is the fact that they use a very simple encoding scheme to represent a solution, which allows for the implementation of an unrestricted crossover operator. The comparison also shows that, on average cases, MHG and MRG perform indifferently well in terms of both the time needed to find a schedule and its execution time. However, MRG does provide coverage of schedules in situations where MHG fails to perform. Furthermore, during the simulation, MRG outperforms MHG in a significant percentage of the cases. This suggests that the percentage of DAGs with optimal schedules outside the "region" covered by MHG is significant. An interesting extension of this work would be to allow different weight combinations apply to different sub-graphs of the given DAG. We believe that might further improve the quality of solutions.

References

[1] A. Y. Zomaya, *Parallel and Distributed Computing Handbook*. McGraw-Hill, 1996.

[2] B. Hellstrom and L. Kanal, Asymmetric mean-field neural networks for multiprocessor scheduling. *Neural Networks*, Vol. 5, pp. 671-686, 1992.

[3] T. L. Adam, K. M. Chandy and J. R. Dickson. A comparison of list schedules for parallel processing systems. *Comm. ACM*, Vol. 17, pp. 685-690, 1974.

[4] E. Hou, N. Ansari and H. Ren, A genetic algorithm for multiprocessor scheduling. *IEEE Transactions Parallel and Distributed Systems*, Vol. 5, no. 2, pp. 113-120, Feb 1994.

[5] R. Corrêa, A. Ferreira and P. Rebreyend, Scheduling multiprocessor tasks with genetic algorithms. *IEEE Transactions on Parallel and Distributed Systems*, Vol. 10, No. 8, pp. 825-837, 1999.

[6] C. L. Chen, C.S.G. Lee and E.S.H. Hou, Efficient scheduling algorithms for robot inverse dynamics computation on a multiprocessor system. *IEEE Transactions System, Man, Cybernetics*, Vol. 18, pp. 729-743, 1998.

[7] Y. K. Kwok and I. Ahmad, Dynamic critical-path scheduling: An effective technique for allocating task graphs to multiprocessors. *IEEE Transactions Parallel and Distributed Systems*, Vol. 7, No. 5, pp. 506-521, 1996.

[8] H. Holland. *Adaptation in Natural and Artificial Systems*. University of Michigan Press, 1975.

[9] David J. Stracuzzi, Some methods for the parallelization of genetic algorithms. *http://ml www.cs.umass.edu/~stracudj/genetic/dga.html.*

[10] Andy AuYeung, Iker Gondra and H. K. Dai, Multi-heuristic list scheduling genetic algorithm for task scheduling. *The Eighteenth Annual ACM Symposium on Applied Computing*, pp. 721-724, 2003.

[11] N. J. Nilsson, Artificial Intelligence: A New Synthesis. Morgan Kaufmann Publishers, Inc., 1998.

[12] K. Hwang, Advanced Computer Architecture Parallelism, Scalability, Programmability. McGraw-Hill, 1993.

[13] S. J. Russell and P. Norvig, Artificial Intelligence: A Model Approach. Prentice Hall, 2002.

[14] G. J. Nutt, Centralized and Distributed Operating System. Prentice Hall, 1992.

[15] D. L. Galli, Distributed Operating System: Concepts and Practice. Prentice Hall, 2000.

[16] G. R. Andrews, Foundations of Multithreaded, Parallel and Distributed Programming. Addison Wesley Longman, Inc., 2000.

[17] R. W. Conway, W. C. Maxwell and L. W. Miller, Theory of Scheduling. Addison Wesley Longman, Inc., 1967.

[18] D. S. Hochbaum, Approximation Algorithms for NP-Hard Problems. PWS Publishing Company, 1997.

[19] D. Fogel, Evolutionary Computation: Towards a New Philosophy of Machine Intelligence. IEEE Press, 1999.

[20] A. Grama, V. Kumar and A. Gupta, An Introduction to Parallel Computing: Design and Analysis of Algorithms. Addison Wesley Longman, 2003.

Scheduling to be competitive in supply chains

Sabyasachi Saha and Sandip Sen

Department of Math & CS
University of Tulsa
Tulsa, Oklahoma, USA
sahasa,sandip@ens.utulsa.edu

Abstract. Supply chain management system has an important role in the globalized trading market, where sub-contracts and sub-tasks are given to other manufacturer or suppliers through competitive auction based contracts. The suppliers can decide their scheduling strategy for completing the contracted tasks depending on their capacity, the nature of the contracts, the profit margins and other commitments and expectations about future contracts. Such decision mechanisms can incorporate task features including length of the task, priority type of the task, scheduling windows, estimated arrival probabilities, and profit margins. Task scheduling decisions for a supplier should be targeted towards creating a schedule that is flexible about accommodating tasks with urgency and in general, those tasks which produce larger profits to the supplier. In previous work, different scheduling heuristics like first fit, best fit and worst fit have been evaluated. We believe that a more opportunistic, comprehensive and robust scheduling approach can significantly improve the competitiveness of the suppliers and the efficiency of the supply chain. We present an expected utility-based scheduling strategy and experimentally compare its effectiveness with previously used scheduling heuristics.

1   Introduction

With the increasing popularity of e-markets business-to-business ($B2B$) trading has received an unprecedented boost. Online trading has become the trend of the day. To be ahead of the competition in this scenario, complexity of business strategies is increasing proportionally. Business enterprises have traditionally viewed their organizations to be modular entities, consisting of different functional groups like manufacturing, sales, marketing, finance, etc. Agent-based systems have been deployed to optimize the functional modules and business strategies of an enterprise. Stable and profitable $B2B$ relationships are predicated on the effectiveness of the supply chains linking these organizations. We believe that optimizing supply chain management is a key step in developing more profitable businesses. Globalization, however, has made supply chains enormously complex to maneuver. With suppliers, retailers and manufacturers distributed geographically, integration of these entities for an efficient supply chain management system is a very challenging task. In [3] we find how software agents following *structured conversations* achieve effective coordination in a supply chain. Meeting deadlines of production, supply and transportation is key to a successful business. In this paper we focus on this critical aspect, and identify different scheduling strategies that can effectively accommodate the market dynamics.

Here we use simulations to show the success of different scheduling algorithms in processing dynamically arriving task requests. We have designed a utility based strategy which tries to capture the seasonal variation in the arriving tasks pattern. The tasks which the manufacturers have to complete differ in the priority or the deadline before which they must be completed. Tasks of different priorities represent the dynamics of market demand. We use a contracting framework for connecting the suppliers to the manufacturers: manufacturers announce contracts for tasks with given specifications (deadline and processing time); suppliers bid on these tasks with prices; and the contract is allocated by an auction mechanism to the highest bidder.

A supplier who can meet deadlines when others cannot can demand more from the manufacturer for their services. We investigate the effect of different scheduling algorithms on generating profits under varying task mixes and group composition. For experimentation, we simulate a three-level supply chain with primary and secondary manufacturers and a group of suppliers. Supplier scheduling strategies are chosen from a set of reasonable heuristics for scheduling dynamically arriving tasks. We analyze the ability of different strategies to produce more flexible schedules. The more flexible a schedule, the more readily it can accommodate late arriving and high priority tasks. If only few suppliers can retain such flexibility, they can correspondingly demand more from the manufacturers because of reduced competition. Correspondingly, we analyze a price adjustment mechanism by which suppliers *up their ante* when they win contracts and reduce their bids when they fail to procure them. This allows for flexible suppliers to benefit monetarily. Our aim is to identify scheduling policies that benefit suppliers in increasing profitability by allowing them to accommodate demands that others can not.

2   Scheduling strategies

We evaluate four scheduling strategies using which the suppliers will schedule assigned tasks. The goal is to allocate a task t of length len and deadline d on the calendar C given the current set of scheduled tasks T in C. The following scheduling strategies have distinct motivations:

First fit: This strategy searches forward from the current time and assigns t to the first empty interval of length len on the calendar. This produces a compact, front-loaded schedule.

Best fit: This strategy searches the entire feasible part of the schedule (between the current time and deadline d) and assigns t to one of the intervals such that there is minimal empty slots around it. This strategy produces clumped schedules, but the clumps need not be at the front of the schedule.

Worst fit: This strategy searches the entire feasible part of the schedule (between the current time and deadline d) and assigns t to one of the intervals such that there is maximum empty slots around it.

Expected Utility Based (EU): All the strategies discussed above schedule the tasks in a greedy way. They bid for a task whenever it can schedule it. In reality, generation of tasks has a seasonal pattern. Sometimes it happens that more profitable and high priority jobs can come later. *EU* agents use the knowledge of the task arrival pattern and whenever a task arrives it estimates whether it is worthwhile to bid on that task. If the task arrival pattern is not known, they can learn it within a short period of time and use the knowledge, provided the pattern does not change drastically every time. Unfortunately, there is no free lunch. There comes a risk along with the lure of achieving more profit, that some slots may remain unused. According to this algorithm, it will bid for a task if the expected utility of scheduling it is positive. The expected utility is negative when there is an expectation of generating more profit from later-arriving tasks. In those cases, it will not bid for this task. We will discuss the expected utility in the next section.

3   Expected Utility of scheduling a task

Every agent has an estimation of the utility for each task. Suppose a task has arrived and its utility for an agent is, say C, depending on the task's priority *prt* and length *len*. The agent finds the number of empty slots, *es* available each day up to the deadline of the task. For each of these days, the agent generates all possible task length combinations that can fill those empty slots. It also generates all possible task length combinations that can fill the remaining empty slots, if it schedules the task at hand. We will choose to schedule the task now if the corresponding utility dominates the utilities from all other ways of filling up the empty slots without scheduling the current task. Let, $L_{i,j,k}$ be a combination of tasks where i_j is the number of tasks of length j and all the tasks in the combination add up to the total empty slots in that day, i.e. $\sum_j j*i_j = es$. $P_{i,j,k}$ is the probability that at least i_j number of tasks of length j and a priority k will arrive at a later time on this day. $P_{i,j,k}$ is calculated from the task distribution. The utility of the task of length j and priority k is $C_{j,k}$. Priority of a task is 0 and 1 for a low and high priority respectively.

So, the expected utility of scheduling the current task is

$$EU = C + \sum_{k>prt} \left\{ \frac{1}{M} \sum_{L_{i,j,k} \in sch} \{P_{i,j,k} \sum_j (i_j \times C_{i_j})\} \right.$$

$$\left. - \frac{1}{T} \sum_{L_{i,j,k} \in notsch} \{P_{i,j,k} \sum_j (i_j \times C_{i_j})\} \right\} \times winchance$$

Where, *sch* is the collection of those task combinations $L_{i,j,k}$ such that

$$\sum_j j * i_j = es - len$$

and *notsch* is the collection of those combinations $L_{i,j,k}$ such that $\sum_j j * i_j = es$
M and T are $|sch|$ and $|notsch|$ respectively. Here *winchance* is the defined as
$winchance = (1 - \frac{emp'}{total})$, where emp' is the number of empty slots till last period
and *total* is the total number of slots considered till last period (*winchance*
initially has a value of 1). The expression for expected utility adds the utility
of the current task to the sum of the average utility over all possible ways of
scheduling the remaining free slots minus the average utility of all possible ways
of scheduling the currently empty slots with tasks of higher priority.

Given the task distribution (discussed in the Section 4) the probabilities $P_{i,j,k}$
can be calculated by using multinomial cumulative probability function.

4   Simulation Framework

In our simulations, we consider each period to consist of 5 days, each day having
six slots. *Deadline* is the latest time by which the task has to be done. A high
priority task has to be finished on the day of its arrival. Ordinary tasks have
to be finished by the fifth day after its arrival. We vary the arrival rate of different task types with the percentage of priority tasks in different simulations.
Each of the tasks are generated and allocated to a manufacturer depending on
whether the manufacturer can accomplish the task by using one of its suppliers.
Selection of one supplier over another by a manufacturer is performed through a
first-price sealed-bid auction procedure. We have used a 3-level supply chain. In
level one, there is one *main manufacturer*, level two is occupied by six *secondary
manufacturers* and at level three there are twelve *suppliers*. Suppliers under one
secondary manufacturer form a *supplier window*. *Supplier windows* of two secondary manufacturers may be overlapping. The supply chain structure is shown
in Figure 1. We have later changed the number of suppliers to show the impact
of presence of too many similar kind of suppliers in the environment. For the
experiments reported in this paper, we have set the supplier window size to be
four and each supplier can get contracts from exactly two different secondary
manufacturers. A task assigned to the main manufacturer has two properties:
(a) *priority*, and (b) *length*, which is the number of time units (slots) the task
requires to be completed. A task must be scheduled on consecutive slots of the
same day. Higher priority tasks are more profitable. Here we assume that initially, utility of an ordinary task is equal to the length of the task and the utility
of a high priority task is two times the length of the task.

In our simulation, the tasks are generated stochastically. We use different
type of task distributions to verify the robustness of the different algorithms.
Table 4 shows the first type of task generation. The row number gives the length
of the task. The column $(2i - 1)$ shows the probability of generating high priority

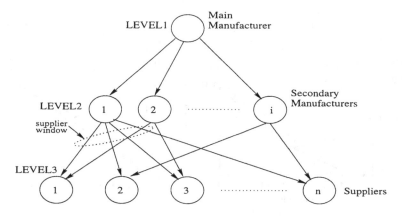

Fig. 1. Supply chain structure.

task in the ith day and the column $2i$ shows the probability of generating a low priority task . Here high priority tasks arrive in the later part of the week. In another type, tasks are generating from a uniform distribution.

Table 1. Probabilities of Generating Tasks

len	Probabilities									
1	.001	.14	.005	.005	.002	.01	.03	.01	.04	.01
2	.001	.14	.005	.005	.002	.01	.03	.001	.04	.01
3	.001	.15	.005	.005	.001	.005	.02	.02	.03	.01
4	.001	.15	.005	.005	.001	.005	.02	.02	.03	.01

Suppliers at level 3 of the supply chain can employ any one of the four scheduling algorithms mentioned in Section 3. The suppliers have the same initial bid price. Only those suppliers in the supplier window can bid who can accommodate the task announced by a secondary manufacturer in their schedule. This implies that contract cancellation is not allowed which precludes use of leveled commitment protocols [11]. The supplier with the minimum bid price wins the task from the secondary manufacturer. Winning a task gives the supplier incentive to increase its bid price for that type ($[priority, length]$ combination) of task. A corresponding decrease in the bid prices occurs if a supplier loses a bid. The bid price cannot go below the initial bid price. When a supplier wins a bid, it increases its *wealth* by an amount equal to its bid price. Initially, wealth of all suppliers is set to zero.

5   Experimental Results

To describe our experimental results we use the following conventions: suppliers using the worst fit algorithm are represented by WF, those using first fit by FF,

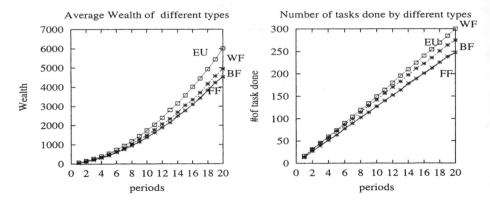

Fig. 2. Average Wealth earned (left) and number of tasks done (right) by different type of suppliers (3 supplier of each type).

those using best fit by BF and those using expected utility based strategy by EU.

In the first set of experiments, we have 3 suppliers of each type and the task distribution is as mentioned in the table 1. Here we study the variation in the wealth earned by suppliers using different types of scheduling algorithms. The result is shown in Figure 2 (left). The average wealth earned by EUs are much better compared to the other three scheduling types. Since the EU strategy is not myopic, it waits for the more profitable high priority tasks, ignoring ordinary tasks that arrived earlier, and make more profits throughout the period of execution of 100 days. The EUs benefit from earning more than the other types of suppliers by winning the bids for these high-profit tasks. Other, myopic suppliers, do not consider passing on arriving tasks, and hence fill up slots by committing to tasks arriving early in the run, and therefore fail to bid for high priority tasks that arrive later. The EUs are followed in performance by WFs. Since the worst fit algorithm produces an evenly-loaded schedule, suppliers using this type of scheduling can better accommodate the high priority tasks (and ordinary tasks too) that arrive more frequently later in the contracting period. We note that at the end of the run, the prices for the high priority task charged by EU suppliers increase considerably. In this run the Figure 2 (right) shows that the EU not only achieves more profit, but also accommodates more tasks compared to the other scheduling strategies. The reason is that in the task distribution, more high priority tasks are of smaller length. In this metric too, the WF strategy comes in second. They accommodate more tasks compared to BF and FF.

In the second set of experiments, we alter the task distribution and generate the tasks such that all task types and their arrival dates are equally likely. Here, tasks per week is decreased to 100 instead of 150, as used in the first simulation. Here also the EU suppliers earned more wealth as shown in Figure 3 (left). The reason is that though all of the suppliers is accommodating some high priority

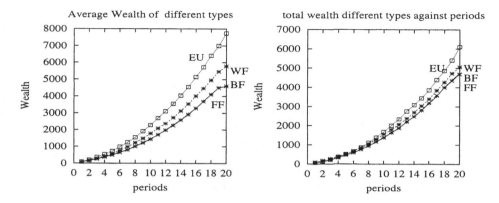

Fig. 3. Average Wealth earned by different type of suppliers with uniform task distributions (left) and learning task distribution (following Table 1) (right). There are 3 suppliers of each type.

tasks, the other strategies also contract a number of ordinary tasks. The *EU*s, however, focus more on the high priority tasks. So, we show that the *EU* suppliers are robust enough to do better under different type of task distributions.

In the third set of experiments, we assume that the *EU* suppliers do not know about the task distribution. So, they try to estimate the distribution in the first two iterations and bid for all tasks. From the third period onward, however, it uses the expected utility based strategy as it has an estimate of task distributions by this time. In Figure 3 (right) we can see that *EU*s utility is less in the first two periods. But the *EU*s dominate the other strategies in the later periods. In this experiment set the *EU*s continues to estimate, in every period, the task arrival probability distribution with the task arrival data of last two periods using relative frequency. So, this *EU* scheduling strategy works well even when the task generating distribution is not known *a priori*.

Figure 4 shows that if the number of *EU* suppliers increase, the per supplier utility decreases for *EU* suppliers. In this experiment we increase the number of the *EU*s and keep number of suppliers under a secondary manufacturer equal and change tasks proportionately. There is a significant decrease in the wealth earned by the *EU* suppliers individually. This decrease, as the number of *EU*s increase, is due to the accumulation of too many "smart" people (*EU*s) in the population. When there is only few *EU* supplier, it enjoys the advantage of more consistently contracting the high priority tasks. It can increase the bid price, and hence the profit, to a greater extent since it may be the only bidder or among very few bidders for a secondary manufacturer (other suppliers do not have a vacant schedule for the task at all). With more EU bidders, there is more contention for high priority tasks, which reduces their individual profitability. Though they are doing better than the other strategies, the differences in performances decreases because they will also go for a large number of ordinary tasks after the first iteration.

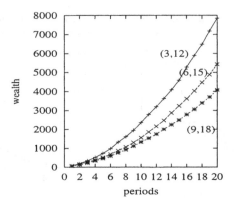

Fig. 4. Average wealth varies as the number of *EU* increases,in the figure the number of *EU* and total number of suppliers are mentioned in the brackets.

We note that, in these experiments, we have supplied a relatively large number of tasks. Over each period, i.e., for 5 day intervals, we use at least 100 tasks for 12 suppliers. If the number of tasks per period is very small, the profits of different suppliers or the number of tasks they perform, do not differ significantly. This is because then scheduling strategies do not have a significant effect as most of the slots remain empty.

6   Related work

Scheduling in the context of supply chain management has long been an active area of research among the operations research community. Typically, the operations research approach was to search for optimum schedules from a single objective function [8]. Finding the optimal schedule, however, becomes an intractable problem in complex real-world situations. Faced with such real-world problems, researchers adopted the heuristic scheduling approach [9] that allows the scheduler to locally evaluate alternative solutions.

Scheduling problems have also been the focus of AI researchers. Scheduling by expert systems has been one of the premiere approaches in AI [4], that attempts to capture the decision-making ability of a human expert using situation-specific "if-then" rules contained in a knowledge base. Iterative techniques, including genetic algorithm and simulated annealing, that generate high-fidelity solutions using neighborhood based search form another successful AI approach to scheduling.

More recently the problem of supply chain management has drawn the attention of multi-agent systems (MAS) researchers [1]. In terms of MAS, a supply chain is conceptualized as a group of collaborative autonomous software agents [7]. Distributing the corporate-wide business management system into autonomous problem solving agents, managers are able to better coordinate and schedule processes. Negotiation-based supply chain management has been accounted in [5], where a *virtual* supply chain evolves by the negotiating software agents when an order arrives. The *blackboard architecture* is a proven methodology for integrating multiple knowledge sources for problem solving [6]. Using this architecture, a mixed-initiative agent-based architecture for supply chain planning and scheduling is implemented in MASCOT [10].

The contract-bid-award framework used in our simulation is similar to the contract-net framework that has been used widely in multi agent systems research to solve the "connection problem", i.e., finding people who can do the job that one needs to get done [2, 12].

Scheduling for increased flexibility for accommodating late-arriving tasks has been studied in the context of distributed meeting scheduling in the form of search biases [13]. The supply-chain setting do not involve the concurrent timing agreement requirement of the meeting scheduling domain. On the other hand the meeting scheduling domain does not involve the competitive bidding framework.

7 Conclusions and future work

We have experimentally evaluated the relative effectiveness of different scheduling heuristics in accommodating dynamically arriving mix of tasks in the context of *B2B* trading. We note that when a significant number of high priority tasks are present, the expected utility based strategy and the worst-fit strategy provides greater flexibility. On the contrary, when such tasks are rare and also there are longer tasks in the mix, first fit or best-fit strategies may be equally good.

Additionally we have studied the competitive advantage to be gained by suppliers in upping their price when they can schedule effectively to win task contracts. We need to augment this work by adaptively selecting the scheduling strategy from observed job mixes. We also plan to study swapping of resources based on side-payments. This allows for effective coalition to develop to better respond to uncertain market needs.

Acknowledgments This work has been supported in part by an NSF award IIS-0209208.

References

1. *Working Notes of the 1999 Autonomous Agents Workshop on Agent-based Decision-support for Managing the Internet-enabled Supply-chain* (May 1999).
2. BAKER, A.: Complete manufacturing control using a contract net: A simulation study. In *Proceedings of the 1988 International Conference on Computer Integrated Manufacturing* (May 1988), pp. 100–109.
3. BARBUCEANU, M., AND FOX, M. S.: Coordinating multiple agents in the supply chain. In *Proceedings of the Fifth Workshop on Enabling Technologies: Infrastructure for Collaborative Enterprises* (1996), pp. 134–142.

4. BENSANA, E., CORREGE, M., BEL, G., AND DUBOIS, D.: An expert-system approach to industrial job shop scheduling. In *Proceedings of the International Conference on Robotics and Automation* (1986), pp. 1645–1650.

5. CHEN, Y., PENG, Y., FININ, T., LABROU, Y., COST, R., CHU, B., SUN, R., AND WILLHELM, R.: A negotiation-based multi-agent system for supply chain management. In *Working Notes of the ACM Autonomous Agents Workshop on Agent-based Decision-support for Managing the Internet-enabled Supply-chain* (May 1999).

6. ERMAN, L. D., HAYES-ROTH, F., LESSER, V. R., , AND REDDY, D. R.: The hearsay-ii speech-understanding system: Integrating knowledge to resolve uncertainty. *Computing Surveys 12*, 2 (June 1980), 213–253.

7. JENNINGS, N. R., FARATIN, P., NORMAN, T. J., O'BRIEN, P., AND ODGERS, B.: Autonomous agents for business process management. *International Journal of Applied Artificial Intelligence 14*, 2 (2000), 145–189.

8. JOHNSON, S. M.: Optimal two- and three-stage production schedules with setup times included. *Naval Research Logistics Quarterly 1*, 1 (1954), 61–68.

9. KURTULUS, I., AND DAVIS, E. W.: Multi-project scheduling: Categorization of heuristic rules performance. *Management Science 28*, 2 (1982), 161–172.

10. SADEH, N., HILDUM, D., KJENSTAD, D., AND TSENG, A.: Mascot: An agent-based architecture for coordinated mixed-initiative supply chain planning and scheduling. In *Working notes of the ACM Autonomous Agents Workshop on Agent-Based Decision Support for Managing the Internet-Enabled Supply Chain* (May 1999).

11. SANDHOLM, T., AND LESSER, V.: Leveled commitment contracting: A backtracking instrument for multiagent systems. *AI Magazine 23*, 3 (2002), 89–100.

12. SANDHOLM, T. W., AND LESSER, V. R.: Issues in automated negotiation and electronic commerce: Extending the contract net framework. In *First International Conference on Multiagent Systems* (Menlo Park, CA, 1995), AAAI Press/MIT Press, pp. 328–335.

13. SEN, S., AND DURFEE, E.: Unsupervisesd surrogate agents and search bias change in flexible distributed scheduling. In *First International Conference on Multiagent Systems* (Menlo Park, CA, 1995), AAAI Press/MIT Press, pp. 336–343.

Contract Net Protocol for Cooperative Optimisation and Dynamic Scheduling of Steel Production

[1]Ouelhadj D., [2]Cowling P. I., [3]Petrovic S.

[1,3] Automated Scheduling, Optimisation and Planning Research Group, School of Computer Science and IT, University of Nottingham, Nottingham NG8 1BB, UK, [1]Email: dxs@cs.nott.ac.uk, [3]Email: sxp@cs.nott.ac.uk.

[2]Modelling Optimisation Scheduling and Intelligent Computing Research Group, Department of Computing, University of Bradford, Bradford BD7 1DP,UK, Email: P.I.Cowling@Bradford.ac.uk.

Abstract. This paper describes a negotiation protocol proposed for inter-agent cooperation in a multi-agent system we developed for optimisation and dynamic integrated scheduling of steel production. The negotiation protocol is a two-level bidding mechanism based on the contract net protocol. The purpose of this protocol is to allow the agents to cooperate and coordinate their actions in order to find globally near-optimal robust schedules, which are able to optimise the original production goals whilst minimising the disruption caused by the occurrence of unexpected real-time events. Experimental results show the performance of this negotiation protocol to coordinate the agents in generating good quality robust schedules. This performance is evaluated in terms of stability and utility measures used to evaluate the robustness of the steel production processes in the presence of real-time events.

Keywords. steel production, multi-agents, cooperation and coordination, contract net protocol, dynamic scheduling

1 Introduction

Steel production involves a variety of processes [1, 2] including continuous casters, hot strip mills and furnaces. Continuous casters cast the molten steel into slabs with different widths and chemical composition. The hot strip mill rolls these slabs in order to produce the output of the primary steelmaking production process, steel coils. The scheduling systems of the caster and the hot strip mill have very different objectives and constraints and operate in an environment where there is a substantial quantity of real-time information concerning production failures and customer requests. Note that there are usually several continuous casters that can supply a single hot strip mill with slabs. Due to the complexity of the steel production environment involving multiple distributed production processes, in our previous work, we proposed a multi-agent architecture for integrated dynamic scheduling of the hot strip mill and continuous casters [3, 4]. Each steel production process is represented by an agent, including the continuous caster agents (CCA), the hot strip mill agent (HSMA), the slabyard agent (SYA) and the user agent (UA). The HSMA performs the robust scheduling of the hot strip mill. The CCA performs the robust scheduling of the continuous caster. Robust scheduling focuses on building predictive/reactive schedules to minimise the effects of disruption while optimising some measures of performance for the realised schedule. The SYA communicates with the CCA(s) to convey details of the slabs requested by the HSMA, and maintains information on slabs already produced which are currently cooling down. The UA manages and announces the orders to produce, and allows outside input of data for dynamic changes of order conditions. These agents communicate by exchanging asynchronous messages formatted in XML (Extensible Mark-up Language). An experimental prototype was developed as a multi-thread application in Microsoft Visual C++/MFC.

In order for the agents to find global feasible schedules, the agents must cooperate and coordinate their local actions. The most widely used cooperation and coordination method is the Contract-Net Protocol (CNP). The CNP is a high level protocol for achieving efficient cooperation introduced by Smith [13] based on a market-like protocol. The CNP has been extensively used for inter-agent cooperation in dynamic production scheduling. Yams [9] is one of the earliest agent-based manufacturing system where part agents negotiate with resource agents to assign tasks to the resource agents using the CNP. Shaw [11] developed a dynamic scheduling system in a cellular manufacturing system, where the manufacturing cell agent could sub-contract work to other cells through a bidding mechanism to schedule the tasks. Lin and Solberg [5] used the CNP for inter-agent cooperation in a shop floor. In their model part agents enter the system with certain currency and negotiate with resource agents via a bidding mechanism. When a resource agent is in failure, it informs the corresponding part agent, and the latter proceeds to a renegotiation process on the operations in failure with the resource agents. Sousa and Ramos [14] used the CNP for dynamic scheduling in manufacturing systems. The task agents negotiate the operations of the task with the resource agents using the CNP. When a resource agent detects a malfunction, it sends a machine fault message to the task agents that have contracted its operations. The task agents renegotiate the operations in failure with other resource agents capable of performing the operations. Ouelhadj et al. [7] proposed a multi-agent autonomous architecture for dynamic scheduling in flexible manufacturing systems where resources agents negotiate using the CNP for a global schedule. Following the occurrence of real-time events, the resource agent of the resource in failure sends a failure message to the contractor resource agents specifying the operations in failure. On receiving the message, the corresponding resource agents renegotiate the operations in failure with other resource agents. Maturana and Norrie [6] proposed a mediator architecture, Metaphor I, for dynamic scheduling of virtual enterprises combining mediation and sub-tasking using the CNP. Mediator agents are used to coordinate the resource agents using the CNP. A resource breakdown is simulated by introducing a breakdown period into the resource. Each job allocated within the halt-period is rescheduled to other available time slots found in the malfunctioning resource) or in a different resource. The same idea was adopted by Shen et al. [12] with Metaphor II, for integrating the manufacturing enterprise's activities. The manufacturing resource agents are coordinated by appropriate mediators by combining the mediation mechanism and the CNP. They defined several rescheduling mechanisms for different real-time events such as: rush orders, order cancellation, machine breakdown, etc. Recently leveled commitment contracts were proposed as an extension of the CNP for increasing the economic efficiency of contracts between self-interested agents in the presence of incomplete information about future events. Sandholm [10] described a leveled commitment contracting protocol for automated contracting in distributed manufacturing by giving the possibility for each agent to decommit from the contract by simply paying a decommitment penalty to the other contract party.

The research presented in this paper extends our previous work concerning the integration of the dynamic scheduling of the continuous caster and the hot strip mill in steel production using multi-agent systems. It focuses on inter-agent cooperation for optimal dynamic robust scheduling of steel production. Section 2 describes the negotiation protocol developed for cooperative optimisation and robust scheduling. Section 3 presents the cooperative robust scheduling in the presence of real-time events. The experimental results are presented in Section 4. Conclusions are presented in Section 5.

2 The negotiation protocol for cooperative optimisation and robust scheduling

The HSMA and CCA(s) present different objectives and constraints and no agent possesses a global view of the entire agency, therefore cooperation is essential to allow the agents to adjust their local schedules in order to achieve global objectives and to react to the presence of real-time events. In this paper, we propose a negotiation protocol based on the contract net protocol for inter-agent cooperation. The negotiation protocol is a two-level bidding mechanism involving negotiation at HSMA-SYA level and SYA-CCA(s) level. At the HSMA-SYA negotiation level, the HSMA requests the supply of slabs from the SYA. At the SYA-CCA (s) negotiation level, the SYA requests the production of slabs not available in the slabyard from the CCA(s). In the CNP no agent takes any responsibilities before reaching mutual agreements on the tasks, which can cause the degradation of the global coordination performance. By attaching the commitment duration to the negotiation messages, the performance of the CNP is improved especially for dynamic environments where desired tasks and available resources may be continuously changing. The commitment duration specifies the time windows by which the agents must respond to a given negotiation message. The negotiation protocol is a three-step process, which begins after the session has been initiated by the

UA, which sends a request message to the HSMA to produce a collection of coil orders. The order describes the grade (carbon and aluminium composition) and physical (weight, width, length, and thickness) properties of the coils and their due dates. The steps of the negotiation protocol are described above.

2.1 Task announcement

Upon receiving the request message from the UA, the HSMA generates a schedule of coils which satisfies the hot strip mill's constraints and maximises its objective function. Then, negotiation at the HSMA-SYA level starts. The HSMA issues an announcement message (HSMA-announcement) to the SYA to supply the slabs necessary to produce the coils, which have been scheduled. Due to hot strip mill hard constraints, the HSMA generates a number of sub-schedules (coffins) [4], and announces each sub-schedule separately to the SYA. The HSMA-announcement message describes the following information:

> *Sender: HSMA,*
> *Receiver: SYA,*
> *Type: task announcement,*
> *Task specification: the list of slabs to produce with the description of their grade, dimensions, and the requested production ready time,*
> *Bid-reception deadline: the time by which the SYA must respond with a bid.*

Upon receiving the HSMA-announcement message, the SYA analyses the announcement and negotiation at the SYA-CCA level starts to identify the CCA to produce the non-available slabs in the slabyard. The SYA sends the SYA-announcement message to the CCA(s) to request the casting of slabs, which are not currently available in the slabyard. The SYA-announcement message describes the following information:

> *Sender: SYA,*
> *Receiver: CCA,*
> *Type: task announcement,*
> *Task specification: the list of slabs requested by the HSMA which are currently not available in the slabyard with the description of their grade, dimensions, and the requested production ready time,*
> *Bid-reception deadline: the time by which the CCA must respond with a bid.*

2.2 Bidding

A bid represents an offer to execute the task specified in the SYA-announcement concerning the production of the slabs non-available in the slabyard. Each CCA will inspect the SYA-announcement message and will decide whether or not it should respond with a bid, considering the quality of the caster schedule resulting from adding the requested slabs to its current engagements. If it chooses to respond, it will develop its locally optimised scheduling and sends a CC-bid message to the SYA. The CC-bid message describes the following information:

> *Sender:* CCA,
> *Receiver:* SYA,
> *Type:* bid,
> *Bid specification:* the slabs to produce with the proposed production dates,
> *Bid-acceptance deadline:* the time by which the SYA must respond with
> a contract.

After receiving the bids before the bid-reception deadline, the SYA evaluates the CCA-bids and selects the best bid based on the earliest production date. The SYA sends a SYA-bid to the HSMA with the information about the available slabs in the slabyard and production dates of the slabs to be produced by the CCA(s) proposed in the selected bid. This bid is used at the HSMA-SYA negotiation level. The SYA-bid message describes:

> *Sender:* SYA,
> *Receiver:* HSMA,
> *Type:* bid,
> *Bid specification:* slabs available in the slabyard plus the slabs from the accepted CCA bid with the proposed production dates,
> *Bid-acceptance deadline:* the time by which the HSMA must respond with a contract.

2.3 Contracting

After the HSMA has received the bid from the SYA, it can respond before the bid-acceptance deadline with the following alternatives:

- Accept the bid and sends the award message to the SYA. The SYA on its turn sends the award message to the appropriate CCA before the bid-acceptance deadline. The production of slabs then can start.

- Accept a subset of the slabs in the bid and renegotiates the production of slabs with unsatisfied production dates. The HSMA restarts a re-negotiation session with a new announcement message, which specifies the slabs with unsatisfied production dates and an additional allowance tolerance on the width. Width tolerance is a relaxation on the width hard constraint, since a coil can be made from slabs having a variety of different physical dimensions. The renegotiation process iterates cyclically until the HSMA states that the solution matches the requirements of its schedule within acceptable tolerances.

A failure to send a contract message before the bid-accept deadline means the HSMA or the SYA is rejecting the bid.

3 Cooperative robust rescheduling

In the proposed multi-agent architecture, both the hot strip mill and the continuous caster are subject to various real-time events. Among the most frequent real-time events that can occur on the continuous caster are heat of molten steel may arrive with wrong chemical composition. These real-time events can affect the hot strip mill production process by causing delayed delivery of slabs or failure to meet prescribed quality control standards of the slabs.

On the occurrence of real-time events, the CCA and the HSMA react locally to the real-time events and cooperate in order to define a globally feasible robust schedule. In order for the CCA and the HSMA to react locally to real-time events, we defined several domain-specific rescheduling strategies, as well as utility, stability and robustness measures. The CCA and the HSMA reschedule so as to maximise robustness. Robustness combines the maximisation of utility and the minimisation of stability. Utility measures the change in the value of the schedule objective function following the schedule revision. It is expressed by the difference between the value of the objective function of the new schedule after reacting into the real-time events and the objective function of the predictive schedule before taking into account real-time events. Stability measures the deviation from the original predictive schedule caused by schedule revision to quantify the undesirability of making large changes to the initial schedule unless absolutely necessary. For the CCA the stability is expressed by the weighted sum of the absolute difference between the original starting production time of each slab in the original schedule, and the new starting production time after the occurrence of the real-time event. On the HSMA, the stability is expressed by the sum of the absolute difference between the original completion time of each coil of the original schedule and the new completion time after the occurrence of the real-time event. For a detailed description of the utility, stability, robustness measures and rescheduling strategies refer to [8] for the CCA, and to [3, 4] for the HSMA.

The rescheduling strategies of the HSMA are: Do-nothing (NOT), Simple Replacement (SR), Closed Schedule Repair (CSR), Open Schedule Repair (OSR), Hybrid Open Schedule Repair (HOSR), Partial Reschedule (PR), and Complete Rescheduling (CR).

The rescheduling strategies of the CCA are: Insert-at- End Schedule Repair (IESR), Insert-Heat Schedule Repair (IHSR), Shift Schedule Repair (SHSR), Swap Schedule Repair (SWSR), Hybrid-shift-swap Schedule Repair (HBSR), and Complete rescheduling (CR).

When the CCA applies a rescheduling strategy to react to the real-time events heat with wrong chemical composition specification, it does a least commitment scheduling by proposing an alternative schedule which specifies the new production dates for the slabs in the current schedule. These new production dates are sent to the SYA in order to inform it on the new production dates. The SYA on receiving the message analyses the new production dates, and starts renegotiation at the SYA-CCA level with other CCA (s) to find alternative CCA that can produce the delayed slabs. If the SYA cannot manage to find the appropriate CCA to produce on time the slabs in failure it will send an alert message to the HSMA on the delayed slabs which will not be available on time. The HSMA on receiving the message reacts to the real-time events by applying the rescheduling strategies so as to maximise robustness to find alternative slabs from the slabyard to be inserted into the gap. The HSMA sends a new announcement message to the SYA to request in real-time the remaining slabs of its current schedule found after reacting to the delayed slabs. These slabs should be available in the slabyard or in the process of casting.

4 Experimental results

To show the performance of the negotiation protocol proposed to generate feasible global schedules in the presence of real-time events, we conducted several simulation experiments. In the simulation, the multi-agent architecture involves two CCA(s), a HSMA, a SYA and a UA. We carried out 5 runs for various instances of data obtained from a steel manufacturer. For each run, 5 real-time events were generated on the CCA(s). Each real time-event specifies up to 5 heats of molten steel with the wrong aluminium and carbon composition, which affects up to 4 slabs per heat. Each real-time event on the CCA can affect the production of up to 20 slabs on the HSMA. For each real-time event, both the HSMA and the CCA evaluate the best rescheduling strategy, which maximises the robustness to react to the real-time events. Note that low values of the stability measure yield low schedule disruption, and high values of the utility measure give a good value of the objective function.

Figure 1 presents a Gantt chart of a simulation run to illustrate the negotiation and re-negotiation sessions. The HSMA generates its optimal schedule of 150 orders using tabu search and requests the SYA to supply it with the slabs. On receiving the message, the SYA requests the CCA1 and CCA2 to produce 100 slabs not available in the slabyard. The CCA1 and CCA2 generate their local schedules using tabu search which is a sequence of 24 molten steel heats, and submit bids with different production dates of the slabs to the SYA. The SYA selects the CCA1 bid and submits its bid to the HSMA. We suppose in this example that the HSMA is satisfied with the production dates proposed and then it sends a contract to the SYA. The SYA on its turn sends a contract to CCA1. During production, real-time event1 occurred on the CCA1, which specifies that heat 2 and heat 3 are out of carbon and aluminium ranges which affects 9 slabs in total. The CCA1 reacts to the real-time events by applying HSR strategy and submits an alternative bid to the SYA which specifies the revised schedule. The SYA notices that 9 slabs will be late. Then it requests the CCA2 to produce the late slabs. The CCA2 bid cannot satisfy the required production dates. So the SYA sends an alert message to the HSMA on late slabs. The HSMA reacts to the real-time event by applying the CSR strategy and requests the revised schedule by the SYA. On the occurrence of real-time event 2 which affects 3 slabs of heat 6, the revised schedule of CCA1 does not affect the production dates of the slabs. The SYA on receiving the new

CCA1 bid on the revised schedule sends a contract message to the CCA1 since the slabs will still be produced on time by the CCA1.

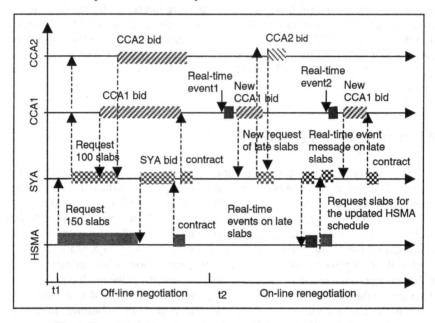

Fig. 1. Gantt Chart of an example of negotiation and renegotiation sessions

Figure 2 and figure 3 show the performance of the utility and stability measures for various problem instances on both the CCA and the HSMA obtained by applying the schedule repair and complete rescheduling strategies. Each point is the new value of utility and stability after reacting to a real-time event. In figure 2, the experiment results show that usually schedule repair strategies (NOT, SR, CSR, OSR, HCSR, HOSR, HCSR, HOSR, PR) maintain a very good stability of the system and at the same time they can achieve a very good utility comparable to the one of complete rescheduling. Consequently, schedule repair strategies define more robust schedules to react to real-time events. Complete rescheduling is competitive with schedule repair strategies in terms of utility measure, but does not dominate the schedule-repair strategies, which attain similar utility, but for better stability. Similar results are observed on the CCA. Figure 3 shows that schedule repair strategies IESR, SWSR, SHSR, HBSR maintain a good stability measure compare to complete reschedule except IHSR because of the large number of heats added to the schedule. HBSR outperforms the other rescheduling strategies in both the utility and stability measures.

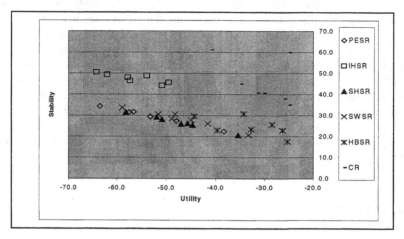

Fig. 2. Performance of the utility and stability measures of the CCA

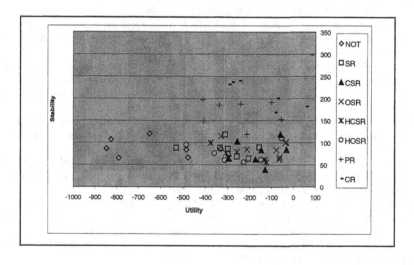

Fig. 3. Performance of the utility and stability measures of the HSMA

5 Conclusion

This paper has presented an inter-agent cooperation protocol for optimisation and dynamic scheduling in steel production. The cooperation protocol is a two level bidding mechanism based on the contract net protocol. The negotiation is a three-step process, which involves announcing, bidding and contracting at the HSMA-

SYA and SYA-CCA(s) levels. In the presence of real-time events, the CCA and the HSMA react locally to real-time events using the utility, stability, and robustness measures and rescheduling strategies. The proposed negotiation protocol is able to deal with the presence of real-time events by renegotiation in order to maintain robust global schedules. The experimental results showed the performance of the negotiation protocol in increased performance of utility and stability measures of the schedule-repair strategies compare to complete rescheduling. We have provided evidence that local autonomy and cooperation capabilities of multi-agent systems yield system robustness against failures. Moreover, using this flexible cooperation architecture allows the agents to be very simply added or removed from the system.

6 References

1. Cowling P I (1995) Optimisation in Steel Rolling. In Optimisation in Industry, John Wiley & Sons, Chichester, England.
2. Cowling P I, Rezig W (2000) Integration of Continuous Caster and Hot Strip Mill Planning for Steel Production. Journal of Scheduling, 3:185-208.
3. Cowling P I, Ouelhadj D, Petrovic S (2001) A multi-agent architecture for dynamic scheduling of steel hot rolling. In the Proceedings of the Third International ICSC World Manufacturing Congress, Rochester, NY, USA.
4. Cowling P I, Ouelhadj D, Petrovic S (2003) A multi-agent architecture for dynamic scheduling of steel hot rolling. To appear in the Journal of Intelligent Manufacturing.
5. Lin G Y, Solberg J J (1992) Integrated shop floor control using autonomous agents. IIE Transactions, 24:57-71.
6. Maturana F, Shen Weiming, Norrie D (1999) MetaMorph: an adaptive agent-based architecture for intelligent manufacturing. International Journal of Production Research, 37: 2159-2173.
7. Ouelhadj D, Hanachi C, Bouzouia B (1999) A Multi-contract net protocol for dynamic scheduling in flexible manufacturing systems. In the Proceedings of the IEEE International Conference on Robotics and Automation, Detroit, USA.
8. Ouelhadj D, Cowling P I, Petrovic S (2003) Utility and Stability Measures for Agent-Based Dynamic scheduling of Steel Continuous Casting. Accepted for publication in Proceedings of IEEE Conference on Robotics and Automation, Taiwan, USA.
9. Parunak H V (1987) Manufacturing experience with the contract net, In M. N. Huhns, ed., Distributed Artificial Intelligence, Pitman, pp 285-310.
10. Sandholm Tuomas W (2000) Automated contracting in distributed manufacturing among independent companies. Journal of Intelligent Manufacturing, 11:271-283.
11. Shaw J M (1988) Dynamic scheduling in cellular manufacturing systems: a framework for Network decision making. Journal of Manufacturing Systems, 7:83-94.
12. Shen Weiming, Maturana F, Norrie D H (2000) Metaphor II: an agent-based architecture for distributed intelligent design and manufacturing. Journal of Intelligent Manufacturing, 11:237-251.
13. Smith R (1980) The contract net protocol: high level communication and control in distributed problem solver. IEEE Transactions on Computers, 29:1104-1113.
14. Sousa P, Ramos C (1999) A distributed architecture and negotiation protocol for scheduling in manufacturing systems. Computers in Industry, 38:103-113.

Part X

**Special Session on
Peer-to-Peer Computing**

Technical Session on
Agent Enabled Peer-to-Peer Systems

Session Organizers:

Prithviraj(Raj) Dasgupta, University of Nebraska, Omaha, USA

Vana Kalogeraki, University of California, Riverside, USA

Over the past few years multi-agent systems have emerged as an active research field that combines principles from different disciplines including artificial intelligence, distributed systems, computer networks and security. Multi-agent systems have been applied to several domains to assist humans in hazardous and high-precision tasks. The performance of multi-agent systems can be significantly improved if software agents can co-ordinate their interactions to collaborate and even compete with one another. Peer-to-peer(P2P) systems provide a suitable paradigm for software agents to collaborate as peers sharing information, tasks and responsibilities with each other. A P2P system is a distributed network of nodes which are capable of sharing resources with one another. Intelligent software agents can be used to implement and enhance existing P2P protocols. Simultaneously, co-ordination mechanisms from P2P systems can be applied to multi-agent systems to improve their efficiency.

In this technical session we have a collection of diverse papers that describe research covering architectural aspects, intelligent algorithms and applications of software agent enabled peer-to-peer systems. The paper by Dasgupta titled "A Peer-to-Peer System Architecure for Agent Collaboration" describes improvements to basic P2P protocols using intelligent agents. Montresor and Baboglu's paper describes P2P algorithms inspired by social insects such as ants and termites while the paper by Rodic and Engelbrecht discusses coordination between robot teams based on a social interaction model. Distributed task management under soft real time constraints is the topic of the paper by Chen and Kalogeraki and the paper by Samaras and Evripidou describes UbAgent - a P2P application for ubiquitous and pervasive computing.

Review Committee
Prof. Prithviraj Dasgupta, University of Nebraska, Omaha, USA; Prof. Vana Kalogeraki, University of California, Riverside, USA; Prof. Manolis Koubarakis, Technical University of Crete, Greece; Prof. G. Manimaran, Iowa State University, USA; Prof. Hanh Pham, State University of New York, New Paltz, USA; Prof. Sandip Sen, University of Tulsa, USA

A Peer-to-Peer System Architecture for Multi-Agent Collaboration

Prithviraj(Raj) Dasgupta

Department of Computer Science
University of Nebraska, Omaha, NE 68182.

E-mail: pdasgupta@mail.unomaha.edu
Phone: (402) 554 4966 Fax: (402) 554 3284

Summary. A peer-to-peer(P2P) network comprises a collection of nodes that can cooperate and collaborate with each other in a de-centralized and distributed manner. A node in a P2P network can access information present in the network using peer discovery followed by a search and retrieval phase. At present, most P2P systems employ TCP/IP based message communication to implement the operations in a P2P network. In this paper, we propose the use of mobile software agents to implement the protocols in a P2P system. Mobile software agents are autonomous, economic in terms of size and bandwidth consumption, and can operate remotely without the continuous supervision of a central server. Our research indicates that mobile software agents provide a suitable paradigm for implementing P2P systems that is both scalable and robust.

Keywords: Peer-to-peer systems, multi-agent systems, mobile agents, intelligent node and resource discovery.

1 Introduction

An agent is a software entity that can operate autonomously without continuous supervision by a central authority. Intelligent software agents[9] can also learn from their environment, react to changes in the environment and take decisions based on their beliefs. With the rapid growth of the Internet over the past decade, software agents called bots are being used extensively to aid online users with different tasks including intelligent Web searches, comparing prices of items from different online shops, and customizing Web sites according to user preferences. However, most of these Internet bots perform their tasks individually and possibly independently of each other. Most of the information that is available online is duplicated across many sites. Due to this vast redundancy in online information, it is quite likely that many of the Internet bots gather and analyze information that has already been collected

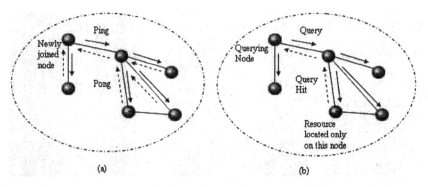

Fig. 1. Messages exchanged in the (a) node discovery and (b) resource discovery protocols in a P2P network.

and, perhaps analyzed as well previously by another bot. Therefore, we envisage that a significant performance improvement can be obtained in agent enabled online processing if Internet agents can share information, analysis of that information and even computational load with one another.

For online agents to interact with each other, every agent should possess more or less similar functionalities so that it can discover and utilize information and resources from other agents. In such a situation Internet agents need to collaborate as peers sharing information, tasks and responsibilities with each other. In this research, we investigate the architecture of a multi-agent system using the peer- to-peer (P2P) computing model[6, 8]. The traditional P2P model employs TCP/IP networking techniques to implement the P2P protocols. In this paper, we first describe the architecture of a P2P system that employs intelligent agents to improve the efficiency of the computations between the nodes of the peer network. Then, we illustrate an implementation scenario for our system to implement dynamic pricing algorithms employed by online sellers.

2 Peer-to-Peer (P2P) Network Protocols

A P2P system is a decentralized and distributed network of nodes that is capable of sharing and distributing resources between themselves. The primary objective of a node in a P2P network is to search and acquire resources available on other nodes in the network and, simultaneously allow other nodes to access resources present on the node itself. For this every node uses a node discovery protocol to determine other nodes in the network and a resource discovery protocol to determine resources on other peer nodes. As shown in Figure 1, in the node discovery protocol a node uses a *ping* message to probe and discover other nodes in the network. A *pong* message is a response to a ping message and it contains information about the replying node including the Internet address, available bandwidth, and networking resources. After a node becomes aware of its peer nodes it uses the resource discovery protocol to discover resources that are possibly present on those peers. The resource discovery protocol comprises the *query* message that is forwarded to successive peers until the resource is discovered or the lifetime of the message expires. If the resource is found on a peer node a *queryHit* message is sent back to the node originating the query. The requesting node and the providing node then decide on the exchange protocol for the resource.

The nodes in a P2P network join and leave the network in an ad-hoc manner. Because of the dynamic nature of the availability of peer nodes and their resources, a node in the P2P network should perform the following tasks continuously as long as it remains connected:

- Determine changes in the network configuration due to new peer nodes that join the network and unresponsive or unreachable nodes that leave the network
- Identify resources on remote peer nodes that the user at the current node is interested in, monitor the availability of those resources and selectively acquire the resources.
- Share the resources present on the current node with other peer nodes.

Intelligent software agents provide a suitable paradigm to implement the tasks at every node of a P2P network in an efficient manner as shown in Figure 2. Once programmed to perform a set of tasks, an intelligent agent can continue to perform those tasks without supervision. Mobile intelligent agents can also selectively migrate to other peer nodes to discover and monitor resources on those nodes. In the next section, we describe some agents that are fundamental for supporting the P2P protocols. We then propose some improvements to the basic P2P protocols that can be obtained by using intelligent software agents.

3 Agents as P2P Mediators

A software agent based P2P system must support the peer discovery and resource discovery protocols in a P2P network. To achieve this, a node should create agents to discover other peer nodes and discover and download resources from the peer nodes as summarized in Figure 2.

Figure 3 above shows the node discovery protocol in an agent enabled P2P system. The stationary controller agent creates a mobile reconnaissance agent and provides a list of addresses that the reconnaissance agent should visit. The address list or itinerary should contain the address of at least one node. Subsequent addresses can be learnt by the reconnaissance agent as it visits other nodes in the P2P network. The reconnaissance agent visits the different nodes contained in its itinerary and obtains node related information from the information agent present on those nodes. At the end of its itinerary, the reconnaissance agent reverts to its origin and presents the node addresses it

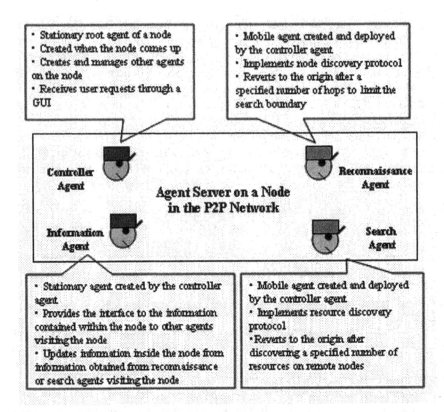

- Stationary root agent of a node
- Created when the node comes up
- Creates and manages other agents on the node
- Receives user requests through a GUI

- Mobile agent created and deployed by the controller agent.
- Implements node discovery protocol
- Reverts to the origin after a specified number of hops to limit the search boundary

Controller Agent

Reconnaissance Agent

Agent Server on a Node in the P2P Network

Information Agent

Search Agent

- Stationary agent created by the controller agent
- Provides the interface to the information contained within the node to other agents visiting the node
- Updates information inside the node from information obtained from reconnaissance or search agents visiting the node

- Mobile agent created and deployed by the controller agent.
- Implements resource discovery protocol
- Reverts to the origin after discovering a specified number of resources on remote nodes

Fig. 2. Different types of agents inside a peer node and their functionalities in an agent enabled P2P system.

has visited and network related parameters obtained from those nodes to the controller agent.

The resource discovery protocol shown in Figure 4 proceeds in a similar manner. The controller agent now creates a search agent and provides the addresses obtained by the reconnaissance agent to the search agent. The search agent also contains the name(s) of the resources the user wishes to locate. The search agent visits each node contained in its itinerary and queries the information agent on that node. At the end of its itinerary, the search agent reverts to its origin and presents the information about the resources obtained from the nodes it has visited to the controller agent.

The agents described in 2 support the basic P2P protocols. However, they do not significantly enhance the performance of the system over a traditional P2P network. In the next section we describe the functionality of some intelligent agents that can be used to make the P2P system more efficient, reliable and easier to use.

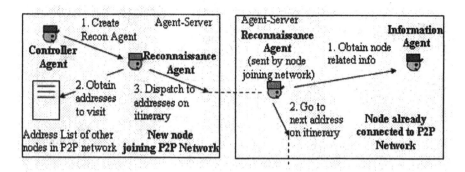

Fig. 3. Node Discovery Protocol using a Reconnaissance Agent in an agent enabled P2P system.

4 Improving P2P Protocols Using Intelligent Agents

Intelligent agents can be used to improve the efficiency and performance of a P2P system in various ways by interacting with other agents in the P2P system, dynamically learning network parameters from peer nodes and by selectively downloading resources according to relevance to user queries. Here, we propose some enhancements to the computations within a P2P system that can be achieved by using intelligent agents:

4.1 Intelligent Search: Query-Classifier Agent

A query request in the traditional P2P protocol is multicast to all the peer nodes to which the node originating the query is connected. To increase the efficiency of the search process, the originating node can maintain a search history from all the peer nodes to which it has already sent queries for resources. The search history can be constructed and even divided into categories by implementing a query classifier agent that identifies peer nodes by their resources. The query classifier agent can use the results returned after the resource discovery phase to determine the availability of resources on peer nodes and the relevance of those resources to the query terms. Later search queries starting from the same node on similar query terms or for similar categories, utilize the ranking of the peer nodes before starting the search and makes the resource discovery phase more efficient.

4.2 Efficient Download: Download Monitor Agent

The availability of peer nodes, their resources and the number of connections each node supports varies dynamically in a P2P network as nodes enter, leave and connect to each other in an ad-hoc manner. Therefore, there is no guarantee that the resources located during the resource discovery phase are still available when the actual download starts. The network related parameters

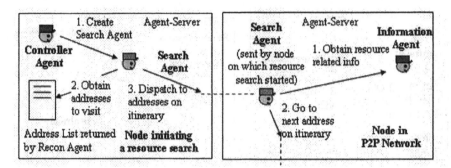

Fig. 4. Resource Discovery Protocol using a Search Agent in an agent enabled P2P system.

of a peer node such as bandwidth and connection speed can also vary dynamically as other nodes connect and disconnect within the system. To increase the reliability of resource download, the user at the node downloading the resource can define threshold parameters for the download such as the minimum download speed and allowable download time. A download monitor agent can be used to select a peer node that currently returns the best values for the download parameters. The download monitor agent can supervise the download process and ensure that the download parameters remain within the specified values. If the parameters fall beyond their threshold, or, if the peer node providing the resource leaves the P2P network while the download is in progress download monitor agent can inform the originating node and determine other peer nodes that might be suitable for a subsequent download of the same resource. If the resource is not available on any other peer node, the download monitor agent can reattempt the aborted download if the node rejoins the P2P network or improves its download parameters.

4.3 Enhancing Resource Discovery: Gossiping Agent

Search agents foraging for resources in the network can improve the efficiency of their search by sharing and exchanging information about the availability and location of resources with other search agents. A search agent is created by a node in response to a user query and the search agent is disposed after the results of the query have been submitted at the node originating the query. User queries are generated at random, and therefore, it is unlikely that search agents from different nodes will remain active at the same time. Therefore, the information exchange between search agents has to be mediated asynchronously by another agent called a gossiping agent. When a search agent arrives on a peer node, the gossiping agent at that node requests information about other nodes already visited by the search agent and resources available on those nodes. The gossiping agent simultaneously allows the search agent to access the gossip it has already collected. Since the information about peer nodes and resources revealed by a search agent can be unauthentic, therefore, we have termed that information unreliable "gossip". A search agent can use a probabilistic belief factor to decide whether or not to believe in the gossip it hears from a gossiping agent.

4.4 Intelligent Peer Discovery: Trail Marker Agent

A reconnaissance agent leaves behind a trail containing the addresses of the peer nodes that it has already visited. Other reconnaissance agents can pick up this trail. Trails are stored in a table indexed by the query category.

4.5 Smart User Interaction: Interface Agents

Like every other software application, a P2P system is likely to be used by a wide variety of users ranging from casual to dedicated users. An interface agent can be used to improve the interaction of the P2P system with its users by creating slightly different versions of interfaces for different types of users. The interface agent can display the appropriate interface depending on the user characteristics. For example, a user who is comfortable with graphics is displayed a GUI that contains more graphics and less text as compared to the default GUI. As another example, an interface that contains few navigational instructions can be used to interact with veteran users of the system while an interface that contains various usage and navigational hints can be used for novices. The interface agents improve the efficiency of the system by reducing the time required by human users to learn and use the software.

4.6 Communities with different service levels: Service Monitor Agent

When a node joins the P2P network, it can specify certain threshold parameters for nodes that it wishes to connect to. These parameters might include available bandwidth, connection speed and number of connections supported. The node joining the P2P network can create a service monitor agent to initiate connections only to those peer nodes that satisfy the threshold parameters. This enables creation of communities of peer nodes, within a large P2P network, with different levels of services possibly with tiered operational charges. A user of the P2P system requiring a certain level of service can then create a node to join appropriate service-level within the P2P network.

4.7 Securing Critical Resources: Security Agents

Security agents can be used to enforce security protocols and trust mechanisms between the peers and are located on the node that they protect. In the future, we envisage mobile security agents that police the entire network

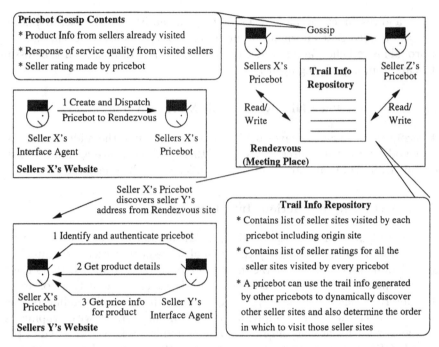

Fig. 5. Pricebot Collaboration Scenario using a agent enabled P2P network.

visiting different nodes to enforce security policies. In a P2P system, security of the nodes and the resources contained on those nodes is important as a malicious node can send rogue agents to sabotage other nodes in the P2P system. Rogue agents can over-populate a certain node resulting in a denial of service attack. They can also spread incorrect gossip to malign the reputation of properly functioning nodes. Therefore, we need to have a trust based mechanism where a node only allows agents that have originated from nodes that it trusts to access its resources. Reconnaissance or search agents sent by un-trusted or partially trusted nodes are provided partial information about a resource and are requested to establish a trust based contract before they can obtain complete information about the resource.

5 Implementation Scenario: Pricebot Collaboration using a Multi-Agent based P2P system

To study the results of our proposed system we have selected the domain of online pricing used by Web based sellers. Online stores selling items over the Internet have matured over the years from text based catalogs of the stores' inventories to intelligent sites that keep track of customer preferences and even charge competitive prices to attract customers from other competing merchants. To achieve this, online sellers frequently employ software agents called pricebots. A pricebot can gather and analyze information from competitors on behalf of the seller, while the seller continues its primary task of selling items to customers. The pricebot also uses a pricing algorithm to dynamically calculate the appropriate price for an item under the current market conditions and reports the price back to the seller[3]. The performance of pricebots can be improved significantly if pricebots from different sellers interact with each other sharing information and perhaps, computational load. For this all pricebots should possess more or less similar functionalities so that every pricebot is capable of discovering and utilizing the information and resources from other pricebots. A scenario for pricebot collaboration within an online seller community is illustrated in Figure 5. In the future, we intend to develop a multi-agent based P2P system that allows pricebots created by different online sellers to cooperate and exchange information with each other so that they can determine the optimal price to be charged for an item by a seller under dynamic market conditions more efficiently.

6 Related Work

P2P systems have become an area of active research and development since the popularity of online file and resource sharing services such as Freenet[4], Gnutella[5], and SETI@home[7]. P2P systems such as Pastry[6], Tapestry[11], PlanetP[2] and Chord[8] propose a scalable, distributed object location and routing algorithm for wide-area peer-to-peer applications and proactively move the data to reduce the distance traveled by the messages in the network. Researchers[10] have also proposed parameters such as interests and local knowledge to improve the overall performance of P2P networks. However, the systems mentioned above are mainly concerned with developing an infrastructure based on message communication between different peers. The Anthill framework[1] uses agent technology to support the development of peer-to-peer systems. Mobile agents or ants move across the nodes in a P2P network to accomplish distributed tasks. However, Anthill uses mobile agents mainly to discover and download resources. Our focus in this paper has been to investigate the interaction and collaboration between intelligent agents in a agent community based on the P2P model to improve the efficiency of agents working individually.

7 Conclusions and Future Work

The P2P model offers several design and implementation challenges. A node in a P2P network needs appropriate techniques for determining peers that possesses useful resources, algorithms for trading those resources between the peers and a trust model for sharing resources. Here we have described the architecture of a multi-agent based P2P system and discussed some of the design issues related to implementing and enhancing a P2P system using mobile intelligent agents. Our future work will also address economic issues such as incentives to nodes that share their resources with peers and preventing free-riders from exploiting the system. We envisage that with the rapid development and deployment of online and real time P2P systems and mobile agents that travel across the Internet, the convergence of these two research areas will generate much research and commercial interest.

References

1. O. Babaoglu, H. Meling and A. Montresor, "Anthill: a Framework for the Development of Agent- Based Peer-to-Peer Systems," Department of Computer Science, University of Bologna, Technical ReportUBLCS-2001-09, November 2001.
2. F. Cuenca-Acuna, C. Peery, R. Martin and T. Nguyen, "Planet P: Using Gossiping to Build Content Addressable Peer-to-Peer Information Sharing Communities," Technical Report DCS- TR-487, Department of Computer Science, Rutgers University, NJ, May 2002.
3. P. Dasgupta, N. Narasimhan, L. Moser, P.M. Melliar Smith, "MAgNET: Mobile Agents for Networked Electronic Trading", IEEE transactions on Knowledge and Data Engineering, Special Issue on Web Technologies, vol. 24, no. 6, July/August 1999, pp 509-525.
4. Freenet, URL http://www.freeproject.org
5. Gnutella Inc., URL http://www.gnutella.com
6. A. Rowstron and P. Druschel, "Pastry: Scalable, decentralized object location and routing for large-scale peer-to-peer systems," Proceedings of the 18th IFIP/ACM International Conference on Distributed Systems Platforms, Heidelberg, Germany, November 2001, pp. 329-350.
7. SETI@Home URL http://setiathome.ssl.berkeley.edu
8. I. Stoica, R. Morris, D. Karger, M. Frans Kaashoek and H. Balakrishnan, "Chord: A scalable peer- to-peer lookup service for Internet applications," Proceedings of the Conference on Applications, Technologies, Architectures, and Protocols for Computer Communications, San Diego, CA, 2001, pp. 149-160.
9. G. Weiss (editor) "Multiagent Systems," MIT Press, 1999.
10. B. Yang and H. Garcia-Molina, "Improving search in peer-to-peer networks," Proceedings of the 22nd International Conference on Distributed Computing Systems (ICDCS'02), Vienna, Austria, July 2002, pp. 5-14.
11. B. Y. Zhao, J. D. Kubiatowicz, and A. D. Joseph, "Tapestry: An infrastructure for fault resilient wide-area location and routing," Technical Report UCB CSD-01-1141, University of California, Berkeley, 2001.

A Soft Real-Time Agent-based Peer-to-Peer Architecture

Fang Chen[1] and Vana Kalogeraki[2]

[1] Department of Computer Science and Engineering, University of California, Riverside, Riverside, CA, 92521, fchen@cs.ucr.edu
[2] Department of Computer Science and Engineering, University of California, Riverside, Riverside, CA, 92521, vana@cs.ucr.edu

Summary. As computers become more pervasive and communication technologies advance, a new generation of peer-to-peer (P2P) networks are increasingly becoming popular for real-time communication, ad-hoc collaboration and resource sharing in large-scale distributed systems. In this paper we present an agent-based peer-to-peer architecture that provides soft real-time guarantees to distributed tasks in peer-to-peer systems. The architecture exploits the urgency of the tasks, the objects at the peers and the resource utilization of the nodes, to dynamically determine an efficient schedule for the distributed activities. The mechanism uses only local knowledge and is entirely distributed and therefore scales well with the size of the system.

1 Introduction

As computers become more pervasive and communication technologies advance, a new generation of peer-to-peer (P2P) systems has become very attractive to provide a single integrated platform for performing large scale distributed computations. Collaborative applications that dynamically interact and share resources (such as CPU cycles, storage and files) [7, 10, 21, 19], are increasingly being deployed in wide-area environments.

A peer-to-peer system creates a virtual - point-to-multipoint - network (overlay) of many peers built on top of the physical infrastructure. The basic characteristic is that there is a group of nodes with the same type of interests connected over the same communication system. The success of the peer-to-peer systems today, has been demonstrated primarily by exchanging objects of "small" size (such as files, instant messages, news and email) in collaborative environments [7, 10, 19]. To support more complicated distributed applications (such as large scale simulations or streaming of media services) that need end-to-end quality of service (QoS) guarantees across multiple nodes such as minimum latency and jitter, several problems remain to be solved.

For example, distributed applications are characterized by potentially variable data rates and sensitivity to losses due to the transmission of data between different locations in local- and wide-area networks. and the concurrent scheduling of multiple activities with different timing constraints and QoS requirements. To provide end-to-end QoS to the distributed applications, one needs to provide predictable computation time, minimum delay and small latency. Providing such end-to-end QoS guarantees to distributed applications requires careful orchestration of the applications and the system resources (processor, network), as remote interactions can lead to excessive utilization and poor quality of service.

The performance of peer-to-peer systems can be significantly improved by employing software agents to coordinate the peer-to-peer interactions. Agents for example can collect information (such as resource utilization loads) from their peers and help to meet the requirements of the distributed applications (*e.g.,* timing requirements) and balance the load on the system resources. This is particularly important due to the characteristics of the environment in which they operate. In dynamic environments, requests arrive concurrently and asynchronously from multiple sources and compete for the same computing resources, thus making it hard to predict the needs of the applications or the availability of the resources in advance.

In this paper we propose an agent-based peer-to-peer architecture that provides soft real-time guarantees to distributed applications. Our architecture manages the applications and the system resources in an integrated way; monitors the utilization of the resources transparently, manages the connections between the peers and establishes new connections in response to changing processing and networking conditions; and schedules distributed invocations dynamically over multiple peers to meet the end-to-end QoS requirements of the applications.

The problem is more difficult in peer-to-peer systems, because the global state of the peer-to-peer system is changing much faster than it can be communicated to the peers. Furthermore, the number and availability of the resources can change at run-time, as new peers join or leave the system. Therefore, the exact structure of the system cannot be known by any single resource manager at any given point in time. These have a direct impact on the scalability, efficiency and performance of the distributed applications.

Our architecture has the following key advantages:

1. *Scalability:* Ability to accommodate hundreds of thousands of nodes and to scale well with respect to the increasing size of the system. It requires little administration cost as each node maintains a small number of peers and manages only the objects in its local data store.
2. *Performance:* Ability to maximize the probability of meeting the soft real-time deadlines of the distributed applications and to minimize end-to-end latency by considering the urgency of the applications when scheduling the invocations across multiple peers.

3. *Adaptability:* Ability to adapt dynamically to changes in the resource availability, system topology and application behavior as it dynamically bal-
ances the processing load across multiple peers.

This paper is organized as follows. Section 2 presents the system model and
Section 3 describes our system architecture. In Section 4 we discuss related
work and Section 5 concludes the paper.

2 System Model

The system model refers to the organization of the peer-to-peer network and
the characteristics of the individual peers, the peer-to-peer communication
protocol, the characteristics of the distributed applications and the scheduling
strategy. We will discuss each of them in detail.

2.1 Peer-to-Peer Network

A peer-to-peer system is modeled as a logical network of nodes (peers) in which
the nodes can be static or mobile in the network. Each node keeps a small
number of connections to other peers; the number of connections is typically
limited by resources at the peer. The node keeps the number of known peers
in a list which is updated dynamically; these are peers with which the node is
currently connected or had previous connections to and is likely to connect to
in the future. For example, the number of connections available by a powerful
workstation could be ten, while for a mobile device cannot exceed two. A
node becomes a member of the system, by establishing a connection with at
least one peer currently in the system. The advantage is that the peers can
join and leave the system dynamically without explicit knowledge of their
memberships or need for group membership algorithms.

A node decides locally whether it should accept a connection to a remote
peer, based on its own incentives, such as if it has enough bandwidth on its
communication links or if it shares a trusted relationship with that peer. Furthermore, if a node discovers that a peer frequently produces good results
to its requests [16], it can attempt to move closer to it in the network by
connecting directly to that peer. If a node (temporarily or permanently) dis-
connects from the system, its peers can choose another node to connect to. As
a result, the topology of the resulting network could be arbitrary with various
connectivity degrees between the peers, which makes managing the system a
difficult problem. The connections are typically initiated by one peer, but they
are symmetric, in that, either peer can send messages or choose to disconnect
at any time.

2.2 Peer-to-Peer Protocol

The P2P protocol is used for establishing connections and propagating requests among a large number of peers. Nodes in the peer-to-peer system communicate by exchanging the following messages:

- *ping* : to discover server peers in the network. A node that receives a ping message can reply with a pong message.
- *pong*: a peer node accepts the connection by replying with a pong message. In addition, it piggybacks information about the node, such as its geographic location and the type of connection it supports.
- *query* : this is the primary mechanism for executing object invocations across multiple processors. If a value needs to be returned, the node will reply with a query_reply message.
- *query_reply* : this is the reply message that responds to a previous request, carrying along the return value of the invocation.

Each query and query_reply message carries a number of parameters in the message header, including a *unique_msg_id* ($UUID$) that distinguishes each message from others, a *descriptor_id* that uniquely identifies the sender of the message, a *laxity* value that represents the urgency of the request and a *time_to_live(ttl)* field that measures the number of times the message will be forwarded among the peers before it expires.

2.3 Application Tasks

Each application consists of a number of application tasks. An application task is defined as a sequence of invocations of objects located across multiple nodes in the peer-to-peer system. A task is executed by a single client thread (such as a user operation) and completes when the client thread finishes execution. Multiple tasks originating from different client threads and multiple sources can be executed in sequence or in parallel.

The object is the smallest scheduling entity. Tasks invoke methods on objects, and objects are scheduled based on the scheduling parameters carried along with the tasks. The advantage is that tasks enable end-to-end scheduling in that they carry scheduling parameters from one node to another yielding a wide area scheduling strategy that requires only local computations. As multiple tasks execute, they compete for the same computing resources. Tasks can be preempted only by other tasks that execute with higher priority.

Each task is characterized by its deadline, which is the time within which the task should complete and the estimated execution time which is the estimated amount of time required for the task to execute. We assume that these parameters can be provided by the application user in advance, or defined automatically by the system as it executes.

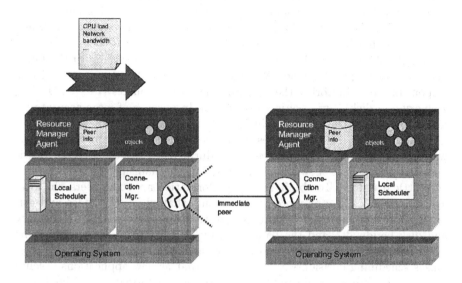

Fig. 1. The agent-based peer-to-peer architecture.

2.4 Scheduling Objective

The objective of our scheduling algorithm is two-fold: First, we want to implement a soft real-time scheduling strategy that schedules distributed tasks in peer-to-peer systems. Second, we want to maximize the number of tasks that meet their deadlines and dynamically balance the load on the system resources. By exploiting the physical location, resource capabilities and objects available at the peers we can give higher priority to tasks with higher urgency.

3 Our Architecture

Figure 1 shows our agent-based peer-to-peer architecture. There is a Connection Manager on each node, responsible for managing the connections at the node and maintaining current peer information. Because of the dynamic nature of the network, the Connection Manager has only a limited view of the system, in terms of the peers in the network and the resource availability. For example, it knows the physical (IP address) and logical address (1 hop away) of its immediate peers, but it learns about other peers through its immediate peers. For example, it learns about remote peers by looking at the information carried with the message invocations. As it discovers new nodes (*e.g.,* new peers joining the system), it decides whether to make new connections and drop existing ones. Periodically, the Connection Manager monitors the current bandwidth on its communication links by sending ping messages to its peers. The Connection Manager chooses to disconnect from a peer when it discovers that it over-uses its resources.

The local Resource Manager agent works in concert with both the Local Scheduler and the Connection Manager. This is a stationary agent which is responsible for monitoring the usage of the local resources (*i.e.*, CPU load, memory and disk) during object executions. In absence of more accurate information about the tasks, profiling mechanisms can be employed to monitor local and remote execution times of the objects as invoked by the tasks as well as the load imposed on the resources during the executions. In our previous work [9] we have shown that such profiling mechanisms can be used with very little overhead. To obtain resource information about its peers, the local agent can deploy one or more mobile agents that visit peer nodes and collect resource management information, in terms of CPU utilization, memory and disk bandwidth and the bandwidth on their communication links. The agent uses this information to estimate what is the load of its peers when it makes scheduling decisions. Note, however, that as new applications execute and tasks are created and completed, the resource information collected from remote nodes may be out of date. Therefore, the Resource Manager agent has to decide the frequency with which it will generate mobile agents and the nodes they should visit.

The Local Scheduler on the node is responsible for specifying a local ordered list (schedule) for the tasks which defines how the objects will be invoked. The Scheduler implements the least laxity scheduling algorithm (LLS) which has been shown to be efficient in distributed systems [4]. The initial laxity value of the task is computed as a difference between the deadline and the estimated time for the task to complete. The local agent measures the execution time of the task as the sum of the execution times of all the objects invoked by the task. This includes the processing time of the objects in each node and the communication time to propagate the object invocation into a remote node and receive a reply back. The Scheduler uses this information to determine the laxity value of the tasks.

3.1 Soft Real-Time Scheduling Algorithm

The Scheduler on each node does not have a priori information of the objects at the peers or the user requests, especially because the structure and the resource availability of the system changes so rapidly that its state cannot be communicated into a single node. Furthermore, as the objects are allocated and reallocated in the system, the execution times of the tasks may vary under different resource utilizations.

For such dynamic environments, soft real-time algorithms are more applicable to meet the timeliness requirements of the tasks. The goal in soft real-time scheduling is to maximize the number of tasks that meet their deadlines and guarantee smooth operation in the system. This is different from hard real-time systems where timeliness is satisfied by meeting the deadlines of all the tasks; if one task misses its deadline, the system cannot operate correctly. A soft real-time system, on the other hand, can survive some deadline misses. In addition, it allows the task to continue execution even after its deadline has passed, as long as it does not delay the execution of other tasks in the system.

The Local Scheduler implements the least laxity scheduling algorithm which has been proven efficient in distributed systems [4, 9]. In least laxity scheduling, each application task is associated with a laxity value which represents the measure of urgency for the task and determines the priority of the task in the system. The laxity value for each task is computed as:

$$Laxity_t = Deadline_t - EstCompTime_t$$

where $Deadline_t$ is the time within which the task must complete and $EstCompTime_t$ is the estimated time for the task to execute. The estimated computation time of the task is computed as the sum of the computation times of the objects invoked by the task and the corresponding communication times. The task with the smallest laxity value has the highest priority for the system.

As new tasks are introduced, objects are invoked and existing tasks finish execution, each Local Scheduler makes decisions as follows (Figure 2): (a) when a new task arrives for execution locally at the node, (b) when the executing object completes its execution, (c) when an object makes a remote invocation to an object on a peer node, and (d) when the executing task misses its deadline (negative laxity value). In each of these cases, the Scheduler will evaluate the urgency of the objects queued for execution. If, for example, a new task is more urgent than the task currently executing on the node, the executing task will be suspended and a new thread carrying the new task will be dispatched for execution. The laxity value of the task will also determine the priority of the thread that will carry the task execution. The interrupted task will be inserted back in the local queue and will be ordered based on its scheduling priority.

Figure 3 demonstrates the execution of our least-laxity scheduling algorithm for a task that executes objects across multiple peers. As the task executes, the objects invoked by the task are scheduled according to the laxity value of the task. For example, at time 1200 milliseconds, another task with a higher priority (smaller laxity value) arrives at the node, therefore, the current task is interrupted and the new task is dispatched for execution. The interrupted task resumes execution when the new task finishes and its priority is higher than all the tasks scheduled for execution at the node. A task with a negative laxity value indicates that the task has missed its deadline, even if the deadline has not passed yet.

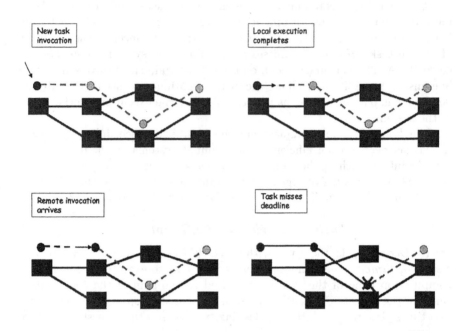

Fig. 2. The Local Scheduler makes scheduling decisions in the above four cases.

4 Related Work

Research into scheduling has been dominated by hard real-time systems, but some useful results are available for soft real-time distributed systems. Several researchers [4, 8, 20] have shown that least laxity scheduling is an effective algorithm for scheduling real-time distributed systems. Abeni *et al* [1] use probabilistic models to analyze the task arrival and service to estimate the deadline guarantee of a given task set. Most of these strategies [12] have shown that efficient dynamic scheduling algorithms require knowledge of real-time task information including the task's deadline, resource requirements and worst-case execution time.

Peer-to-peer systems have received much attention recently for sharing information and resources in large-scale systems [6, 7, 13, 14]. Many of these research efforts have focused on employing local search strategies [2, 11, 23] to improve the efficiency and performance of these systems. P2P systems have inspired work on building persistence utility infrastructures [10, 17, 19, 22]. These employ location and routing algorithms [18] to guarantee a bounded number of messages, while retaining the scalability and self-organizing properties of the peer-to-peer networks.

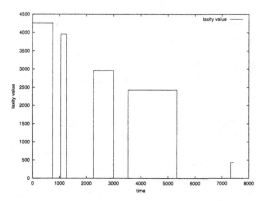

Fig. 3. Laxity value of a task that executes objects across multiple nodes in the peer-to-peer system.

Recent work has proposed integrating peer-to-peer networking and computing in a common agent-based framework [3, 5, 15]. Most of this work however focuses on dynamic resource discovery in peer-to-peer networks. Our work is different in that it proposes a soft real-time scheduling architecture for agent-based peer-to-peer systems.

5 Conclusions

Peer-to-peer systems are increasingly becoming popular for enabling direct and real-time communication among many participants in wide-area environments. Such systems allow individual computers to share data (e.g., files) and computing resources (e.g., CPU, storage) directly, without the need for dedicated servers. In this paper we presented an agent-based peer-to-peer architecture for providing end-to-end soft real-time guarantees to distributed applications. The system employs local agents, Schedulers and Connection Managers that monitor the behavior of the objects, measure the usage of the resources and manage the connections to the peers. The scheduling algorithm is driven by the least laxity scheduling algorithm that uses the urgency of the applications to schedule distributed object invocations and allows the system to balance the load across multiple peers and to adapt to changes in the application behavior or the processing and networking conditions dynamically.

References

1. Abeni, L. Buttazzo, G., "QoS guarantee using probabilistic deadlines", *Proceedings of 11th Euromicro Conference on Real-Time Systems. Euromicro RTS'99*
2. L. A. Adamic, R. M. Lukose, A. R. Puniyani, B. A. Huberman, "Search in power-law networks", *Phys. Rev. E, 64 046135 (2001).*

3. The anthill project, http://www.cs.unibo.it/projects/anthill/
4. M. L. Dertouzos and A. K.-L. Mok, "Multiprocessor on-line scheduling of hard real-time tasks," *IEEE Transactions on Software Engineering*, vol. 15, no. 12 (December 1989), pp. 1497-1506.
5. C. R. Dunne, "Using Mobile Agents for Network Resource Discovery in Peer-to-Peer Networks," *SIGecom Exchanges*, vol. 2, no. 3, pp 1-9 (2001).
6. The Freenet Homepage, http://freenet.sourceforge.net
7. The Gnutella Homepage, http://www.gnutella.wego.com/
8. Hildebrandt, J. Golatowski, F. Timmermann, D., "Scheduling coprocessor for enhanced least-laxity-first scheduling in hard real-time systems", Proceedings of 11th Euromicro Conference on Real-Time Systems. Euromicro RTS'99
9. V. Kalogeraki, "Resource Management for Real-Time Fault-Tolerant Distributed Systems," *Ph.D. Thesis*, Department of Electrical and Computer Engineering, University of California, Santa Barbara
10. J. Kubiatowicz, D. Bindel, Y. Chen, S. Czerwinski, P. Eaton, D. Geels, R. Gummadi, S. Rhea, H. Weatherspoon, W. Weimer, C. Wells and B. Zhao, "OceanStore: An Architecture for Global-Scale Persistent Storage," *Proceedings of ASPLOS*, Cambridge, MA (2001).
11. The Limewire Homepage, http://www.limewire.com/
12. G. Manimaran and c. R. Murthy, " An efficient dynamic scheduling algorithm for multiprocessor real-time systems," *Proceedings of the IEEE 20th Real-Time Systems Symposium*, Phoenix, AZ (December 1999), pp. 315-326.
13. D. Milojicic, V. Kalogeraki, R. Lukose, K. Nagaraja, J. Pruyne, B. Richard, S. Rollins and Z. Xu, "Peer-to-Peer Computing", *HP Technical Report, HPL-2002-57*.
14. The Napster Homepage, http://www.napster.com/
15. B. J. Overeinder, E. Posthumus, and F. M. T. Brazier, "Integrating Peer-to-Peer Networking and Computing in the Agentscape Framework," *Second International Conference on Peer-to-Peer Computing*, Linkping, Sweden.
16. M. Ramanathan, V. Kalogeraki and j. Pruyne, "Finding Good Peers in Peer-to-Peer Networks," *Proceedings of the International Parallel and Distributed Computing Symposium*, Fort Lauderdale, Florida (April 2002).
17. S. Rathnasamy, P. Francis, M. Handley and R. Karp, "A scalable content-addressable network," *Proceedings of the ACM SIGCOMM*, San Diego, CA (August 2001).
18. S. C. Rhea and J. Kubiatowicz, "Probabilistic Location and Routing", *Proceedings of the INFOCOM 2002*, New York (2002)
19. A. Rowstron and P. Druschel, "Storage management and caching in PAST, a large-scale persistent peer-to-peer storage utility", *Proceedings of the 18th SOSP*, Toronto, Canada (2001).
20. F. Sandrini, F. D. Giandomenico, A. Bondavalli and E. Nett, "Scheduling solutions for supporting dependable real-time applications," *Proceedings of the IEEE Third International Symposium on Object-Oriented Real-Time Distributed Computing*, Newport Beach, CA (March 2000), pp. 122-129.
21. The Seti@Home Home page, http://setiathome.ssl.berkeley.edu
22. I. Stoica, R. Morris, D. Karger, M. Frans Kaashoek and H. Balakrishnan, "Chord: A scalable peer-to-peer lookup service for Internet applications", *Proceedings of the ACM SIGCOMM Conference*, San Diego, CA (August 2001).
23. B. Yang and H. Garcia-Molina, "Comparing Hybrid Peer-to-Peer Systems", *Proceedings of Very Large Databases*, Rome, Italy (September 2001).

Social Networks as a Coordination Technique for Multi-Robot Systems

Daniel Rodic[1] and Andries P. Engelbrecht[1]

[1] Department of Computer Science, University of Pretoria, Pretoria 0002, Republic of South Africa

Abstract

The last decade saw a renewed int erest in the robotics research field and a shift in research focus. In the eighties and early nineties, the focus of robotic research was on finding optimal robot architectures, often resulting in non-cognitive, insect-like entities. In recent years, processing power available to autonomous agents has improved and that has allowed for more complex robot architectures. The focus has shifted from single robot to multi-robot teams. The key to the full utilisation of multi-robot teams lies in cooperation. Although a robot is a special case of an agent, many existing multi-agent cooperation techniques could not be directly ported to multi-robot teams. In this paper, we overview mainstream multi-robot coordination techniques and propose a new approach to coordination, based on models of organisational sociology. The proposed coordination model is not robot specific, but it can be applied to any multi-agent system without any modifications.

Keywords: Coordination, Multi-Robot Teams, Self-organisation, Socio-economic models, Multi-Agent Systems

1. Introduction

Until recently, robots have been seen as a novelty. Today, the variety of robotic applications is growing at a tremendous rate and the trend will carry on in future, as the progress in technology opens new possibilities for applications. A single robot system is not an optimal solution for all applications. The growing range of existing and envisaged tasks benefits from applications of multi-robot teams, such as search and rescue tasks, mapping of hazardous/hostile environments and space exploration/colonisation.

There are numerous reasons to prefer the use of multi-robot systems over single robot and some of them are enumerated below:

? Multi-robot teams can accomplish certain tasks faster than a single robot. Examples of such tasks are mapping and foraging tasks [3][10].

? Multi-robot teams are more fault-tolerant than a single robot. If implemented properly, the coordination mechanism must allow for reorganisation due to the failure of a member of the team.

? Multi-robot teams can accomplish tasks that are not achievable by a single robot, for an example, box-pushing tasks [13].

Despite reported progress in coordination techniques for multi-robot teams, the issue of coordination and cooperation is still a very open one. The mainstream approaches are overviewed and a new one is proposed in this paper.

2.   Swarm Robotics   – Coordination Perspective

Before considering multi-robot systems inspired by insects (commonly referred to as swarm robotics systems), it is useful to consider some fundamental differences between real, biological agents in insect colonies and those in higher, typically mammalian societies. The differences are discussed from the cognitive and the social perspectives. A full comparison is outside the scope of this paper and only the characteristics related to cooperation and coordination are considered.

Ability to learn. Insects have a much lower level of self-awareness and their individual learning ability is often non-existent. Insects typically do not develop memories and do not learn from past experience. Mammals do.

Communication methods. Insects have much simpler communication mechanisms that prohibit exchange of complex messages. Mammal societies usually employ complex means of communication. The other related issue is the localised nature of insect communications. They usually communicate by touch and/or chemical reactions [2], and sound is rarely employed. Mammals can communicate over greater distances.

Individualism. In insect colonies, most agents are homogenous. In other words, the insect colonies are an anonymous society where agents of the same type are indistinguishable. In contrast kinship and other social relationships are of extreme importance to mammalian societies.

Cooperation is a form of positive interaction between the agents and it is a process of working together to achieve a common goal. Cooperation usually requires some kind of coordination and/or negotiation mechanism. A view expressed by Mataric [11] is that

"...any cooperative behaviours that require negotiation between agents depend on direct communication in order to assign particular tasks".

Direct communication, when a specific agent is identified and addressed, is not possible in insect societies, due to the lack of individualism and their limited communication mechanisms. However, it would be wrong to say that insects do not cooperate, they do, but through much simpler mechanisms based on interaction with their environment, using the principle of stigmergy. The stigmergy principle is derived from observations of social insect colonies, such as bees and ants. The process of stigmergy is described as:

"The production of a certain behaviour as a consequence of the effect produced in the local environment by previous behaviour" [2].

Due to the lack of individualism, hierarchies and social relations between agents in insect colonies are virtually non-existent. In mammalian societies, hierarchies and social relations play a fundamental role in the organisation of such a society. Cooperation models are often based on hierarchical and role-based models. Insect colonies, as a cooperation model, were and still are very attractive for the applications in robotics. The main advantages of swarm robotics are that the insect-like robots are fairly simple to construct; cooperation between such robots should be an emergent property of such a system. Swarm robot teams are usually fault-tolerant (to a degree, because if a sufficient number of agents fails then the whole team might fail).

The subsumption architecture [3] (also often referred to as reactive architecture) can be seen as an example of an insect inspired robotic system, and some of the characteristics are discussed below. Many researchers are taking this architecture as a starting point and evolving it into behaviour based robotics [12].

There are a few fundamental, underlying concepts on which reactive architectures are built:

? **Situatedness** – an agent is situated in its environment and it interacts with the environment directly, without building a world model

? **Embodiment** – Brooks [4] argues that the only way to make sure that an agent can function in the real world is if it has a physical body.

? **Intelligence** – The approach for intelligence modelling adopted in reactive architectures is the bottom-up approach. Reasoning is also deemed unnecessary as it relies on a symbolic world model.

? **Emergence** – The idea is that from an agent's interaction with its environment, intelligent behaviour will emerge

Coordination in such insect-like multi-robot systems was initially done through interaction with the environment, using the principle of stigmergy[2]. The stigmergy-based coordination mechanism was very limited and, although emergent cooperative behaviour was observed [17], it has imposed limitations on cooperation methods. The need for a more capable communication mechanism, even in insect like societies, has been recognised relatively early in research and it has lead to various communication-capable behaviour-based multi-agent systems [10]. A summary of the differences between the agent models is given in the table below:

Table 1. Differences between agent models

Characteristic of an Agent	Insect Colonies	Mammal Societies
Communication	Sparse, localised	Complex
Individualism	No	Yes
Learning ability	No	Yes

3.   Markets and Hierarchies – Traditional Coordination

Agents in a Multi-Agent System (MAS) can exhibit cooperative or competitive behaviours. In robotics, the cost of building a robot is relatively high thus the competitive approach is often not desirable. It is important to note that cooperative behaviour does not exclude market based competitive approaches, such as the auctioning coordination technique. The traditional coordination mechanisms for cooperation are the market-based approach and the hierarchical approach.

3.1   Market Based Approach

Markets are based on voluntary exchange of commodities between parties at some price and market based coordination is based on the same premise. Markets have many properties, but from the point of view of its applicability to robotics, the following properties are of main interest:

? **Self-organisational property**: markets are self-organising through a pricing mechanism. This property is highly desirable as it helps with the social approach to MAS design, where agents are viewed as a self-organising society.
? **Demand-supply relationship** supply and demand are inseparable and self-regulating. This relationship assumes the existence of two entities: a buyer and a seller.
? **Scalability**: in theory there is no limit to the number of participants in a market.

Ideally, when the market functions properly, there is an equilibrium price that is the fairest cost of the transaction and coordination is nearly optimal. The idea of using market based coordination in MAS in particular, and in AI in general, has lead to the development of various auction-based algorithms. One of the most widely used coordination mechanisms used in AI is the Contract Net Protocol

(CNP). CNP is based on auctioning [18] and it assumes existence of a buyer, a seller and a price.

With regard to robotics, auction based coordination has been mainly applied to simulated multi-robot teams. An example of such an application is given in [5]. Recently, the first real embodied agent systems that use auction based coordination have appeared [8]. It may be too early to judge the success of such coordination based MAS in real embodied agent applications, but there are some concerns with the future of a purely auction based approach. Firstly, the method for awarding a bid must be determined. It is usually a metric or fitness function that is used to determine a winning bidder. In case of a task that has not been done before it is uncertain how to determine such a function. Secondly, the auctioning mechanism relies on objective information from the bidding agents. The information that is submitted is not necessarily accurate, either due to the inaccuracies of its sensors (if the bid award mechanism relies on self evaluation, as in MURDOCH [8]) or due to some other problem, such as the "eager bidder" scenario. Thirdly, it is unclear how a purely auction-based approach can handle a scenario when a task exceeds the capabilities of each individual bidder. One of the main strengths of MAS is to allow for solving a problem that exceeds each individual agent's capability.

3.2   Hierarchical Approach

Probably the simplest way of coordinating agents is by establishing a relatively strict hierarchical architecture that prescribes roles for each agent. Such an approach often assumes a globally coordinated multi-robot system. The coordination task is done either by a specialised agent [7] or by an agent that has been temporally assigned the coordination role [9]. The hierarchical approach often uses a symbolic based planner that can provide the optimal solution based on its symbolic environment model. There are numerous potential problems with such an approach and more can be found in [20]. Other potential shortcomings of the hierarchical approach are that it is somewhat contrary to the idea of autonomous agents and that it prohibits optimal load balancing due to the extreme specialisation of each agent in the system. The hierarchical approach can also lead to easier specialisation of agents; as they can have different physical and logical characteristics. One of the extreme specialisations is that of MACTA [1], where a deliberative agent is a desktop PC and the other team members are physical robots. Specialisation has a positive side: the agents can be designed according to the role they perform and that, in turn, should lower the complexity and cost. The side effect of specialisation is that redundancy is reduced: one robot cannot easily replace the one that has failed due to the physical and logical differences between them. If a centralised coordination mechanism is employed, the role of a reliable communication channel is crucial. In case of a failure of either the global planner or the communication channel, the hierarchical multi-robot system will fail – hence the single point of failure syndrome. In certain environments, communication channels can be limited and unreliable (due to the physical properties of the environment) and it is questionable if a hierarchical approach would be effective in such applications.

4.   Social Networks

Robots in multi-robot teams can be viewed as members of a society. Zini, for example, defines MAS as "...a society of agents" [21]. With that in mind, a multi robot -system can be defined as:

"...a social and cognitive entity with a relatively identifiable boundary, that functions on a relatively continuous basis through the coordination of loosely interdependent, cognitive and autonomous agents" [19].

Traditionally, societies are organised according to socio-economic structures such as markets, hierarchies and networks. Networks have been identified as social organisational structures only relatively recently, with one of the first definitions of social networks given by Mitchell [20]:

"A network is generally defined as a specific type of relation linking a defined set of persons, objects or events"

The membership to a social network is not exclusive: a member can belong to multiple social networks. A social network can be seen as a social group. Examples in human society abound: a person can be a member of a sport club, a university study group and a family. In human societies, social networks are present in everyday interactions but they are not always simple to express and quantify. Key questions applicable to social networks are how the relations that define them are formed and how they are maintained [19]. In more complex animal societies, concepts of kinship and trust form the fundamental role in the creation of such networks. For the purpose of multi-robot teams, different criteria must be used and these are presented in the next section. The societies with multiple, well-established social networks have one great advantage: they are flexible enough to allow the best team for the task to be selected, using the most appropriate social relation, that in turn defines the social group. For example, if a task involves participation in a sport, the member belonging to the sport club that practices that sport should be used to form a team. Affinities between social groups can also play a significant role. If there is an uncertainty about the task, and no social group satisfies demands of that task, then a group with the highest affinity for such a task should be selected. The existence of such affinity relationships between social groups is very important when there is uncertainty (lack of detailed information) about a task or in the case where the best candidates for a task are already allocated to some other task. These features (that can be seen as robustness and fault tolerance) make social networks an attractive tool for coordination between multi-robot teams.

5.   Social Networks in Multi-Robot Teams

Social networks in INDABA [16] are utilised as a coordination and optimisation tool and they play a crucial role in team selection and formation as well as in task allocation. INDABA is a hybrid multi-robot architecture that utilises a decentralised coordination system. The question of formation and maintenance of social networks is answered through a reward mechanism that strengthens or weakens social relations, based on the successful execution of an allocated task. Coordination in INDABA is achieved through a cooperative problem solving process. Current literature divides the cooperative problem solving into four stages [6]:

? **Potential Recognition**: where an agent investigates which agents are capable of executing the task, and tasks are allocated.
? **Team Formation**: where agents start to share a common goal, and self organise to form teams to achieve this common goal.
? **Plan Formation**: where an agent (manager) or a whole team decides on the division of a goal into subtasks and on the allocation of those subtasks.
? **Plan Execution**: where agent or, in the case of a team, agents execute their allocated subtasks.

INDABA proposes five stages with the addition of an additional task:
? **Task Success Evaluation**: where the manager awards points based on the full or partial success of the allocated task. These points are used to reinforce social networks and their affinity towards the task allocated.

For the purpose of the discussion on the role of social networks in INDABA, plan formation and plan execution are of little importance and they are not discussed in this paper. The rest of this section overviews potential recognition, team formation and task success evaluation processes as proposed in INDABA.

5.1   Potential Recognition

The potential recognition and task allocation functions are implemented as an auctioning house mechanism, based on Contract Net [18]. If a task is very complex, the auctioning agent will try to subdivide the task into smaller tasks.

Contract Net works particularly well in the situation of a "market economy" where there are competing agents, willing and capable to perform a task. Although some authors [6] believe that Contract Net works only in these conditions, INDABA allows for extension of Contract Net by allowing teams to be formed upon the consensus obtained through negotiations between agents.

In other words, INDABA provides the cleanliness and efficiency of Contract Net, but it also caters for more complex problem domains where there is uncertainty about tasks and suitability of each agent for a specific task. This is achieved by maintaining social networks that are based on reward points and affinity relations for each agent. By means of maintaining affinity relations for each agent, INDABA provides a mechanism for the development of social networks. Social networks, in turn, provide specialisation amongst the agents, as opposed to other frameworks that require all agents' capabilities to be known and determined up-front.

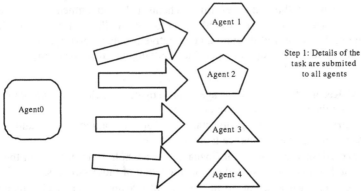

Fig. 1. Agent0 propagates task details

In figure 1, *agent3* and *agent4* are members of a social group that has a strong affinity to the type of task that is submitted, but the capabilities of each agent in that group are not sufficient enough to execute the task in isolation. *Agent1* is a member of a group that has almost total affinity to the submitted task and each member of that group can execute the submitted task in isolation. Each of the agents submits a bid based on its internal evaluation mechanism. The evaluation mechanism takes into account the affinity of the agent to the particular task, based on its history. It also takes into account the risk factor involved. Some agents will have a more conservative bidding evaluation mechanism than others. Ideally an agent will respond to a bid if it determines that it can successfully execute the allocated task, and it bids, using some cost. In order to counter the well-known problem of "eager bidder", the INDABA auctioning mechanism takes into account the agent's credibility, based on previous track record. Ultimately, the auctioning agent, *agent0*, or in terms of Contract Net, the manager, must form his own beliefs about the capabilities, reliability and cost of each bidder. This belief is influenced by the historic performance of such a bidder, as well as its trust points. The manager must also perform the initial breakdown of the task into subtasks, if need be. This breakdown is achieved by utilising broad plans stored in a plan library. Part of the manager's functionality is to decide on the trust points allocated to each task, based on a previous history of successes and failures if the task is not given, as an initial condition, any trust points. All task types have some predetermined default settings that get refined throughout the system's execution.

5.2 Team Formation

When the task exceeds the capability of a single agent, then cooperation between agents is required. Some agents might subdivide the task by forming teams. In the figure 2, *agent1* is fully capable of executing the task on its own, while *agent3*, while it has affinity for the required task, it cannot complete the task on its own.

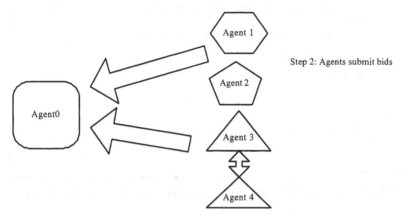

Fig. 2. Team formation initiated by agent3

In the team scenario, the primary agent allocates points and its affinity to organise increases. The cost of the bid is the accumulated cost of the whole team who are involved in that bid. Distribution of the reward is negotiated between the primary agent (confirming its role as a team leader) and the team members, taking into account the strength of the social relationship between them.

The manager now allocates the task to one or multiple bidders, based on some evaluation function (such as a combination of the minimum bid and the confidence in the agent that submitted the bid). Task execution takes place and upon completion (successful or unsuccessful) the task success evaluation process takes place.

5.3 Task success evaluation

If the execution of the task was successful, the affinity bond between the agents that have participated in the same team will be reinforced. In the case of failure, the opposite will take place. Task success evaluation is a function that awards agents that have successfully completed their allocated tasks, with or without the help of other agents. The manager keeps a "score board" that keeps the history of an agent's performance. This scoreboard must differentiate between various task types in order to keep track of the agent's affinity to a particular task type.

The potential recognition and team formation process are described in agent pseudo-code in figure 3:

```
If not Auctioning agent
Receive task attributes
For all social groups that agent is a member of
    If affinity of social group to task > threshold
        If group leader
                Divide task into subtasks
                Collect subtasks bids from other agents in group
                Submit collective bid to auctioning agent
        Else if
                Receive subtask details from group leader
                Evaluate subtask effort
                Send bid to group leader
        End Else If
    End If
End for
Else if
        Send all agents task details
        Collect Bids
        Award Bid
End Else if
```

Fig. 3. Pseudo-code for INDABA potential recognition and team formation processes

6.   Conclusions and Future Work

This paper overviewed the three main coordination approaches used in multi-robot teams. The first, based on swarm robotics, has properties of self-organisation and fault tolerance, but it suffers from the lack of a learning mechanism and it does not, in its standard form, cater for heterogeneous teams. The next two approaches, market based and hierarchies both have their advantages and disadvantages. For example, markets can suffer from a number of conditions that lead to market failure and hierarchies suffer from the consequences of having a centralised planning system. This paper proposed a novel coordination method that introduces social networks as a coordination tool. In INDABA[16], these are coupled with a market based auctioning mechanism. This coordination mechanism allows for the applicability of the market approach even in conditions when traditional markets fail. The model has been tested on test scenarios and it will be implemented in virtual and real robot teams.

References:

1.   Aylett RS; Coddington AM; Barnes DP, Ghanea-Hercock RA (1997) What does a planner need to know about execution? in Steel S, Alami R (eds): Recent advances in AI planning, Springer, pp26-38

2. Beckers R, Holland OE, Deneubourg JL (1994) From local actions to global tasks: stigmergy and collective robotics. In Brooks R, Maes P (eds) Proceedings of fourth international workshop on artificial ife, MIT Press, Boston, pp181-189

3. Brooks RA (1986) A robust layered control system for a mobile robot. IEEE Journal of Robotics and Automation, vol 2 (1):pp14-23

4. Brooks RA (1991) Intelligence without reason. AI Memo 1293, MIT Artificial Intelligence Laboratory, MIT

5. Dias MB, Stentz A (2000) A free market architecture for distributed control of a multirobot system. In Pagello et al. (eds), Proceedings of the 6th international conference on intelligent autonomous systems IAS6, IOS Press, pp115-122

6. Dignum F, Dunin-Keplicz,B, Verbrugge R (2000) Agent theory for team formation by dialogue. In. Castelfranchi C, Lesperance Y (eds), Seventh workshop on agent theories, architectures, and languages, Boston, pp141-156

7. Fox D, Burgard W, Kruppa H, Thrun S (2000) A probabilistic approach to collaborative multi-robot exploration. Autonomous Robots vol 8 (3):325-344

8. Gerkey BP, Mataric MJ (2002) Sold!: Auction methods for multirobot coordination. IEEE Transactions on Robotics and Automation, vol 18 (5):758-768

9. Lima P, Ventura R, Aparicio A, Custodio L (1999) A functional architecture for a team of fully autonomous cooperative robots. RoboCup, pp378-389

10. Mataric MJ (1994) Interaction and intelligent behaviour. Ph.D. thesis, MIT

11. Mataric MJ (1995) Issues and approaches in design of collective autonomous agents, Robotics and Autonomous Systems 16:321-331

12. Mataric MJ (1999) Behaviour Based Robotics. In: MIT Encyclopaedia of Cognitive Science. MIT Press, Boston, pp74-77

13. Mataric MJ, Nilsson M, Simsarian K (1995) Cooperative multi-robot box-pushing. In IEEE/RSJ IROS, pp556--561

14. Mitchell JC (1969) The concept and use of social networks. Social networks in urban situations, Manchester University Press, Manchester pp1-50

15. Panzarasa P, Jennings NR (2001) The organisation of sociality: A Manifesto for a new science of multi-agent systems. In Proceedings of the tenth European workshop on multi-agent systems

16. Rodic D (2002) INDABA – Proposal for INtelligent Distributed Agent Based Architecture. Technical Report, University of Pretoria

17. Sen S, Sckaran M, Hale J (1994) Learning to Coordinate without Sharing Information. In Proceedings of the Twelfth National Conference on Artificial Intelligence,. Seattle, pp426- 431.

18. Smith RG (1980) The Contract Net Protocol: High-Level Communication and Control in Distributed Problem Solver. IEEE Transactions on Computers C29 vol 12:1104-1113

19. Thompson G (1991). Markets, Hierarchies & Networks: In Thompson G, Frances J, Levacic, Mitchell J (eds) The Coordination of Social Life, Sage Publications, London

20. Wooldridge M, Jennings NR (1995) Intelligent Agents: Theory and Practice. The Knowledge Engineering Review, 10 (2):115-152

21. Zini F (2001) CaseLP: A Rapid Prototyping Environment for Agent Based Software, Ph.D. thesis, University of Genoa

Biology-Inspired Approaches to Peer-to-Peer Computing in BISON *

Alberto Montresor and Ozalp Babaoglu

Department of Computer Science, University of Bologna, Mura Anteo Zamboni 7, 40127 Bologna, (Italy), E-mail: {montresor,babaoglu}@CS.UniBO.IT
Tel. +39 051 2094973 Fax +39 051 2094510

Summary. BISON is a research project funded by the European Commission that is developing new techniques and paradigms for the construction of robust, self-organizing and self-repairing information systems as ensembles of autonomous agents that mimic the behavior of some natural or biological process. In this paper we give a brief overview of BISON, discuss some preliminary results for peer-to-peer systems, describe the ongoing work and indicate future research directions that appear to be promising.

1 Introduction

Modern information systems that are distributed are gaining an increasing importance in our every day lives. As access to networked applications become omnipresent through PC's, hand-held and wireless devices, more and more economical, social and cultural transactions are becoming dependent on the reliability and availability of distributed services.

As a consequence of the increasing demands placed by users upon networked environments, the complexity of modern distributed systems has reached a level that puts them beyond our ability to design, deploy, manage and keep functioning correctly through traditional techniques. Part of the problem is due to the sheer size that these systems may reach, with millions of users and interconnected devices. The other aspect of the problem is due to the extremely complex interactions that may result among components even when their numbers are modest. Our current understanding of these systems is such that even minor perturbations (e.g., a software upgrade, a failure) in some remote corner of the system will often have unforeseen, and at times catastrophic, global repercussions. In addition to being fragile, many situations (e.g., adding/removing components) arising from their highly dynamic

---

* This work was partially supported by the Future & Emerging Technologies unit of the European Commission through Project BISON (IST-2001-38923).

environments require manual intervention to keep information systems functioning.

In order to deal with the scale and dynamism that characterize modern distributed systems, we believe that a paradigm shift is required that includes self-organization, adaptation and resilience as intrinsic properties rather than as afterthought. With this motivation, we have initiated BISON (*Biology-Inspired techniques for Self Organization in dynamic Networks*), an international research project funded by the European Commission. BISON draws inspiration from natural and biological processes to devise appropriate techniques and tools to achieve the proposed paradigm shift and to enable the construction of robust and self-organizing distributed systems for deployment in highly dynamic modern network environments.

2 Overview of the BISON Project

Nature and biology have been a rich source of inspiration for computer scientists. *Genetic algorithms* [4], *neural networks* [8] and *simulated annealing* [10] are examples where different natural processes have been mimicked algorithmically to successfully solve important practical problems. More recently, *social insects* [2] have been added to this list, inspiring biomimetic Algorithms to solve combinatorial optimization problems. *Immune networks* [11] represent a recent frontier in this area.

These processes are examples of *complex adaptive systems (CAS)* [7] that arise in a variety of biological, social and economical phenomena. In the CAS framework, a system consists of large numbers of *autonomous agents* that individually have very simple behavior and that interact with each other in very simple ways. CAS are characterized by total lack of centralized coordination. Despite the simplicity of their components, CAS typically exhibit what is called *emergent behavior* that is surprisingly complex and unpredictable [7]. Furthermore, the collective behavior of a well-tuned CAS is highly adaptive to changing environmental conditions or unforeseen scenarios, is robust to deviant behavior (modeling failures) and is self-organizing towards desirable configurations.

Parallels between CAS and advanced information systems are immediate. BISON will exploit this fact to explore techniques and ideas derived from CAS in order to derive robust, self-organizing and self-repairing solutions to problems arising in dynamic networks through ensembles of autonomous agents that mimic the behavior of some natural or biological process. In our opinion, exploitation of CAS techniques will enable developers to meet the challenges arising in dynamic network settings and to obtain desirable global properties like resilience, scalability and adaptability without explicitly programming them into the individual agents. This represents a radical shift from traditional algorithmic techniques to that of obtaining desired system properties as a result of emergent behavior that often involves evolution, adaptation, and learning.

The dynamic network architectures that will be studied during the project include *peer-to-peer* (P2P) and *grid* computing systems [14, 3], as well as *ad-hoc networks* (AHN) [6]. P2P systems are distributed systems based on the concept of resource sharing by direct exchange between *peer* nodes, in the sense that all nodes in the system have equal role and responsibility [14]. Exchanged resources may include content, as in popular P2P document sharing applications, or CPU cycles and storage capacity, as in computational and storage grid systems. P2P systems exclude any form of centralized structure, requiring control to be completely decentralized. In AHN, heterogeneous populations of mobile, wireless devices cooperate on specific tasks, exchanging information or simply interacting informally to relay information between themselves and the fixed network [17]. Communication in AHN is based on multi-hop routing among mobile nodes. Multi-hop routing offers numerous benefits: it extends the range of a base station; it allows power saving; and it allows wireless communication, without the use of base stations between users located within a limited distance of one another.

We have chosen these domains for their practical importance in future distributed computing technologies as well as their potential for benefiting from our results. P2P/Grid systems can be seen as dynamic networking at the application level, while AHN results from dynamic networking at the system level. In both cases, the topology of the system typically changes rapidly due to nodes voluntarily joining or leaving the network, due to involuntary events such as crashes and network partitions, or due to frequently changing interconnection patterns.

The use of CAS techniques derived from nature in the context of information systems is not new. Numerous studies have abstracted principles from biological systems and applied them to network-related problems, such as routing. However, much of the current work in this area can be characterized as *harvesting* — combing through nature, looking for a biological system or process that appears to have some interesting properties, and applying it to a technological problem by modifying and adapting it through an enlightened trial-and-error process. The result is a CAS that has been empirically obtained and that appears to solve a technological problem, but without any scientific explanation of why.

BISON proposes to take exploitation of CAS for solving technological problems beyond the harvesting phase. We will study a small number of biology-inspired CAS, applied to the technological niche of dynamic networks, with the aim of elucidating principles or regularities in their behavior. In other words, BISON seeks to develop a rigorous understanding of why a given CAS does or does not perform well for a given technological problem. A systematic study of the rules governing good performance of CAS offers a bottom-up opportunity to build more general understanding of the rules for CAS behavior. The ultimate goal of the BISON project is then the ability to synthesize a

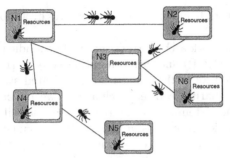

Fig. 1. An overlay network of nests in Anthill

CAS that will perform well in solving a given technological task based on the accumulated understanding of its regularities when applied to different tasks. In addition to this ambitious overall objective, BISON has more concrete objectives for obtaining robust, self-organizing and self-repairing solutions to important problems that arise in dynamic networks at both the system layer and the application layer.

BISON has officially started on January 1st, 2003, and will continue for three years. Given the interdisciplinary nature of the problem ahead, the BISON consortium brings together expertise from a wide range of areas including core disciplines (physics, mathematics, biology) and "user" disciplines (information systems, telecommunications industry). The BISON consortium consists of the Department of Computer Science at the University of Bologna (Italy), Telenor (the leading telecommunications operator in Norway), Department of Methods of Innovative Computing, University of Dresden (Germany), Dalle Molle Institute for Artificial Intelligence (IDSIA, Switzerland) and the Santa Fe Institute (USA).

3 The Anthill Framework

One of the goals of BISON is to develop a framework for the the design, analysis and evaluation of novel peer-to-peer applications based on ideas such as multi-agent systems borrowed from CAS [4]. This work will be based largely on the *Anthill* [1] framework that is being developed at Bologna as an open-source project. Anthill provides a "testbed" for studying and experimenting with CAS-based P2P systems in order to understand their properties and evaluate their performance. Although the current prototype of Anthill has been designed for P2P systems, it will serve as the basis for evaluation frameworks of other dynamic networks that will be studied in BISON.

An Anthill system is composed of a self-organizing overlay network of interconnected *nests*, as illustrated in Fig. 1. Each nest is a middleware layer capable of hosting resources and performing computations. The overlay network is characterized by the absence of any fixed structure, as nests come and go and discover each other on top of a communication substrate. Nests interact with local instances of one or more *applications* and provide them with a set of *services*. Applications are the interface between users and the P2P network, while services have a distributed nature and are based on the collaboration among nests. An example application might be a file-sharing system, while a service could be a distributed indexing service used by the file-sharing application to locate files.

An application performs *requests* and listens for *replies* through its local nest. Requests and replies constitute the interface between applications and services. When a nest receives a request from the local application, an appropriate service for handling the request is selected from the set of available services. This set is dynamic, as new services may be installed by the user. Services are implemented through *ants*, which are autonomous agents capable of traveling within the nest network. In response to a request, one or more ants are generated and assigned to a particular task. Ants may explore the nest network and interact with nests that they visit in order to accomplish their goal. Anthill does not specify which services a nest should provide, nor does it impose any particular format on requests. The provision of services and the interpretation of requests are delegated to ants.

Nests offer their resources to visiting ants through one or more *resource managers*. Example resources could be files in a file-sharing system or CPU cycles in a computational grid, while the respective resource managers could be a disk-based file system or a job scheduler. Resource managers typically enforce a set of policies for managing their (inherently limited) resources.

Ants do not communicate directly with each other; instead, they communicate indirectly by leaving information related to the service they are implementing at the appropriate resource manager found in the visited nests. For example, an ant implementing a distributed lookup service may leave routing information that aids subsequent ants in directing themselves toward regions of the network that are more likely to contain the searched key. This form of indirect communication, used also by real ants, is known as *stigmergy* [5].

A Java prototype of the Anthill runtime environment based on JXTA [9] has been developed [16]. JXTA, an open-source P2P project promoted by Sun Microsystems, aims to establish a programming platform for P2P systems by identifying a small set of basic facilities necessary to support P2P applications and providing them as building blocks for higher-level services. The benefits of basing our implementation on JXTA are several. For example, JXTA allows the use of different transport layers for communication and deals with issues related to firewalls and NAT.

In addition to the runtime environment, Anthill includes a simulation environment to help developers analyze and evaluate the behavior of their P2P systems. Simulating different P2P applications requires developing appropriate ant algorithms and a corresponding request generator characterizing user interactions with the application. Each simulation study is specified using XML by defining a collection of component classes and a set of parameters for component initialization. For example, component classes to be specified include the simulated nest network, the request generator to be used, and the ant algorithm to be simulated. Initialization parameters include the duration of the simulation, the network size, the failure probability, etc. This flexible configuration mechanism enables developers to build simulations by assembling pre-defined and customized component classes, thus simplifying the process of evaluating ant algorithms.

Unlike other toolkits for multi-agent simulation [12], Anthill uses a single ant implementation in both the simulation and actual run-time environments, thus avoiding the cost of re-implementing ant algorithms before deploying them. This important feature has been achieved by a careful design of the Anthill API and by providing two distinct implementations for it to be used in simulation and deployment.

Having developed a first Anthill prototype, we are now in the process of testing the viability of our ideas regarding biology-inspired approached to P2P by developing common P2P applications over it [1, 13]. As an example, we have developed Messor [13] which is a load-balancing application for grid computing that supports concurrent execution of highly-parallel, time-intensive computations where the workload may be decomposed into a large number of independent jobs. Messor exploits the computational power offered by a network of Anthill nests by assigning a set of jobs constituting a computation to a dispersed set of nests. To balance the resulting load among the nests, Messor uses a biology-inspired algorithm. Several species of ants are known to group objects (e.g., dead corpses) in their environment into piles so as to clean up their nests. It is possible to obtain this emergent behavior (randomly dispersed objects being transformed into neat piles) in a completely decentralized manner through virtual ants that act autonomously according to the following three rules [15]: (i) wander around randomly, until an object is encountered; (ii) if carrying an object, drops it and continue to wander randomly; (iii) if not carrying an object, pick it up and continue to wander randomly. Despite their simplicity, a colony of these "unintelligent" ants is able to group objects into large clusters, independent of their initial distribution in the environment.

The Messor algorithm is inspired by "inverting" the above virtual ant behavior such that an artificial ant drops an object it may be carrying only after having wandered about randomly "for a while" without having encountered other objects. Colonies of such ants try to disperse objects uniformly over their environment rather than clustering them into piles. As such, they could form the basis for a distributed load balancing algorithm.

Figure 2 illustrates the load balancing achieved by Messor over time. The results were obtained in a network of 20 Messor ants and 100 nests that are initially idle and form a ring-structured overlay network. At time zero, 10,000 jobs are inserted at a single nest. The different histograms depict the load observed at the 100 nests (linearly ordered along the x-axis) after 0, 5, 10, 15, 20, and 50 iterations of the algorithm. At each iteration, the 20 ants perform

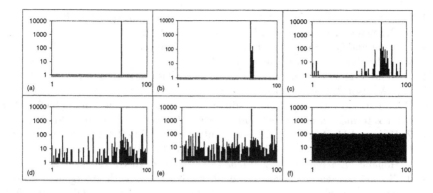

Fig. 2. Load distribution after 0, 5, 10, 15, 20, 50 iterations in 20-ant Messor.

a single step of the Messor algorithm: they randomly move from one node to another, possibly moving jobs with them from overloaded nests to under-loaded nests. As the figure illustrates, after 15-20 iterations most of the nests in the network have some work to do, and after 50 iterations, the load is perfectly balanced. The first few iterations serve for the ants to explore their neighborhoods in the network and create new connections beyond the initial ring structure so that future steps of the algorithm can transfer load to more remote parts of the network.

4 Conclusions

BISON and Anthill are relatively new projects and results obtained so far are limited to a few "proof-of-concept" demonstration applications. Neverthe-less, preliminary simulation results confirm our intuition that biology-inspired techniques for solving challenging problems in dynamic networks is indeed a promising approach. Our preliminary results exhibit performances that are comparable to solutions obtained through traditional techniques. Yet, the methodology followed to obtain our algorithms is completely different from previous approaches, as we mimic the behavior of natural systems, thus in-heriting their resilience and self-organizational properties for free.

Additional information on Anthill and BISON can be found at their re-spective web sites http://www.cs.unibo.it/projects/anthill/, http://www.cs.unibo.it/bison/.

References

1. Ö. Babaoğlu, H. Meling, and A. Montresor. Anthill: A Framework for the Development of Agent-Based Peer-to-Peer Systems. In *Proc. of the 22th Int. Conf. on Distributed Computing Systems*, Vienna, Austria, July 2002.

2. E. Bonabeau, M. Dorigo, and G. Theraulaz. *Swarm Intelligence: From Natural to Artificial Systems*. Oxford University Press, 1999.
3. I. Foster and C. Kesselman, editors. *The Grid: Blueprint for a Future Computing Infrastructure,*. Morgan Kaufmann, 1999.
4. D. Goldberg. *Genetic Algorithms in Search, Optimization and Machine Learning*. Addison-Wesley, 1999.
5. P. Grasse. La reconstruction du nid et les coordinations interindividuelles chez bellicositermes natalensis et cubitermes sp. *Insectes Sociaux*, 6:41–81, 1959.
6. Z. Haas and M. Pearlman. The zone routing protocol (ZRP) for ad-hoc networks (InternetDraft). Mobile Ad-hoc Network (MANET) Working Group, Aug. 1998.
7. J. Holland. *Emergence: from Chaos to Order*. Oxford University Press, 1998.
8. J. Hopfield. Neural networks and physical systems with emergent collective computational abilities. *Proc. of the Natl. Acad. Sci. USA*, 79:2554–2558, 1982.
9. Project JXTA. http://www.jxta.org.
10. S. Kirkpatrick, C. Gelatt, and M. Vecchi. Optimization by simulated annealing. *Science*, 220(4598):671–680, 1983.
11. E. Klarreich. Inspired by Immunity. *Nature*, 415:468–470, Jan. 2002.
12. N. Minar, R. Burkhart, C. Langton, and M. Askenazi. The Swarm Simulation System, A Toolkit for Building Multi-Agent Simulations. Technical report, Swarm Development Group, June 1996. http://www.swarm.org.
13. A. Montresor, H. Meling, and O. Babaoğlu. Messor: Load-Balancing through a Swarm of Autonomous Agents. In *Proc. of the 1st Workshop on Agent and Peer-to-Peer Systems*, Bologna, Italy, July 2002.
14. A. Oram, editor. *Peer-to-Peer: Harnessing the Benefits of a Disruptive Technology*. O'Reilly, Mar. 2001.
15. M. Resnick. *Turtles, Termites, and Traffic Jams: Explorations in Massively Parallel Microworlds*. MIT Press, 1994.
16. F. Russo. Design and Implementation of a Framework for Agent-based Peer-to-Peer Systems. Master's thesis, University of Bologna, July 2002.
17. C. Toh. *Ad Hoc Mobile Wireless Networks*. Prentice Hall, 2002.

UbAgent: A Mobile Agent Middleware Infrastructure for Ubiquitous/Pervasive Computing

George Samaras and Paraskevas Evripidou and Department of Computer Science, University of Cyprus, {cssamara, skevos}@ucy.ac.cy

Abstract: Pervasive computing can be summarized as the process to automatically gather, organize, process, and analyses data in order to provide sophisticated services that enhance and improve human activities, experience, learning and knowledge anytime, anywhere and from any device. UbAgent's middleware infrastructure provides an efficient way of handling the back-end processing of Pervasive Computing: accessing information from distributed databases, computation processing, and location specific services. UbAgent is a mobile agent based middleware for Pervasive computing. It is build on top of commercially available mobile agent systems. The second tier components are the TRAcKER a location management system, the Taskhandler framework and a Unified Message System for Mobile agents. At tier three we have PaCMAn, a web-based meta computer, the DBMS-agent system for distributed information retrieval and the DVS system for the creating of personalized views. At present UbAgent deals with the back-end of Pervasive applications; provides the services and it does not deal with real life user interface, i.e. to interpret the user needs and wishes.

Introduction

Pervasive computing is the trend towards increasingly ubiquitous, connected computing devices in the environment [1]. Pervasive/Ubiqitous services are not about browsing the Web on a cell phone or a PDA. It is about providing personalized services that are highly sensitive to the immediate environment and needs of the user. Real world will become a Marketplace for Services, Products and Information. Project UbAgent is developing a Mobile Agent based middleware for the Ubiquitous/Pervasive Computing. This middleware focuses in providing access to information, scalable computing, and location-based services. Mobile agents appear to be a promising platform for Ubiquitous/Pervasive computing because they can retrieve information, perform heavy computation, can function well within wireless and wireline networked environments, they can operate both synchronously and asynchronously. Java Mobile agents are a promising platform for Pervasive computing because of their ubiquitous nature. They can operate in a variety of devices from Java enable devices such as phones and PDAs to the largest computers. Pervasive computing will rely in large number of embedded devices for the user interface. Java was originally developed as a universal language for appliances with embedded processors, thus it should it should be easily migrate to such devices.

The UbAgents design philosophy:

- **Portability**: Various components of UbAgent have been implemented on the major research and commercial mobile agent systems. Furthermore UbAgent separates the mobility specific code form the Task specific code, thus making the task specific code even more portable.
- **Modularity/Reusability**: Task specific code is separated from the mobile shell. Introduction of a Mobile Agent messaging methodology that its mapped onto the native messaging system of the various mobile agent platforms that we use in UbAgent.
- **Performance**: UbAgent components have been optimized to perform better than competing technologies.
- **Support for low bandwidth and intermittent connectivity**: UbAgents can be launched from a laptop or a palmtop computer during a short connection. The UbAgents will continue roaming around performing their tasks and then wait until the communication link is again available to return home with their task result
- **Flexibility/Dynamicity**: UbAgents can be used to dynamically deploy specialized and specific infrastructures on a need basis.

The UbAgent's middleware infrastructure is build around three standalone components developed at the University of Cyprus (see figure 1): (i) Tracker: Location Management system [2](ii) PaCMAn, Mobile-agent based Web Metacomputer [3], and (iii) Mobile-agent based framework for distributed information retrieval [4]. Tracker is a universal and MASIF compliant location management middleware for Mobile Agents. Tracker is hierarchical and distributed; it manages the location of mobile agents and allows agents to locate other agents. Thus, providing the necessary support for location and context based services. In the area of web-based meta-computing we have developed PaCMAn. PaCMAn (Parallel Computing with Java Mobile Agents) launches, anywhere in the Internet, multiple Java-mobile Agents that communicate and cooperate to solve problems in parallel.

In the area of distributed information retrieval we have already developed prototypical systems for mobile-agent based distributed access to databases and for creating personalized views and data warehousing for the mobile client.

Pervasive computing can be summarized as the process to automatically gather, organize, process, and analyses data in order to provide sophisticated services that enhance and improve human activities, experience, learning and knowledge anytime, anywhere and from any device. UbAgent's middleware infrastructure provides an efficient way of handling the back-end processing of Pervasive Computing accessing information from distributed databases, computation processing, and location specific services.

We will demonstrate UbAgent in action in Pervasive computing using the following scenario *form "The Semantic Web" by Tim Berners-Lee, James Hendler and Ora Lassila that appear in Scientific America [5]:*

> *"The entertainment system was belting out the Beatles' "We Can Work It Out" when the phone rang. When Pete answered, his phone turned the sound down by sending a message to all the other local devices that had a volume control. His sister, Lucy, was on the line from the doctor's office: "Mom needs to see a specialist and then has to have a series of physical therapy sessions. biweekly or something. I'm going to have my agent set up the appointments." Pete immediately agreed to share the chauffeuring. At the doctor's office, Lucy instructed her Semantic Web agent through her handheld Web browser. The agent promptly retrieved information about Mom's prescribed treatment from the doctor's agent, looked up several lists of providers, and checked for the ones in-plan for Mom's insurance within a 20-mile radius of her home and with a rating of excellent or very good on trusted rating services. It then began trying to find a match between available appointment times (supplied by the agents of individual providers through their Web sites) and Pete's and Lucy's busy schedules".*

The semantic web technologies deal only with what we call front end interface of the above example. By front end interface we mean the interface between the user's desires, implicit and/or explicit, and the pervasive computing infrastructure. By back-end we refer to the computational needs for information retrieval from structure and semi-structure sources and the necessary computing power needed to provide the services.

The *Agent* launched by *Lucy* has to access the doctor's systems database in order to get information about the prescribed treatment. This requires that the agent launched by *Lucy* will somehow granted permission to access the doctor's system and to also have the ability to perform the required queries in the database. Next the *agent* has to check several lists of providers that can provide the required treatment. This requires access to one or more databases. It then applies multiple queries to the sub-list of providers that can offer the needed treatment (20-mile radius of her home and with a rating of excellent or very good on trusted rating services). The gathering and processing of this information is handled by the Ub-Agent's DBMS-agent framework. A number of mobile agents are sent out to visit the relevant database servers and retrieve the information. The agents also have the capability to combine their intermediate results in order to provide the needed information, the list of providers that fulfill all the requirements set up by Lucy.

After that the agent tries to find a match between available appointment times (supplied by the agents of individual providers through their Web sites) and *Pete's* and *Lucy's* busy schedules. Here again the agent need more queries to multiple distributed databases and possibly create personalized views for the mobile agents in order to find all the providers that passed all the other criteria and have appointment times that can fit the schedule of *Pete* and *Lucy*.

The next step is to find the best schedule for all three of them *Mom, Lucy* and *Pete*. This is a scheduling with constrains problem that might need substantial computing power in order to find an optimal result. Where are we going to perform all these computations? UbAgent's web-based meta-computer enters the picture. It uses available resources in the Internet to produce one or multiple schedules

UbAgent's Architecture

UbAgents employs tier architecture. At tier one we have the fundamental infra-structure, the JVM and the mobile agent platforms. At tier two we have the Ub-Agents core technologies, the location management, the taskhandlers and the communication infrastructures. At tier three we have the UbAgents systems for web based meta-computing and distributed information retrieval PaCMAn and DBMS-Agent. Pervasive applications built on top of UbAgents belong to Tier four. In this section we will deal with tier one and tier two technologies. Tier three and Tier four are described in the next two chapters.

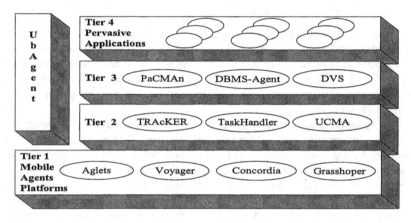

Fig. 1. UbAgent Tier Architecture

Mobile-Agent Platforms

Mobile agents are processes dispatched from a source computer to accomplish a specified task [6]. Each mobile agent is a computation along with its own data and execution state. After its submission, the mobile agent proceeds autonomously and independently of the sending client. When the agent reaches a server, it is delivered to an agent execution environment. Then, if the agent possesses necessary authentication credentials, its executable parts are started. To accomplish its task, the mobile agent can transport itself to another server, spawn new agents, or interact with other agents. Upon completion, the mobile agent delivers the results to the sending client or to another server. By letting mobile hosts submit agents, the burden of computation is shifted from the resource-poor burden of computation is shifted from the resource-poor mobile hosts to the fixed network. Mobility is inherent in the model; mobile agents migrate not only to find the required resources but also to follow mobile clients. Finally, mobile agents provide the flexibility to adaptively shift load to and from a mobile host depending on bandwidth and other available resources. Mobile-agent technology is suitable for wireless or dial-up environments

In our work we have utilized most of the commercial Java-based mobile agent platforms (i.e., Aglets [7], Concordia [8], Grasshopper [9] and Voyager [10]. In retrospect all the platforms are build around the same principles (i.e., object mobility) offering similar functionalities without essential to our work differences. We have implemented parts of our work in all these platforms but recently, and without loss of generality, we focused our implementation on the Voyager platform.

Dynamic Mobile Agents and Task Handlers

Pervasive computing environments are very dynamic by nature. It is also often needed to maintain their functionality for a long duration of time but with different responsibilities assigned to the various participants. Participants may switch roles; for example, a site might be a buyer in one transaction, a seller in another and a mediator in a third. This requires a more dynamic and flexible framework. In our framework, we generalize the notion of mobile agent by separating the mobile shell from the specific task code of the target application [3]. We achieve this with the introduction of TaskHandlers, which are Java objects capable of performing the various tasks. TaskHandlers are dynamically assigned to our agents. Each agent can carry with it any number of TaskHandlers and it can dynamically choose when and where to use them.

A TaskHandler Library is a collection of Java objects that are serializable, and thus can travel along with our mobile agents. The TaskHandler is an object that is in the disposal of an agent to use when it needs to perform a specific computational task. We have implemented a number of Task Handlers to perform specific tasks, such as database querying. When the need arise for a database query, the agent utilizes the DBMS-TaskHandler from our TaskHandler Library. The Task-Handlers can be used to dynamically change the role of an agent.

Tracker

In any mobile agent system, the ability to communicate with agents in real-time, as agents move from one network node to another, is essential for retrieving any data or information that they have collected, and for supporting co-ordination and co-operation among them. Communication with mobile agents presupposes the ability to locate (i.e., find the node and execution environment in which they currently reside) [11]. Locating agents efficiently is thus an issue central in any mobile agent system or environment that utilizes such entities. However, the support provided by current existing Java-based mobile agent platforms is limited in three areas: (1) does not scale to large number of agents, (2) can handle well high mobility and (3) they lack proper support for Internet-wide distribution of their agents. There is also a need for a system that will provide location management across platforms; i.e., it will be able to locate agents independently of their native implementation platform.

TRAcKER, is a distributed location management middleware which has the ability to manage the location of mobile agents that travel autonomously across the Internet. Being MASIF compliant [12] and based on JAVA, TRAcKER is independent of the semantics of any particular. Java based mobile agent platform and thus able to integrate with each one of them seamlessly. Major advantages of this middleware include: (a) the ability to manage the location of any mobile agent and (b) allowing agents to locate other agents, independently of their native execution environment. TRAcKER is designed in such a way as to incorporate, test and benchmark any different location mechanisms [11]. This makes TRAcKER very flexible and able to dynamically adjust to various location management mechanisms; it can support hierarchical location protocol and two hierarchical cellular-like mechanisms. Thus, TRAcKER may adopt dynamically the most appropriate mechanism for the current environment, e.g., the distribution of mobile agents or their mobility pattern. We experimentally evaluated several such these mechanisms under different mobility patterns and we studied an exhaustive variety of such algorithms and concluded that a combination of the above two type of approaches is the most optimal [2].

The location of all agents is maintained in a common for all agents node (called TRAcKERRegistry). The TRAcKER-enabled agents notify, depending on the location algorithm used, the current and destination node/TRAcKERRegistry of their whereabouts.

- The mobile shell has a JobSlice variable that contains the name of the TaskHandler to use and a TaskHandler variable to reference the newly created object.
- The following code creates the TaskHandler: *MyMainTask* = (TaskHandler) (Class.forName (*MainTask*)).newInstance();
- *MainTask*: A string variable that contains the name of the the TaskHandler to use (given by a JobSlice).
- *MyMainTask*: A reference to the TaskHandler object.
- The following code starts the TaskHandler: new Thread(*MyMainTask*).start();

Fig. 2. The relationship betwenen mobile shell, taskhandler and Jobslice

Unifying messaging system

The four mobile agent platforms that we use in our work have different mechanism for communication between the mobile agents. In order to achieve true interoperability between the core components of UbAgent we need to define a common way for exchanging messages, here we briefly describe the messaging method of each platform and them we present UbAgents approach in providing a unified method for message exchange between mobile agent platforms.

In the Aglet Platform we have the proxy notion. Any client that wants to send a request/message to an agent must know the identity and its exact position of that agent. Using that he will get the proxy for that agent, which will use to send a message. In the Concordia platform we have the notion of the **event manager;** an agent can be registered to an event manager and then receive messages from it. In Voyager technology, we also have the **proxy** notion. The difference with Aglets lays in the way the proxy operates in Voyager. Any client that wants to send a request to an agent must look up the agent in the name service of the node in which the agent is registered.

For UbAgent we have adopted a messaging scheme that is based on the Aglets and Grasshopper. One of the most straightforward schemes of communication between agents (or RPC in general) is the exchange of messages. Voyager doesn't have a formal messaging protocol and instead it uses RMI as its primary communication scheme. However, in Voyager we could easily build a mechanism to support a more formal communication protocol between agents. More specifically we can adopt the communication paradigm of Aglets and Grasshopper by just implementing a set of methods. We design and developed of a Java class that implement the communication paradigm. The following methods have been implemented:

1. *sendMessage.* This method has as its purpose to send message to other agents. It should take as input the message itself, as well as the name and location of the agent to which we want to send the message. The actual sending of the message (in Voyager) would be the call of another method in the target agent – the method *handleMessage.* Finally via TRAcKER it is possible to add functionality to this method and omit as input for this method the location of the target agent.

2. *handleMessage.* This method has as its purpose to handle the received, from other agents, messages. It should never be called by the owner agents but only by other agents. More specifically by their *sendMessage* method.

This is just the core of a simple one-way communication scheme between the various agents. Through object orientation we can easily adopt and adjust the actual messages; we just inherit those methods and extend them to include new, task specific, messages. In order to advance this scheme to include two-way messaging we can either extend these two methods, or better yet, can write two new methods that follow the same mode of operation as the one-way methods.

UbAgent's meta-computing support: PaCMAn

At any given time, there are vast numbers of computers connected to the Internet that are idle. The collective computational power of these computers dwarfs the computational power of all the world's high performance computers put together. PaCMAn aims in harness these otherwise idle resources into a collection of meta-computers each dynamically configure at real-time to solve specific problems submitted by users of PaCMAn. Such a system can provide the computing power needed for pervasive applications anywhere, anytime.

The PaCMAn (**Pa**rallel **C**omputing with **M**obile **Ag**ents) meta-computer [3] launches multiple mobile agents that cooperate and communicate to solve problems in parallel. Each agent supports the basic communication and synchronization tasks of the classical parallel worker assuming the role of a process in a parallel processing application. It consists of three major components:

- **Broker**: It keeps track of the system configuration. All participating clients and servers and their capabilities are registered with the broker. The broker also monitors the systems resources and performs network traffic/load forecasting.
- **Server:** It is a Java enabled computer (has installed the Java runtime environment) that runs the PaCMAn daemon, in order to host incoming PaCMAn mobile agents.
- **Client:** It is a Java enabled computer that runs the PaCMAn daemon in order to launch/host PaCMAn mobile agents.

A machine has to be explicitly registered in order to part in the PaCMAn meta-computer. Owners of the machines can make them available to PaCMAN at times they do not use them. Users have the ability to make their computer unavailable explicitly by turning off the PaCMAn daemon or implicitly when the local load reaches a predetermined threshold.

The PaCMAn mobile agent can assume various roles such as:

- **Mobile-Worker (MW)**: This is the workhorse of the PaCMAn system. It performs all the necessary computation, communication and synchronization tasks. The MW initializes and uses the appropriate TaskHandler.
- **Mobile-Coordinator/Dispatcher(MCD)**: The MCD does not participate in the computation. Instead, it performs task-specific coordination among the mobile-workers. This is done by dynamically assigning tasks to the MWs through the use of TaskHandlers. This is done by dynamically assigning tasks to the MWs through the use of JobSlices and also by providing information about the status of other MWs. For example, in a computationally intensive application, the MCD can produce a large number of JobSlices (much larger than the number of the MWs), effectively dividing the entire task into a number of subtasks, and dynamic assign a JobSlice to each MW whenever it requests it. This mode of operation is appropriate for application where the load cannot be balanced statically, such as the Prime number generator.

The PaCMAn model for using computer cycles from idle machines can be effectively used for Pervasive computing with different business models. For example an accounting system can be setup that supports micro-payments so that any client that uses the system resources will have to pay according to usage. Similarly the account of each server used will be credited with the appropriate payment. Of course it could also adopt the free software trend and the free Wi-Fi connectivity trend and make its resources available to anyone for free.

The collection of information through data mining has an integral role in Pervasive computing. Data mining algorithms have extremely large runtimes even for modest amount of data. The proliferation of the Web and the globalization of business have given rise to huge amounts of data been collected and stored in large-scale distributed data warehouses. Thus, the need for efficient data mining is ever more important under the current conditions. The obvious solution is to employ parallel and distributed processing techniques for data mining. In our approach, we go one step further and employ mobile agents that travel to the data sources and communicate and coordinate among themselves to perform data mining. For example we utilize the PaCMAn meta-computer to extract association rules for points of sales data from geographically distributed stores. A more user centric use of data mining for pervasive computing might for a customer to initiate a request as to the most probable time that a particular product, let say a Palm Pilot, might go on sale and how much the discount will be. Based on the results of such search he can decide to buy that particular item now or waiting for the expected sale.

UbAgents framework for distributed information retrieval

Mobile Agents for Web Databases Access (The DBMS-Agent): Web-based access to databases is an integral part of pervasive type of systems and applications. The real challenge is the formation of smart, lightweight, flexible, independent and portable Java DBMS client programs that will support database connectivity over the Internet from anywhere and at anytime. The traditional methodologies [13] overload the client and offer limited flexibility and scalability. The UbAgent approach is based on using mobile agents, between the client program and the server machine, to provide database connectivity, processing and communication, and consequently eliminate the overheads of the existing methodologies. Our database framework is comprised of a set of such Java based agents that cooperate to efficiently support Web database connectivity. The main agent, called DBMS-agent, acquires its database capabilities dynamically not at the client but at the server. The other agents of the framework assist this dynamic acquisition.

In particular, in most current approaches to Web database connectivity, an applet at the client machine downloads from the remote SQL server and initiates a JDBC driver, and then handles a complex set of JDBC interfaces. The UbAgent DBMS-applet creates and fires a mobile agent (or agents if necessary) that travels directly to the remote SQL server. At the SQL server, the mobile agent initiates a local JDBC driver, connects to the database and performs any queries specified by the sending client. When the mobile agent completes its task at the SQL server, it dispatches itself back to the client machine directly into the DBMS-applet from where it was initially created and fired. Since our mobile agents possess database capabilities, they are called *DBMS-agents*. The DBMS-agent is independent of the various JDBC driver implementations. The DBMS mobile agent can not (and is not supposed to) be aware of which JDBC driver to load when it arrives at an SQL server. Upon arrival at the SQL server's context, the DBMS-agent is informed of all available JDBC drivers and corresponding data sources. The DBMS-agent is then capable of attaching itself to one or more of these vendor data sources.

Our framework promotes a much more efficient way of utilizing the JDBC API and the JDBC driver API and eliminates the overheads of the various conventional approaches. It supports lightweight, portable and autonomous clients as well as operation on slow or expensive networks. The implementation of the framework shows that its performance is comparable to, and in many cases outperforms, the performance of current approaches. In fact, in wireless and dial-up environments and for average size transactions, a client/agent/server adaptation of the framework provides a performance improvement of approximately a factor of ten.

Furthermore, this new form of Web-based database access supported by the "UbAgent DBMS-Agent Framework" is [3]: (a) flexible: it can be set up dynamically and efficiently, (b) scalable: its extension to support multidatabase systems not only maintained but also increased its performance benefits, and (c) robust: the statistical analysis performed found the DBMS-agent framework more stable than the current JDBC-based database connectivity.

Dynamic View System for Wireless and Mobile Clients: Large-scale Internet queries applied directly to the source databases are often quite expensive. Such queries are even more expensive for wireless clients due to the severe limitations of the wireless links that are in general characterized by high communication costs, limited bandwidth and high latency [14]. The efficient creation and materialization of personalized views provides an attractive alternative. To this end, we have proposed a new UbAgent component, called the Dynamic View System (DVS), that allows the automatic definition, creation and maintenance of such views. Views are defined using queries over a number of database relations, called base relations. Views can be materialized (the result of the query is stored) or abstract (the query is stored and its result is re-calculated any time the query is posed). DVS's multitier architecture is based on mobile agents [16] and is built upon the DBMS-agent sub-system. Through DVS, wireless clients identify Web databases of interest, access their metadata, set up dynamically via mobile agents the necessary view infrastructure and create personalized views encapsulated in mobile agents (i.e., the View-agent). The system allows these views to be shared by other mobile clients.

DVS also supports materialize view maintenance in the case of updates at the sources. View maintenance can be achieved in various ways [15]. The proposed infrastructure, via the notion of Task-Handlers, can be dynamically adapted to serve any existing approach. Each agent can carry with it various protocols, as

TaskHandlers, and it can dynamically choose (or be directed to) when and where to use them.

DVS brings data close to the clients thus eliminating the overhead of transmitting data over slow networks. In addition, it allows the sharing of views among the clients of the system. Accessing remote views and databases is done both synchronously and asynchronously [2]. With asynchronous communication, clients can pose their query, disconnect and connect later to receive the results. Analogously, busy servers can postpone processing of selected submitted queries. Furthermore, the implementation of DVS with mobile agents allows dynamic code deployment and relocation of views (a view within a View-Agent can be dynamically transferred to another site).

Developing Pervasive Applications with UbAgent

The current approach adopted in developing pervasive applications is first to create a large collection of services in the form of taskhandlers, each tested and fine tuned individually. Pervasive applications are set up by using a number of these taskhandlers. We have a large collection of information retrieval, creation and processing of mobile views and several computational codes. The scenario of section 2 can be implemented with our existing collection of taskhandlers.

In addition to the taskhandlers developed by our team UbAgent's modular approach enable us to easily adopt and use java codes developed by others. Actually some of the codes used to test PaCMAn were given to us by colleagues. At present we would like to incorporate into our taskhandler library code for the following: (1) Support for Intelligence/semantic web services, (2) Support for Web Services, (3) Support for sensors, (4) Computational code, (5) Security & Authentication

Our future plans is to create a meta-level intelligent system that will interpret the users needs and automatically assemble and schedule the necessary services in the form of taskhandlers.

Conclusions

UbAgent provides a mobile-agent based middleware for Pervasive/Ubiquitous computing. It has a layer architecture that is build on top of commercially available mobile agent platforms. It core technologies is the TRAcKER, a location management system for mobile agents, the Taskhandler, a framework for creating dynamic mobile agents that can change role son the fly and a unifying messaging system that acts as bridge between the messaging systems of the mobile agent platforms. On the higher tier we have PaCMAn, DBMS-agent and DVS systems. The four tier is the pervasive applications which are build on top of the major systems of the system of the third tier. All the components of UbAgents has been developed, tested and fine tuned as stand alone applications. Special emphasis was given in order to achieve better performance than competing approaches.

UbAgent current emphasis is the back-end of Pervasive applications; provides the services and it does not deal with real life interface with user's needs and demands. The user needs to explicitly request the services and UbAgent retrieves and process the relevant information and performs the computations needed to deliver the service to the user

References

[1] M Satyanarayanan, "Pervasive Computing Vision and challenges," IEEE Personal Communications, Vol 6, no 8, Aug. 2001, pp. 10-17
[2] G. Samaras, C. Spyrou, E. Pitoura, M. Dikaiakos, "Tracker: A Universal Location Management System for Mobile Agents", European Wireless, Feb 2002.
[3] P. Evripidou., G. Samaras, E. Pitoura, P. Christoforos., "The PacMan Metacomputer: Parallel Computing with Java Mobile Agents", Journal FGCS special issue on JAVA in High Performance Computing, 2001.
[4] S. Papastavrou, G. Samaras, and E. Pitoura, *"Mobile Agents for WWW Distributed Database Access"*, Proc. 15th International Data Engineering Conference (IEEE-ICDE'99) , Sydney, Australia, March 1999
[5] Tim Berners-Lee, J. Hendler and O. Lassila *"The Semantic Web" Scientific America*
[6] D. B. Lange and M. Oshima. Seven Good Reasons for Mobile Agents. *Communications of the ACM*, 42(3):88-91, March 1999
[7] Aglets Workbench, by IBM Japan Research Group. Web site:http:/aglets.trl.ibm.co.jp
[8] D. Wong, N. Paciorek, T. Walsh, J. DiCelie, M. Young and B. Peet. Concordia: An Infrastructure for Collaborating Mobile Agents. *Lecture Notes in Computer Science*, 1219, 1997. <http:// www.meitca.com/HSL/Projects/ Concordia/>.
[9] M. Breugst, I. Busse, S. Covaci and T. Magedanz. Grasshopper A Mobile Agent Platform for IN Based Service Environments. *IEEE IN Workshop, Bordeaux*, France, May 10-13, 1998.
[10] ObjectSpace Voyager [tm] Technical Overview. Web Site <http://www.objectspace.com/voyager/whitepapers/Voyager TechOview. pdf>
[11] E. Pitoura, and Samaras, G., *"Locating Objects in Mobile Computing"*. IEEE Transactions on Knowledge and Data Engineering Journal (TKDE) 2001.
[12] Mobile Agent System Interoperability Facilities Specification. OMG TC Document orbos 97/10/05 http://www.omg.org/cgi-bin/doc?orbos/97-10-05
[13] Roedy Greem. Article: Java Access to SQL Databases. Canadian Mind Products, 1997.
[14] E. Pitoura and G. Samaras, *"Data Management for Mobile Computing"*, Kluwer Academic Publishers, ISBN 0-7923-8053-3, 1998.
[15] Nick Roussopoulos: Materialized Views and Data Warehouses. KRDB 1997.
[16] Constantinos Spyrou, George Samaras, Evaggelia Pitoura, Evripidou Paraskevas *"Mobile Agents for Wireless Computing: The Convergence of Wireless Computational Models with Mobile-Agent Technologies"*, Journal of ACM/Baltzer Mobile Networking and Applications (MONET), special issue on Mobility in Databases & Distributed Systems", (In Print)

Part XI

**2003 International Workshop on
Intelligence, Soft computing and the Web**

International Workshop on Intelligence, Soft Computing and the Web (ISCW'03)

This workshop is concerned about ways in which soft computing techniques can enhance the current usage of the Web. Another main issue being considered is how (or whether??) soft computing techniques can contribute to more intelligent use of the web.

Papers exploring new directions and ideas in the following areas are invited for this workshop.

Neural networks, fuzzy systems, evolutionary computation, probabilistic reasoning and hybrid systems

Data mining, Web mining, text mining, text and image retrieval

Software agents, intelligent agents

Web intelligence

Soft computing in a Web environment

Business intelligence, E commerce

Data bases, data warehouses, OLAP and the Web

Dynamic, changing environments and intelligent control

This workshop is organized by the Soft Computing Research Group, School of Business Systems, Faculty of Information Technology, Monash University.

Organizing Chair's

Dr. Damminda Alahakoon
School of Business Systems,
Monash University,
PO Box 63 B, Victoria 3800, Australia.
Phone +61 3 9905 9662 Fax +61 3 9905 5159
Email: damminda.alahakoon@infotech dot monash dot edu dot au

Dr. Leon S.L. Wang
Associate Professor
Department of Computer Science
New York Institute of Technology
1855 Broadway
New York, New York 10023
E-mail: slwang@nyit.edu

Enhanced Cluster Visualization Using the Data Skeleton Model

R. Amarasiri, L. K. Wickramasinghe and L. D. Alahakoon

(rasika.amarasiri, kumari.wickramasinghe,
damminda.alahakoon@infotech.monash.edu.au)
School of Business Systems, Monash University, Australia

Abstract: The Growing Self Organizing Map (GSOM), which is an extended version of the Self Organizing Map (SOM), has significant advantages when dealing with dynamically or incrementally changing data structures. The Data Skeleton Model (DSM) is a technique developed using the GSOM to identify clusters and relate them to the input data sequence. In this paper, we have developed the DSM as a visualization tool to automate the cluster identification process. We also demonstrate the advantage of using the Spread Factor (SF) in the GSOM for clustering data.

1 Introduction

Growing Self Organizing Map (GSOM) [1-4] has a flexible structure which self organizes to represent the input data stream more accurately. Therefore cluster identification with GSOM is easier by observing the direction of the growth of the network. The concept of Spread Factor (SF) is used to control the spread of the map. By using a high value for SF data analyst has the capability of going for very finer clusters if he wishes to.

Identifying the paths along which the GSOM grew develops the Data Skeleton Model (DSM). DSM can be considered as an automated cluster identification tool. A visualization tool is proposed in this paper, which enhances the cluster identification process using GSOM.

GSOM grows from boundary nodes. If such a node is selected to grow, all its free neighboring positions will be grown with nodes irrespective of calculating the exact position of the new node. This method has a disadvantage when it comes to DSM as it generates some isolated clusters. The paper presents remedial actions to overcome this problem.

Section 2 of this paper presents an introduction to GSOM and DSM. Development of data skeleton as a visualization tool is described in section 3 in detail. It also contains the proposed techniques to overcome isolated cluster issues. Section 4 contains some simulations and experiments that were done to demonstrate the results. Conclusion is in section 5.

2 The Growing Self Organizing Map and The Data Skeleton Model

2.1 The Growing Self Organizing Map

The GSOM [1-4] is a self-generating unsupervised neural network algorithm, which is an extended version of the self-organizing map (SOM) [6-9]. Unlike SOM, the GSOM doesn't start with the full-blown network. It has a minimum number of starting nodes (usually four) and generates new nodes only when it's required using a heuristic to identify such a need. By using the value called Spread Factor (SF), the data analyst has the ability to control the growth of the GSOM.

All the starting nodes of the GSOM are boundary nodes, i.e. each node has the freedom to grow on its own direction at the beginning. (Fig. 1) New Nodes are grown from the boundary nodes. Once a node is selected for growing all its free neighboring positions will be grown new nodes.

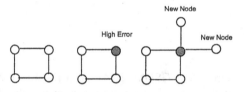

Fig. 1. New node generation in the GSOM

The GSOM process is as follows.
1) Initialization phase.
   a) Initialize the weight vectors of the starting nodes (usually four) with random numbers between 0 and 1.
   b) Calculate the growth threshold (GT) for the data set of dimension D according to the spread factor (SF) using the formula $GT = -D \times \ln(SF)$
2) Growing phase.
   a) Present input to the network.
   b) Determine the weight vector that is closest to the input vector mapped to the current feature map (winner), using Euclidean distance (similar to the SOM). This step can be summarized as: find q' such that

$$|v - w_{q'}| \le |v - w_q| \, \forall q \in N$$, where v, w are the input and

weight vectors, respectively, q is the position vector for nodes, and

N is the set of natural numbers.

c)   The weight vector adaptation is applied only to the neighborhood of the winner and the winner itself. The neighborhood is a set of neurons around the winner, but in the GSOM the starting neighborhood selected for weight adaptation is smaller compared to the SOM (localized weight adaptation). The amount of adaptation (learning rate) is also reduced exponentially over the iterations. Even within the neighborhood, weights that are closer to the winner are adapted more than those further away. The weight adaptation can be described by

$$w_j(k+1) = \begin{cases} w_j(k), j \notin N_{k+1} \\ w_j(k) + LR(k) \times (x_k - w_j(k)), j \in N_{k+1} \end{cases} \quad \text{where}$$

the learning rate $LR(k)$, $k \in N$ is a sequence of positive parameters converging to zero as $k \to \infty$. $w_j(k), w_j(k+1)$ are the weight vectors if the node j before and after the adaptation, and N_{k+1} is the neighborhood of the winning neuron at $(k+1)^{\text{th}}$ iteration. The decreasing value of $LR(k)$ in the GSOM depends on the number of nodes existing in the network at time k.

d)   Increase the error value of the winner (error value is the difference between the input vector and the weight vectors).

e)   When $TE_i \leq GT$(where TE_i is the total error of node i and GT is the growth threshold). Grow nodes if i is a boundary node. Distribute weights to neighbors if i is a non-boundary node.

f)   Initialize the new node weight vectors to match the neighboring node weights.

g)   Initialize the learning rate (*LR*) to its starting value.

h)   Repeat steps b) – g) until all inputs have been presented and node growth is reduced to a minimum level.

3) Smoothing phase.

a)   Reduce learning rate and fix a small starting neighborhood.

b)   Find winner and adapt the weights of the winner and neighbors in the same way as in growing phase.

The GSOM adapts its weights and architecture to represent the input data. Therefore, in the GSOM, a node has a weight vector and two-dimensional coordinates that identify its position in the net, while in the SOM, the weight vector is also the position vector.

2.2 The Data Skeleton Model

GSOM generates nodes in accordance with the input data stream. The Data Skeleton Model (DSM) is developed by identifying the paths along which the GSOM grew. This model has been proposed as a cluster identification tool in [5]. The process of developing the DSM is described later. Before we can describe the process some terms need to be defined.

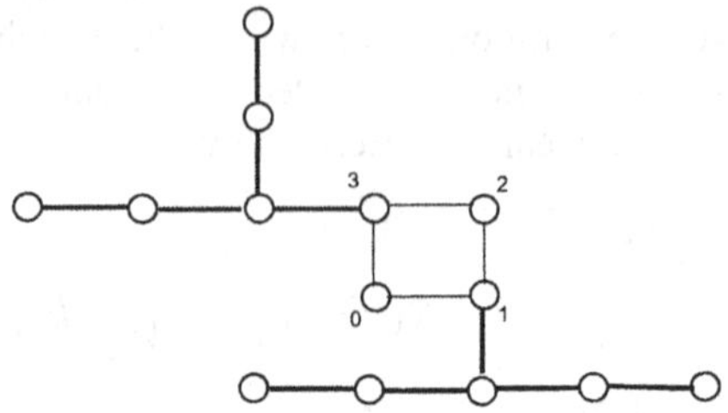

Fig. 2. Path of spread plotted on the GSOM

GSOM starts with four initial nodes and grows nodes according to the input data stream. The Path of Spread is obtained by joining these nodes in the chronological order of their growth. A sample path of spread is illustrated in Fig. 2.

After the GSOM has been grown, it is calibrated using a different data set than that of the training data set. The calibration data map on to different nodes on the GSOM, and those nodes that have input data mapped on to them are referred to as Hit Points or Hit Nodes.

Once the POS is identified, it can be seen that some of the hit points do not occur in the POS. These points are linked to the closest points in a POS. When every POS is joined to the initial four nodes, and the additional hit points are linked to the POS, the Data Skeleton for the data set is generated.

When external hit points are connected to the POS, if the point on the POS that is linked is not a hit point, it will become a junction. The distance (Euclidian distance in weight vector value) between two consecutive hit points, which could include junctions, is called a path segment.

The algorithm for data skeleton building and cluster separation is given below. The algorithm assumes that there is a fully generated GSOM to be used for this purpose.

1) Skeleton modeling phase:
   a) Plot the *node numbers* for each node of the GSOM. The node numbers (0,...N) are assigned to the nodes as they are generated. The initial four nodes have the numbers 0 ...3. The node numbers represent the chronological order of node generation in the GSOM.
   b) Join the four initial nodes to represent the *base* of the skeleton.
   c) Identify the nodes, which initiated new node generation, and link such parent nodes to the respective child nodes. This linking process is carried our in the chronological order of node generation. The node numbers of nodes, which initiate growth, are stored for this purpose during the growing phase of the GSOM creation.
   d) Identify the paths of spread (POS) from the links generated in (b)
   e) Plot the hit points on the GSOM.

    f)   Complete the data skeleton by joining the used nodes not on the POS, to the respective POS as sub-branches.

    g)   Identify the junctions on the POS.

2)   Cluster separation phase

    a)   Identify the path segments by using the POS, hit points and the junctions.

    b)   Calculate the distance between all neighboring junctions on the skeleton using Euclidian metric as the distance measure.

    c)   Delete path segments starting from the largest distance value.

    d)   Repeat c until the data analyst is satisfied with the separation of clusters in the GSOM.

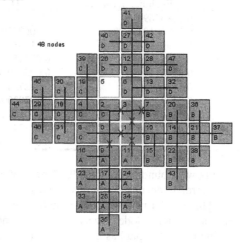

Fig. 3. Data Skeleton of a GSOM with 4 clusters. The number in each cell is the node number and the character is the mapped class name.

The above algorithm results in a separation of the clusters using the data skeleton, such that the data analyst can observe the separation. We believe that this visualization of the clusters provides a better opportunity for the analyst to decide on the ideal level of clustering for the application.

3 Development of Data Skeleton into a Visualization Tool

As described earlier, the Data Skeleton Model can be used as a visualization tool for identifying clusters in input data. In order to do this, we extended the simulation software for GSOM developed by Arthur Hsu from the Mechatronics Research Group of University of Melbourne, to generate the actual data skeleton by applying the method described in 2.2.5. A simple visualization of the data skeleton generated is illustrated in Fig. 3. It was for an artificial two-dimensional data set having clusters in the four corners of the grid. A partial list of segment lengths for the skeleton in the descending order are given in Table 1, which can be used to identify the order in which the skeleton breaks to form the clusters.

When testing the data skeleton, we found out that the data skeleton did not cluster that well and some isolated clusters were being generated. This was obvious when the SF for the GSOM was increased. For example, in the GSOM and skeleton illustrated in Fig. 3, nodes 7, 8 and 11 will be clustered first although they should have been belonged to the larger clusters that are closer to them.

Table 1: Sorted Partial Weight Distances of Skeleton in Fig. 3

Segment	Euclidian Distance	Difference
1 to 0	0.14867699	
3 to 1	0.14260101	0.0060759783
3 to 7	0.12991986	0.012681156
0 to 9	0.055865165	0.07405469
1 to 11	0.04981446	0.006050706
2 to 3	0.0470604	0.0027540587
3 to 6	0.04003031	0.0070300885
0 to 2	0.030630061	0.00940025
1 to 10	0.023284368	0.007345693
0 to 8	0.02090912	0.0023752488

This unusual clustering was found to be a result of the way the GSOM was being grown. For each hi error node in the boundary, all possible neighbors were grown from the node. This results in un-necessary nodes also being grown. For example, in the previous example, node 7 was grown with node 6, which was caused by a data point in the cluster that has been grown from node 6. But actually node 7 maps on to some data in the cluster starting from node 10

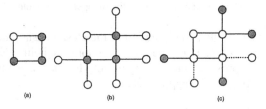

(a)   (b)   (c)

Fig. 4. Node growth of GSOM with node elimination

The problem of isolated data clusters was a minor issue in the visualization tool as it still showed those clusters being separate and the main clusters were very easily identified. To overcome the problem of these isolated clusters, we propose two methods and the selection of the most suitable method is ongoing research.

3.1 Node elimination method

The first method that we propose for the elimination of the isolated initial clusters is to do a simple alteration to the initial growing phase of the GSOM that will result in some new node elimination. The process is as follows and is illustrated in Fig. 4

(a) Iterate through the whole training data set identifying all hi-error nodes in the boundary of the already available nodes. The process can be stopped before all the data in the training data set have been evaluated if we find that all the boundary nodes are hi-error nodes, as further processing would not result in any new discoveries.

(b) Grow nodes as for the original GSOM algorithm for all the hi-error nodes and distribute their weights appropriately, using the same weight distribution algorithms as the GSOM.

(c) Apply the whole training dataset to the new map and identify nodes that don't get any hits and are on the boundary. Since we apply the whole data set to the map, if there are no hits on a specific boundary node, it is very unlikely that those nodes will have any hits in the future and also very unlikely to grow nodes from them. Therefore, we eliminate those boundary nodes from the current map.

(d) Repeat steps (a) to (c) until no new nodes are grown or until a specified time period has reached.

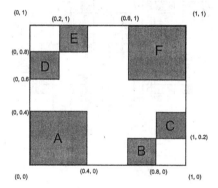

Fig. 5. Data patterns used for testing the data skeleton

Initial experiments on this method eliminated most of the isolated nodes that could have come from the middle of the skeleton, but isolated clusters from the initial nodes still remained. To eliminate this problem, we propose the second method, which is changing the starting number of nodes.

3.2 Method with 9 initial nodes

The second method that we propose is to increase the initial nodes to 9 in a 3x3 matrix. This will allow more options for the growing directions and if we apply the previous method of growing to this, we would be able to eliminate the isolated small clusters completely. This method has not been tested for its results at the time of writing and the results of this method would be available in future publications.

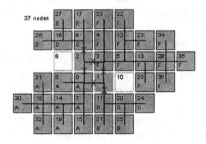

Fig. 6. Data skeleton for the 6 cluster data set with SF=0.2

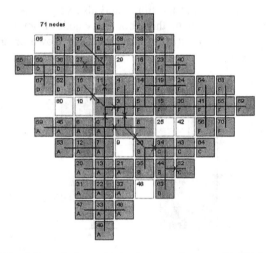

Fig. 7. Data skeleton for the 6 cluster data set with SF=0.6

4 Simulations and Experiments to Demonstrate the Results

As described in [5], the spread factor of the GSOM plays an important factor in spreading out the clusters into detail. To illustrate these phenomena, we created the data skeleton for a two-dimensional dataset that had data clusters arranged as illustrated in Fig. 5. The results of the data skeleton developed with the spread factor set to 0.2 and 0.6 are illustrated in fig. 6 and Fig. 7 respectively. This clearly illustrates how clusters B, C and D, E are separated from each other to form independent clusters when the spread factor is increased.

As proof to show that the method works for real data, we used the wine data set obtained from [10] which consists of a chemical analysis of wines grown in the same region in Italy but derived from three different cultivars. The resulting data skeleton is illustrated in Fig. 8, and shows that the wines are grouped together by the GSOM. The data skeleton can identify most of the clusters properly but the growing pattern of the GSOM again has caused group 2 and 1 to be overlapped into the same cluster.

5 Conclusions

The data skeleton model as presented in this paper can be used to visually identify clusters in a data set. It also showed that weaknesses in the GSOM's growing algorithm hinders the cluster identification process and we hope that the two improved methods of growing the GSOM will result in better clustering as well as improved automated cluster identification. The fact that the data clustering can be applied to real data and not only for simulated data shows that the method has potential as a good data mining tool.

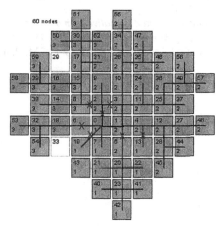

Fig. 8. Data skeleton for the wine data set

Acknowledgement

The GSOM simulation software developed by Arthur Hsu from the Mechatronics Research Group of University of Melbourne was extended to develop the actual data skeleton visualization package by the authors.

References

[1] Alahakoon L. D. and Halgamuge S. K. Knowledge Discovery with Supervised and Unsupervised Self Evolving Neural Networks, *In Proceedings of the International Conference on Information-Intelligent Systems*, pp 907-910, 1998.

[2] Alahakoon L. D., Halgamuge S. K. and Sirinivasan B. A Self Growing Cluster Development Approach to Data Mining, *In Proceedings of the IEEE conference on Systems, Man and Cybernetics*, pp 2901-2906, 1998.

[3] Alahakoon L.D., Halgamuge S. K. and Sirinivasan B. A Structure Adapting Feature Map for Optimal Cluster Representation, *In Proceedings of the International Conference on Neural Information Processing*, PP 809-812, 1998.

[4] Alahakoon L. D., Halgamuge S. K. and Sirinivasan B. Dynamic Self Organizing Maps With Controlled Growth for Knowledge Discovery, *In IEEE Transactions on Neural Networks, Special Issue on Knowledge Discovery and Data Mining*, 2000.

[5] Alahakoon D, Halgamuge SK, Sirinivasan B Mining a Growing Feature Map by Data Skeleton Modeling. In: Kandel A, Last M, Bunke H (eds) *Data Mining and Computational Intelligence*, Heidelberg, New York, Physica-Verl, pp 217-250 2000.

[6] Kohonen T. Analysis of Simple Self-Organizing Process, *Biological Cybernetics*, 44:139-140, 1982.

[7] Kohonen T. Self-Organized formation of Topological Correct Feature Maps, *Biological Cybernetics*, 43:59-69, 1982.

[8] Kohonen T. *Self-Organization and Associative Memory*, Springer, Verlag, 1989.

[9] Kohonen T. *Self-Organizing Maps*, Springer Verlag, 1995.

[10] E. Koegh, C. Blake and C. J. Mertz, UCI Repository of Machine Learning Databases, 1999.

Generating Concept Hierarchies for Categorical Attributes using Rough Entropy

Been-Chian Chien[1], Su-Yu Liao[1], and Shyue-liang Wang[2]

[1] Institute of Information Engineering, I-Shou University, Kaohsiung 840, Taiwan,
R.O.C. cbc@isu.edu.tw, m9003036@stmail.isu.edu.tw
[2] Department of Computer Science, New York Institute of Technology, 1855
Broadway, New York 10023 slwang@nyit.edu

Abstract. Extracting knowledge from a large amount of data is one of the important research topics in knowledge discovery. A concept hierarchy is a kind of concise and general form of concept description that organizes relationships of data and expresses knowledge as a tree-like or partial ordering structure. In this paper, we propose an approach to generate concept hierarchies automatically for a given data set with nominal attributes based on rough entropy. The proposed method reduces the number of attributes after each process of generating concept level. We give two experiments to show that the proposed algorithm is intuitive and efficient for knowledge discovery in databases.

1 Introduction

Extracting knowledge from large databases is one of the most important research topics in both area of machine learning and databases. The objective of KDD is to examine how to extract knowledge automatically from huge amounts of data. The discovered knowledge from a large practical database can be applied to many applications, such as decision-making, marketing, information management, and so on. For different applications, the different types of knowledge lead to lots of methodologies in the development of data mining techniques [2]. Regardless of the mining approaches involved, we can imagine the various types of discovered knowledge as the concepts learned from databases based on different viewpoints.

For representing knowledge as concepts explicitly, a clear and meaningful concept description is required. The concept hierarchy is a kind of concise and general form of concept description widely used. A concept hierarchy defines a sequence of mappings from a set of low-level concepts to higher-level, more general concepts on one or a set of attribute domains. Such mappings may organize the set of concepts in partial order according to a general-to-specific ordering and form a tree-like structure (a hierarchy, a taxonomy). The most general concept can be an universal concept, whereas the most specific concepts correspond to the specific values of attributes in the database. Each node in the tree of concept hierarchy represents a concept which helps to express knowledge and data relationships in a database as a concise, high level term [3].

Many knowledge representations are variants of concept hierarchies in different ways of expression. For transaction databases, as an example, association rules are the set of rules that describe the relative relationships in a partial ordering hierarchy of large item sets. Decision tree is another type of concept hierarchy used for generating classification rules and classifying the categories of data. The other type of concept hierarchy, like clusters, is a one-level concept grouping data with high similarity into classes. It is reasonable for different users that distinct concept hierarchies can be constructed on the same attribute(s) based on users' viewpoints or preferences.

For constructing different types of concept hierarchies, many methods for numerical or categorical data have been developed. We summarize the methods of generating concept hierarchies briefly as follows. The first method is to specify concept hierarchies by users, experts or data analysts according to a partial or total ordering of attributes explicitly at the schema level. A user or expert can easily define a concept hierarchy by semantics of schema or his/her need. Another method is to define a portion of a concept hierarchy by clustering explicit data in databases. However, it is essentially unrealistic to define an entire concept structure by explicit value enumeration manually in a large database. Such method thus focuses on automatic generation of concepts by grouping data by clustering techniques. The related researches include Cluster/2 by Michalski and Stepp [7], COBWEB by Fisher [1] etc. The third method is that given a set of attributes, the system can then try to generate the partial ordering of attributes automatically so as to construct a meaningful concept hierarchy [2]. Without knowledge of data semantics, a concept hierarchy can be automatically generated based on the number of distinct values in attributes at the lowest level of hierarchy; the higher levels include generated concepts in the hierarchy. Some researches on generating concept hierarchies in a single numerical attribute [3] and finding the partial ordering in a set of categorical attributes [5] have been proposed. However, the problem for finding a complete hierarchical ordering from an arbitrary set of categorical attributes in a large database efficiently and meaningfully is still interesting and important.

The purpose of this paper is to construct a meaningful concept hierarchy by their partial ordering relationships automatically among attributes in databases. We introduce an information measure called rough entropy based on the rough set theory to determine the levels of attributes and generate the concept hierarchy for the given database. It can be proved that the values of rough entropy are monotonously increasing in accord with the generalization in higher levels of concept hierarchy. We then propose an efficient algorithm to determine the partial ordering of attributes and construct a meaningful complete concept hierarchy. The proposed algorithm makes use of rough entropy to find a weakly relevant attribute, group the most relevant concepts and remove the weak relevant attributes recursively to generate the concept hierarchy for a given set of attributes in a large database. Two experiments are given to show the effectiveness and simplicity in the results of generated concept hierarchies

The paper is organized as follows: Section 2 introduces the concepts of rough set theory and rough entropy. In Section 3, we presented the proposed algorithm based on rough entropy for generating concept hierarchies on categorical attributes. In Section 4, we show the performance of the proposed method by experimental results of two examples databases. Finally, conclusions and future work are discussed in Section 5.

2 Rough Entropy

The rough set theory was first proposed by Zdzislaw Pawlak in 1982 [9]. Rough set is used to deal with the identification of common attributes in data sets. This theory provides a powerful foundation to reveal and discover important structures in data and to classify complex objects. An attribute-oriented rough sets technique can reduce the computational complexity of learning processes and eliminate the unimportant or irrelevant attributes so that the knowledge can be learned from large databases efficiently. In an information system, some categories of objects cannot be distinguished in terms of the available attributes. They may be just defined roughly or approximately. The idea of rough sets is based on the establishment of equivalence classes on the given data set U. We will give the formal definitions of the rough set theory as follows: Let $S = (U, A)$ denote an information system where U means a non-empty finite set of objects with a non-empty finite set of attributes A. For all $a \in A$ and each $B \subseteq A$, we define a binary indiscernibility relation $R_A(B)$ as follows:

Definition 1: Let $x, y \in U$, we say that objects x and y are indiscernible if the equivalence relation $R_A(B)$ is satisfied on the set U,
$$R_A(B) = \{(x, y) \mid x, y \in U, \forall a \in B, a(x) = a(y)\}.$$

Definition 2: $apr = (U, R_A)$, is called an approximation space. The object $x \in U$ belonging to one and only one equivalence class. Let $[x]_B$ denote an equivalence class of $R_A(B)$ and $[U]_B$ denote the set of all equivalence classes $[x]_B$ for $x \in U$. We have
$$[x]_B = \{ y \mid x \, R_A(B) \, y, \, \forall x, y \in U \} \text{ and } [U]_B = \{[x]_B \mid x \in U \}.$$

Definition 3: Let $S = (U, A)$, $x_i \in U$ and $B \subseteq A$. The rough entropy of information based on a subset of attributes B is
$$E(B) = -\sum_{i=1}^{n} |[x_i]_B| \, log_2 \frac{1}{|[x_i]_B|},$$
where $|[x_i]_B|$ denotes the cardinality of the equivalence class of $[x_i]_B$ and n is the number of objects in U.

Theorem 1: Let $S = (U, A)$, $B, B' \subseteq A$. If $[x_i]_B \subseteq [x_i]_{B'}$, then $E(B) \leq E(B')$.
Proof. The values of rough entropy

$$E(B) = -\sum_{i=1}^{n} |[x_i]_B| \log_2 \frac{1}{|[x_i]_B|},$$

$$E(B') = -\sum_{i=1}^{n} |[x_i]_{B'}| \log_2 \frac{1}{|[x_i]_{B'}|}.$$

Since $[x_i]_B \subseteq [x_i]_{B'}$, we have $|[x_i]_B| \le |[x_i]_{B'}|$,

$$\sum_{i=1}^{n} |[x_i]_B| \log_2 |[x_i]_B| \le \sum_{i=1}^{n} |[x_i]_{B'}| \log_2 |[x_i]_{B'}|,$$

$$-\sum_{i=1}^{n} |[x_i]_B| \log_2 \frac{1}{|[x_i]_B|} \le -\sum_{i=1}^{n} |[x_i]_{B'}| \log_2 \frac{1}{|[x_i]_{B'}|}.$$

Hence,

$$E(B) \le E(B').$$ □

Corollary 1: Let $S = (U, A)$, $B, B' \subseteq A$. If $B' \subseteq B$, then $E(B) \le E(B')$.
Proof. For B' and B, the two subsets of A, if $B' \subseteq B$, we have $[x_i]_B \subseteq [x_i]_{B'}$ for all x_i
 $\in U$. As shown in Theorem 1, since $[x_i]_B \subseteq [x_i]_{B'}$, we have $E(B) \le E(B')$. □

3 The Generating Algorithm of Concept Hierarchies

3.1 Notations

The symbols used in our proposed method are defined in the following.

U	: A database with categorical attributes.
n	: The number of objects of U.
A	: The finite set of categorical attributes in U.
m	: The number of attributes in A.
B	: A set of reduced attributes, $B \subseteq A$.
B_j	: The j-th set of reduced attributes, $B_j \subseteq A$.
$|B|$	: The number of attributes in B.
A_j	: The j-th attribute of A, $A_j \in A$, $1 \le j \le m$.
$|[x_i]_B|$	: Let $B \subseteq A$, the cardinality of equivalence classes of $[x_i]_B$, $x_i \in U$.
C_k	: The k-th concept.

Except the above symbols, we define a property of compatible inclusion for the partitions of set of reduced attributes, as follows:

Definition 4: Let $S = (U, A)$ and $x \in U$. Assume that $B, B' \subseteq A$ and $|B| = |B'|$, we say that $[U]_B$ is compatible included in $[U]_{B'}$ if the partitions of U on B and B' satisfy $[x]_B \subseteq [x]_{B'}$, for all $x \in U$, denoted as $[U]_B \prec [U]_{B'}$.

3.2 The Concept Hierarchy Generating Algorithm

After the definitions in Section 3.1. We give the algorithm for generating a meaningful concept hierarchy with respect to a given data set. The algorithm is described in seven main steps, as follows:

Input: A database U with the attributes A.

Output: A concept hierarchy H for $S = (U, A)$.

Step 1: Initialize the values of $m' = m$ and $k = 1$.

Step 2: Generate the lowest level of concept hierarchy H_m with the set of attributes
   A. That is, $H_m = [U]_A$. Let the initial reduced set of attributes $B = A$.

Step 3: Let $|B|$ be the number of attributes in the reduced set of attributes B, we set
   $m' = |B| - 1$. Find $P_{m'}(B)$, which is the set of all possible subsets with m' at-
   tributes in the reduced set of attributes B.

Step 4: Compute the rough entropy $E(B_j)$ for $B_j \in P_{m'}(B)$.

Step 5: Generate a higher level of concept hierarchies $H_{m'}$ by the following three
   sub-steps:

   5.1) Find $B' = max\{E(B_j)\}$, for $B_j \in P_{m'}(B)$; then generate new concepts C_k
       in $H_{m'}$ with partitions of $[U]_{B'}$. The C_k is the concept that merge the
       concepts C_i in the lower level $H_{m'+1}$ belonging to the partition of $[x]_{B'}$.

   5.2) For the remainder concepts in $H_{m'+1}$, if there exists $B_j \in P_{m'}(B)$ and B_j
       $\neq B$, the property $[U]_{B'} \prec [U]_{B_j}$ is satisfied. We then generate another
       new concepts C_k in $H_{m'}$. The C_k is also the concept that merge the
       concepts C_i in the lower level $H_{m'+1}$ belonging to the partition of $[x]_{B_j}$.

   5.3) For the others C_i in the lower level $H_{m'+1}$, if $C_i \equiv [x]_{B_j}$, the C_i is un-
       changed and added into $H_{m'}$.

Step 6: The new reduced set of attributes $B = B' \cap (\underset{[U]_{B'} \prec [U]_{B_j}}{\cap} B_j)$.

Step 7: If $B \neq \varnothing$, go to Step 3; otherwise, output all levels of the concept hierarchy
   $H_0, ..., H_m$.

The time complexity of constructing an equivalence relation is shown to be
$O(mn^2)$, where m and n are number of attributes and objects, respectively [10].
However, the time of finding partition classes is obviously bounded by sorting algorithm. The time complexity of the proposed algorithm is mainly dependent on
the partitioning of equivalence class $[U]_B$. Thus, the time complexity for n objects
with m attributes is computed in $O(mT(n))$ to obtain $[U]_B$ for $|B|=m$, $|U| = n$ and
$T(n)$ is $O(nlogn)$. The number of attributes is decreased after a new concept level
is generated. Hence, the overall time complexity of the algorithm is bounded by
$O(m^2(T(n)))$. If a good predefined hashing function for our proposed algorithm, we
can obtain almost near $O(mn)$.

3.3 An Example

We give an example to explain the above algorithm more clearly. Assume that the
example used in Table 1 shows a data set containing eight objects and four attrib-

utes. The symbolic categories in the four attributes, respectively, are: *gender*: {*male, female, boy, girl*}, *income*: {*high, low*}, *insurance*: {*yes, no*}, *status*: {*single, married, divorce, yet*}.

Table 1. The example dataset

	gender	in-come	insur-ance	status		gender	in-come	insur-ance	status
x_1	male	low	no	single	x_5	boy	low	yes	yet
x_2	male	low	no	divorce	x_6	girl	low	yes	yet
x_3	female	high	yes	divorce	x_7	female	low	no	married
x_4	female	high	yes	married	x_8	male	high	yes	single

Input: The dataset in Table 1.
Output: The concept hierarchy *H*.
Step 1: Initialize the values of $m' = 4$ and $k = 1$.
Step 2: Generate the lowest level of concept hierarchy with the set of attributes A.
   Let $B = \{gender, income, insure, status\}$, $H_4 = [U]_A$.
   $[U]_{\{gender, income, insure, status\}} = \{[x_1]_B, [x_2]_B, [x_3]_B, [x_4]_B, [x_5]_B, [x_6]_B, [x_7]_B, [x_8]_B \}$.
   Thus $H_4 = C_1 \cup C_2 \cup C_3 \cup C_4 \cup C_5 \cup C_6 \cup C_7 \cup C_8$.
Step 3: We set $m' = |B| - 1 = 3$. Find $P_3(B)$, which is the set of all possible subsets with 3 attributes in the reduced set of attributes B.
   $P_{m'}(B) = \{B_j \mid B_j \subseteq B, |B_j| = m'\} = \{B_1, B_2, B_3, B_4\}$,
   $\{B_1\} = \{gender, income, insurance\}$, $\{B_2\} = \{gender, income, status\}$,
   $\{B_3\} = \{gender, insurance, status\}$, $\{B_4\} = \{income, insurance, status\}$.
Step 4: Compute the rough entropy $E(B_j)$ for $B_j \in P_3(B)$.
   $E(\{B_1\}) = 8$, $E(\{B_2\}) = 0$, $E(\{B_3\}) = 0$, and $E(\{B_4\}) = 4$.
Step 5: Generate a higher level of concept hierarchy H_3 by the following three sub-steps:
   5.1) $E(B_1)$ is the maximum of $E(B_j)$ for $B_j \in P_3(B)$. Thus $C_9 = C_1 \cup C_2$, $C_{10} = C_3 \cup C_4$.
   5.2) For the remainder concepts in H_4, there is a subset of attributes $B_j' \in P_3(B)$, $B_j' = \{income, insurance, status\}$ have the property of compatible inclusion with B_1. Thus $C_{11} = C_5 \cup C_6$ is generated in H_3.
   5.3) For the others C_i in the lower level H_4, because the C_i is unchanged and added into H_3. Thus $H_3 = C_7 \cup C_8 \cup C_9 \cup C_{10} \cup C_{11}$.
Sept 6: The new reduced set of attributes $B = B_1 \cap B_j'$, $B = \{income, insurance\}$.
Step 7: Because $B = \{income, insurance\}$, $|B| = 2$, then go to Step 3.
The final result of the concept hierarchy is shown in Figure 1.

4   The Experimental Results

In this section, we give two databases to evaluate the propose algorithm.

Experiment 1. Table 2 shows an artificial data set consisting of 16 objects and five attributes. These attributes have two or five values which are used by Murphy and Smith in psychological experiments [8]. The experiment result is shown in Figure 2.

Table 2. The 16 object used in Experiment 1

	A_1	A_2	A_3	A_4	A_5		A_1	A_2	A_3	A_4	A_5
x_1	1	1	1	1	1	x_9	2	4	3	1	1
x_2	1	1	1	1	2	x_{10}	2	4	3	1	2
x_3	1	1	1	2	1	x_{11}	2	4	3	4	1
x_4	1	1	1	2	2	x_{12}	2	4	3	4	2
x_5	1	2	2	3	1	x_{13}	2	5	4	5	1
x_6	1	2	2	3	2	x_{14}	2	5	4	5	2
x_7	1	3	2	3	1	x_{15}	2	5	5	5	1
x_8	1	3	2	3	2	x_{16}	2	5	5	5	2

For such data set, we first generate the lowest level of concept hierarchy H_5 with the set of attributes A. That is, $H_5 = [U]_A$. For all $x_i, x_j \in U$, $[x_i]_A \neq [x_j]_A$, there are 16 concepts generated in level H_5. Then we produced the equivalence classes by removing one of attributes and compute the rough entropy. The results are listed in Table 3. We chose the attribute with the maximum rough entropy as criterion of generating concept hierarchy. Thus the new concepts in H_4 with the partition of $[U]_{\{A1, A2, A3, A4\}}$, $C_{17} = C_1 \cup C_2$, $C_{18} = C_3 \cup C_4$, $C_{19} = C_5 \cup C_6$, $C_{20} = C_7 \cup C_8$, $C_{21} = C_9 \cup C_{10}$, $C_{22} = C_{11} \cup C_{12}$, $C_{23} = C_{13} \cup C_{14}$, and $C_{24} = C_{15} \cup C_{16}$ are generated. Two phenomena should be noted here. First, the rough entropies of the equivalence classes consist of the A_2, A_3, A_4 and A_5 attributes are zero, it indicated that A_1 is the key attribute. The other is that attribute A_5 is considered dispensable in the set A first. Theorem 1 can be used to explain this observed phenomenon. A_1 and A_5 attributes have two values each only. Assume $B' = \{A_2, A_3, A_4, A_5\}$ and $B = \{A_1, A_2, A_3, A_4\}$. We observed that each x in B is contained in B' (i.e. $[x]_R \prec [x]_{B'}$), and $E(B')$ is greater than $E(B)$. Obviously, the result is just the same as Theorem 1 states.

Table 3. The equivalence classes containing four attributes in Experiment 1.

j	B_j	$E(B_j)$	j	B_j	$E(B_j)$
1	$\{A_1, A_2, A_3, A_4\}$	32	4	$\{A_1, A_3, A_4, A_5\}$	8
2	$\{A_1, A_2, A_3, A_5\}$	16	5	$\{A_2, A_3, A_4, A_5\}$	0
3	$\{A_1, A_2, A_4, A_5\}$	8			

Second, as before Step 3 to Step 7, we produced the equivalence classes by removing one of the attributes and compute the rough entropies of the set of reduced attributes. The results are listed in Table 4. The new concepts in H_3 with the partition of $[U]_{\{A1, A2, A3\}}$, C_{25} and C_{26} are generated. We note that the compatible inclusion occur in this case. For the remainder concepts in H_4, the set of reduced attributes $\{A_1, A_2, A_4\}$ and $\{A_1, A_3, A_4\}$ have the property of compatible inclusion with $\{A_1, A_2, A_3\}$ (i.e. $[U]_{\{A1, A2, A3\}} \prec [U]_{\{A1, A2, A4\}}$, $[U]_{\{A1, A2, A3\}} \prec [U]_{\{A1, A3, A4\}}$). We can then generate another new concept in H_3. Since the concepts C_{23} and C_{24} consist of the attributes A_1, A_2, and A_4, C_{19} and C_{20} consist of the attributes A_1, A_3, and A_4, the new concepts C_{27} and C_{28} are generated. The new set of reduced attributes is $\{A_1\}$.

Table 4. The equivalence classes containing three attributes in Experiment 1.

j	B_j	$E(B_j)$	j	B_j	$E(B_j)$
1	$\{A_1, A_2, A_3\}$	80	3	$\{A_1, A_3, A_4\}$	56
2	$\{A_1, A_2, A_4\}$	56	4	$\{A_2, A_3, A_4\}$	32

Finally, we select A_1 attributes to generate a higher level of concept hierarchy H_1. The new concepts are $C_{29}=C_{25} \cup C_{28}$ and $C_{30} = C_{26} \cup C_{27}$.

The above experiment shows that the generality measure is effective and it is helpful for us to discover a meaningful structure of concept hierarchy. The interest of this data set stems from the fact that it exhibits an underlying hierarchical structure of three levels with two, four, and eight classes, respectively [6].

Table 5. The weather dataset

data	outlook	temperature	humidity	windy	play
x_1	sunny	hot	high	false	no
x_2	sunny	hot	high	true	no
x_3	overcast	hot	high	false	yes
x_4	rainy	mild	high	false	yes
x_5	rainy	cool	normal	false	yes
x_6	rainy	cool	normal	true	no
x_7	overcast	cool	normal	true	yes
x_8	sunny	mild	high	false	no
x_9	sunny	cool	normal	false	yes
x_{10}	rainy	mild	normal	false	yes
x_{11}	sunny	mild	normal	true	yes
x_{12}	overcast	mild	high	true	yes
x_{13}	overcast	hot	normal	false	yes
x_{14}	rainy	mild	high	true	no

Experiment 2. The second experiment uses the weather data set with 14 objects as shown in Table 5. $A = \{outlook, temperature, humidity, windy, play\}$. The decision attribute is $\{play\}$.

In this experiment, the process starts from the generation of the lowest concept level H_5 with the set of attributes A. The hierarchies as depicted in Figure 3, the result of the algorithm divides the weather data into two groups, one is suitable for playing, and the other is not. The result of the algorithm is the same as resultant classification.

From the experimental results, we observed that generated concept hierarchies by the proposed method performs fairly well. Experiments with several data sets demonstrate that the proposed approach is able to detect weak attributes and without significantly hurting the performance of the approach. The results are correspond with human's intuitive recognition.

5   Conclusions

Knowledge discovery from a large database is an important research topic of late years. The concept hierarchy is an explicit representation of knowledge and can be widely used in many applications. In this paper, we present an algorithm for generating concept hierarchies based on rough set and rough entropy. The algorithm can constructs a meaningful concept hierarchy by their partial ordering relationships among attributes in databases automatically. The experimental results demonstrate that the proposed method is feasible and can be applied to other issues of knowledge discovery, such as classification, clustering etc. In the future, we are trying to cope with the data having missing values.

References

1.   Fisher, D. (1987) Improving Inference through Conceptual Clustering. In: Proc. of 1987 AAAI Conference. Seattle Washington, pp.461-465
2.   Han, J., Kamber, M. (2001) Data Mining: Concept and Techniques, Morgan Kaufmann publishers
3.   Han, J., Yongjian, F. (1994) Dynamic Generation and Refinement of Concept Hierarchies for Knowledge Discovery in Databases. In: AAAI Workshop on Knowledge Discovery in Databases
4.   Hong, J., Mao, C. (1991) Incremental Discovery of Rules and Structure by Hierarchical and Parallel Clustering. In: Piatetsky-Shapiro, G., Frawley, W.J.(eds.): Knowledge Discovery in Databases. AAAI/MIT Press 177-193
5.   Hu, X. (1995) Knowledge Discovery in Databases: Attribute-Oriented Rough Set Approach. PhD thesis, University of Regina, Canada
6.   Luis, T., Javier, B. (2001) Generality-Based Conceptual Clustering with Probabilistic Concepts. IEEE Transactions on Pattern Analysis and Machine Intelligence, Vol. 23(2). 196-206
7.   Michalski, R.S., Stepp, R. (1983) Automated Construction of Classifications: Conceptual Clustering Versus Numerical Taxonomy. IEEE Transaction Pattern Analysis and Machine Intelligence, 396-410
8.   Murphy, G. L., Smith, E.E. (1982) Basic Level Superiority in Picture Categorization, J. Verbal Learning and Verbal Behavior, Vol. 21, 1-20
9.   Pawlak, Z. (1982) Rough set. International Journal of Computer and Information Sciences, 341-356
10.   Sever, H., Raghavan, V.V., Johnsten, T.D. (1998) The Status of Research on Rough Sets for Knowledge Discovery in Databases. In: ICNPAA-98: Second Internal Conference On Nonlinear Problems in Aviation and Aerospace, Daytona Beach, FL

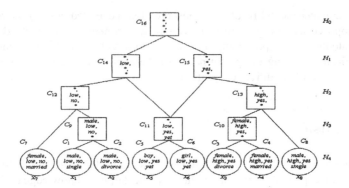

Fig. 1. The result of the Example Table 1

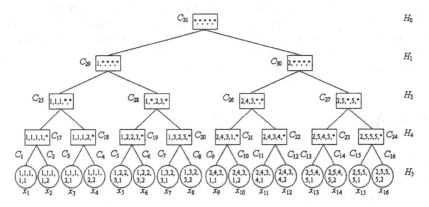

Fig. 2. The concept hierarchy in Experiment 1

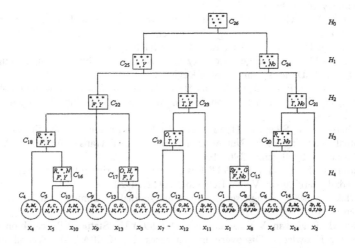

Fig. 3. The concept hierarchy in Experiment 2

Learning from Hierarchical Attribute Values by Rough Sets

Tzung-Pei Hong[1], Chun-E Lin[2], Jiann-Horng Lin[3], Shyue-Liang Wang[4]

[1] Department of Electrical Engineering, National University of Kaohsiung
   Kaohsiung, 811,Taiwan, R.O.C. **tphong@nuk.edu.tw**
[2] Institute of Information Management, I-Shou University
   Kaohsiung, 840, Taiwan, R.O.C. **m9022006@stmail.isu.edu.tw**
[3] Institute of Information Management, I-Shou University
   Kaohsiung, 840, Taiwan, R.O.C. **jhlin@isu.edu.tw**
[4] Department of Computer Science, New York Institute of Technology
   New York, 10023, U.S.A. **slwang@nyit.edu**

Abstract. The rough-set theory has been widely used in dealing with data classification problems. Most of the previous studies on rough sets focused on deriving certain rules and possible rules on the single concept level. Data with hierarchical attribute values are, however, commonly seen in real-world applications. This paper thus attempts to propose a new learning algorithm based on rough sets to find cross-level certain and possible rules from training data with hierarchical attribute values. It is more complex than learning rules from training examples with single-level values, but may derive more general knowledge from data.

Keywords: machine learning, rough set, certain rule, possible rule, hierarchical value.

1. Introduction

Recently, the rough-set theory has been used in reasoning and knowledge acquisition for expert systems [3]. It was proposed by Pawlak in 1982 [12] with the concept of equivalence classes as its basic principle. Several applications and extensions of the rough-set theory have also been proposed. Examples are Lambert-Torres *et al.*'s knowledge-base reduction [7], Zhong *et al.*'s rule discovery [18], Lee *et al.*'s hierarchical granulation structure [8] and Tsumto's attribute-oriented generalization [17]. Because of the success of the rough-set theory in knowledge acquisition, many researchers in the machine-learning fields are very interested in this research topic since it offers opportunities to discover useful information from training examples.

Most of the previous studies on rough sets focused on deriving certain rules and possible rules on the single concept level. Hierarchical attribute values are, however, usually predefined in real-world applications. Deriving rules on multiple concept levels may thus lead to the discovery of more general and important knowledge from data. It is, however, more complex than learning rules from training examples with single-level values. In this paper, we thus propose a new learning algorithm based on rough sets to find cross-level certain and possible rules from training data with hierarchical attribute values. Boundary approximations, instead of upper approximations, are used to find possible rules, thus reducing some subsumption checking. Some pruning heuristics are also used in the proposed algorithm to avoid unnecessary search.

2.   Review of the Rough-Set Theory

The rough-set theory, proposed by Pawlak in 1982 [12][14], can serve as a new mathematical tool for dealing with data classification problems. It adopts the concept of equivalence classes to partition training instances according to some criteria. Two kinds of partitions are formed in the mining process: lower approximations and upper approximations, from which certain and possible rules can easily be derived.

Formally, let U be a set of training examples (objects), A be a set of attributes describing the examples, C be a set of classes, and V_j be a value domain of an attribute A_j. Also let $v_j^{(i)}$ be the value of attribute A_j for the i-th object $Obj^{(i)}$. When two objects $Obj^{(i)}$ and $Obj^{(k)}$ have the same value of attribute A_j, (that is, $v_j^{(i)} = v_j^{(k)}$), $Obj^{(i)}$ and $Obj^{(k)}$ are said to have an indiscernibility relation (or an equivalence relation) on attribute A_j. Also, if $Obj^{(i)}$ and $Obj^{(k)}$ have the same values for each attribute in subset B of A, $Obj^{(i)}$ and $Obj^{(k)}$ are also said to have an indiscernibility (equivalence) relation on attribute set B. These equivalence relations thus partition the object set U into disjoint subsets, denoted by U/B, and the partition including $Obj^{(i)}$ is denoted $B(Obj^{(i)})$.

The rough-set approach analyzes data according to two basic concepts, namely the lower and the upper approximations of a set. Let X be an arbitrary subset of the universe U, and B be an arbitrary subset of attribute set A. The lower and the upper approximations for B on X, denoted $B_*(X)$ and $B^*(X)$ respectively, are defined as follows:

$$B_*(X) = \{x \mid x \in U, B(x) \subseteq X \}, \text{ and}$$

$$B^*(X) = \{x \mid x \in U \text{ and } B(x) \cap X \neq \varnothing \}.$$

Elements in $B_*(x)$ can be classified as members of set X with full certainty using attribute set B, so $B_*(x)$ is called the lower approximation of X. Similarly, elements in $B^*(x)$ can be classified as members of the set X with only partial certainty using attribute set B, so $B^*(x)$ is called the upper approximation of X.

3.   Hierarchical Attribute Values

Most of the previous studies on rough sets focused on finding certain rules and possible rules on the single concept level. However, hierarchical attribute values are usually predefined in real-world applications and can be represented by hierarchy trees. Terminal nodes on the trees represent actual attribute values appearing in training examples; internal nodes represent value clusters formed from their lower-level nodes. Deriving rules on multiple concept levels may lead to the discovery of more general and important knowledge from data. A simple example for attribute *Transport* is given in Figure 1.

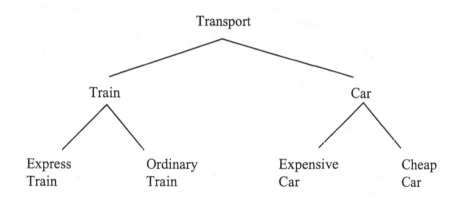

Figure 1: An example of predefined hierarchical values for attribute *Transport*

In Figure 1, the attribute *Transport* falls into two general values: *Train* and *Car*. *Train* can be further classified into two more specific values *Express Train* and *Ordinary Train*. Similarly, assume *Car* is divided into *Expensive Car* and *Cheap Car*. Only the terminal attribute values (*Express Train*, *Ordinary Train*, *Expensive Car*, *Cheap Car*) can appear in training examples.

The concept of equivalence classes in the rough set theory makes it very suitable for finding cross-level certain and possible rules from training examples with hierarchical values. The equivalence class of a non-terminal-level attribute value for attribute A_j can be easily found by the union of its underlying terminal-level equivalence classes for A_j. Also, the equivalence class of a cross-level attribute value combination for more than two attributes can be derived from the intersection of the equivalence classes of its single attribute values.

4.   The Algorithm

In this paper, we will propose a rough-set-based learning algorithm for deriving cross-level certain and possible rules from training examples with hierarchical attribute values. It is stated below.

A rough-set-based learning algorithm for training examples with hierarchical attribute values:
Input: A data set U with n objects, each of which has m decision attributes with
   hierarchical values and belongs to one of c classes.
Output: A set of multiple-level certain and possible rules.
Step 1: Partition the object set into disjoint subsets according to class labels.
   Denote each subset of objects belonging to class C_l as X_l.
Step 2: Find terminal-level elementary sets of single attributes; that is, if an
   object $obj^{(i)}$ has a terminal value $v_j^{(i)}$ for attribute A_j, put $obj^{(i)}$ in the
   equivalence class from $A_j = v_j^{(i)}$.
Step 3: Link each terminal node for $A_j = A_j^{t_i}$, $1 \le i \le /A_j^t/$, in the taxonomy tree
   for attribute A_j to the equivalence class for $A_j = A_j^{t_i}$, where $/A_j^{t_i}/$ is the
   number of terminal attribute values for A_j.
Step 4: Set l to 1, where l is used to represent the number of the class currently
   being processed.
Step 5: Compute the terminal-level lower approximation of each single
   attribute A_j for class X_l as:
$$A_{j*}^t(X_l) = \{ A_j^t(x)/x \in U, \ A_j^t(x) \subseteq X_l \},$$
   where $A_j^t(x)$ is the terminal-level equivalence class including object x
   and derived from attribute A_j.
Step 6: Compute the terminal-level boundary approximation of each single
   attribute A_j for class X_l as:
$$A_j^{t*}(X_l) = \{ A_j^t(x)/x \in U, \ A_j^t(x) \cap X_l \ne \phi, A_j^t(x) \not\subset X_l \}.$$
Step 7: Compute the non-terminal-level lower and boundary approximations of
   each single attribute A_j for class X_l from the terminal level to the root
   level in the following substeps:

(a) Derive the equivalence class of the i-th non-terminal-level attribute value $A_j^{nt_{ki}}$ for attribute A_j on level k by the union of its underlying terminal-level equivalence classes.

(b) Put the equivalence class of a non-terminal-level attribute value $A_j^{nt_{ki}}$ in the k-level lower approximation for attribute A_j if all the equivalence classes of the underlying attribute values of $A_j^{nt_{ki}}$ are in the $(k+1)$-level lower approximation for attribute A_j.

(c) Put the equivalence class of a non-terminal-level attribute value $A_j^{nt_{ki}}$ in the k-level boundary approximation for attribute A_j if at least one equivalence class of the underlying attribute values of $A_j^{nt_{ki}}$ is in the $(k+1)$-level boundary approximation for attribute A_j.

Step 8:   Set $q = 2$, where q is used to count the number of attributes currently being processed.

Step 9:   Compute the lower and the boundary approximations of each attribute set B_j with q attributes (on any levels) for class X_l from the terminal level to the root level by the following substeps:

(a) Skip all the combinations of attribute values in B_j which have the equivalence classes of their any value subsets already in the lower approximation for X_l.

(b) Derive the equivalence class of each remaining combination of attribute values by the intersection of the equivalence classes of its single attribute values.

(c) Put the equivalence class $B_j(x)$ of each combination in substep (b) into the lower approximation class X_l if $B_j(x) \subseteq X_l$.

(d) Put the equivalence class $B_j(x)$ of each combination in substep (b) into the boundary approximation class if $B_j(x) \cap X_l \neq \emptyset$ and $B_j(x) \not\subseteq X_l$.

Step 10:   Set $q=q+1$ and repeat 8 to 10 until $q>m$.

Step 11:   Derive the certain rules from the lower approximations.

Step 12:   Remove certain rules with condition parts more specific than those of some other certain rules.

Step 13:   Derive the possible rules from the boundary approximations and calculate their plausibility values as:

$$p(B_j(x)) = \frac{\left|B_j(x) \cap X_l\right|}{\left|B_j(x)\right|},$$

where $B_j(x)$ is the equivalence class including x and derived from attribute set B_j.

Step 14:   Set $l = l + 1$ and repeat Steps 5 to 14 until $l > c$.

Step 15:   Output the certain rules and possible rules.

5. An Example

In this section, an example is given to show how the proposed algorithm can be used to generate certain and possible rules from training data with hierarchical values. Assume the training data set is shown in Table 1.

Table 1 contains ten objects U=$\{Obj^{(1)}$, $Obj^{(2)}$, ..., $Obj^{(10)}\}$, two decision attributes A={Transport, House}, and a class attribute C={Consumption Style}.

Table 1: The training data used in this example

	Transport	Residence	Consumption Style
$Obj^{(1)}$	Expensive car	Villa	High
$Obj^{(2)}$	Cheap car	Single house	High
$Obj^{(3)}$	Ordinary train	Suite	Low
$Obj^{(4)}$	Express train	Villa	Low
$Obj^{(5)}$	Ordinary train	Suite	Low
$Obj^{(6)}$	Express train	Single house	Low
$Obj^{(7)}$	Cheap car	Single house	High
$Obj^{(8)}$	Express train	Single house	High
$Obj^{(9)}$	Ordinary train	Suite	Low
$Obj^{(10)}$	Express train	Apartment	Low

The possible values of each decision attribute are organized into a taxonomy, as shown in Figures 1 and 2.

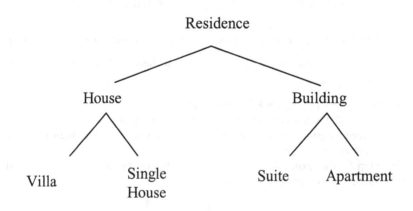

Residence

House   Building

Villa   Single House   Suite   Apartment

Figure 2: Hierarchy of Residence

In Figures 1 and 2, there are three levels of hierarchical attribute values for attributes *Transport* and *Residence*. The roots representing the generic names of attributes are located on level 0 (such as "Transport" and "Residence"), the internal nodes representing categories (such as "Train") are on level 1, and the terminal nodes representing actual values (such as "Express Train") are on level 2. Only values of terminal nodes can appear in training examples. Assume the class has only two possible values: {High (H), Low (L)}. The proposed algorithm then processes the data in Table 1 as follows.

Step 1: Since two classes exist in the data set, two partitions are found as follows:

$$X_H = \{Obj^{(1)}, Obj^{(2)}, Obj^{(7)}, Obj^{(8)}\}, \text{ and}$$
$$X_L = \{Obj^{(3)}, Obj^{(4)}, Obj^{(5)}, Obj^{(6)}, Obj^{(9)}, Obj^{(10)}\}.$$

Step 2: The terminal-level elementary sets are formed for the two attributes as follows:

$$U/\{Transport^t\} = \{\{(Obj^{(1)})\}, \{(Obj^{(3)})(Obj^{(5)})(Obj^{(9)})\},$$
$$\{(Obj^{(4)})(Obj^{(6)})(Obj^{(8)})(Obj^{(10)})\},$$
$$\{(Obj^{(2)})(Obj^{(7)})\}\};$$
$$U/\{Residence^t\} = \{\{(Obj^{(1)})(Obj^{(4)})\}, \{(Obj^{(2)})(Obj^{(6)})(Obj^{(7)})(Obj^{(8)})\},$$
$$\{(Obj^{(3)})(Obj^{(5)})(Obj^{(9)})\}, \{(Obj^{(10)})\}\}.$$

Step 3: Each terminal node is linked to its equivalence class for later usage.

Steps 4 and 5: The terminal-level lower approximations for class X_H are first calculated as:

$$Transport^t_*(X_H) = \{\{(Obj^{(1)})\}, \{(Obj^{(2)})(Obj^{(7)})\}\}, \text{ and}$$

$$Residence^t_*(X_H) = \varnothing.$$

Step 6: The terminal-level boundary approximations for class X_H are calculated as:

$$Transport^{t*}(X_H) = \{(Obj^{(4)})(Obj^{(6)})(Obj^{(8)})(Obj^{(10)})\}, \text{ and}$$
$$Residence^{t*}(X_H) = \{\{(Obj^{(1)})(Obj^{(4)})\}, \{(Obj^{(2)})(Obj^{(6)})(Obj^{(7)})(Obj^{(8)})\}\}.$$

Step 7: The non-terminal-level lower and boundary approximations of single attributes for class X_H are computed as:

$$Transport^{nt}_*(X_H) = \{(Obj^{(1)})(Obj^{(2)})(Obj^{(7)})\}, \text{ and}$$

$$Transport^{nt*}(X_H) = \{(Obj^{(3)})(Obj^{(4)})(Obj^{(5)})(Obj^{(6)})(Obj^{(8)})(Obj^{(9)}) (Obj^{(10)})\}.$$

Steps 8 and 9: The lower and the boundary approximations of each attribute set with two attributes for class X_H on the terminal level are found as:
$$\{Transport^t, Residence^t\}_*(X_H) = \varnothing, \text{ and}$$

$$\{Transport^{t}, Residence^{t}\}^{*}(X_H) = \{(Obj^{(6)})(Obj^{(8)})\}.$$

Steps 10 and 11: All the certain rules are then derived from the lower approximations. Results for this example are shown below:

If Transport is Car then Consumption Style is High.

Steps 12 and 13: All the possible rules are derived from the boundary approximations. The plausibility measure of each rule is also calculated in this step. The resulting possible rule with their plausibility values are shown below:

If Transport is Express Train then Consumption Style is High,
with plausibility = 0.25.

Steps 14 and 15: All the certain rules and possible rules are then output.

6.   Conclusions and Future Works

In this paper, we have proposed a new learning algorithm based on rough sets to find cross-level certain and possible rules from training data with hierarchical attribute values. The proposed method adopts the concept of equivalence classes to find the terminal-level elementary sets of single attributes. These equivalence classes are then easily used to find the non-terminal-level elementary sets of single attributes and the cross-level elementary sets of multiple attributes by the union and the intersection operations. Lower and boundary approximations are then derived from the elementary sets from the terminal level to the root level. Boundary approximations, instead of upper approximations, are used in the proposed algorithm to find possible rules, thus reducing some subsumption checking. Lower approximations are used to derive maximally general certain rules. Some pruning heuristics are also used to avoid unnecessary search. The rules derived can be used to infer a new event with both terminal and non-terminal attribute values.

REFERENCES

1.   Buchanan BG, Shortliffe EH (1984) Rule-Based Expert System: The MYCIN Experiments of the Standford Heuristic Programming Projects. Massachusetts: Addison-Wesley

2. Chmielewski MR, Grzymala-Busse JW, Pererson NW, Than S (1993) The rule induction system LERS – A version for personal computers. Foundation of Computation Decision Science Vol 18, No 3, pp 181-212

3. Grzymala-Busse JW (1988) Knowledge acquisition under uncertainty: A rough set approach. Journal of Intelligent Robotic Systems. Vol 1, pp 3-16

4. Hirano S, Sun X, Tsumoto S (2002) Dealing with multiple types of expert knowledge in medical image segmentation: A rough sets style approach. The 2002 IEEE International Conference on Fuzzy system. Vol 2, pp 884 -889

5. Hong T.P, Wang TT, Wang SL (2000) Knowledge acquisition from quantitative data using the rough-set theory. Intelligent Data Analysis. Vol 4, pp 289-304

6. Hong TP, Wang TT, Chien BC (2001) Mining approximate coverage rules. International Journal of Fuzzy Systems. Vol 3, No 2, pp 409-414

7. Lambert-Torres G, Alves DA, Silva AP, Quintana VH, Borges DA, Silva LE (1996) Knowledge-base reduction based on rough set techniques. The Canadian Conference on Electrical and Computer Engineering. pp 278-281

8. Lee CH, Swo SH, Choi SC (2001) Rule discovery using hierarchical classification structure with rough sets. The 9th IFSA World Congress and the 20th NAFIPS International Conference. Vol 1, pp 447 -452

9. Lingras PJ, Yao YY (1998) Data mining using extensions of the rough set model. Journal of the American Society for Information Science. Vol 49, No 5, pp 415-422

10. Michalski RS, Carbonell JG, Mitchell TM (1983) Machine Learning: An Artificial Intelligence Approach. Vol 1, California: Kaufmann Publishers

11. Michalski RS, Carbonell JG, Mitchell TM (1984) Machine Learning: An Artificial Intelligence Approach. Vol 2.

12. Pawlak Z (1982) Rough set, International Journal of Computer and Information Sciences. Vol 11, No 5, pp 341-356

13. Pawlak Z (1991) Rough Sets: Theoretical Aspects of Reasoning about Data. Kluwer Academic Publishers

14. Z Pawlak (1996) Why rough sets? The Fifth IEEE International Conference on Fuzzy Systems. Vol 2, pp 738 -743

15. Tsumoto S (1998) Extraction of experts decision rules from clinical databases using rough set model. Intelligent Data Analysis. Vol 2, pp 215-227

16. Tsumto S (1998) Knowledge discovery in medical databases based on rough sets and attribute-oriented generalization. The 1998 IEEE International Conference on Fuzzy Systems. Vol 2, pp 1296 -1301

17. Yao YY (1999) Stratified rough sets and granular computing. The 18th International Conference of the North American. pp 800 -804

18. Zhong N, Dong JZ, Ohsuga S, Lin TY (1998) An incremental probabilistic rough set approach to rule discovery. The IEEE International Conference on Fuzzy Systems. Vol 2, pp 933 -938

Delivering Distributed Data Mining E-Services

Shonali Krishnaswamy

School of Computer Science and Software Engineering
900 Dandenong Road, Monash University, Caulfield East, Victoria –3145, Australia.
shonali@csse.monash.edu.au

Abstract. The growing number of commercial Internet-based data mining service providers is indicative of the emerging trend of data mining application services. It validates the recognition that knowledge is a key resource in strategic organisational decision-making. The trend also establishes that the Application Service Provider (ASP) paradigm is seen as a cost-effective approach to meet the business intelligence needs of small to medium range organisations that are the most constrained by the high cost of niche software technologies. An important issue in a service context is the optimization of performance in terms of throughput and response time. The paper presents a *hybrid* distributed data mining (DDM) model that improves the overall response time.

1 Introduction

There are a growing number of Application Service Providers (ASPs) like digiMine™ (http://www.digimine.com) and Information Discovery™ (http://www.datamine.aa.psiweb.com), who offer commercial data mining e-services [8]. The increasing focus on data mining e-services can be attributed to the recognition of data mining as an important technology in aiding strategic decision making coupled with the commercial viability of the ASP paradigm that underlines the cost benefits of "renting" software [1]. The option of Internet-delivery of data mining services is emerging as attractive for small to medium range organisations, which are the most constrained by the high cost of data mining software, and consequently, stand to benefit by paying for software usage without having to incur the costs associated with buying, setting-up and training. An important issue in a service environment is the performance in terms of response time and throughput. Thus, it is important to perform data mining in a manner whereby the utilisation of resources is optimal and cost-efficient and implies task allocation strategies that minimise the overall response time. The optimal utilisation of resources has a significant bearing the successful operation of a data mining e-services as it leads to the ability to accept more jobs and increase revenue. The e-services concept of differentiated services where services are customised to specific requirements of individual client also requires efficient task allocation techniques. While traditionally this notion of task allocation and resource utilisation is not incorporated into data mining systems, as it is not considered to be of significant bearing, a consequence of the e-services paradigm is that it requires such capabilities in data mining systems. In this paper, we present our *hybrid* distributed data mining (DDM) architectural model and experimentally show how it improves the response time when compared with the traditional DDM models that use either mobile agents or client-server approaches [4].

This paper is organised as follows. Section 2 presents the hybrid DDM model. Section 3 presents experimental evaluation of the hybrid DDM model. Section 4 concludes the paper.

2 Hybrid Model for Data Mining E-Services

In this section, we propose a novel hybrid DDM, which integrates the traditional client-server and mobile agent architectural models for developing DDM systems. The hybrid DDM model, shown in figure 1 is driven by an optimiser that uses estimates of the DDM response time in order to choose a task allocation strategy that minimises the response time.

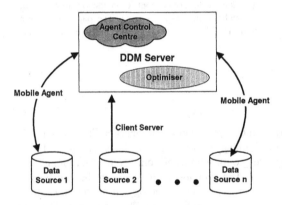

Figure 1 Hybrid DDM Architectural Model

Thus by integrating the client-server and mobile agent models, the hybrid model endeavours to minimise the response time. The hybrid model by virtue of being able to support both the client-server and mobile agent models for distributed mining has the ability to use a combination of the two strategies for performing a task. This assumes added significance in view of the studies by [2, 12], which have shown that a mix of client-server and hybrid approaches can lead to improved communication times for certain distributed applications.

This operation of the hybrid DDM model is shown in the example scenario illustrated in figure 2. The scenario represents a distributed e-services environment and involves a client (based at location 1) who requires two datasets to be mined ("Data 1" and "Data 2") and has available two computational resources ("Server 1" and "Server 3") that it can make available for mining. One of the datasets (Data 1) is located on Server 2 that is not available for mining. A service provider (based at location 2) has the data mining software and has the ability to mine using its own computational resources (i.e. three servers) or dispatch mobile data mining agents to the client's site. The three servers may be geographically distributed, thereby making the communications costs variable between the three servers.

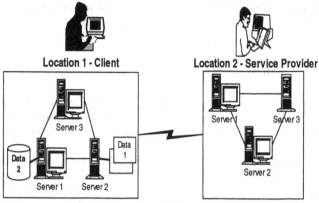

Figure 2 A Scenario for Optimisation

There are three possible options that the client can choose:

1. The mining can be done locally using only the client's computational resources. Obviously, mobile agents will have to be employed in order to perform the task at the client's site since the client possesses no data mining software. However the optimisation issue will involve using the most appropriate allocation/transfer of the data to the locally available computational resources.

2. The mining should be done remotely using only the service provider's computational resources. In this case the client-server model will have to be used and the data has to be transferred to the service provider's servers and the optimisation issue will involve using these resources in the most suitable manner.

3. The client has no preference for the location and the service provider has the option of choosing the combination that is optimal for the given response time requirement. In this scenario, the optimisation question involves choosing the most appropriate combination of mobile agents and client-server models to achieve the optimal response time.

We have informally presented an overview of our hybrid DDM strategy to support minimising the response time. The hybrid model has the following principal characteristics: it combines the best aspects of the mobile agent model and the client-server approach by integrating an agent framework with dedicated computational resources. It brings with it the advantage of combining the concept of dedicated data mining resources (and thus alleviating the issues associated with lack of control over remote computational resources in the mobile agent model) and the ability to circumvent the communication overheads associated with the client-server approach. This gives the model the option of applying either approach to a particular DDM task. The hybrid model involves an optimisation component (i.e. the "optimiser"), that uses *a priori* estimates of the DDM response time for various alternative scenarios and choose the most cost-effective one. Thus, the optimiser provides the infrastructure necessary to build the cost-estimates and determine the strategy to be applied on a case-by-case basis. The key to the successful operation is the ability to estimate the DDM response time. The estimation in turn requires the identification of the cost components and formalisation of a cost model for DDM response time for different scenarios. We now formalise the cost components of the DDM response time and state the operation of the hybrid model in terms of the cost components.

2.1 Cost Components of the DDM Response Time

In this section we specify the different cost components of the response time in distributed data mining e-services. The response time of a task in a distributed data mining e-service broadly consists of three components: *communication, computation and knowledge integration.*

Communication: The communication time is largely dependent on the operational model. It varies depending on whether the task is performed using a client-server approach or using mobile agents. In the client-server model the communication time is principally the time taken to transfer data from distributed servers to a high performance machine where the data mining is performed. In the mobile agent model, the communication time revolves around the time taken to transfer mobile agents carrying data mining software to remote datasets and the time taken to transfer results from remote locations for integration.

Computation: This is the time taken to perform data mining on the data sets and is a core factor irrespective of the operational model.

Knowledge Integration: This is the time taken to integrate the results from the distributed datasets.

The response time for distributed data mining is as follows:

$$T = t_{dm} + t_{com} + t_{ki} \tag{1}$$

In Eq. 1 above, T is the response time, t_{dm} is the time taken to perform data mining, t_{com} is the time involved in communication and t_{ki} is the time taken to perform knowledge integration. The modelling and estimation of the knowledge integration (t_{ki}) variable is dependent on the size and contents of the results obtained from the distributed datasets. Given that the primary objective of data mining is to discover hitherto unknown patterns in the data [3], we consider the *a priori* estimation of the time taken to perform knowledge integration to be outside the scope of this paper (since knowledge integration depends on the characteristics of the results of the data mining process). Having identified the cost components of the DDM response time, we now formalise the overall estimation cost for different models and scenarios.

2.2 Cost Matrix for Representing the Composite Response Time

We now present a cost matrix for computing the composite DDM response time estimates for different strategies. The response time for distributed data mining as presented in Eq. 1 consists of three components including communication (due to either transfer of mobile agents and/or transfer of data), computation (performing data mining) and knowledge integration. The following discussion focuses on the communication and computation components and does not consider the knowledge integration component. The strategy is to compute estimates for the individual cost components and then uses the estimates to determine the overall response times for different strategies. The cost matrix to calculate the composite response time for different DDM strategies is denoted by *CM* and is represented as a two-dimensional *m* x *n* matrix, where *m* is the number of available servers and *n* is the number of datasets. A fundamental feature of the cost matrix that makes it applicable for both the mobile agent and client-server models is that we incorporate location information on the datasets and servers.

The elements of the cost matrix represent the estimated response time and are defined as follows:

1. Let *m* be the number of servers.
2. Let $S = \{S_1, S_2, ..., S_m\}$ be the set of servers. A server S_j can either be located at the service provider's site or at the client's site. Therefore, let S^{SP} be the set of servers located at the service providers site and let S^C be the set of servers located at the client's site. The following properties are true for the sets S, S^{SP}, S^C.

    ❑ $S^{SP} \cup S^C = S$; the set of servers available at the client's site and the set of servers available at the service provider's site summarily constitute the total set of available servers. The obvious corollaries are $S^{SP} \subseteq S$ and $S^C \subseteq S$.

    ❑ $S^{SP} \cap S^C = \phi$; thus a server either belongs to the client or the service provider. It cannot belong to both.

    ❑ $S^{SP} = \phi$ is valid and indicates that the client is not willing to ship the data across and $S^C = \phi$ is also valid and indicates that the client's computational resources are unavailable or inadequate.

    In order to specify the location of a server we use the following notation: $S_j \in S^{SP}$ and $S_l \in S^C$ where $l, j = 1, 2, ..., m$. This distinction is necessary for specification of how the response time has to be estimated in the cost matrix.
3. Let *n* be the number of datasets and let $DS = \{ds(1), ds(2), ..., ds(n)\}$ represent the labelling of the datasets.
4. Let $ds(i)_{S_j}$ represent the location of a dataset labelled $ds(i)$, $i=1, 2, ..., n$ at server S_j, where $ds(i) \in DS$ and $j = 1, 2, ..., m$. Thus datasets are uniquely identified and multiple datasets at locations can be represented.

Let $cm_{ij} \in CM$ be the estimated response time for taking a dataset located at the server j and mining it at the server i, where $1 \le i \le m$ and $1 \le j \le n$. The value of cm_{ij} is computed as follows:

$$cm_{ij} = \begin{cases} MA_{S_X \to S_i} + TR_{S_j \to S_i}^{ds(k)_{S_j}} + W_{S_i} + DM_{S_i}^{ds(k)_{S_j}} , i \neq j, S_X \in S^{SP}, S_j \in S^C, S_i \in S^C \\ MA_{S_X \to S_i} + W_{S_i} + DM_{S_i}^{ds(k)_{S_j}} , i = j, S_X \in S^{SP}, S_j \in S^C, S_i \in S^C \\ TR_{S_j \to S_i}^{ds(k)_{S_j}} + W_{S_i} + DM_{S_i}^{ds(k)_{S_j}} , i \neq j, S_i \in S^C, S_i \in S^{SP} \end{cases}$$

In the above equation:

- $MA_{S_x \to S_i}$ is the time to transfer a mobile data mining agent from server S_X (which is a server of the service provider) to server S_i

- $TR_{S_j \to S_i}^{ds(k)_{S_j}}$ is the time to transfer a dataset $ds(k)_{S_j}$ located at server S_j to server S_i

- $DM_{S_i}^{ds(k)_{S_j}}$ is the time to mine dataset $ds(k)_{S_j}$ located originally at server S_j at server S_i

- W_{S_i} is the wait time at server S_i required for the completion of previous tasks and is generally more significant when $S_i \in S^{SP}$ (i.e. server S_i is located at the service provider's site).

As presented above there are three formulae for estimating the response time for different scenarios in the cost matrix. The first one is for the case where the server Si where the mining is to be performed is at the client's site but does not contain the dataset $ds(k)_{S_j}$ (which as indicated is located at the server Sj). Hence there is a need to transfer the data from its original location Sj to the server Si to perform the data mining. Further, the client's site would not have the data mining software and a mobile agent needs to be transferred from the service provider's site (represented as S_X). The second formula is for the case where the server where the mining is to be performed is at the client's site and the dataset is located on the same server (i.e. i = j). In this case, the mobile agent needs to be transferred but there is no need to transfer the data. The third formula is for the case where the server where the mining is to be performed is located at the service provider's site (i.e. Si $\in$ S^{SP}) and therefore the data has to be shipped across from the client's site to perform mining. The three formulae map to the three scenarios outlined earlier, namely, that of mining at the client's site, mining at the service provider's site and using both sites.

We have modelled cost formulae and estimation algorithms to estimate the individual cost components in [6, 9]. The analysis of the cost matrices to determine the minimum response time reduces to the well-known Generalised Allocation Problem (GAP) [10]. We apply an approximate technique that has two phases to determine the task allocation strategy from the cost matrix. We first use the polynomial-time algorithm proposed by [11] to determine the initial allocation of tasks and then apply a Tabu search [5] to improve the initial allocation by exchanges. We now present experimental evaluation of the performance of the hybrid model.

3 Experimental Evaluating the Hybrid DDM Model

The objective of the experimental evaluation is to analyse the performance of the hybrid DDM model when compared with the current predominant models for constructing distributed data mining systems, namely, the client-server and mobile agent models. The comparison focuses on the overall response times obtained by using each approach to perform DDM tasks. We used our *a priori* estimation techniques to compute the cost for each model – mobile agent, client-server and hybrid. We note that we have the experimentally shown the accuracy of our *a priori* estimates [6, 9]. Therefore irrespective of the DDM model that is used, the estimates provide an accurate prediction of the response time. This facilitates a valid comparison of the different DDM models using the estimates as the costing strategy for task allocation. In order to experimentally evaluate the performance of the hybrid model we have a developed a DDM system – Distributed Agent-based Mining Environment (DAME). DAME is an implementation of the hybrid model presented in this paper and has been developed in Java. It has an optimiser and cost calculators to estimate the DDM response time for different scenarios. It uses Aglets™ to provide the mobile agent infrastructure for dispatching mobile data mining agents and the Weka package [13] of data mining algorithms. For details of the DAME system, readers are referred to [7].

The experiments were conducted using a network of four distributed machines located on two different campuses and by varying the following parameters:

1. We assigned different roles to the servers for different experimental runs. That is for a given run, certain servers were assigned the role of "Service Provider" and certain servers were assigned the role of "Client". As discussed in section 2, the cost matrix computes the composite response time for performing a given task at server based on its location at either the service provider's site or the client's site.
2. We varied the data mining tasks, the datasets and the number of datasets assigned per server.
3. We applied different "Wait Times" to different servers to simulate the unavailability of servers for a certain period that is typically brought about by the either the execution of previous tasks or by lack of computational resources.

For each experimental run, the task was allocated to servers for the following strategies:

1. *Mobile Agents Only.* In this case mobile agents are sent from one of the nominated "Service Provider" servers to the location of the datasets. The datasets are not moved and remained at the client's sites *in their original location* (i.e. a dataset originally located at the client's site *c* remains at *c*.
2. *Mobile Agents with Local Data Transfers.* In this case, mobile agents are dispatched as before, but the task allocation is used to perform local transfers of the datasets within the client's servers. That is, if there are two servers located at the client's site and one dataset is to be mined, the dataset may be mined at either of the two servers, depending on whichever provides the faster response time. The cost matrix and the task allocation strategy is used to determine the precise allocation.
3. *Client-server.* In this case, the datasets are moved to the service provider's servers. However, if there are two or more servers nominated as service provider servers, then the server that can provide the faster response time is allocated the task using the cost matrix.
4. *Hybrid.* In this case, both mobile agent and client-server approaches are used to determine the task allocation strategy that results in the minimum response time.

The evaluation of the cost matrix for each of the above cases for a sample experimental run involving three servers nominated as being client sites and one server nominated as being service provider sites is shown in figure 3.

Datasets

Servers	ds1@c3	ds2@c2	ds3@c1
SP1	49	34	32
C1	12	10	10
C2	12	8	9
C3	11	7	7

a) Cost Matrix for Hybrid DDM Model

Datasets

Servers	ds1@c3	ds2@c2	ds3@c1
SP1	49	34	32
C1	12	10	10
C2	12	8	9
C3	11	7	7

b) Cost Matrix for Mobile Agent Model with Local Data Transfers

Datasets

Servers	ds1@c3	ds2@c2	ds3@c1
SP1	49	34	32
C1	12	10	10
C2	12	8	9
C3	11	7	7

c) Cost Matrix for Mobile Agent Model

Datasets

Servers	ds1@c3	ds2@c2	ds3@c1
SP1	49	34	32
C1	12	10	10
C2	12	8	9
C3	11	7	7

d) Cost Matrix for Client Server Model

Figure 3 Evaluating the Cost Matrix to Determine Task Allocation

The shaded regions in the matrices show the part of the matrix that is used to determine the allocation of tasks. For example, figure 3a shows the cost matrix that is used to determine the task allocation in the hybrid model and it can be seen that all the servers are taken into account. In contrast, it can seen from the shaded region of the cost matrix of figure 3b that in the mobile agent model such an evaluation is not performed and the datasets are mined at their original locations. Similarly figure 3d shows that in the client-server model only the service provider's servers are used. Figure 3b shows the region of the cost matrix that is evaluated to determine the minimum response time strategy for the case where the mobile agent model is used in conjunction with local data transfers (i.e. all the servers at the client's site are considered in the task allocation process).

The comparison between the response times of the client-server, mobile agent and hybrid DDM models is shown in figure 4. The figure clearly indicates that the response time achieved by the hybrid DDM model is always lower than or equal to the minimum response time of the other two models. The comparison between the hybrid DDM model and the mobile agent model with local data transfers is shown in figure 5. This model is the closest in performance to the hybrid DDM model. However, the hybrid DDM model outperforms even this model, which consists of mobile agents with local optimisation. It must be noted that the concept of local data transfers is also based on the model of task allocation based on *a priori* estimates to reduce the overall response time.

From our experiments we note that the mobile agent model performs better than the client-server model except where the wait times are considerably higher in the servers located at the client's site and the mobile agent model with local data transfers always performs better than or equal to the mobile agent model. In summary, the experimental results validate our hypotheses that *a priori* estimation is an effective basis for costing alternate DDM strategies (if the estimates are accurate) and that the hybrid DDM model lead to better overall response time than either the client-server or mobile agent models.

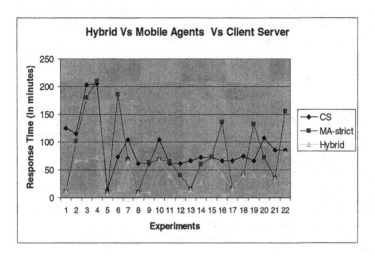

Figure 4 Comparative Evaluation of the Hybrid Model with the Mobile Agent and Client-server Models

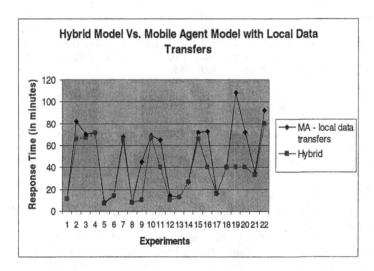

Figure 5 Comparative Evaluation of the Hybrid Model with the Mobile Agent Model Including Local Data Transfers

4 Conclusions

This issue of performance in application services is important and one that needs to be addressed. This paper takes a first step in improving the response time for data mining e-services. The hybrid model presented in this paper is applicable to distributed, data intensive application services in general..

References

1. Clark-Dickson, P., (1999), *"Flag-fall for Application Rental"*, Systems, (August), pp.23-31.
2. Chia, T., and Kannapan, S., (1997), *"Strategically Mobile Agents"*, First International Workshop on Mobile Agents (MA '97), Berlin, Germany, Lecture Notes in Computer Science (LNCS) 1219, Springer Verlag, pp. 149-161.
3. Fayyad, U, M., Piatetsky-Shapiro, G., and Smyth, P., (1996), *"From Knowledge Discovery to Data Mining an Overview"*, Advances in Knowledge Discovery and Data Mining, (eds) Fayyad, Piatetsky-Shapiro, Smyth and Uthurusamy, AAAI / MIT Press, pp. 1-37.
4. Fu, Y., (2001), *"Distributed Data Mining: An Overview"*, Newsletter of the IEEE Technical Committee on Distributed Processing, Spring 2001, pp.5-9.
5. Glover, F., and Laguna, M., (1997), *"Tabu Search"*, Kluwer Academic Publishers, Massachusetts, USA.
6. Krishnaswamy, S., Loke, S, W., and Zaslavsky, A., (2002), *"Supporting the Optimisation of Distributed Data Mining by Predicting Application Run Times"*, Proceedings of the Fourth International Conference on Enterprise Information Systems (ICEIS 2002), April 3-6, Ciudad Real, Spain, pp. 374-381.
7. Krishnaswamy, S., Loke, S, W., Zaslavsky, A., (2002), *"Towards Anytime Anywhere Data Mining Services"*, Proceedings of the Australian Data Mining Workshop (ADM'02) at the 15th Australian Joint Conference on Artificial Intelligence, (eds) S.J. Simoff, G.J. Williams, and M. Hegland. Canberra, Australia, December 2002, pp. 47 - 56, Published by the University of Technology Sydney, ISBN 0-9750075-0-5.
8. Krishnaswamy, S., Zaslavsky, A., and Loke, S, W., (2003), *"Internet Delivery of Distributed Data Mining Services: Architectures, Issues and Prospects"*, Chapter 7 in the book: Architectural Issues of Web-enabled Electronic Business, Murthy, V.K. and Shi, N. (eds.), pp. 113 - 127, Idea Group Publishing.
9. Krishnaswamy, S., Zaslavsky, A., and Loke, S, W., (2002), *"Techniques for Estimating the Computation and Communication Costs of Distributed Data Mining"*, Proceedings of International Conference on Computational Science (ICCS2002) – Part I, Lecture Notes in Computer Science (LNCS) 2331, Springer Verlag. pp. 603-612.
10. Martello, S., and Toth, P., (1981), *"An Algorithm for the Generalised Assignment Problem"*, Operational Research'81, Amsterdam, pp. 589-603.
11. Martello, S., and Toth, P., (1990), *"Knapsack Problems – Algorithms and Computer Implementations"*, John Wiley and Sons Ltd, England, UK.
12. Straßer, M., and Schwehm, M., (1997), *"A Performance Model for Mobile Agent Systems"*, in Proceedings of the International Conference on Parallel and Distributed Processing Techniques and Applications (PDPTA'97), (eds) H. Arabnia, Vol II, CSREA, pp. 1132-1140.
13. Witten, I, H., and Eibe, F., (1999), *"Data Mining: Practical Machine Learning Tools and Techniques with Java Implementations"*, Morgan Kauffman.

Maintenance of Discovered Functional Dependencies: Incremental Deletion

Shyue-Liang Wang[1], Wen-Chieh Tsou[2], Jiann-Horng Lin[2], Tzung-Pei Hong[3]

[1] Department of Computer Science, New York Institute of Technology
New York, 10023, U.S.A.
slwang@nyit.edu
[2] Institute of Information Management, I-Shou University
Kaohsiung, 840, Taiwan, R.O.C.
m9022005@stmail.isu.edu.tw, jhlin@isu.edu.tw
[3] Department of Electrical Engineering, National University of Kaohsiung
Kaohsiung, 811, Taiwan, R.O.C.
tphong@nuk.edu.tw

Abstract. The discovery of functional dependencies (FDs) in relational databases is an important data-mining problem. Most current work assumes that the database is static, and a database update requires rediscovering all the FDs by scanning the entire old and new database repeatedly. Some works consider the incremental discovery of FDs in the presence of a new set of tuples added to an old database. In this work, we present two incremental data mining algorithms, top-down and bottom-up, to discover all FDs when deletion of tuples occurred to the database. Based on the principle of monotonicity of FDs [2], we avoid re-scanning of the database and thereby reduce computation time. Feasibility and efficiency of the two proposed algorithms are demonstrated through examples.

Keywords: *data mining, functional dependency, incremental discovery.*

1. Introduction

Functional dependencies are relationships among attributes of a database relation. A functional dependency states that the value of an attribute is uniquely determined by the values of some other attributes. It is often used as a guideline for the design of relational schema that are conceptually meaningful and free of update anomalies. It therefore plays a crucial role in enforcing integrity constraints for logical database design. Due to its importance in database technology, mining of functional dependencies from databases has been identified as an important database analysis technique. It has received considerable research interests in recent years.

To find all functional dependencies from a given database relation r, we need to search for all possible FDs and test their validity. There are essentially two approaches for searching for all possible FDs: top-down searching [1,4-9], and bottom-up searching [2,3,7]. The top-down approach starts with the set of trivial FDs and adds those FDs that are satisfied by r. It can be further classified into depth-first search and level-wise (breadth-first) search. An efficient algorithm (TANE) proposed by Huhtala etc [4-6] for discovering functional and approximate dependencies is a level-wise top-down approach. The bottom-up approach starts with the set of all possible FDs and remove those FDs that are contradicted by r. An efficient algorithm (FDEP) proposed by Flach etc [2,3] for discovering functional and multi-valued dependencies is a bottom-up inductive approach. To test the validity of a given possible FD, pairwise comparisons between any two tuples, decision trees method and tuple partition method have been proposed. However, most data mining techniques of determining functional dependencies assumes that the database is static, and a database update requires rediscovering all the FDs by scanning the entire old and new database repeatedly.

In [2,10], incremental data-mining algorithms to discover FDs, when new tuples are added to database, were proposed. Based on the property that FDs are monotonic in the sense that $r_1 \leq r_2$ implies $FD(r_1) \geq FD(r_2)$, the algorithms find the new FDs incrementally and thereby reduce computation time.

In this work, we present an efficient two data-mining algorithms to incrementally discover all the cover of a database when a set of tuples are deleted from the database. Feasibility and efficiency of the two proposed algorithms are also demonstrated.

The rest of our paper is organized as follows. Section 2 reviews the functional dependency and the incremental discovery of FDs. Section 3 presents the incremental searching algorithms for the discovery of functional dependencies when a set of tuples is deleted from the database. Section 4 gives an example for the proposed algorithms. A conclusion is finally given at the end of the paper.

2.   Problem Description

This section reviews the definition of FD and gives the problem statement of discovering FDs when deletion of tuples occurred to database.

2.1   Mining of Functional Dependency

Functional dependencies are relationships among attributes of a relation. A FD states that the value of an attribute is uniquely determined by the values of some other attributes. A FD on conventional relational database is defined as follows.

Definition 1: Let R be a relation scheme and r be a relation over R. A functional dependency, expressed as $X{\rightarrow}A$, $X{\subseteq}R$ and $A{\in}R$, holds or is valid in r if, for all pairs of tuples t_1 and $t_2 \in r$,

$t_1[X] = t_2[X]$ implies $t_1[A] = t_2[A]$,

where $t_i[X]$ denotes the values of the attribute set X belonging to tuple t_i.

For a given relation r defined on scheme R, the set of all FDs, $X{\rightarrow}A$, is denoted by $FD(r)$. If $X{\rightarrow}A$ holds in r, then for any $X{\subseteq}Y$, $Y{\rightarrow}A$ also holds in r. We call $X{\rightarrow}A$ is more general than $Y{\rightarrow}A$ (or $Y{\rightarrow}A$ is more specific than $X{\rightarrow}A$). A set of FDs forms a *COVER* for $FD(r)$ if $FD(r)$ is the deductive closure of this set. A *COVER* is non-redundant, if for any two elements of cover $X \rightarrow A$ and $Y \rightarrow A$, it implies $X{\not\subseteq}Y$. For two relations, r_1 and r_2, defined on the same scheme R, the sets of FDs are monotonic in the sense that $r_1 \subseteq r_2$ implies $FD\!\left(r_1\right) \supseteq FD\!\left(r_2\right)$ [2].

Example 1   Consider the relation r_1 defined on scheme R={A,B,C,D,E} in Figure 1. The size of r_1 is 8 and the cover of r_1 contains 13 FDs.

Row id	A	B	C	D	E
1	1	a	x	p	i
2	1	b	v	q	k
3	2	b	x	r	m
4	2	b	x	p	m
5	2	c	v	s	i
6	3	c	x	t	i
7	3	d	v	p	i
8	3	d	z	u	m

Cover of r_1
AC->B,AC->E,AD->B,
AD->C,AD->E,BD->A,
BD->C,BD->E,BC->A,
BC->E,CDE->A,CDE->B
ABE->C

Fig. 1. An example relation r_1 and its cover of FDs

2.2   Incremental Discovery of Functional Dependency

A database update is an insertion, deletion, or modification of tuples to the database. Any FD data mining technique on static database can be used to discover the new FDs. However, when the size of update is small compared to the size of old database, incremental discovery of FDs should be considered in order to save computation efforts. In this work, we consider the case that given the cover of a database when deletion of tuples occurred to the database find a cover of the updated database as efficient as possible. It can be formally described as follows:

Input: a relation r_1 on R
         a non-redundant *cover* of r_1
         a set of deleted tuples r_2 on R

Output: a non-redundant cover of the set of FDs satisfied by r_1 - r_2

Example 2 Consider the relation r_1 in Figure1 and the relation r_2 in Figure 2. The cover of r_1 - r_2 shown in Figure 3 can be discovered incrementally from cover of r_1.

Row id	A	B	C	D	E
7	3	d	v	p	i
8	3	d	z	u	m

Fig. 2. An example relation r2 deleted from relation r1 in Fig. 1.

Right Hand Side	A	B	C	D	E
Functional	BC->A	AC->B	D->C	φ	AB->E
Dependency	BD->A	AD->B	AB->C		AC->E
	DE->A	AE->B	AE->C		AD->E
		DE->B			BC->E
					BD->E

Fig. 3. Cover of (r_1 - r_2)

The searching of FDs in the updated (deleted) database consists of two main phases. The first phase searches for FDs that are more general them the FDs in the cover of old database. The second phase searches for FDs that are not related to FDs in the old cover. In each phase, all candidate FDs for each attribute can be organized as a lattice structure, such as Figure 4 for attribute A.

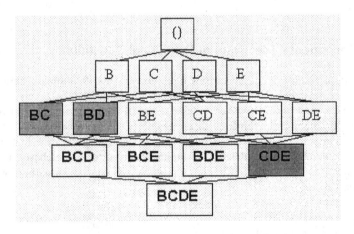

Fig. 4. The FD lattice for right hand side attribute A

In the searching for FDs, two pruning strategies can be applied in order to reduce the number of candidates. The first strategy is that if a node in the lattice is an FD, then all its descendant nodes are FDs. The second strategy is that if a node in the lattice is not an FD, then all its ancestor nodes are not FDs. Based on these two pruning strategies, we propose two incremental data-mining algorithms to discover a cover of FDs when deletion occurred to the database.

3.   Proposed Algorithms

This section describes the two proposed incremental searching algorithms of functional dependencies when deletion occurred to the database.

3.1   Top-Down Approach

Input: a relation r_1 on R
            a non-redundant *cover* of r_1
            a set of deleted tuples r_2 on R
Output: a non-redundant cover of the set of FDs satisfied by $r_1 - r_2$

STEP 1:   Group the FDs in the given cover of r_1 according to the same right hand size (RHS) of each attribute

STEP 2:   Sort the FDs in each group according to the size of left hand size (LHS) of FDs

STEP 3:   Set cover of $(r_1 - r_2) = \phi$

STEP 4:   For each group
            4.1   If group is NULL then goto Step 4.5
            4.2   *New1FD* = ϕ ;   //For more general FDs
            4.3   For each FD in the group
                    4.3.1   If FD is not a descendant of *New1FD*
                                    If there exists ancestor FD
                                            Then find the highest ancestors, put
                                            into *New1FD*, and mark the FD
                                                    //Specific FD
                            else
                                            mark the FD
                                                    //Specific FD
            4.4   Find those NonFDs that are most general and not descendent of New1FD and the group, put them in UnknowFD

//new candidate FD not in cover(r1)

4.5 $New2FD = \phi$ //new FD

4.6   For each candidate FD in *UnknowFD*,

    4.6.1   If candidate FD is not a descendent of *New2FD* Find its maximum ancestor FDs and put them into *New2FD*

4.7   cover(r_1 - r_2) = cover(r_1 - r_2) + *New1FD* + *New2FD* + *group(RHS) - marked FD*

    //*group(RHS)* is all FDs in the group

STEP 5: cover of $(r_1 - r_2)$ = cover(r_1 - r_2)   //all covers in each group

3.2   Bottom-Up Approach

Input: a relation r_1 on R
    a non-redundant *cover* of r_1
    a set of deleted tuples r_2 on R

Output: a non-redundant cover of the set of FDs satisfied by r_1 - r_2

STEP 1: Group the FDs in the given cover of r_1 according to the same RHS of each attribute

STEP 2: Sort the FDs in each group according to the size of left hand size (LHS) of FDs

STEP 3: Set cover of $(r_1 - r_2) = \phi$

STEP 4: For each group

    4.1   If group is NULL then goto Step 4.5

    4.2 $New1FD=\phi$ //For more general FDs

    4.3   For each FD in the group

        4.3.1 If FD is not a descendant of *New1FD*
            If there exists ancestor FD
                Then find the highest ancestors, put into *New1FD*, and mark the FD   //Specific FD

        else

                mark the FD
                //Specific FD

    4.4   Find those NonFDs that are most specific and not descendent of New1FD and the group, put them in UnknowFD
        //new candidate FD not in cover(r1)

    4.5 $New2FD = \phi$ //new FD

4.6   For each candidate FD in *UnknowFD*
    4.6.1   If candidate FD is not a descendent of *New2FD*
        Find its maximum ancestor FDs and put them into *New2FD*

4.7   cover(r_1 - r_2) = cover(r_1 - r_2) + *New1FD* + *New2FD* + *group(RHS)* - *marked FD*
        // *group(RHS)* is all FDs in the group

STEP 5: cover of (r_1 - r_2) = cover(r_1 - r_2)   //all covers in each group

4. Example

This section gives an example to show how the proposed algorithms can be used to discover FDs incrementally. For simplicity, we show only the bottom-up approach. The top-down approach works similarly.

Input: a relation r_1 on R in **Fig. 1.**
        a non-redundant *cover* of r_1 in **Fig. 1.**
        a set of deleted tuples r_2 on R in **Fig. 2.**
Output: a non-redundant cover of the set of FDs satisfied by r_1 - r_2, in Figure 3

STEP 1: Group the FD's in the given cover according to the same RHS of each attribute. The result is as follows.

Right Hand Side	A	B	C	D	E
Functional Dependency	BC->A	AC->B	AD->C	ϕ	AC->E
	BD->A	AD->B	ABE->C		AD->E
	CDE->A	CDE->B	BD->C		BC->E
					BD->E

STEP 2: Sort the FD's in each group according to size of LHS. The result is as follows.

Right Hand Side	A	B	C	D	E
Functional Dependency	BC->A	AC->B	AD->C	ϕ	AC->E
	BD->A	AD->B	BD->C		AD->E
	CDE->A	CDE->B	ABE->C		BC->E
					BD->E

STEP 3: Set cover of $(r_1 - r_2) = \phi$

STEP 4: For each group with RHS = A
          4.1   groupA is not NULL
          4.2   *New1FD*$= \phi$
          4.3   For *BC->A* in the groupA
                4.3.1 *BC->A* is not a descendant of *New1FD* and
                            There exists no ancestor FD
          4.3   For BD->A in the groupA
                4.3.1 *BD->A* is not a descendant of *New1FD* and
                            There exists no ancestor FD

STEP 3: Set cover of $(r_1 - r_2) = \phi$

STEP 4: For each group with RHS = A
          4.1   groupA is not NULL
          4.2   *New1FD*$= \phi$
          4.3   For *BC->A* in the groupA
                4.3.1 *BC->A* is not a descendant of *New1FD* and
                            There exists no ancestor FD
          4.3   For BD->A in the groupA
                4.3.1 *BD->A* is not a descendant of *New1FD* and
                            There exists no ancestor FD
          4.3   For *CDE->A* in the groupA
                4.3.1 *BC->A* is not a descendant of *New1FD* but
                            Its maximum ancestor FD is *DE->A*
                            put *DE->A* into *New1FD*, and mark *CDE->A*
          4.4   *BE->A* is the most specific candidate and not descendent of
                    New1FD and the group, put *BE->A* in *UnknowFD*
          4.5   *New2FD* $= \phi$
          4.6   For *BE->A* in *UnknowFD*,
              4.6.1   It has no ancestor FD
          4.7   cover$(r_1 - r_2) = \phi$ +*{DE->A}* + ϕ +*{BC->A,BD->A,CDE->A}*
                    - *{CDE->A}*
                            ⋮
                            ⋮

                FDs in each group can be obtained similarly

STEP 5: The resulting cover$(r_1 - r_2)$ is shown Figure 3

In the following, we summarize the candidate FDs that are required to test their validity, using the example in this section. For both top-down and bottom-up approaches, there are only 38 candidate FDs need to be validated in total. In contrast, there are 75 candidates to be validated if exhaustive searching is adopted. This indicates that there exists computation reduction in the proposed approaches.

5. Conclusion

In this paper, we have presented two incremental mining algorithms to discover functional dependencies when a set of tuples is deleted from a database relation. We have also presented examples for the proposed algorithms and compared their efficiency. The comparisons indicate that there exists computation saving by the proposed approaches. However, further experiments need to be carried out to distinguish the efficiency of each approach.

Although the proposed mining algorithm works well, it is just a beginning. ĪŠhĩẽldŝ ŝ̃ĩdĩ m̃ẽũãũhp̃ẽẽọr̃kõt̃ọp̃ãẽĩ x̃dõṽs̃ẽ õĩñ th incremental and non-incremental discovery of functional dependencies have not yet been studied. Other types of database activities, such as updates, need to be considered. Utilizing database management system functions such as SQL commands to incrementally discover functional dependencies might deserve further investigation, as it will facilitate the processes in the database reverse engineering. Discovering other types of data dependencies incrementally may require further studies as well.

REFERENCES

1. Bell S, Brockhausen P (1995) Discovery of Data Dependencies in Relational Databases. Tech. Rep. LS-8, Report-14, University of Dortmund, Apr.

2. Flach PA (1990) Inductive Characterization of Database Relations. ITK Research Report, November

3. Flach PA, Savnik I (1999) Database Dependency Discovery: A Machine Learning Approach. AI Communications, 12(3), November, pp 139-160

4. Huhtala Y, Karkkainen J, Porkka P, Toivonen H (1997) Efficient Discovery of Functional and Approximate Dependencies Using Partitions.(Extended Version) Report C Report C-1997-79, November

5. Huhtala Y, Karkkainen J, Porkka P, Toivonen H (1998) Efficient Discovery of Functional and Approximate Dependencies Using Partitions. Proceedings of IEEE International Conference on Data Engineering, pp 392-410

6. Huhtala Y, Karkkainen J, Porkka P, Toivonen H (1999) TANE: An Efficient Algorithm for Discovering Functional and Approximate Dependencies. The Computer Journal, Vol 42, No 2, pp 100-111

7. Mannila H, Räihä KJ (1994) Algorithms for Inferring Functional Dependencies. Data & Knowledge Engineering, 12(1), pp 83-99

8.   Mannila H, Toivonen H (1997) Levelwise Search and Borders of Theories in Knowledge Discovery. Data Mining and Knowledge Discovery, 1 (3), pp 241-258

9.   Schlimmer JC (1993) Efficiently Inducting Determinations: A Complete and Systematic Search Algorithm that Using Optimal Pruning. In G. Piatetsky-Shapiro, editor, Proceedings of the Tenth International Conference on Machine Learning, Morgan Kaufmann, pp 284-290

10.   Wang SL, Shen JW, Hong TP (2001) Incremental Data Mining of Functional Dependencies. Proceedings of the Joint 9th IFSA World Congress and 20th NAFIPS International Conference, Vancouver, Canada, pp 1322-1326

Filtering Multilingual Web Content Using Fuzzy Logic

Rowena Chau and Chung-Hsing Yeh

School of Business Systems, Faculty of Information Technology, Monash University, Clayton, Victoria 3800, Australia

Abstract.

An emerging requirement to sift through the increasing flood of multilingual textual content available electronically over the World Wide Web has led to the pressing demand for effective multilingual Web information filtering. In this paper, a content-based approach to multilingual information filtering is proposed. This approach is capable of screening and evaluating multilingual documents based on their semantic content. As such, correlated multilingual documents are disseminated according to their corresponding themes/topics to facilitate both efficient and effective content-based information access. The objective of alleviating users' burden of information overload is thus achieved. This approach is realized by incorporating fuzzy clustering and fuzzy inference techniques. To illustrate, development of a Web-based multilingual online news filtering system by applying this approach is also presented.

1 Introduction

Within the multilingual information community, there are users who need to keep track of the global information development that is relevant to some specific areas of interest. With the explosive growth of the WWW over the globe, voluminous text streams containing documents in multiple languages have been made electronically available. To mediate information access in an overloading multilingual environment, multilingual information filtering, which is capable of automatically disseminating relevant multilingual documents in response to a stable and specific information need, is highly desirable. Multilingual information filtering aims at relieving such users of the burden of screening large amount of multilingual information manually. Benefits of multilingual information filtering are not limited to multilingual users. Monolingual users who need to examine multilingual documents will also find it useful. Multilingual information filtering will help to reduce translation effort by identifying only the relevant documents to be translated [6].

As a process of automatic text detection and delivery, multilingual information filtering involves sifting through incoming multilingual text streams, removing irrelevant documents, and delivering only documents that are relevant to a topic (or a set of topics) in a constant manner. Multilingual information filtering is considerably challenging. Given a topic and a document stream, a multilingual information filtering system constructs a topic profile and a filtering function (selection rule) which will make a decision to either select or reject each document from the stream as the document arrives. Current monolingual information filtering approaches use topic profile represented as a bag of terms. Documents containing the identical set of terms are considered relevant. In a multilingual environment, the same topic is represented by different sets of terms in different languages. Translating every term of the topic profile into a unique form in every existing language appears to be the most straightforward solution. However, errors in word sense selection during such word-based translation can adversely affect filtering performance when polysemous terms are encountered. Therefore, a syntactic-oriented approach to multilingual information filtering is basically infeasible. Some content-based approach capable of representing and comparing multilingual text based on their semantic context is necessary.

In this paper, a novel approach to facilitate content-based multilingual information filtering based on fuzzy set theory is proposed. This approach is capable of screening and organizing multilingual documents based on their semantic content. As such, correlated multilingual documents are associated with their corresponding themes/topics to facilitate both efficient and effective content-based information access. The objective of alleviating users' burden of information overload is thus achieved. This paper is organized as follows: First, the content-based approach to multilingual information filtering is explained in Section 2. This is followed by details about the generation of universal topic profiles and their application towards content-based multilingual information filtering in Section 3 and Section 4 respectively. In Section 5, development of a Web-based multilingual online news filtering system by applying this approach is presented. Finally, a conclusive remark is given in Section 6.

2   The Content-Based Approach

On constantly detecting and delivering multilingual documents relevant to a set of themes/topics, multilingual information filtering involves two major tasks, namely, topic profiling and text categorization. Due to the feature incompatibility problem contributed by the vocabulary mismatch phenomenon across multiple languages, documents describing the same topic in different languages are represented by different sets of features (i.e. terms) in separate feature spaces. This language-specific representation has made multilingual text incomparable. Therefore, monolingual information filtering techniques that rely on shared syntactic terms will not work for multilingual information filtering. To overcome this problem, a content-based multilingual information filtering approach is proposed. The idea is: by generating a set of universal content-based topic profiles encapsulating all feature variations among multiple languages, semantically-relevant multilingual documents can then be categorized regardless of the syntactic terms they contain.

Information filtering concerns the relevance of documents against topics. In a multilingual environment, synonymous terms are used by documents in different languages to describe the same topic. To determine the topical relevance of these multilingual documents, universal content-based topic profiles, which are linguistically comprehensive as well as semantically rich, are necessary. Each topic profile should be capable of capturing the semantic content of its corresponding topic while encapsulating the syntactic variation among term usage. To effectively modelling a set of domain-specific topics, a novel content-based topic profiling technique is developed.

Using the co-occurrence statistics of a set of multilingual terms extracted from a parallel corpus, fuzzy clustering is applied to group semantically-related multilingual terms to form topic profiles. A parallel corpus is a collection of documents containing identical text written in multiple languages. Statistical analysis of parallel corpus has been suggested as a potential means of providing effective lexical basis for multilingual text management applications, such as cross-lingual text retrieval [5]. This has been successfully applied in research for building translation-based cross-lingual text retrieval models [2,7,4,3]. In this paper, the basic idea of applying fuzzy clustering for universal multilingual topic profiling is based on the intuition that translated versions of the same text are linguistic variants of the same topic(s). Hence, multilingual terms co-occurring within corresponding translated documents are semantically related to the same topic(s). Therefore, by grouping topic-related multilingual terms extracted from a training parallel corpus using fuzzy clustering, a set of content-based topic profiles, which are represented by clusters of multilingual terms, are automatically generated. As such, these topic profiles will be available as a kind of topic schema to filter multilingual document by facilitating content-based multilingual text categorization.

A text categorization algorithm performs the crucial task of information filtering by determining the topical relevance of each document. To enable multilingual information filtering, a content-based multilingual text categorization algorithm is developed using fuzzy reasoning. The algorithm regards the topical relevance of a multilingual document as the aggregation of the semantic relevance of all its constituent terms towards the corresponding topic. As terms are relevant to various topics with different degrees simultaneously, content-based multilingual text categorization is essentially a fuzzy inference process. Given by the fuzzy clustering result during the process of universal topic profiling, degrees of semantic relevance between multilingual terms and topics has already been encoded in the topic profiles. Hence, to compose the overall topical relevance of a document, a conclusion is deduced by combining topical relevance of its index terms with the application of the composition rule of fuzzy inference.

3 Content-Based Universal Topic Profiling

Universal topic profiling concerns the construction of multilingual topic profiles representing the semantic content of interesting topics, which are described universally by documents in multiple languages. Towards this end, a fuzzy clustering algorithm, known as fuzzy c-means (FCM) is applied. The purpose is to reveal the semantic content of a set of topics by generating corresponding fuzzy clusters of topically related multilingual terms. Each of the resulting fuzzy multilingual term cluster is then regarded as a topic profile representing a topic. Every cluster comprises topically related terms in multiple languages. Having effectively captured the linguistic diversity of a topic in every language, this topics profile is thus universal. Hence, it can be applied for detecting relevant multilingual documents regardless of the language of an incoming text.

Due to the intrinsic vagueness of natural language, semantic content of topic profiles tend to be overlapping in terms of the multilingual terms they comprise. Thus, crisp clustering algorithm like k-means [8] that generates partition such that each term is assigned to exactly one cluster is inadequate for representing the real term-topic relationship. In this aspect, fuzzy clustering methods that allow objects (multilingual terms) to be classified to more than one cluster with different membership values is preferred. With the application of fuzzy c-means, the resulting fuzzy multilingual term clusters, which are overlapping by admitting partial membership, will provide a more realistic representation of the multilingual topic profiles.

The fuzzy c-means algorithm developed by [1] aims at minimizing the objective function

$$J(X,U,v) = \sum_{i=1}^{c} \sum_{k=1}^{n} (\mu_{ik})^m d^2(v_i, x_k) \qquad (1)$$

under the constraints

$$\sum_{k=1}^{n} \mu_{ik} > 0 \quad \text{for all } i \in \{1, \ldots, c\} \qquad (2)$$

and

$$\sum_{i=1}^{c} \mu_{ik} = 1 \quad \text{for all } k \in \{1, \ldots, n\} \qquad (3)$$

where $X = \{x_1, \ldots, x_n\} \subseteq R^p$ is the set of objects, c is the number of fuzzy clusters, $\mu_{ik} \in [0,1]$ is the membership degree of object x_k to cluster i, v_i is the prototype (cluster center) of cluster i, and $d(v_i, x_k)$ is the Euclidean distance between prototype v_i and object x_k. The parameter $m > 1$ is called the fuzziness index. For $m \to 1$, the clusters tend to be crisp, i.e. either $\mu_{ik} \to 1$ or $\mu_{ik} \to 0$; for $m \to \infty$, we have $\mu_{ik} \to 1/c$. Usually, $m = 2$ is chosen.

As a procedure of objective function optimization, FCM is most suitable for finding the optimal groupings of objects that best represent the structure of the data set. By minimizing the sum of within group variance, strength of associations of objects are maximized within clusters and minimized between clusters. In this aspect, FCM is particularly useful in text management applications, such as term clustering, where intrinsic topical structure and semantic relationship among terms of a given term set must be revealed in order to gain knowledge for better document categorization and filtering.

Based on the hypothesis that semantically related multilingual terms representing similar concepts tend to co-occur with similar inter- and intra-document frequencies within a parallel corpus, fuzzy c-means is applied to sort a set of multilingual terms into clusters (topics) such that terms belonging to any one of the clusters (topics) should be as similar as possible while terms of different clusters (topics) are as dissimilar as possible in terms of the concepts they represent.

To realize this idea for generating a set of universal content-based topic profiles by means of fuzzy multilingual term clustering, a set of multilingual terms are first extracted from a parallel corpus of N parallel documents. To apply FCM for clustering these multilingual terms into clusters representing different topics, each term is then represented as an input vector of N features where each of the N parallel document is regarded as an input feature with each feature value representing the frequency of that term in the nth parallel document. The universal topic profiling algorithms is given as follows:

1 Initialize the membership values μ_{ik} of the k multilingual terms t_k to each of the i clusters (topics) for $i = 1, \ldots, c$ and $k = 1, \ldots, K$ randomly such that:

$$\sum_{i=1}^{c} \mu_{ik} = 1 \quad \forall k = 1,\ldots, K \quad (4)$$

and

$$\mu_{ik} \in [0,1] \quad \forall i = 1,\ldots c \,; \forall k = 1,\ldots K \quad (5)$$

2 Calculate the cluster prototype (cluster centres) v_i using these membership values μ_{ik} :

$$v_i = \frac{\sum_{k=1}^{K} (\mu_{ik})^m \cdot t_k}{\sum_{k=1}^{K} (\mu_{ik})^m}, \quad \forall i = 1,\ldots,c \quad (6)$$

3 Calculate the new membership values μ_{ik}^{new} using these cluster centers v_i:

$$\mu_{ik}^{new} = \frac{1}{\sum_{j=1}^{c} \left(\frac{\|v_i - t_k\|}{\|v_j - t_k\|} \right)^{\frac{2}{m-1}}}, \quad \forall i = 1,\ldots,c \,; \forall k = 1,\ldots, K \quad (7)$$

4 If $\left\|\mu^{new} - \mu\right\| > \varepsilon$, let $\mu = \mu^{new}$ and go to step 2. Else, stop. ε is a convergence threshold.

Finally, after the completion of the universal topic profiling process, all existing topic profiles represented by a $c \times K$ fuzzy partition matrix S is obtained such that

$$
S = \begin{array}{c} \\ v_1 \\ v_2 \\ v_3 \\ \\ \\ \\ v_c \end{array}
\begin{array}{cccccc}
t_1 & t_2 & t_3 & & & t_K \\
\begin{bmatrix} \mu_{11} & \mu_{12} & \mu_{12} & \cdot & \cdot & \cdot & \mu_{1K} \\
\mu_{21} & \mu_{22} & \mu_{23} & \cdot & \cdot & \cdot & \mu_{2K} \\
\mu_{31} & \mu_{32} & \mu_{33} & \cdot & \cdot & \cdot & \mu_{3K} \\
\cdot & & & & & & \cdot \\
\cdot & & & & & & \cdot \\
\cdot & \cdot & \cdot & \cdot & \cdot & \cdot & \cdot \\
\mu_{c1} & \mu_{c2} & \mu_{c3} & \cdot & \cdot & \cdot & \mu_{cK} \end{bmatrix}
\end{array}
\tag{8}
$$

where μ_{ik} for $i = 1, \ldots, c$ and $k = 1, \ldots, K$ is the membership value of multilingual terms, $t_{k \in \{1,2,\ldots,K\}}$, in cluster i. Each row of the matrix refers to a topic profile defining a topic with its corresponding set of multilingual terms. To label a topic profile for every language involved, the term in each language being assigned the highest membership values in a cluster is selected.

As a result, a collection of fuzzy multilingual term clusters, acting as a set of universal topic profiles, is now available for multilingual document categorisation for the purpose of content-based multilingual information filtering.

4   Content-Based Fuzzy Multilingual Text Filtering

In information filtering, determination of a document's topic relevance against a set of topic profiles is essentially a text categorization task. The objective is to infer the degree of interestingness of this document for a who is interested in that particular topic. By its nature, this inference is based on knowledge whose meaning is not sharply defined. Conventional approaches such as classical logic which requires high standards of precision for exact reasoning is thus ineffective, because of its inability to grip with the linguistic fuzziness involved. On the other hand, fuzzy logic [9], through the use of approximate reasoning, allows the standards of precision to be adjusted to fit the imprecision of the information involved. Fuzzy reasoning is an inference procedure that deduces conclusion from a set of fuzzy IF-THEN rules by combining evidence through the compositional rule of inference. Briefly, a fuzzy IF-THEN rule is a fuzzy conditional proposition assuming the form:

"IF x is A, THEN y is B."

where "x is A" and "y is B" are fuzzy propositions, and A and B are linguistic values defined by fuzzy sets on universe of discourse X and Y respectively. By the composition rule of inference, individual conclusions inferred from IF-THEN rules are aggregated to compose an overall conclusion.

In the context of our content-based multilingual information filtering approach, this text categorization task is modelled as a fuzzy reasoning process using the composition of fuzzy relations. The intuitive meaning of the fuzzy inference is that: IF there exists a term t_k in a document that is relevant to a topic c_i to a degree of $\mu_{ci}(t_k)$, THEN this document is also relevant to topic c_i at least to a degree equal to $\mu_{ci}(t_k)$.

To do so, the partition matrix S (as presented by equation 8) given by the universal topic profiling process is interpreted as a fuzzy relation associating all multilingual terms and existing topics. A piece of multilingual text txt_{ML} is expressed as a fuzzy set such that

$$txt_{ML} = \left\{ \mu_{txt_{ML}}(t_k) \right\} \qquad \text{for all } t_k \in T \tag{9}$$

where

$$\mu_{txt_{ML}}(t_k) = \begin{cases} 1 & \text{for } t_k \in txt_{ML} \\ 0 & \text{for } t_k \notin txt_{ML} \end{cases} \tag{10}$$

Based on the composition rule of fuzzy inference, the fuzzy content-based text categorisation function g is defined as

$$g(txt_{ML}, S) = txt_{ML} \circ S \tag{11}$$

where $\circ$ is the max-min composition operator.

This fuzzy content-based text categorization function assigns each piece of multilingual text a series of topic relevance weights with respect to all existing topics such that

$$txt_{ML} = \left\{ \mu_{c1}(txt_{ML}), \mu_{c2}(txt_{ML}), \mu_{c3}(txt_{ML})...\mu_{cn}(txt_{ML}) \right\} \tag{12}$$

where

$$\mu_{ci}(txt_{ML}) = \max_{t_k \in txt_{ML}} min\left(\mu_{txt_{ML}}(t_k), \mu_{ci}(t_k) \right) \tag{13}$$

As a result, every piece of multilingual text can now be categorised to all existing topics regardless of languages. Multilingual texts originally represented in separate feature spaces defined by terms in different language can now be compared and filtered based on a common set of linguistic comprehensive universal topic profiles. The problem of feature incompatibility is thus overcome. Moreover, our text categorisation algorithm does not just make a binary decision of relevant or irrelevant. It assigns a degree of topical relevance to each document. These topical relevance degrees provides additional information to the users by indicating how well it match a topic of the user's information interest. This information is particularly useful for a user of a multilingual information filtering system when time is not allowed to examine a large numbered of relevant documents being delivered within a short period of time, and the user just wants to limit his/her focus onto the most relevant ones.

5 Web-Based Multilingual Online News Filtering

To illustrate this approach for filtering multilingual Web content, an application to filter multilingual online news is presented. In a prototype system, the fuzzy multilingual information filtering approach described in this paper is applied to realise a Web-base multilingual online news filtering service. The purpose of this service is to automatically disseminate multilingual news by topics regardless of the languages in which the news stories are being written. An overview of this system is depicted in Figure 1.

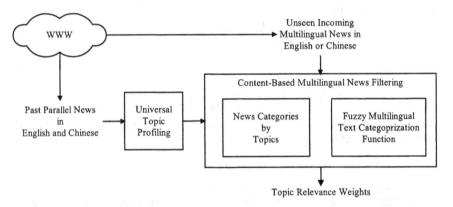

Fig. 1. Development of a Web-based Multilingual Online News Filtering System

First, to gather a set of training parallel document for the process of content-based universal topic profiling, a collection of past parallel online news in both English and Chinese are collected from the Web. To generate a set of content-based topic profiles indicating the major news categories, a set of multilingual (both English and Chinese) terms are extracted from the training parallel documents using a bilingual wordlist, and then the universal topic profiling algorithm for clustering topic-related multilingual terms is applied. As such, a set of universal topic profiles is made available for a multilingual Web news filtering service to categorise unseen incoming news in both English and Chinese base. Given a set of topic relevance weights by the fuzzy multilingual text categorization function, every piece of unseen incoming news in either English or Chinese can then be categorised to a topic to which it has been assigned the highest topic relevance weight. Consequently, multilingual online news, regardless of languages, can then be intelligently disseminated based on the semantic content it exhibits rather than the syntactic string patterns (i.e. keywords) it contains.

6 Conclusion

Multilingual information filtering is important for users who want to keep track of the global information development that is relevant to his/her areas of information interests. In this paper, a content-based approach to multilingual information filtering is proposed. The key to its effectiveness is the generation of the universal topic profiles. These topic profiles overcome the feature incompatibility problem by capturing the semantic content of their corresponding topics while encapsulating the syntactic variation among term usage in multiple languages. As such, multilingual documents can then be filtered based on their topical relevance regardless of the syntactic features it contains. To generate the topic profiles, a universal topic profiling algorithm using fuzzy clustering us developed. To facilitate the delivery of relevant multilingual documents by topics, a content-based text categorisation algorithm based on fuzzy inference is proposed. By enabling topic-focused content-based multilingual information filtering, the objective of alleviating the user's burden of information access in an overloading multilingual environment is thus achieved. Realisation of this idea is illustrated with an application to the development of a Web-based multilingual online news filtering service for intelligent content-based multilingual news dissemination.

References

1. Bezdek JC (1981) Pattern Recognition with Fuzzy Objective Function Algorithms, Plenum Press, New York
2. Carbonell JG, Yang Y, Frederking RE, Brown RD, Geng Y and Lee D (1997) Translingual information retrieval: a comparative evaluation. In: Pollack ME (ed.) IJCAI-97 Proceedings of the 15th International Joint Conference on Artificial Intelligence, 708-714
3. Chau R. and Yeh C-H. (2001) Construction of a fuzzy multilingual thesaurus and its application to cross-lingual text retrieval. In: Zhong N (ed.): Web Intelligence: Research and Development, First Asia-Pacific Conference, WI 2001, Maebashi City, Japan, October 23-26, 2001, Proceedings. Lecture Notes in Artificial Intelligence. Springer-Verlag. Germany, 340-345.
4. Chen J, Nie J-Y. (2000) Parallel Web Text Mining for Cross-Language IR. In: Proceedings of RIAO-2000: "Content-Based Multimedia Information Access", Paris.
5. Croft WB, Broglio J and Fujii H. (1996) Applications of multilingual text retrieval. In: Proceedings of the 29th Annual Hawaii International Conference on System Sciences. 5: 98-107
6. Oard. DW (1996) Adaptive Vector Space Information filtering for Monolingual and Cross-Language Applications. PhD thesis, University of Maryland, College Park.

7. Littman ML, Dumais ST and Landaur TK (1998) Automatic cross-language informa-
    tion retrieval using latent semantic indexing. In: Grefenstette G (ed) Cross-Language
    Information Retrieval. Kluwer Academic Publishers, Boston.
8. McQueern J (1967) Some methods for classification and analysis of multivariate ob-
    servations. In: Proceedings of the Fifth Berkeley Symposium on Mathematical Statis-
    tics and Probability,281-297
9. Zadeh LA. (1965) Fuzzy sets. Information and control 8: 338-353.

A Comparison of Patient Classification Using Data Mining in Acute Health Care

Eu-Gene Siew[1], Kate A. Smith[1], Leonid Churilov[1], Jeff Wassertheil[2]

[1] School of Business Systems, Monash University, Victoria 3800, Australia
[2] Peninsula Health, PO Box 52, Frankston, Victoria 3199, Australia

Abstract. Patients' diagnoses are used currently as a basis for resource consumption. There are other alternative forms of groupings: one approach is to group patients according to common characteristics and infer their resource consumption based on their group membership. In this paper, we compare the effectiveness of the alternative forms of patient classification obtained from data mining with the current classification for an objective assessment of the average difference between the inferred and the actual resource consumption. In tackling this prediction tasks, classification trees and neural clustering are used. Demographic and hospital admission information is used to generate the clusters and decision tree nodes. For the case study under consideration, the alternative forms of patient classifications seem to be better able to reflect the resource consumption than diagnosis related groups.

1 Introduction

Patient groupings are currently used by the Victorian government for funding purposes. Case Mix funding for public hospitals is a system based on diagnosis related groups (DRG). The basic premise of Case Mix is that those patients with similar DRG will consume similar amounts of resources [1].

The role of DRG, which are used by Case Mix as a predictor of resource consumption, has been overestimated. According to the Victorian Auditor General's report on Case Mix implementation in Victoria, Australia, there have been problems in funding hospitals [2].

Even though Case Mix has problems it continues to be used. It provides an objective measure, a benchmark for improve and a basis for funding. However, there are alternative patient classifications and the information obtained could be incorporated into DRG.

This study uses data mining techniques to homogeneously group acute patients based on a case study done in a medium-sized hospital in Victoria, Australia. The focus as explained in more detail in Section 2, is to compare patient classifications obtained from data mining with that of the current patient classification, using patient diagnosis groups to verify the hypothesis that using DRG alone is not a good indicator of resource consumption.

Because directly measuring consumption is difficult, costly and time consuming, LOS is used as a resource consumption proxy according to Ridley et al. [3]. Ridley et al. have also used classification trees as an alternative method for grouping iso-resource consumption based on LOS.

Walczak et al. [4] use neural networks to predict resource consumption. They too use LOS as a proxy for resource consumption. Neural nets prediction was made on two different types of emergency patients: pediatric trauma and acute pancreatitis. It was found that neural networks are able to predict LOS to an accuracy of within 3 to 4 days.

The other alternative for predicting LOS, is by first classifying homogenous patient groups and inferring their resource consumption based on their group membership. Siew et al. [5] propose clustering as an alternative method of grouping patients, utilising both demographic and hospital admission information. Once the clusters have formed, the average length of stay (LOS) of each cluster can be inspected, and used as a proxy for the predicted resource consumption of that group of patients.

Siew et al. [6] also look at the different learning paradigms with a different patient base classification system from data mining techniques. There were many clusters produced by Kohonen's self organising map (SOM), an unsupervised learning paradigm, similar, in many ways, to the groups produced by Classification and Regression Trees (CART), a supervised learning paradigm.

This paper is organised into 4 sections. The first presents the structure of the paper. The second deals with an explanation of data mining techniques. The third presents an analysis of the case study data. The final section outlines the analysis of the results.

2 Framework

The first part of this paper deals with getting the predicted LOS; the second deals with comparisons to see how accurate the prediction is.

For predicting LOS, one methods is to homogeneously group patients using data mining techniques from the case study data (referred to here as observed data). Two data mining techniques are used: CART and SOM. Once this is done, we infer the LOS prediction of the individual patients based on the average LOS of their group membership.

The alternative LOS prediction is from the DRG. Two methods are used to predict LOS using DRG: one uses the average LOS of the patient's group membership and infers that to be the predicted LOS. We refer to that as the "average DRG LOS." The other method uses the target LOS that the Victorian government set. Each year the government publishes cost weights that are used to calculate the funding for public hospitals. Central to the computation of the cost weights is the average LOS for each individual DRG group, which is referred to here as the "notional" LOS. The figure below illustrates this process.

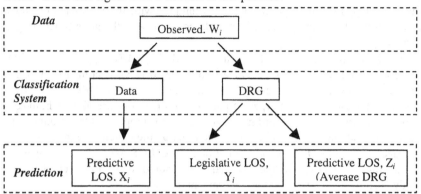

Fig. 1. Framework Outline

Let, W_i = actual LOS of individual patient i

X_i = average LOS prediction of patient i's group based on data mining techniques.

Y_i = average LOS prediction of patient i's DRG group

Z_i = expected LOS of patient i's DRG group obtained from the Victorian Government funding calculation

A three-way comparison of the data mining-generated classifications' prediction, the DRG generated prediction and the observed data (individual patient LOS data) shows how well the two different LOS predictions stack up to reality. The mean absolute error (MAE) is used by averaging the absolute difference between the predicted LOS to the actual individual patient LOS. The smaller the MAE, the more accurate the prediction. MAE calculation is produced for the data mining techniques prediction as shown in the equations below. Equation 1 is for CART or SOM.

$$MAE_{Xi(CART / SOM)} = \frac{\sum_{i=1} |W_i - X_i|}{N} \tag{1}$$

where N are the total number of patients in the entire case study data.

Similarly, MAE calculation for notional DRG is given in E.q 2.

$$MAE_{yi} = \frac{\sum_{i=1} |W_i - Y_i|}{N} \tag{2}$$

MAE calculation for average DRG LOS is given in E.q 3.

$$MAE_{Zi} = \frac{\sum_{i=1} |W_i - Z_i|}{N} \tag{3}$$

Comparisons can be made between:

- data mining prediction of LOS using CART ($MAEx_{i(CART)}$) and SOM ($MAEx_{i(SOM)}$), and that of the notional DRG LOS($MAEy_i$). A comparison from the predictive LOS using data mining techniques with the notional LOS tells how well the data mining techniques' LOS compare with the government target results.
- data mining prediction of LOS using CART ($MAEx_{i(CART)}$) and SOM ($MAEx_{i(SOM)}$), and that of the average DRG LOS($MAEz_i$). A comparison of the predictive LOS from data mining techniques to the average DRG LOS shows how well the two different types of patient classifications are able to homogeneously group patients according to resource consumption. One grouping is made from data mining techniques and the other by the patients' diagnoses.
- average DRG LOS($MAEz_i$) and the notional DRG LOS($MAEy_i$). By comparing the average DRG LOS to the notional DRG would show how well the case study hospital is able to meet the government targets.

3 Data Mining Approach

The aim of our clustering approach is to see how well the available inputs are able to predict LOS without actually using LOS as an input. Demographic and initial admission information is used as the set of inputs to the clustering process. Once the clusters have formed, each is inspected for its average LOS. Any new patient arriving at the hospital can be expected to stay for a number of days equivalent to the average LOS of the cluster to which they are closest.

The neural clustering approach used in this paper is based on Kohonen's Self Organising Map (SOM) [7]. Self organisation refers to a type of neural network which classifies data and discovers relationships within the data set without any guidance during learning (unsupervised learning). The basic principle of identifying those hidden relationships is that, if input patterns are similar, they should be grouped together.

A software package called Neuroshell 2 was used to model the data. The neural networks were forced into a certain number of groups from 5 to 12 groups to see how many actual subdivisions were formed.

In contrast to the unsupervised approach described above, classification and regression trees (CART) can be used to learn to predict LOS in a supervised manner. CART is a supervised learning data analysis tool that explores the effects of the independent variables on the dependent variable: LOS. At each stage of the tree generation process, the CART algorithm chooses the best independent variable that allows it to split the data into two groups on the basis of reducing the variation of LOS. It keeps splitting in twos with new independent variables until no further splits are possible. After finding the maximal tree, CART prunes the trees until it finds the best, based both on predictive accuracy and on a penalty applied to large trees. The inputs to the CART algorithm are the same as those listed for the unsupervised approach, while the dependent variable is LOS.

4 Case Study

The data was obtained from Frankston Hospital, a medium sized health unit in a region in the south-eastern suburbs of Melbourne. It is part of a network of hospitals which serves up to nearly 290,000 people, with a seasonal influx of visitors of up to 100,000[8].

The data consists of about 16,200 records of admitted patients from 1 July, 2000 to 30 Jun, 2001. It contains information about demographics and admission information about patients and their LOS.

Age, gender and initial admission information are known when the patient arrives. Most of those admitted to Frankston Hospital during this time were elderly patients (as shown in Table 1). Elderly patients tend to consume more than the "average" amount of resources compared to other patients in the same DRG groupings. The distribution of male and female patients is almost equal. The LOS of patients is determined by subtracting the time when patients are admitted from the time they are discharged.

Patients can be admitted to emergency (EMG), children (CHW), emergency department ward (EDW), cardiac care (C/C), short stay unit (SSU), post natal (PNA), intensive care unit (ICU) and to various geographical areas of the hospital. EDW differs from EMG in that EDW patients are sent for observation but consumes some emergency department resources because patients are looked after by emergency staff.

Table 1 shows provides insight into the characteristics of the data. All the data appears to be distributed widely. For example, the average LOS of patients is 3 days but its standard deviation is 5 days and 20 hours. Middle-aged patients tend to be the average age group for the hospital and there is an equal number of males and females (the mean is 0.49).

5 Results

This section presents summary descriptions of the groups obtained from CART and then the clusters obtained from the SOM.

The tree generated by CART is shown in Figure 3, and a summary of the terminal nodes is shown in Table 4. Terminal Node 17 represents patients admitted immediately on emergency. They represent the largest of all groups, but have the lowest LOS. Terminal Node 1 represents the youngest patients which is the third largest group. Most of these patients are children and they tend to have the second shortest stay.

There are three major groups of patients with high LOS: terminal groups 9 and 13 to 16. Although they are different from each other, as shown in the tree, they do share similar features. They tend to have older groups of patients compared to rest and their average age lies between 60 and 70 years.

Clustering from SOM information is presented in Table 3 which shows that there is a young age cluster compared to the rest and that there is a good division of LOS even when LOS is not used as an input. There were important features that are similar regardless of the number of cluster. There is a children's ward where the average age is 6 years. This cluster represent the second lowest LOS.

The next group is from the initial admission to emergency group. They tend to have the lowest LOS of all groups. However the SOM tend to break this emergency group into smaller clusters based on the age and the last discharge wards as the number of forced cluster is increased. For instance when the SOM is forced into 12 cluster there are 4 clusters of this emergency group whereas there is only 1 cluster when SOM is forced into a 5 cluster. Their LOS of these different clusters in emergency group shows similar LOS.

The cluster of patients who stay in EDW has an average of 7-8 days. They tend to have an average age of 66 to 68. There is another cluster of patients that stay in the 4th and 5th south and west wing of the hospital. They tend to have a high LOS about 7 days.

Both SOMs and CART are similar in many ways. Both classify important groups such as emergency and children's groups. They also came up with the same grouping when classifying the 4th and 5th wing group and EDW groups. The difference between the two, as stated earlier, is that since CART is able to use LOS as a predictor, it is able to increase its accuracy in picking the groups according to their different LOS. Although it can be more accurate it may not be relevant for prediction purposes since LOS information is not available beforehand.

This section presents a comparison between the two different learning paradigms and DRG by MAE. New patients can be expected to stay for a number of days equivalent to the average LOS of the subdivision to which they belong. It is on this basis that the comparison is made with the techniques and DRG. Table 2 shows MAE for the comparison mentioned in the framework section.

From Figure 2, it can be seen that the larger the subdivision, the smaller the MAE becomes. To illustrate: if one subdivides equal to the number of records in the whole data, the mean LOS of each group would be equal to the LOS of each individual records. Thus, the MAE in this case is zero.

The lowest of all the MAE is CART. Since it uses LOS as a target variable when splitting the nodes, it is able to choose the subdivisions that have the lowest LOS subject to the minimum number each terminal node can be.

The neural clustering approach seems to follow the trend that the larger the subdivision the smaller the MAE. The graph shows that, although DRG groups have the largest subdivision, they also have one of the largest MAE of all groupings. The highest MAE is cluster 5 of SOM followed by the notional DRG and then the average DRG LOS. It is not surprising that the notional DRG has a higher MAE than the average DRG LOS, since the "legislative" LOS are based on the average of selected hospitals throughout Victoria whereas the DRG groupings' LOS is obtained from the Frankston data.

If DRG is a good indicator of resource consumption then it might be expected that the larger subdivisions of DRG would have one of the lowest MAE. This is not the case, as shown in Figure 2. The lowest subdivision from clusters 6 to 12 has a lower MAE than DRG. This means that the added subdivisions for the DRG adds complexity without having any benefit for predicting resource consumption.

It also seems that the case study hospital is unable to meet the government targets. MAE of notional DRG is much higher than the average DRG LOS. This may means that there is a need for more improvement to the current system of Case Mix.

6 Conclusion

This paper argues that, for this set of data, DRG alone is not a good indicator of resource consumption in hospital. Even when there is a high number of subdivisions for the DRG cases, it still produce a higher MAE than SOM or CART. There is even a problem with using the notional DRG: in the current case study, the hospital was unable to meet the government targets. The high MAE notional DRG also shows that imposing targets based on the average of selected hospitals throughout Victoria may not be applicable to the local situation.

The clustering done from the data mining techniques could be used to supplement the DRG. Homogenous groups, as shown, contain certain information and certain characteristics that can have some impact on resource consumption.

References

1. Sanderson H, Anthony P, Mountney L (1998) Casemix for All. Radcliffe Medical Press, Oxon, UK
2. Auditor-General of Victoria (1998) Acute health care services under casemix. A case of mixed priorities. Special Report No 56
3. Ridley S, Jones S, Shahani A, Brampton W, Nielsen M, Rowan K (1998) Classification Trees. A possible method for iso-resource grouping in intensive care. Anaesthesia 53:833-840
4. Walczak, S, Pofahl, WE, Scorpio, RJ (2002) A decision support tool for allocating hospital bed resources and determining required acuity of care. Decision Support System 34:445-446
5. Siew E-G, Smith KA, Churilov L, Ibrahim M (2002) A Neural Clustering Approach to Iso-Resource for Acute Healthcare in Australia. Proceedings of the 35th Annual Hawaii International Conference on System Science
6. Siew, E-G, Smith KA, Churilov L, Wassertheil J (2002) A comparison of supervised and unsupervised approaches to iso-resource grouping for acute healthcare in Australia. Proceedings of the International Conference on Fuzzy Systems and Knowledge Discovery 2: 601-605
7. Deboeck G, Kohonen T (1998) Visual Explorations in Finance with Self Organizing Maps. London: Springer-Verlag
8. Peninsula Health Annual Report (2000)

Appendix: Results in Tables and Figures

Table 1. Data characteristics

Gender	Age					LOS	
	Mean	Quartile 1	Quartile 2	Quartile 3	Mean	Std. Dev.	
0.49	50	28	52	74	3.03	5.84	

Table 2. MAE of all the predictive LOS groupings

Name of Grouping	Number of groups formed	Mean Absolute Error
Notional DRG	476	2.637
Average DRG LOS	476	2.187
CART	12	1.899
Cluster 12	10	2.106
Cluster 11	9	2.109
Cluster 10	8	2.104
Cluster 9	8	2.119
Cluster 8	7	2.120
Cluster 7	6	2.119
Cluster 6	6	2.127
Cluster 5	5	2.200

Table 3. The mean of each cluster groups

Clusters		1	2	3	4	5	6	7	8	9	10	11	12
12	LOS	6.8	9.98	7.9	6.9	5.3	3.45	1.5	2.6	0.2	0.3	0.3	0.3
	Gender	0.4	0.5	0.4	0.47	0.5	0.53	0.5	0.6	0.4	0.5	0.5	0.4
	Age	62	59	66	64	52	49	6	15	14	36	58	81
11	LOS	6.8	8.48	6.9	5.35	3.4	1.53	2.6	0.2	0.3	0.3	0.3	
	Gender	0.4	0.46	0.4	0.5	0.5	0.57	0.6	0.4	0.5	0.5	0.4	
	Age	62	64	64	52	49	6	15	14	36	58	81	
10	LOS	1.5	10.9	6.6	9.98	7.9	6.9	5.2	4.0	0.3	0.3		
	Gender	0.5	0.44	0.4	0.5	0.4	0.47	0.5	0.4	0.5	0.4		
	Age	6	59	62	59	66	64	52	43	28	73		
9	LOS	6.7	8.48	5.5	5.86	1.5	14.1	0.3	0.3	0.3			
	Gender	0.4	0.46	0.5	0.33	0.5	0.43	0.4	0.5	0.4			
	Age	62	64	59	42	6	28	19	47	78			
8	LOS	6.7	8.48	5.5	5.86	1.5	12.4	0.3	0.3				
	Gender	0.4	0.46	0.5	0.33	0.5	0.5	0.5	0.4				
	Age	62	64	59	42	6	33	28	73				
7	LOS	8.4	5.64	1.5	10.7	6.7	0.34	0.3					
	Gender	0.4	0.5	0.5	0.5	0.4	0.5	0.4					
	Age	64	55	6	37	62	28	73					
6	LOS	7.8	6.09	1.7	11.7	7.2	0.34						
	Gender	0.4	0.48	0.5	0.58	0.4	0.5						
	Age	61	53	7	60	60	52						
5	LOS	3.7	7.96	6.7	10.8	0.3							
	Gender	0.5	0.46	0.4	0.38	0.4							
	Age	34	64	62	48	50							

Table 4. T-Node attributes for CART

T-Node	Size	Avg LOS	Gender	Age
1	1711	1.64	0.56	7
2	547	2.86	0.38	35
3	1064	4.93	0.49	37
4	95	7.71	0.58	36
5	413	1.90	0.53	33
6	147	8.49	0.54	32
7	816	3.93	0.56	69
8	50	8.25	0.48	71
9	89	14.63	0.53	67
10	59	6.73	0.42	70
11	665	6.61	0.52	60
12	1840	8.52	0.41	80
13	51	15.24	0.57	70
14	611	10.62	0.47	77
15	20	17.05	0.45	61
16	17	34.87	0.41	64
17	7983	0.37	0.50	51

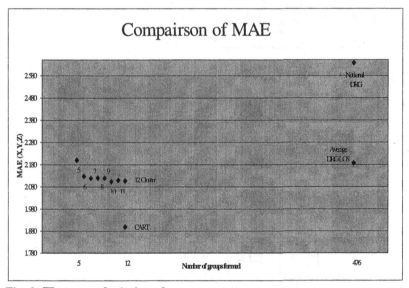

Fig. 2. The mean deviations from average

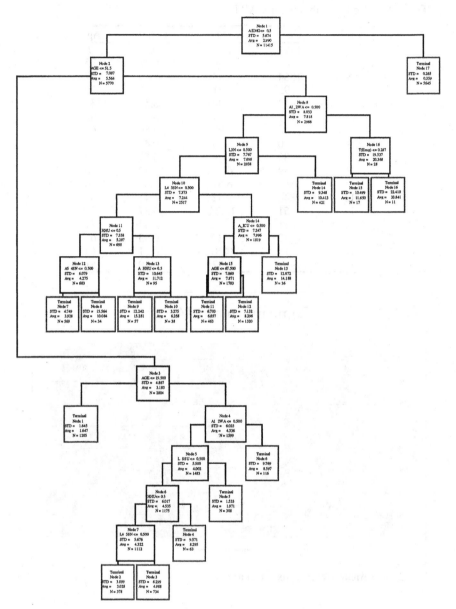

Fig. 3. CART Tree

Criteria for a Comparative Study of Visualization Techniques in Data Mining

Robert Redpath and Bala Srinivasan

School of Computer Science and Software Engineering,Faculty of Information Technology,Monash University,Caulfield East, Victoria, Australia

Abstract. The paper considers the relationship of information visualization tools to the data mining process. The types of structures that may be of interest to data mining experts in data sets are outlined. The performance of a particular visualization technique in revealing those structures and supporting the subsequent steps in the process needs to be based on a number of criteria. The criteria for performing an evaluation are suggested and explained. A division into two main groups is suggested; criteria that relate to interface issues and criteria that relate to the characteristics of the data set. An example application of some of the criteria is made.

Keywords. Information visualization, data mining, comparative criteria, evaluation, multi-dimensional, patterns, visualization software

1 Introduction

Information visualization can be a useful first step in the data mining process. It can act as an exploratory tool to guide the choice of some other algorithmic technique or it can be used to isolate some subset of the larger data set for more detailed analysis. It may also directly reveal structures of interest in the data set under consideration. In order to judge the effectiveness of any particular visualization technique criteria need to be established upon which the success or failure of each technique can be measured.

This paper extends the criteria on which visualization techniques can be compared and categorizes them into two groups ; those that are interface issues and those that are related to the characteristics of the dataset. Each visualization technique has features that affect how the visualization can be used and, consequently, how effective the technique is and therefore criteria relating to these features needs to be considered also. These are the criteria which have been grouped together as *criteria relating to interface issues.*

In section 2 the contributions of this paper are placed in the context of the current literature on data visualization for data mining. Some of the patterns that may exist in data sets and that will determine a number of the criteria upon which the data sets will be judged are outlined in section 3. Section 3 also advances a schema treatment for associations, which causes them to present as clusters in scatterplot visualizations of two or three dimensions. This allows criteria mentioning clusters to be applied to associations as well as clusters. In section 4 the criteria relating directly to the dataset under consideration, and that suggest the features that a standard test dataset could contain in order to evaluate and compare the usefulness of the various visualization techniques, are advanced. A number of other criteria are also suggested which relate to issues of user perception and the ease of use of the interface. In section 5 an example of the application of the criteria to two different visualization techniques is given.

2 Visualization in Data Mining

To evaluate the effectiveness of visualization techniques one approach is to establish standard test data sets. A model for establishing test data sets has been proposed by Keim et al. (Keim et al. 1995). Their aims are to provide a method for judging the effectiveness of visualization systems and also to allow comparative studies of visualization techniques. A taxonomy is suggested for the data structures that visualization systems might need to address. A evaluation tool, which generates visualizations of multidimensional multivariate (mDmV) data test data, was developed by Wong et al. (Wong and Bergeron 1995). The aim of the tool is to evaluate human responses to visualizations of scientific data. The tool developed supports scientific visualization in a number of ways. Test data of limited characteristics determined by the researcher can be generated. It is used against two visualization techniques, which are part of the tool. The tool also keeps track of test sessions and will produce evaluation reports.

A more useful approach would consider a number of more generally used visualization techniques and would try to do a comparison on these techniques. A comparison study has been done by Keim and Kriegel (Keim and Kriegel 1996). The techniques chosen for comparison were those included in the visualization software developed by Keim and Kreigel called *VisDB*. The aim of the study was to evaluate the expressiveness of those different techniques. This expressiveness was defined as the perceptibility of the characteristics of the data. The authors noted that the results must necessarily depend on the human performing the perceptual task as well as the potential of the visualization technique. They did not make use of an evaluation tool, such as those described in Wong(Wong and Bergeron 1995), rather they drew conclusions based upon their own perception of how the different techniques performed with some different data sets. The researchers evaluated the performance of the visualization techniques at different types of tasks. The tasks included identifying functional dependencies, distributions, clusters and outliers. The visualization techniques were also compared on the basis of their ability to handle a maximum number of data items, a maximum number of dimensions, overlapping data instances and finally data items with inherent two or three dimensional data semantics. The conclusion reached was that the different visualization techniques should be used in parallel. In this way the strengths of all can be employed and the weaknesses avoided.

In spite of the progress made it was considered, by Pickett and Grinstein (Pickett and Grinstein 2002) that the evaluation and enhancement of visualizations systems is still largely a matter of qualitative judgment. In their paper *Evaluation of Visualization Systems* a brief discussion on directions that evaluations should take is provided. They suggest that a set of test patterns could be generated for particular domains. They see the problem of generating a generalized set of test patterns in an abstract way as a much more difficult problem. The example, in section 5 addresses this problem by applying the suggested criteria to a cluster in an exploratory way and does this by attempting to identify the existence of a structure against a background of noise instances.

3 Data Mining and the Search For Underlying Patterns

In order to know if a visualization technique is successful or not we must understand the patterns that we are searching for. Section 3.1 provides an overview of the main types of pattern and section 3.2 explains how the particular pattern know as an association needs to be treated if it is to be revealed in a visualization software tool.

3.1 A Taxonomy of Patterns in Data Mining

Data mining aims to reveal knowledge about the data under consideration. This knowledge takes the form of patterns within the data that embody our understanding of the data. Patterns are also referred to as structures, models and relationships. The patterns within the data cannot be separated from the approaches that are used to find those patterns because all patterns are, essentially, abstractions of the real data. The approach used is called an abstraction model or technique. The approach chosen is inherently linked to the pattern revealed. Data mining approaches may be divided into two main groups. These are verification driven data mining and discovery driven data mining.

The name of the pattern revealed in the discovery driven group area is often a variation of the name of the approach. This is due to the difficulty of separating the pattern from the approach. The terms *regression, classification, association analysis* and *segmentation* are used to name the approach used and the pattern revealed. For example *segmentation* reveals segments (also called clusters), *classification* assigns data to classes and *association analysis* reveals associations. An exception to this terminology is the use of the approach called *deviation detection* to reveal a pattern of *outliers* but outliers form a particular kind of segment in any case.

The data will rarely fit an approach exactly and different approaches may be used on the same data. The data will have an inherent structure but it is not possible to describe it directly. Rather a pattern is an attempt to describe the inherent structure by using a particular approach. Patterns are best understood in terms of the approach used to construct them. For this reason the patterns are often discussed in terms of how they are arrived at rather than stating the data has a pattern in some absolute sense.

The taxonomy in table 1, based on Simoudis (Simoudis 1996), classifies approaches to the data mining task and the patterns revealed. It is not expected that all the approaches will work equally well with any data set.

Verification Driven
　　　　　Query and reporting
　　　　　Statistical Analysis

Discovery Driven
　　　　Predictive (Supervised)
　　　　　　Regression
　　　　　　Classification
　　　　Informative (Unsupervised)
　　　　　　Clusters (Segmentation)
　　　　　　Association
　　　　　　Outliers (Deviation detection)

Table 1 Taxonomy of approaches to data mining

3.2 Schema Treatment for Associations

Schema definition is an important part of the task of a data-mining analyst. Consider the situation for associations within a data set. The appropriate schema definition needs to be chosen as part of the knowledge discovery process if useful information is to be revealed by a visualization technique.

An association pattern is useful for market basket analysis (such as "which items do customers often buy together?"). It involves comparing tuples and considering binaries. That is an item is either purchased or not purchased (true or false). If attributes can be organized in a hierarchy of classes this must be considered. For example, ask the question, not do they buy light alcohol beer but rather, do they buy beer. Classifying instances as beer will, potentially, bring many more instances together.

Whether or not a 3 dimensional or 2 dimensional scatter plot visualization will reveal an association rule depends on how the schema of the data set is defined. For example transactions at a shop might be recorded for market basket analysis, which is a common application of Association rules. The schema of instances in the data set would be of the form *(tran-id,{ product-type, product-id})* where the product-type and product-id would be repeated as many times as there were products in the transaction. A visualization of the instances in the data set will not show anything that would indicate the presence of an association (if it were present) unless it is treated in a particular way. The instances would need to be stored with a flag to indicate the presence or not of each product type. The schema definition would need to be of the form *(tran-id, product-type-1, product-type-2, ..., product-type-n)* where the values of true or false would be held for each product type.

Each column will hold true if the product is purchased and false if the product is not purchased. In a scatter plot each axis would have only two possible values, these being *true* and *false.* By viewing scatter plot visualizations, in two dimensions, of product-type-A against product-type-B associations of the type when A is purchased B is also purchased X% of the time will be revealed as a cluster. A scatter plot visualization in three dimensions, with product-type-A, product-type-B and product-type-C mapped to each of the axes, will reveal associations of the type when A and B are purchased C is purchased X% of the time. As the data is categorical (that is with values of either TRUE or FALSE) many points will be co-incident on the graph. The visualization tool needs to be able to deal with this overlap. For example ***DBMiner*** (DBMiner 1998), which uses a variation a 3 dimensional scatterplot technique, would be useful in this context as the size of the cube in a region indicates the number of points in a region.

4 Criteria for the Comparison of Visualization Techniques in Data Mining

In order to evaluate a visualization system, and following on from the suggestions made by Pickett and Grinstein (Pickett and Grinstein 2002), a series of tests needs to be carried out on standard test data sets with different characteristics in which one or another parameter of the system is manipulated to determine whether it has a beneficial or detrimental effect. The different data sets and parameters are not suggested but there choice corresponds to a determination of the criteria which will used to judge the visualization system. Based on this we have made the overriding principle in selecting the criteria, that they are chosen so that they reflect the issues affecting the success or failure of the various techniques in (1) revealing patterns at the exploratory stage of the data mining process and (2) guiding subsequent processing. We have divided the criteria into two groups; *interface issues* and *characteristics of the data set*. The criteria grouped under interface issues could be applied to visualizations in other areas but they are of particular concern in data mining if the process is to be successful. The choice of the criteria grouped under *characteristics of the data set* is justified as summarizing the general features of data that would apply in most domains that data mining would be applied to. The visualization techniques are put forward as general techniques not specific to a particular domain so the features of the data that apply in most domains are the basis for the criteria chosen. While other criteria could be included they represent the most important considerations. Issues of individual user perception are not considered. It is assumed the user understands the role of data mining and that a simple judgment of whether the criteria are satisfied or not can be made by that user.

When judging visualization techniques against the criteria, grouped under the heading *interface issues,* it is necessary to separate effects that are inherent in the visualization technique and those which are due to the particularities of the visualization tool implementing the technique. For instance the incorporation of dynamic controls would relate to a particular implementation.

Interface Issues

Perceptually Satisfying Presentation
When considering whether a visualization technique is perceptually satisfying we are making a judgment about how naturally and clearly features of the data set are revealed. Graphical displays, commonly used and learnt early in a person's education, are easily interpreted. The meaning of what is presented in two-dimensional or three-dimensional graphical displays can also be understood in terms of the three dimensional world we inhabit. When the attributes of a dataset are translated into co-ordinates in a two or three-dimensional space they gain a location which can be understood in terms of that space. The understanding of such displays is therefore much more easily achieved and the display is usually more perceptually satisfying.

Intuitiveness of the Technique
The intuitiveness of a technique relates to how easily what is viewed can be interpreted. If the features of a dataset, such as a cluster or outlier, are easily identified without an extended familiarity with the technique or recourse to a number of displays requiring a considered comparison, the technique may be said to be intuitive. Even two and three-dimensional scatter plots require learning to understand how they are presented and implemented in some particular tools. This need not be a criticism of the technique if the learning task can be achieved in a reasonably short time or achieved in progressive stages with greater facility being gained in the use of the tool at each stage. Thus there is a learning curve associated with each tool and corresponding technique. If, once learnt, the displays

are easily and naturally interpreted the initial novelty of the technique need not be a disadvantage.

Ability To Manipulate Display Dynamically

When a visualization technique is actually implemented as a visualization tool in some specific software it is usually accompanied by a number of features for deliberately controlling the display. Features may exist for querying on particular records or over particular ranges. It may be possible to zoom in on a particular part of the display. Controls may also exist for assigning colours to particular dimensions in the display. There may be features to allow cycling through different dimensions of the data set in the display. The cycling may be necessary to establish relationships over those dimensions. For instance when using a two-dimensional scatterplot if a matrix of scatterplots is not presented the user must cycle through the individual two dimensional scatterplots to establish a relationship or pattern. The features provided depend on the particular implementation rather than the visualization technique.

Ease of Use

The ease of use of the display or visualization technique relates to a combination of factors. These factors include the flexibility of data set format that can be imported. It also relates to how efficiently the data is displayed. If significant delays exist in changing the display the user will have difficulty using the visualization techniques. If the design of the controls is awkward, not obvious or fails to follow common conventions the tool will not be easy to use. These issues relate directly to the architecture and design of the tool rather than the particular visualization technique the tool employs. But unless the tool is easy to use the visualization will not be successful in its goal of revealing patterns in the data.

Characteristics of the Data Set

Size Of Data Set

It is important to know whether the visualization techniques deal with data sets of different sizes. Data sets may range in size from a few hundred instances to many millions of instances. Not all the techniques will successfully deal with large numbers of instances. The concern here is not the capacity of the computer hardware used. Rather the visualization technique may be unable to effectively display and distinguish large numbers of instances. The capability of the visualization techniques to deal with large numbers of instances without overlap and the possible loss of information must therefore be considered.

Support For Multi- Dimensional Data

Some of the visualization techniques are able to display many dimensions in a single display and others have an upper limit of two, three or four dimensions. Simple scatterplots can display only two or three dimensions. If the point plotted has other distinguishing features such as color or is an icon, which relates to further dimensions through some aspect of its shape, a greater number of dimensions can be represented in a single display. Other techniques use multiple windows to display a large number of dimensions or a number of straight line axes as in the case of *parallel co-ordinates*.

Ability to Reveal Patterns in a Data Set

The purpose of the visualization tools is primarily to gain knowledge through the recognition of patterns in the data. The technique must be able to reveal patterns in the data set if they are present. If the visualization is unable to do this it has failed in its basic purpose. It would be desirable to be able to distinguish between different types of pattern. The criteria following consider more particular aspects of the ability to reveal patterns in the data set.

Ability To Reveal Clusters Of Two Or Three Dimensions

Clusters indicate the presence of relationships between attributes. They may be indicative of associations or classes also. For the visualization technique to be useful it is expected that as a minimum requirement two and three-dimensional clusters would be revealed.

Ability To Reveal Clusters Of Greater Than Two Or Three Dimensions

We also wish to know if clusters greater than two or three dimensions can be successfully revealed by the visualization techniques. It needs to be established whether higher dimensional clusters can be revealed by consideration of a single display or whether a more laborious cognitive procedure is required such as cycling through two-dimensional displays.

Number Of Clusters Present

Most patterns manifest as clusters in the visualizations. The visualization techniques must be able to distinguish between clusters if a number of clusters are present. We are concerned as to whether the clusters obscure each other or are clearly revealed as separate clusters.

Amount Of Background Noise

Another important consideration is how the visualization technique performs against a background of *noise* instances. Real data will usually have many instances that do not contribute to any pattern. If presence of background noise, as such instances are termed, obscures what patterns are present the visualization technique is less useful. It is necessary to test the visualization techniques against various levels of background noise to determine the usefulness in the presence of such noise.

Variance Of The Clusters

The instances that contribute to a cluster may be tightly packed or spread out in the region of space where the cluster or pattern appears. Given that there is usually some background noise clusters that are spread out will be more difficult to detect.

It would be interesting to know if some visualization techniques are better than others at dealing with clusters that are more spread out.

Redpath (Redpath 2000) has used these criteria as the basis for the development of a program to generate test data sets. A number of test data sets were generated which varied in respect of the following features; the dimensionality of the clusters in each data set, the number of clusters present in each data set, the amount of noise present in each data set and the variance of the clusters in each data set. An evaluation and comparison of some visualization software was then carried out using these standard test data sets. The visualization software represented variations on the two and three dimensional scatterplot visualization techniques (Chambers et al. 1983) and the parallel co-ordinates visualization technique (Inselberg and Dimsdale 1990).

5 Example Application of the Criteria

In order to make clear the application of the criteria a particular example has been chosen where the dataset possesses a cluster of more than usual variance against a background of noise. It demonstrates that one visualization technique is successful and one is not in this situation. A full range of evaluations for all the criteria can be seen in the work of Redpath (Redpath 2000).

This example is an application of the criteria for *Variance of the Clusters* to a 3 dimensional scatterplot technique and 2 dimensional scatterplot technique. A 3 dimensional scatterplot technique is represented by the software tool DBMiner (DBminer 1998). DBMiner is essentially a 3D scatterplot with the difference that a cube icon represents all the instances within a region of the 3D space. The size of the cube varies according to how many instances are in a region. The parallel co-ordinates technique is represented by the software tool WinViz (WinViz 1997). WinViz represents each instance by a polygonal line that intersects each of the parallel axes at a point corresponding to it's value.

 A test data set has been seeded with a 3 dimensional cluster in a higher dimensional space with minimal noise and a large variance. The aim is to evaluate the ease with which clusters can be identified under these conditions. The data set is fed into DBMiner and the three-dimensional cluster is shown in figure 1. The test data set 6 contains 1100 instances in total. 1000 instances contribute to a three-dimensional cluster in a 6 dimensional space. The cluster is located around the mean position given by the co-ordinates field 1= 65.0 field 2= 55.9, field 3=52.6. The variance of the normal distribution, which determines the values of the instances about the mean, is 2.

In figure 1 it is extremely difficult to identify a three-dimensional cluster. The cubes are slightly larger at the known cluster position but the change in size could easily be attributed to noise. It can be concluded that if the cluster becomes too spread out it cannot be identified.

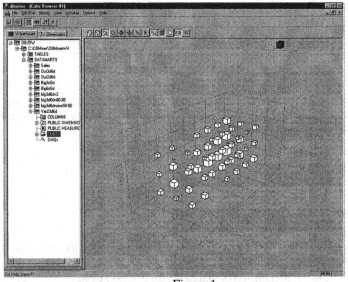

Figure 1
Test data set 6 with 3 dimensional cluster, with spread out (Variance = 2) in DBMiner

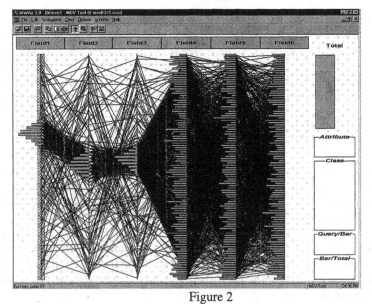

Figure 2
Test data set 6; A 3 dimensional cluster with spread (variance = 2) in WinViz

In figure 2 the same test data is fed into WinViz. The cluster can be easily identified. The variance is indicated by the spread of the probability distribution for each of dimension 1, 2 and 3. It is also apparent that in the spread of the connecting lines, in spite of this variance, that the display still clearly shows a cluster is present.

It can be concluded that when the variance of a cluster becomes large it is difficult to identify a cluster in **DBMiner**. There were a number of slightly larger cubes in the region of the cluster. This was against a background of noise of only 10% of the number of instances present. *WinViz* clearly identified the presence of the cluster under the same conditions.

6 Conclusion

In order to carry out a comparison of visualization techniques suitable criteria for comparison need to be established. We have chosen the following criteria as addressing the relevant issues in determining the success or failure of a particular visualization technique. We have divided the criteria for comparing the techniques into (1) interface considerations and (2) characteristics of the data set.

There are four criteria under interface considerations. Whether the technique is
1. perceptually satisfying
2. intuitive in its use
3. able to manipulate the display dynamically
4. easy to use.

There are four criteria to consider under the characteristics of the data set.
1. The number of instances in the data set.
2. The number of dimensions in the data set.
3. The patterns contained within the data set. These patterns, if clusters, have more specific characteristics
   a. The dimensionality of the clusters present
   b. The number of clusters present.
   c. The variance of the clusters present.
4. The number of noise instances present.

The criteria relating to the characteristics of the data set can be used as the basis to generate a standard set of test data. These standard test data sets could then be used to compare the performance of visualization techniques based on the criteria.

7 References

DBMiner-Software (1998) Version 4.0 Beta; DBMiner Technology Inc., British Columbia, Canada

Chambers JM, Cleveland WS, Kleiner B, Tukey PA (1983) *Graphical Methods for Data Analysis*; Chapman and Hall, New York

Inselberg A, Dimsdale B (1990) *Parallel Coordinates: A Tool for Visualizing Multi-Dimensional Geometry*; Visualization '90, San Francisco, CA, pp 361-370

Keim DA, Bergeron RD, Pickett RM (1995) *Test Data Sets for Evaluating Data Visualization Techniques*; Perceptual Issues in Visualization (Eds. Grinstein G. and Levkowitz H.) ; Springer-Verlag; Berlin, pp 9-21

Keim DA, Kriegel H (1996) *Visualization Techniques for Mining Large Databases: A Comparison*; IEEE Transactions on Knowledge and Data Engineering, Special Issue on Data Mining; Vol. 8, No. 6, December, pp 923-938

Pickett RM, Grinstein GG (2002) *Evaluation of Visualization Systems*; Information Visualization in Data Mining and Knowledge Discovery, Ed. Fayyad U. Grinstein G.G. and Wierse A., Morgan Kaufmann, San Francisco, pp 83-85

Redpath R(2000) *A Comparative Study of Visualization Techniques for Data Mining*; A thesis submitted to the School of Computer Science and Software Engineering of Monash University in fulfillment of the degree Master of Computing, November 2000

Simoudis E (1996) *Reality Check for Data Mining,* IEEE Expert Vol. 11, No. 5:October, pp 26-33.

WinViz-Software (1997) for Excel 7.0 Version 3.0; Information Technology Institute, NCB, Singapore

Wong P, Bergeron RD (1995) *A Multidimensional Multivariate Image Evaluation Tool* Perceptual Issues in Visualization (Ed. Grinstein, G. and Levkowitz, H.) Springer-Verlag Berlin, pp 95-108

Controlling the Spread of Dynamic Self Organising Maps

L. D. Alahakoon

School of Business Systems, Monash University, Clayton, Victoria, Australia.
damminda.alahakoon@infotech.monash.edu.au

Summary. The Growing Self Organising Map (GSOM) has recently been proposed as an alternative neural network architecture based on the traditional Self Organising Map (SOM). The GSOM provides the user with the ability to control the map spread by defining a parameter called the Spread Factor (SF), which results in enhanced data mining as well as hierarchical clustering opportunities. In this paper we highlight the effect of the spread factor on the GSOM and contrast this effect with grid size change (increase and decrease) in the SOM. We also present experimental results to describe our claims regarding the differance in the GSOM to the SOM.

1 Introduction

The Growing Self Organising Map (GSOM) has been proposed as a dynamically generating neural map which has particular advantages for data mining applications [4], [1]. Several researchers have previously developed incrementally growing SOM models [6], [8], [5]. These models have similarities as well as differances from each other, but attempt to solve the problem of pre-defined , fixed structure SOMs. A parameter called the Spread Factor (SF) in the GSOM provides the data analyst with control over the spread of the map. The ability to control the spread of the map is unique to the GSOM and could be manipulated by the data analyst to obtain a progressive clustering of a data set at different levels of detail. The SF can be assigned values in the range 0..1 which are independant of the dimensionality of the data set. Thus the SF provides a measure of the *level of spread* across differant data sets. The spread of the GSOM can be increased by using a higher SF value. Such spreading out can continue until the analyst is satisfied with the *level of clustering* achieved or until each node is identified as a separate cluster.

A data analyst using the SOM experiments with different map sizes by mapping the data set to grids (generally two dimensional) of different sizes. In this paper we highlight the difference between such grid size change in the traditional SOM usage with the controlled map spreading in the GSOM. We describe the SF controlled *growth* of the GSOM as a technique for more representative feature map generation where the inter and intra cluster relationships are better visualised. The fixed structure SOM forces distortion on it's mapping due to the inability of expanding when required.

The traditional SOM does not provide a measure for identifying the size of a feature map with it's *level of spread*. Therefore the data analyst using the SOM can only refer to the length and the width of the grid to relate to map size. We show that the shape or size of the SOM cannot be meaningfully related to the spread of the data without accurate knowledge of the distribution of the data set. The use of the SF in the GSOM provides a single indicator with which to dictate map spread and also the ability to relate a map to a particular level of spread.

The identification of clusters in a map will always depend on the data set being used as well as the *needs* of the application. For example, the analyst may only be interested in identifying the most significant clusters for one application. Alternatively, the analyst may become interested in more detailed clusters. Therefore the clusters identified will depend on the *level of significance* required by the analyst at a certain instance in time. In data mining type applications, since the analyst is not aware of the clusters in the data, it is generally required to study the clusters generated at different levels of spread (detail) to obtain an understanding of the data. Attempting to obtain different sized SOMs for this purpose can result in distorted views due to the need of forcing a data set into maps with pre-defined structure. Therefore we describe the number of clusters shown in a map also as a variable which depends on the level of significance required by the data analyst. We present the SF as our technique of providing the lavel of significance as a parameter to the map generation process.

In section 2 we provide a description of the GSOM and SF techniques. Section 3 explains the workings of the spreadout effect in the GSOM with the SF as a parameter. Here we also compare the GSOM and SOM in terms of the spreading out effect. In section 4 we describe experimental results to clarify our claims. Section 5 provides the conclusions to the paper.

2 Controlling the GSOM Spread with the Spread Factor

2.1 Growing Self Organising Map

The GSOM is an unsupervised neural network which is initialised with 4 nodes and *grows* nodes to represent the input data [2], [3]. During the node growth, the weight values of the nodes are *self organised* according to a similar method as the SOM. The GSOM process is as follows:

1. Initialisation Phase
   a) Initialise the weight vectors of the starting nodes (4) with random numbers.

b) Calculate the Growth Threshold (GT) for the given data set according to the user requirements.
2. Growing Phase
   a) Present input to the network.
   b) Determine the weight vector that is closest to the input vector mapped to the current feature map (winner), using Euclidean distance (similar to the SOM). This step can be summarised as:
   Find q' such that $|v - w_{q'}| \leq |v - w_q| \; \forall q \in \mathcal{N}$ where v, w are the input and weight vectors respectively, q is the position vector for nodes and $\mathcal{N}$ is the set of natural numbers.
   c) The weight vector adaptation is applied only to the neighbourhood of the winner and the winner itself. The neighbourhood is a set of neurons around the winner, but in the GSOM the starting neighbourhood selected for weight adaptation is smaller compared to the SOM (localised weight adaptation). The amount of adaptation (learning rate) is also reduced exponentially over the iterations. Even within the neighbourhood weights which are closer to the winner are adapted more than those further away. The weight adaptation can be described by:

$$w_j(k+1) = \begin{cases} w_j(k), j \notin N_{k+1} \\ \\ w_j(k) + LR(k) \times \\ (x_k - w_j(k)), j \in N_{k+1} \end{cases}$$

where the learning rate $LR(k), k \in \mathcal{N}$ is a sequence of positive parameters converging to 0 as $k \to \infty$. $w_j(k), w_j(k+1)$ are the weight vectors of the the node j, before and after the adaptation and N_{k+1} is the neighbourhood of the winning neuron at $(k+1)^{th}$ iteration. The decreasing of $LR(k)$ in the GSOM, depends on the number of nodes existing in the network at time k.
   d) Increase the error value of the winner (error value is the difference between the input vector and the weight vectors).
   e) When $TE_i \leq GT$ (where TE is the total error of node i and GT is the growth threshold),
   Grow nodes if i is a boundary node. Distribute weights to neighbours if i is a non-boundary node.
   f) Initialise the new node weight vectors to match the neighbouring node weights.
   g) Initialise the learning rate (LR) to it's starting value.
   h) Repeat steps b..g until all inputs have been presented, and node growth is reduced to a minimum level.
3. Smoothing Phase
   a) Reduce learning rate and fix a small starting neighbourhood.
   b) Find winner and adapt weights of winner and neighbors same as in growing phase.

Therefore, instead of the weight adaptation in the original SOM, the GSOM adapts it's weights and architecture to represent the input data. Therefore in the GSOM a node has a weight vector and two dimensional coordinates which identify it's position in the net, while in the SOM the weight vector is also called the position vector.

2.2 The Spread factor

As described in the algorithm, the GSOM uses a threshold value called the Growth Threshold (GT) to decide when to initiate new node growth. GT will decide the amount of spread of the feature map to be generated. Therefore if we require only a very abstract picture of the data, a large GT will result in a map with a fewer number of nodes. Similarly a smaller GT will result in the map spreading out more. When using the GSOM for data mining, it might be a good idea to first generate a *smaller* map, only showing the most significant clustering in the data, which will give the data analyst a summarised picture of the inherent clustering in the total data set.

The node growth in the GSOM is initiated when the error value of a node exceeds the GT. The total error value for node i is calculated as :

$$TE_i = \sum_{H_i} \sum_{j=1}^{\mathcal{D}} (x_{i,j} - w_j)^2 \tag{1}$$

where H is the number of hits to the node i and $\mathcal{D}$ is the dimension of the data. $x_{i,j}$ and w_j are the input and weight vectors of the node i respectively. For new node growth :

$$TE_i \geq GT \tag{2}$$

The GT value has to be experimentally decided depending on our requirement for map growth. As can be seen from equation 1 the dimension of the data set will make a significant impact on the accumulated error (TE) value, and as such will have to be considered when deciding the GT for a given application. The Spread Factor SF was introduced to address this limitation, thus eleminating the need for the data analyst to consider data dimensionality when generating feature maps. Although exact measurement of the effect of the SF is yet to be carried out, it provides an indicator for identifying the level of spread in a feature map (across differant dimensionalities). The derivation of the SF has been described in [4]. The formula used is :

$$GT = -D \times \ln (SF)$$

Therefore, instead of having to provide a GT, which would take different values for different data sets, the data analyst has to provide a value SF, which will be used by the system to calculate the GT value depending on the dimension of the data. This will allow the GSOM's to be identified with their spread factors, and will be a basis for comparision of different maps.

3 The Spreading out Effect of the GSOM compared to the traditional SOM

3.1 The Relationship between the Shape of the SOM to the Input Data Distribution

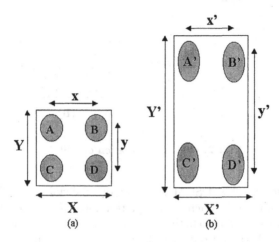

Fig. 1. The shift of the clusters on a feature map due to the shape and size

Figure 1 shows the diagram of a SOM with four clusters A, B, C and D which can be used to explain the spread of clusters due to the change of grid size in a SOM. As shown in Figure 1(a) the SOM has a grid of length and width X and Y respectively. The intra cluster distances are x and y as shown in Figure 1(a). In Figure 1(b) a SOM has been generated on the same data but the length of the grid has been increased (to $Y' > Y$) while the width has been maintained at the previous value ($X = X'$). The intra cluster distances in Figure 1(b) are x' and y'. It can be seen that inter cluster distances in y direction has changed in such a way that the cluster positions have been *forced* into maintaining the proportions of the SOM grid. The clusters themselves have been *dragged out* in y direction due to the intra cluster distances also being forced by the grid. Therefore in Figure 1 $X : Y \approx x : y$ and $X' : Y' \approx x' : y'$. This phenomena can be considered in an intuitive manner as follows [7]:

Considering two dimensional maps, the inter and intra cluster distances in the map can be separately identified in the X and Y directions. We simply visualise the spreading out effect of the SOM as the inter and intra cluster distances in the X and Y directions proportionally being adjusted to fit in with the width and length of the SOM .

The same effect has been described by Kohonen [7] as a limitation of the SOM called the *oblique orientation*. This limitation has been observed and demonstrated experimentally with a two dimensional grid, and we use the same experiment to indicate the limitations of the SOM for data mining.

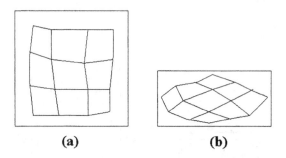

(a)   **(b)**

Fig. 2. Oblique orientation of a SOM

Figure 2(a) shows a 4 × 4 SOM for a set of artificial data selected from a uniformly distributed 2 dimensional square region. The attribute values x, y in the data are selected such that $x : y = 4 : 4$ and as such the grid in Figure 2(a) is well spread out providing an optimal map. In figure 2(b) the input value attributes $x : y \neq 4 : 4$ while the input data *demands* a grid of 4 : 4 or similar proportions. As such it has resulted in a distorted map with a crushed effect. Kohonen has described oblique orientation as resulting from significant differences in variance of the components (attributes) of the input data. Therefore the grid size of the feature map has to be initialised to match the values of the data attributes or dimensions to obtain a *properly* spread out map. For example, consider a two dimensional data set where the attribute values have the proportion $x : y$. In such an instance a two dimensional grid can be initialised with $n \times m$ nodes where $n : m = x : y$. Such a feature map will produce an optimal spread of clusters maintaining the proportionality in the data. But in many data mining applications the data analyst is not aware of the data attribute proportions. Also the data mostly is of very high dimensions, and as such it becomes impossible to decide a suitable two dimensional grid structure and shape. Therefore initialing with an optimal grid for SOM becomes a non-feasible solution.

Kohonen has suggested a solution to this problem by introducing *adaptive tensorial weights* in calculating the distance for identifying the winning nodes in the SOM during training. The formula for distance calculation is

$$d^2[x(t), w_i(t)] = \sum_{j=1}^{N} \psi_{i,j}^2 [\xi_j(t) - \mu_{i,j}(t)]^2 \tag{3}$$

where ξ_j are the attributes (dimensions) of input x, the $\mu_{i,j}$ are the attributes of w_i and $\psi_{i,j}$ is the weight of the jth attribute associated with node i respectively. The values of $W_{i,j}$ are estimated recursively during the unsupervised learning process [7]. The resulting adjustment has been demonstrated using artificial data sets in Figure 3.

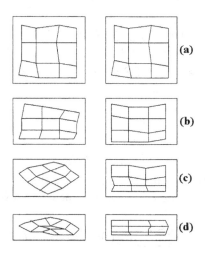

Fig. 3. Solving Oblique orientation with tensorial weights (from [7])

The variance of the input data along the vertical dimension (attribute) versus the horizontal one is varied ($1:1, 1:2, 1:3$, and $1:4$ in Figure 3(a), 3(b), 3(c), 3(d) respectively). The results of the unweighted map is shown on the left and the weighted map on the right in Figure 3.

We interpret the *oblique orientation* as an occurrence due to the map attempting to fit in with a pre-defined network, and resulting in a distorted structure. Tensorial weights method attempts to reduce the oblique orientation but still keeping within the network *borders* thus forcing the shape of the network on the data. This is opposite to the ideal solution since it is the data which should dictate the size and shape of the grid. By changing the size of the grid in the SOM, the map is forced to fit in with a new network size and shape. If the data attributes are not proportionate (in the x and y directions) to the network grid, a distorted final map can occur.

3.2 Effect of the Spread Factor on the GSOM

In GSOM, the map is spread out by using different SF values. According to the formula presented in section 2, a *low* SF value will result in a *higher* growth threshold (GT). In such a case a node will accommodate a higher error value before it initiates a growth. Therefore we can state the spreading out effect (or new node generation) of GSOM as follows :

The criteria for new node generation from node i in the GSOM is

$$E_{i,tot} \geq GT \tag{4}$$

where $E_{i,tot}$ is the total accumulated error of node i and GT is the growth threshold. The $E_{i,tot}$ is expressed as (equation 3.20)

$$E_{i,tot} = \sum_{H_i}\sum_{j=1}^{D}(x_j(t) - w_j(t))^2 \tag{5}$$

If we denote a low SF and high SF values by SF_{low} and SF_{high} respectively, and $\Longrightarrow$ denotes *implies*, then

$$SF_{low} \Longrightarrow GT_{high} \tag{6}$$

$$SF_{high} \Longrightarrow GT_{low} \tag{7}$$

Therefore from equations 5, 6 and 7 we can say that:
when the $SF = SF_{low}$, node i will generate new nodes when

$$E_{i,tot} = \underbrace{\sum_{H_i}\sum_{j=1}^{D}(x_j(t) - w_j(t))^2}_{R_l} \geq GT_{high} \tag{8}$$

Similarly, when the $SF = SF_{high}$, node i will generate new nodes when

$$E_{i,tot} = \underbrace{\sum_{H_i}\sum_{k=1}^{D}(x_k(t) - w_k(t))^2}_{R_s} \geq GT_{low} \tag{9}$$

where $x_j \in R_l$ and $x_k \in R_s$ are two regions in the input data space.

It can be seen that region R_l represents a larger number of hits and accommodates a larger variance in the input space. Similarly region R_s represents a smaller number of hits and a smaller variance. Thus R_l represents a larger portion of the input space and R_s represents a smaller portion. Therefore we can infer that in the case of a low SF value, node i represents a larger region of the input space and with a high SF value, node i represents a smaller region. By generalising i to be any node in the GSOM it can be concluded that, with a small SF value, the nodes in the GSOM represents larger portions of the input space and with a high SF value, the nodes represent smaller portions. Therefore using the same input data, a low SF value will produce a smaller representative map and a high SF value will produce a larger representative map.

In the case of the SOM, the spread of the map is pre-determined by the data analyst and the input data in a certain direction is forced to fit into the available number of nodes. As such, unless the analyst has a method of (or knowledge of) assessing the *proper* number of nodes, a distorted map can occur. With the GSOM, the input values dictate the number of nodes and the spread factor provides, a global control of the spread of the nodes. Therefore the GSOM does not result in the oblique orientation or distorted view of the data as in the case of SOM.

4 Experiments

In this section we present two sets of experimental results to highlight the difference between GSOM with the SF and the traditional SOM. Two artificial data sets with uniform distributions of square and L-shaped two dimensional data are used. The first experiment (Fig 4) show the results from using the SOM on the two data sets. The small dots in the figures represent the input data distribution.The figures show that the shape of the grid effects the level of match to the input data distribution.

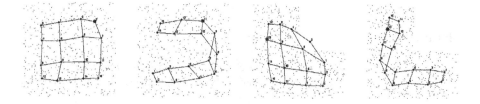

Fig. 4. Square and L-shaped data sets mapped to the SOM

Figure 5 shows the same data sets mapped to the GSOM. Since the GSOM increases the grid size using the spread factor, we have shown two sets of GSOMs with 0.1 and 0.5 spread on the square and L-shaped data. These experiments show that the shape of the gsom grid is decided by the input data distribution.

5 Conclusions

In this paper we have highlighted the advantages of the parameter called the Spread Factor (SF) in the GSOM model. The paper attempts to describe the difference between the GSOM and the traditional SOM in terms of obtaining maps which fit in to the data distributions. In the traditional SOM, the user has to experiment with different grid sizes to arrive at a *best fitting* grid. The

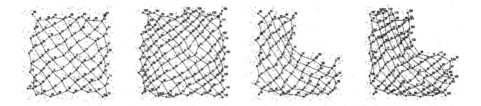

Fig. 5. Square and L-shaped data sets mapped to the GSOM

rows to columns ratio of the grid needs to be initialised to match the data distribution otherwise resulting in a distorted map. We show that the GSOM with the SF takes this burden of pre-defining the *correct* grid size off the user. Therefore we describe the GSOm as a better alternative in generating maps which fit the input data distribution

Acknowledgements

The author acknowledges that the figures 4 and 5 were obtained using the GSOM program developed by Mr. Arthur Hsu, University of Melbourne Australia.

References

1. Alahakoon L. D. *Data Mining with Structure Adapting Neural Networks*. PhD thesis, School of Computer Science and Software Engineering, Monash University, 2000.
2. Alahakoon L.D., Halgamuge S. K. and Srinivasan B. A Self Growing Cluster Development Approach to Data Mining. In *Proceedings of the IEEE Conference on Systems Man and Cybernatics*, pages 2901–2906, 1998.
3. Alahakoon L.D., Halgamuge S. K. and Srinivasan B. A Structure Adapting Feature Map for Optimal Cluster Representation. In *Proceedings of the International Conference on Neural Information Processing*, pages 809–812, 1998.
4. Alahakoon L.D., Halgamuge S. K. and Srinivasan B. Dynamic Self Organising Maps with Controlled Growth for Knowledge Discovery. *IEEE Transactions on Neural Networks*, 11(3):601–614, 2000.
5. Blackmore J. and Miikkulainen R. Incremental Grid Growing: Encoding High Dimensional Structure into a Two Dimensional Feature Map. In Simpson P., editor, *IEEE Technology Update*. IEEE Press, 1996.
6. Fritzke B. Growing Cell Structure: A Self Organising Network for Supervised and Un-supervised Learning. *Neural Networks*, 07:1441–1460, 1994.
7. Kohonen T. *Self-Organizing Maps*. Springer Verlag, 1995.
8. Martinetz T. M. and Schultan K. J. Topology Representing Networks. *Neural Networks*, 7(3):507–522, 1994.

Index of Contributors

Druck und Bindung: Strauss Offsetdruck GmbH